Species Sensitivity Distributions in Ecotoxicology

Environmental and Ecological Risk Assessment

Series Editor
Michael C. Newman
College of William and Mary
Virginia Institute of Marine Science
Gloucester Point, Virginia

Published Titles

Coastal and Estuarine Risk Assessment
Edited by
Michael C. Newman, Morris H. Roberts, Jr., and Robert C. Hale

Risk Assessment with Time to Event Models
Edited by
Mark Crane, Michael C. Newman, Peter F. Chapman, and John Fenlon

Species Sensitivity Distributions in Ecotoxicology
Edited by
Leo Posthuma, Glenn W. Suter II, and Theo P. Traas

Species Sensitivity Distributions in Ecotoxicology

Edited by

Leo Posthuma
Glenn W. Suter II
Theo P. Traas

CRC Press
Taylor & Francis Group
Boca Raton London New York

CRC Press is an imprint of the
Taylor & Francis Group, an **informa** business

CRC Press
Taylor & Francis Group
6000 Broken Sound Parkway NW, Suite 300
Boca Raton, FL 33487-2742

First issued in paperback 2019

© 2002 by Taylor & Francis Group, LLC
CRC Press is an imprint of Taylor & Francis Group, an Informa business

No claim to original U.S. Government works

ISBN-13: 978-1-56670-578-3 (hbk)
ISBN-13: 978-0-367-39648-0 (pbk)

Library of Congress Cataloging-in-Publication Data

Catalog record is available from the Library of Congress

Visit the CRC Press Web site at www.crcpress.com

Visit the Taylor & Francis Web site at
http://www.taylorandfrancis.com

and the CRC Press Web site at
http://www.crcpress.com

Foreword

Different species have different sensitivities to a chemical. This variation can be described with a statistical or empirical distribution function, and this yields a species sensitivity distribution (SSD). The idea to use SSDs in risk assessment originated almost simultaneously in Europe and in the United States. Scientists began to use these distributions for the derivation of environmental quality criteria, challenged by policy makers to make optimal use of single-species toxicity test data for chemicals. This development coincided with the notion that risks cannot be completely eliminated but should be reduced to an acceptable low level.

In 1990, the Organization for Economic Cooperation and Development (OECD) Hazard Assessment Advisory Body organized a workshop in Arlington, Virginia, to discuss these and other approaches for extrapolation of laboratory aquatic toxicity data to the real environment. The extrapolation workshop, together with other workshops on the application of quantitative structure–activity relationships (QSARs) to estimate ecotoxicity data (Utrecht, the Netherlands) and effects assessment of chemicals in sediment (Copenhagen, Denmark), formed the backbone of the *OECD Guidance Document for Aquatic Effects Assessment*, which was published in 1995. This guidance document is applied, for example, in the OECD existing chemicals program.

As head of the OECD Environment, Health and Safety Division, which supported the transatlantic discussions on the use of SSDs in 1990, it is a great pleasure to see that this specific approach in ecotoxicology has been taken up by scientists and is still developing. The fact that it has become so well used in environmental management should not keep us from being critical and demanding about the scientific rationale and validity of the methods used. It is my firm belief that this book contributes to this goal and that it serves as an excellent stimulus to pursue the continued development of SSD-based risk assessment in ecotoxicology.

Rob Visser
Head, Environment, Health and Safety Division

Organization for Economic Cooperation and Development

Preface

AIMS OF THE BOOK

The aims of this book are many, but the most important ones are the following:

- First, the concept that is the subject of the book, *species sensitivity distributions*, is a practical method in ecological risk assessment and in decision-making processes. It is used in the derivation of environmental quality criteria and in ecological risk assessment of contaminated ecosystems. The question is, whether the past adoption of the concept has been a good decision, especially in view of the large investments in preventive and curative actions resulting from decisions based, fully or in part, on application of the concept. The editors, all working in governmental institutes, felt a sense of urgency in the air to summarize the state of the art of the concept, its scientific underpinning, its current uses, and its predictive accuracy, after approximately two decades of convergent evolution on two continents. Eventually, a review of the state of the art should promote better understanding of all issues relevant to the SSD concept and its applications. Therefore, the major aim is a better understanding of the science of ecological risk assessment concerning the use of a practically adopted method.
- Second, the many relevant publications by academic, regulatory, and industrial scientists in North America and Europe have been scattered throughout the literature. Few papers have been published in the easily accessible scientific journals; many are in the "gray literature." Furthermore, most texts explain the issues in various, context-dependent languages, with local jargon added. The secondary aim, necessary to understand the science, is to bring together open and gray literature, and to make the sources available in clear language in this book.
- Third, by compilation and study of the available material and by review of past criticisms of the SSD concept and the solutions offered so far, a final aim becomes apparent. This aim is to suggest paths forward, to suggest solutions for the most relevant criticisms voiced in the past, and to break inertia in the evolution of the SSD concept itself. This should eventually lead to clear views regarding the advantages and limitations of the method for different applications.

THE EVOLUTION OF EDITORIAL RISK

The pursuit of these three aims began in 1998. At a conference in Bordeaux, organized by the European branch of the Society for Environmental Toxicology and Chemistry (SETAC), various Europeans working with the SSD concept were inspired by the local atmosphere to draft the raw outlines of a plan. After approximately 15 years of evolution on two continents, the need was felt to evaluate the SSD concept. The thought simmered for some time. It was brought to the Laboratory for Ecotoxicology at the Dutch National Institute for Public Health and the Environment (RIVM). At RIVM, Herman Eijsackers sowed the seed, and he and Hans Canton cared most for the undisturbed survival and growth of the young plant. In the next year, it grew into a formal RIVM project. RIVM employees were assigned to compile and evaluate the current state of the art, and to formulate ways forward. This was deemed a necessary task for RIVM, since many sites in the Netherlands are exposed at concentrations exceeding the Dutch Environmental Quality Criteria, and the project was expected to help answer the question: "What are the quantitative ecological risks of mixtures of chemical compound concentrations in the environment that exceed the Environmental Quality Criteria?" The efforts were supported by scientific advisory bodies of the RIVM. Soon, the RIVM project became an international project, and the review plan reshaped into a book plan, with international editorship and contributions.

The addition of a North American editor to this effort continued a connection that began at a 1990 OECD workshop on ecotoxicological extrapolation models (OECD, 1992). The most significant result of that workshop was the realization that a common approach was being used in the United States, the Netherlands, and Denmark to extrapolate from single species toxicity test results to biotic communities. Because there was no name for that class of models, the Working Group B rapporteur coined the term *species sensitivity distributions*. That workshop contributed to the subsequent expansion of the use of SSDs from the setting of regulatory criteria into the emerging field of ecological risk assessment. More to the point, it established the contacts and common interests among users of SSDs in North America and Europe that made this volume possible.

ECOLOGICAL AND AUTHORSHIP RISKS

The contributors to this book are specialists on risks, especially risks from chemical compounds in ecosystems. Especially *they* could have been reluctant to contribute to this book in view of various realistic risks associated with it. Nonetheless, they contributed of their own free will.

What risks did authors and editors face?

- First, they faced the risk that they would create a Gordian knot of risk concepts, definitions, and research results, when their goal was to unravel a knotty problem. If you try to imagine how to describe a Gordian knot, or a research plan to unravel it, you can guess how difficult that can be, especially when you want to do it in a scientific way. Where are the rope ends, and how do they causally connect? Those who contribute to a book on such a knotty problem might never be understood by readers or even by the other authors.

- Second, there is the risk that the interpretation of the chosen risk definition (if any) would be strongly context dependent, yielding a hidden knot within a knot. In a scientific context, one can communicate about risks in a purely numerical context, without value judgments. In the societal context of risk-based decision making, however, risk has an aspect of value judgment. The contributors were aware of this extra complication, as they were recruited from those different contexts, so it was courageous to join. Thinkers and practitioners could have easily split, and two volumes rather than one volume could have resulted.
- Third, there is the risk of interminable debate aroused by the published text, as a consequence of the preceding risks. The authors and editors could have chosen to keep the results of their debates among themselves, since the above risks were effectuated in their internal discussions. There might not have been a book at all.
- Fourth, risks are associated with working on the border between science and policy. Scientists may develop methods that have policy implications, which may not be acceptable to policy makers or advocates for industry or the environment. Clearly, the assumption that SSDs are adequate models of the environment is such a case, and work on the book could have been stopped by the employers of the authors or editors.
- Fifth, publicizing controversial technical and conceptual issues may be unwelcome, because SSDs are firmly embedded in the regulatory practices of the United States, the Netherlands, and other nations. Regulators may not want to be told that the scientific foundations of their actions are still questionable or subject to change.
- Sixth, confusion and conflict could have been almost invited by the editors by their wish to bring together two historical lines of SSD evolution (the North American and the European) in a single volume, each with its own context of adopted principles, terminology, and legislation.

AUTHORSHIP RISKS IN PRACTICE

The editors have seen some of these risks in practice. At the first public introduction of the SSD concept in Europe, it was the initiator of the plan for this book who, metaphorically, suggested killing the first messenger. In 1983, Bas Kooijman, from the Netherlands Organization for Applied Scientific Research (TNO), was asked by the Dutch Ministry of the Environment to help resolve the ethical question: "How much toxicity test data for how many species are needed to underpin adequate risk assessment based decisions?" As a result, an initial Dutch TNO report from 1985 and a well-known paper, in *Water Research* in 1987, were published on the risky subject of the derivation of hazardous concentrations for sensitive species. This evolved further when Nico van Straalen from the Vrije Universiteit Amsterdam was invited to give a thought-provoking introductory plenary lecture at a 1995 meeting of the Dutch Provisional Soil Protection Technical Committee (V-TCB). He began this lecture on SSD basics *avant la lettre* by stating that he felt as if he were putting his head on the guillotine, while the audience members were handed a rope to release the blade. The lecture was completed in full health, although the pertinent audience member said in a whisper that he would have liked to pull the rope. This illustrates the risks of the science policy debate on the SSD concept in a nutshell.

POST-WRITING RISKS

Despite these risks, the contributors have not been reluctant. They produced 22 chapters, and no authors left because of inability to describe their strand of the knot. The contributors also have been willing to project themselves into the role and context of their colleagues. The 22 chapters are thus in one book, not two. Although debates have been many, we hope scientific growth has resulted.

On publication of this book, only the post-writing risks remain. There is a need of risk management here. The management of that risk is your task as reader, acting in your own professional environment after reading the book. To help you with this, we have done our best to present the science and applications to you in manageable portions, despite the double Gordian knot. We identified four sections:

I. General Introduction and History of SSDs
II. Scientific Principles and Characteristics of SSDs
III. Applications of SSDs
    A. Derivation of Environmental Quality Criteria
    B. Ecological Risk Assessment
IV. Evaluation and Outlook

By arranging the chapters within these sections, the different focuses of the chapters are presented.

We can help in managing the remaining risks only a bit further, by stating that our discussions profited first from clearly defining the word *risk* when it was used, second from clearly defining or recognizing the context of those involved in the debate, and third from clearly distinguishing the values obtained in risk calculations from value judgments.

All scientific fields can be seen as Gordian knots. For the field of ecological risk assessment, we hope to have cut through some surface layers, and we hope to have freed thereby some useful lengths of rope. This book is the result of the risky effort of many people, who all hope that the field of ecological risk assessment benefits from their efforts.

Leo Posthuma, Glenn W. Suter II, and Theo P. Traas

Acknowledgments

The editors wish to acknowledge the valuable contributions to this book by:

- Olivier Klepper, for starting the process that evolved into this book;
- The **authors**, who volunteered to contribute to this book with a chapter, and who adapted their chapters based on comments of anonymous peer reviewers, section editors, and editors, so as to optimize scientific quality within the chapters, and line of reasoning among chapters in the four sections and throughout the book;
- The **section editors**, who helped to identify highly qualified potential peer reviewers, so that all chapters were read by reviewers representing two types, namely, those expected to be familiar with the environmental policy setting in the continent of the author and those almost completely unfamiliar with that context; the latter helped remove unnecessary jargon;
- The **reviewers**, who performed their peer-reviewing work with enthusiasm, resulting in main-line comments and detailed suggestions on all chapters, which greatly improved the contents of the book.

The reviewers are:

Prof. Dr. Wim Admiraal
Department of Aquatic Ecology
  and Ecotoxicology
University of Amsterdam
Amsterdam, the Netherlands

Dr. Rolf Altenburger
Centre for Environmental Research
  (UFZ-Umweltforschungszentrum)
Leipzig, Germany

Dr. Steve Bartell
The Cadmus Group, Inc.
Oak Ridge, Tennessee, USA

Dr. Jacques J.M. Bedaux
Institute of Ecological Science
Vrije Universiteit
Amsterdam, the Netherlands

Prof. Dr. Hans Blanck
Botanical Institute
Göteborg University
Göteborg, Sweden

Dr. Kym Rouse Campbell
The Cadmus Group, Inc.
Oak Ridge, Tennessee, USA

Dr. Rick D. Cardwell
Parametrix, Inc.
Kirkland, Washington, USA

Dr. Gary A. Chapman
Paladin Water Quality Consulting
Corvallis, Oregon, USA

Dr. Peter Chapman
Jealott's Hill Research Station
Zeneca Agrochemicals
Bracknell, United Kingdom

Dr. Mark Crane
Royal Holloway College
University of London
Egham, United Kingdom

Dr. Michael Dobbs
Bayer Corporation
Agriculture Division
Stilwell, Kansas, USA

Dr. Rebecca A. Efroymson
Environmental Sciences Division
Oak Ridge National Laboratory
Oak Ridge, Tennessee, USA

Dr. Valery E. Forbes
Department of Life Sciences
  and Chemistry
Roskilde University
Roskilde, Denmark

Dr. Florence Fulk
Office of Research and Development
U.S. Environmental Protection Agency
Cincinnati, Ohio, USA

Dr. John H. Gentile
Center for Marine
  and Environmental Analysis
University of Miami
Miami, Florida, USA

Dr. Jeff Giddings
The Cadmus Group, Inc.
Marion, Massachusetts, USA

Dr. Lenwood Hall, Jr.
University of Maryland
Queenstown, Maryland, USA

Dr. Patrick Hofstetter
Harvard School of Public Health
Cincinnati, Ohio, USA

Dr. Udo Hommen
Private Consultant for Ecological
  Modelling and Statistics
Alsdorf, Germany

Dr. Steve Hopkin
School of Animal and Microbial
  Sciences
University of Reading
Reading, United Kingdom

Prof. Dr. Olivier Jolliet
Laboratory of Ecosystem Management
Ecole Polytechnique Fédérale
  de Lausanne
Lausanne, Switzerland

Dr. Lorraine Maltby
Department of Animal
  and Plant Sciences
University of Sheffield
Sheffield, United Kingdom

Dr. Dwayne Moore
The Cadmus Group, Inc.
Ottawa, Ontario, Canada

Prof. Dr. David F. Parkhurst
School of Public
  and Environmental Affairs
Indiana University
Bloomington, Indiana, USA

Dr. David W. Pennington
National Risk Management Research
  Laboratory
U.S. Environmental Protection Agency
Cincinnati, Ohio, USA

Dr. Ad Ragas
Department of Environmental Sciences
Nijmegen University
Nijmegen, the Netherlands

Dr. Hans Toni Ratte
Department of Biology
Aachen University of Technology
Aachen, Germany

Prof. Dr. Sten Rundgren
Department of Ecology
University of Lund
Lund, Sweden

Dr. Bradley E. Sample
CH2M Hill
Sacramento, California, USA

Dr. Wilbert Slooff
Centre for Substances
  and Risk Assessment
National Institute of Public Health and
  the Environment (RIVM)
Bilthoven, the Netherlands

Dr. Eric P. Smith
Department of Statistics
Virginia Polytechnic Institute
  and State University
Blacksburg, Virginia, USA

Dr. Timothy A. Springer
Wildlife International, Ltd.
Easton, Maryland, USA

Mr. Charles E. Stephan
U.S. Environmental Protection Agency
Duluth, Minnesota, USA

Dr. Helen M. Thompson
Environmental Research Team
Central Science Laboratory
York, United Kingdom

Dr. Nelly Van der Hoeven
ECOSTAT
Statistical Consultancy in Ecology,
  Ecotoxicology and Agricultural
  Research
Leiden, the Netherlands

Dr. William H. Van der Schalie
National Center for Environmental
  Assessment
U.S. Environmental Protection Agency
Washington, D.C., USA

Dr. Bert Van Hattum
Institute of Ecological Science
Vrije Universiteit
Amsterdam, the Netherlands

Prof. Dr. Nico M. van Straalen
Institute of Ecological Science
Vrije Universiteit
Amsterdam, the Netherlands

Dr. Donald J. Versteeg
The Procter & Gamble Company
Miami Valley Laboratories
Cincinnati, Ohio, USA

Dr. Jason M. Weeks
Centre for Ecology & Hydrology
Monks Wood
Huntingdon, United Kingdom

In addition, we acknowledge:

- **Marga van der Zwet** (at RIVM), editorial secretary and "Mother Superior" at the Laboratory of Ecotoxicology, who perfectly kept track of all paperwork, and who triggered taking timely action when necessary; without her, the process might have gone out of control;
- **Dick de Zwart** (at RIVM), the electronics polyglot of the book team, who shaped all electronic formats into one, thereby removing the nonscientific transatlantic heterogeneity in file formats, and who shaped

and optimized the appearances of tables and figures and the single reference list;

- **Miranda Mesman** and **Dick de Zwart** for assistance in proofreading of technically edited chapters;
- **Martin Middelburg** at the Studio of RIVM for formatting of various chapter figures;
- **The directors of the Dutch National Institute of Public Health and the Environment (RIVM), especially of the Division of Risks, Environment and Health,** who provided the atmosphere in which scientific ideas on risks of various agents for humans and environment can flourish with both open scientific discussions and an eye on practical use, and who provided funding and all technical means to achieve the goals of this book project;
- The former and current acting Head of the Laboratory for Ecotoxicology, **Herman Eijsackers** and **Hans Canton**, and the Head and Deputy Head of the Centre for Substances and Risk, **Hans Könemann** and **Cornelis van Leeuwen**, who stimulated and gave ample room for planning and executing the work for the book project;
- **Colleagues** who participated in the discussion at the Interactive Poster Session on SSDs, held at the 20th North American Annual Meeting of the Society for Environmental Toxicology and Chemistry (SETAC) in Philadelphia, PA, USA, in 1999;
- The **Society for Environmental Toxicology and Chemistry (SETAC)** and SETAC office personnel, who provided the opportunity to organize an Interactive Poster Session on SSDs at the 20th North American Annual Meeting of SETAC in Philadelphia, PA, USA, in 1999;
- The editors gratefully acknowledge the support of their life partners, **Connie Posthuma, Linda Suter**, and **Evelyn Heugens**.

Development of this book was supported in part by the Dutch National Institute of Public Health and the Environment (RIVM) (www.rivm.nl), within the framework of the strategic RIVM project "Ecological Risk Assessment," RIVM project number S/607501.

About the Editors

Leo Posthuma is currently Research Staff Member in the Laboratory for Ecotoxicology at the Dutch National Institute of Public Health and the Environment (RIVM), where he is involved in the development, testing, and validation of methods for ecological risk assessment. He studied Biology and received a Ph.D. in Ecology and Ecotoxicology from the Vrije Universiteit, Amsterdam, the Netherlands. He has authored and co-authored more than 75 open literature publications, reports, and book chapters, and has acted as book co-editor. His research experience has included phytopathological studies and studies on the evolutionary ecology and population genetics of contaminant adaptation of exposed soil arthropod populations, on community tolerance evolution, on the bioavailability of toxic compounds for terrestrial organisms, on joint effects of compound mixtures, and on stability and resilience of soil ecosystems.

Glenn W. Suter II is currently Science Advisor in the U.S. Environmental Protection Agency's National Center for Environmental Assessment–Cincinnati, and was formerly a Senior Research Staff Member in the Environmental Sciences Division, Oak Ridge National Laboratory, U.S.A. He holds a Ph.D. in Ecology from the University of California, Davis, and has 26 years of professional experience including 20 years of experience in ecological risk assessment. He is the editor and principal author of two texts in the field of ecological risk assessment, and has edited two other books and authored more than a hundred open literature publications. He is Associate Editor for Ecological Risk of *Human and Ecological Risk Assessment*, and Reviews Editor for the Society for Environmental Toxicology and Chemistry (SETAC). He has served on the International Institute of Applied Systems Analysis Task Force on Risk and Policy Analysis, the Board of Directors of the SETAC, an Expert Panel for the Council on Environmental Quality, and the editorial boards of *Environmental Toxicology and Chemistry*, *Environmental Health Perspectives*, and *Ecological Indicators*. His research experience includes development and application of methods for ecological risk assessment, development of soil microcosm and fish toxicity tests, and environmental monitoring. He is a Fellow of the American Association for the Advancement of Science.

Theo P. Traas is currently Research Staff Member in the Centre for Substances and Risk Assessment at the Dutch National Institute of Public Health and the Environment (RIVM). He studied Biology at the Vrije Universiteit, Amsterdam, the Netherlands. His main task is the derivation of environmental risk limits, using species sensitivity distributions and probabilistic food chain models. He is involved in the development, testing, and validation of models for ecological risk assessment. He has authored and co-authored more than 35 open literature publications, reports, and book chapters.

Editors and Principal Authors

Leo Posthuma
RIVM (Dutch National Institute of Public Health and the Environment)
Laboratory for Ecotoxicology
Bilthoven, the Netherlands

Glenn W. Suter II
U.S. Environmental Protection Agency
National Center for Environmental Assessment
Cincinnati, Ohio, USA

Theo P. Traas
RIVM (Dutch National Institute of Public Health and the Environment)
Centre for Substances and Risk Assessment
Bilthoven, the Netherlands

Section Editors

Section I

    Theo P. Traas (RIVM, Bilthoven, the Netherlands)
    Herman J. P. Eijsackers (Alterra Green World Research, Wageningen, the Netherlands)

Section II

    Tom Aldenberg (RIVM, Bilthoven, the Netherlands)
    Dik van de Meent (RIVM, Bilthoven, the Netherlands)
    Glenn W. Suter II (U.S. EPA, Cincinnati, Ohio, USA)

Section III

    Robert Luttik (RIVM, Bilthoven, the Netherlands)
    Dick de Zwart (RIVM, Bilthoven, the Netherlands)

Section IV

    Leo Posthuma (RIVM, Bilthoven, the Netherlands)
    Glenn W. Suter II (U.S. EPA, Cincinnati, Ohio, USA)

Contributing Authors

Belgium
The Procter & Gamble Company, Eurocor, Temselaan 100, 1853 Stroombeek-Bever, Belgium
   Joanna S. Jaworska

Canada
Environment Canada, National Guidelines and Standards Office, Ottawa, Ontario, Canada
   Kathie Adare
   Connie L. Gaudet
   Kelly Potter

Royal Roads University, Victoria, British Columbia, Canada
   Doug Bright

University of Guelph, Centre for Toxicology, Guelph, Ontario, Canada
   Keith R. Solomon
   Peter Takacs

Denmark
National Environmental Research Institute, Department of Terrestrial Ecology, Silkeborg, Denmark
   John Jensen
   Janeck J. Scott-Fordsmand

The Netherlands
Alterra Green World Research, Department of Water and the Environment, Wageningen, the Netherlands
   Theo C. M. Brock
   Paul J. van den Brink

RIKZ (National Institute for Coastal and Marine Management), Middelburg, the Netherlands
   Belinda J. Kater

Pré Consultants, Amersfoort, the Netherlands
   Mark Goedkoop
   Renilde Spriensma

RIVM (National Institute of Public Health and the Environment), Centre for Substances and Risk Assessment, Bilthoven, the Netherlands
   Trudie Crommentuijn*
   Cornelis J. van Leeuwen
   Robert Luttik

Hans Mensink
Dick T.H.M. Sijm
Theo P. Traas
Annemarie P. van Wezel

RIVM (National Institute of Public Health and the Environment), Laboratory for Ecotoxicology, Bilthoven, the Netherlands
Dik van de Meent
Leo Posthuma
Aart Sterkenburg
Dick de Zwart

RIVM (National Institute of Public Health and the Environment), Laboratory for Water and Drinking Water Research, Bilthoven, the Netherlands
Tom Aldenberg

University of Amsterdam, Institute for Biodiversity and Ecosystem Dynamics, Amsterdam, the Netherlands
Mark A. J. Huijbregts*

Vrije Universiteit, Institute of Ecological Science, Amsterdam, the Netherlands
Nico M. van Straalen

Wageningen University, Toxicology Group, Wageningen, the Netherlands
Timo Hamers

United States
The Cadmus Group, Inc., Durham, North Carolina, USA
William J. Warren-Hicks

The Cadmus Group, Inc., Laramie, Wyoming, USA
Benjamin R. Parkhurst

Tetra Tech, Inc., Research Triangle Park, North Carolina, USA
Jonathan B. Butcher

U.S. Environmental Protection Agency, National Center for Environmental Assessment, Cincinnati, Ohio, USA
Glenn W. Suter II

---

* Current affiliation: Ministry of Housing, Physical Planning and the Environment, The Hague, the Netherlands
* Current affiliation: University of Nijmegen, Faculty of Science, Mathematics and Informatics, Department of Environmental Studies, Nijmegen, the Netherlands

U.S. Environmental Protection Agency, Midcontinent Ecology Division, Duluth, Minnesota, USA
Charles E. Stephan

Virginia Institute of Marine Science, Gloucester Point, Virginia, USA
Britt-Anne Anderson
Tyler R. L. Christensen
Scott B. Lerberg
Laurent C. A. Mézin
Michael C. Newman
David R. Ownby
Tiruponithura V. Padma
David C. Powell

Contents

SECTION III   Applications of SSDs
A. Derivation of Environmental Quality Criteria

B. Ecological Risk Assessment

SECTION IV   Evaluation and Outlook

Appendices

Section I

General Introduction and History of SSDs

This section describes the context and history of the development of species sensitivity distributions (SSDs) for use in ecotoxicology. The general introduction shows that SSDs are used for two purposes: the derivation of environmental quality criteria and ecological risk assessment for contaminated ecosystems. It is followed by historical overviews of the partly independent and convergent evolution of the SSD concept on two continents (North America and Europe). The section illustrates the events that have occurred at the interface of science and regulation, homologies and divergence in SSD-based methods, and the need to unite the existing theories and applications.

1 General Introduction to Species Sensitivity Distributions

Leo Posthuma, Theo P. Traas, and Glenn W. Suter II

CONTENTS

Abstract — The species sensitivity distribution (SSD) concept was proposed two decades ago as an ecotoxicological tool that is useful for the derivation of environmental quality criteria and ecological risk assessment. Methodologies have evolved and are applied in various risk management frameworks. Both support and criticisms have been voiced, spread over diverse sources in reports and scientific literature. This chapter introduces the issues and their interrelationships treated in this book. The aims of the book on SSDs are to present (1) the historical context, (2) the basic scientific principles, characteristics, and assumptions, (3) the current practical applications, and (4) an evaluation and outlook regarding the SSD concept and its uses.

1.1 INTRODUCTION

The possible threat of toxic compounds to ecosystems has elicited a request by society to science, to derive "safe" ambient concentrations for protection of ecosystems and methods to assess ecological risks. Although this societal request is difficult to answer for many reasons, one major difficulty is the estimation of effects on diverse species and ecosystems. This book focuses on the variation in species sensitivities to toxicant exposure, and on a specific method to address this variation.

1-56670-578-9/02/$0.00+$1.50
© 2002 by CRC Press LLC

Different ecologists and ecotoxicologists independently designed ecotoxicological assessment systems based on the variance in response among species (Klapow and Lewis, 1979; Mount, 1982; Blanck, 1984; McLaughlin and Taylor, 1985; U.S. EPA, 1985a; Kooijman, 1987). Interspecies variation in sensitivity to environmental pollutants is apparently not only a core problem, but also a basis for finding solutions.

This book focuses on the history, theories, and current practices of the ecotoxicological extrapolation models known as species sensitivity distributions (SSDs). SSDs represent the variation in sensitivity of species to a contaminant by a statistical or empirical distribution function of responses for a sample of species. The emphasis on the issue of "extrapolation" from the single species to the community level that is captured in the SSD model should not mean neglect of environmental factors. That is, there are other relevant factors modulating the predicted risk of contaminants in ecosystems in addition to sensitivity differences, such as variation in biological availability of the compounds and the occurrence of ecological interactions. Therefore, it is often necessary to make additional extrapolations, to improve prediction accuracy of the SSD. The contributors to this book aim to present an overview and evaluation of the use of SSDs in current ecotoxicology, taking into account the importance of the other sources of variation.

1.2 VARIABILITY AND SPECIES SENSITIVITY

Living organisms constitute a vast diversity of taxonomy, life history, physiology, morphology, behavior, and geographical distribution. For ecotoxicology, these biological differences mean that different species respond differently to a compound at a given concentration (i.e., different species have different sensitivities). The acknowledgment that species sensitivities to toxic compounds differ (without attempting to explain the cause) and description of that variation with a statistical distribution function yields SSDs.

The basic assumption of the SSD concept is that the sensitivities of a set of species can be described by some distribution, usually a parametric distribution function such as the triangular, normal, or logistic distribution (Chapters 4 and 5). Nonparametric methods are used as well (Chapter 7). The available ecotoxicological data are seen as a sample from this distribution and are used to estimate the parameters of the SSD. The variance in sensitivity among the test species and the mean are used to calculate a concentration expected to be safe for most species of interest, which can be used to set an environmental quality criterion (EQC). A more recent application is the use of SSDs in ecological risk assessment (ERA).

Since SSDs were originally proposed to derive EQCs in the late 1970s and mid-1980s in the United States and Europe, respectively, their importance in ecotoxicity evaluations has steadily grown. Intensive discussions have taken place on principles, statistics, assumptions, data limitations, and applications (e.g., Hopkin, 1993; Forbes and Forbes, 1993; Smith and Cairns, 1993; Chapman et al., 1998). The history of SSD approaches for North America and Europe is the subject of Chapters 2 and 3. These chapters explain the purposes for which SSDs were originally developed and their expanding use in various regulatory and management contexts. The reader should also be aware that the use of SSDs has spread beyond its two continents of origin to South Africa (Roux et al., 1996), Australia and New Zealand (ANZECC, 2000a,b), and elsewhere. In these new contexts, the concept is expanding both conceptually and technically.

1.3 SSD BASICS

A SSD is a statistical distribution describing the variation among a set of species in toxicity of a certain compound or mixture. The species set may be composed of a species from a specific taxon, a selected species assemblage, or a natural community. Since we do not know the true distribution of toxicity endpoints, the SSD is estimated from a sample of toxicity data and visualized as a cumulative distribution function (CDF, Figure 1.1). This is the integral of an associated probability density function (PDF). The CDF curve follows the distribution of the sensitivity data obtained from ecotoxicological testing, plotting effect concentrations derived from acute or chronic toxicity tests, for example, LC_{50} values and no-observed-effect concentrations (NOECs), respectively. The number of data to construct SSDs varies widely, between no data at all (for many compounds) to more than 50 or 100 sensitivity values (for a few compounds). It is evident that the number of data is highly important for the derivation of the SSD, and for conclusions based on them.

The arrows in the graphs indicate that the SSD concept can be used in a "forward" as well as "inverse" way (Van Straalen and Denneman, 1989; Chapter 4). For the inverse use, such as the derivation of environmental quality criteria, a cutoff percentage p is chosen (to protect $1-p$ percent of species, Y-axis), and the desired "safe"

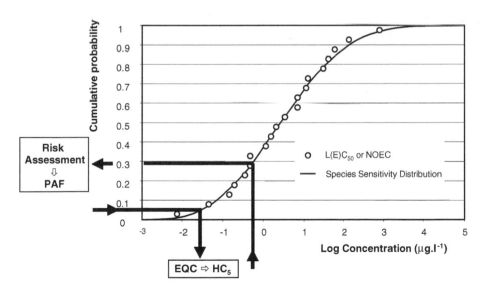

FIGURE 1.1 The basic appearance of SSDs, expressed as a CDF. The dots are input data. The line is a fitted SSD. Forward use (arrows from X → Y) yields the PAF as defined in Chapter 4, or similar estimates of risk as defined by other authors (see also Chapters 5, 15, and 17). Inverse use (arrows from Y → X) yields an EQC at a certain cutoff value, here the hazardous concentration for 5% of the species, HC_5 (e.g., Van Straalen and Denneman, 1989).

concentration (HC_p) is calculated as a result. The 5th percentile of a chronic toxicity distribution has been chosen in the earliest methods as a concentration that is protective for most species in a community, but the value of p is a policy decision, not science. In popular use of the method, the complementary value of p has become known as the 95% (100-p) protection criterion. The forward use, ecological risk assessment, requires estimation of the ambient concentration of a compound at a contaminated site or the concentration predicted to result from a proposed use (X-axis). The potentially affected fraction (PAF) at that concentration can then be estimated using the SSD. If a threshold for significant risk has been identified by policy (e.g., effects on more than 5% of species are unacceptable), any concentration higher than the HC_5 can be considered to pose a significant risk. If variance or uncertainty is estimated, risk may be defined as the probability to harm more than $p\%$ of species. The type of harm is defined by the chosen ecotoxicological endpoint to construct the SSD. These basic concepts can be recognized in all forms of current SSD usage, although the terminology, statistical details, and notation vary.

The SSD method requires three steps: (1) selection of toxicity data, (2) statistical analysis of those data, and (3) interpretation of the output.

1. Toxicity data are collected for species from a community, taxon, or species assemblage. The data set consists of test results with a consistent test endpoint for the pollutant or a mixture of pollutants of concern. The data set should be statistically and ecologically representative of the community or set of species of interest, but in practice the sample of species toxicity data is defined by the available toxicity data rather than by random sampling from the set of interest (Wagner and Løkke, 1991). Some EQC derivation methods require a minimum taxonomic diversity of several genera (e.g., Chapter 12) or families (U.S. EPA, 1985a). Different test endpoints can be used, depending on data availability and the purpose of the exercise. Chronic toxicity data have in practice often been preferred when deriving environmental quality criteria. For the purpose of ecological risk assessment, acute toxicity data are often used because of greater availability, ease of interpretation, or relevance to the duration of exposure. The data set may be subject to quality control measures, averaging within taxa, or modification to normalize for environmental conditions, exposure conditions, or other sources of extraneous variance.

2. Once a data set is assembled, it may be described by a specified statistical distribution such as the normal (Wagner and Løkke, 1991; Aldenberg and Jaworska, 2000), logistic (Kooijman, 1987; Aldenberg and Slob, 1993), or triangular distribution (Erickson and Stephan, 1988). Recently, distribution-free methods based on resampling techniques have been introduced that do not require the specification of the distribution function (Jagoe and Newman, 1997; Newman et al., 2000; Van der Hoeven, 2001).

3. Related to the interpretation of the output, SSDs have mostly been used in the derivation of EQCs, both in the United States and in Europe (see Figure 1.1, inverse use). More recently, SSDs have been used as models of risk to ecological communities or ecosystems as illustrated in

Figure 1.1, forward use (Solomon et al., 1996; Klepper and van de Meent, 1997; Cardwell et al., 1999; Steen et al., 1999; Chapters 15 through 20). In either case, the probability in a CDF (Y-axis) may be interpreted as the probability that an individual species will be affected, or the proportion of the community likely to be affected, encompassing various definitions for risk of contaminants to ecosystems (Chapters 5 and 21).

Criticisms have been voiced about each of these steps, both on the concept of SSDs itself as well as on concepts that are or need to be addressed when SSDs are used, and further on the chance of ecological overinterpretation of the output. For example, Blanck (1984) has already pointed out that a method that is based on single-species toxicity data cannot account for ecosystem-level events such as ecological interactions, implying strict logical limits on the interpretation of risks calculated with the SSDs.

1.4 SSD-RELATED QUESTIONS

A basic assumption in ecological risk assessment is that laboratory-generated single-species toxicity data provide useful information about the communities to be protected. Inherent in the SSD approach is the protection of many species that have not been nor will be tested due to experimental, ethical, or financial restrictions. Statistical extrapolation from a relatively small set of toxicity data to the real world by the use of SSDs and associated extrapolation techniques contains many assumptions and pitfalls, of which the principal authors of the methods were generally aware (U.S. EPA, 1985a; Van Straalen and Denneman, 1989). More questions about the validity of extrapolation techniques based on differences in species sensitivities were voiced shortly after the recommendation of these methods by the OECD (OECD, 1992) and criticism continues to this day (Power and McCarty, 1997; Chapman et al., 1998). Criticisms have addressed a range of issues, from statistical issues, ecotoxicological issues, and ecological issues, to issues related to environmental quality criteria and to ecological risk assessment.

1.4.1 ECOTOXICOLOGICAL ISSUES REGARDING THE INPUT DATA

Ecotoxicological issues focus on the degree to which the responses of a few species in the laboratory represent the responses of the many species exposed in field conditions. Laboratory data sets may be biased toward sensitive or tolerant species and conditions in laboratory tests may be very different from field conditions. Many of these questions revolve around the key issue of bioavailability and exposure routes. An interesting property of SSD methods is that the variation in sensitivity of species can be related to the toxic mode of action and classification of toxicants (e.g., Vaal et al., 1997a,b), which may help to partly overcome data limitations on tested substances. By using toxicity databases such as the U.S. EPA's AQUIRE, a large source of prior information can be accessed and used for the derivation of EQCs or ecological risk assessment of poorly tested substances (Chapter 8) or very small data sets (Luttik and Aldenberg, 1997; Chapter 6).

1.4.2 Statistical Issues

Many statistical issues are related to the assumption that the data set of ecotoxicological endpoints such as NOEC or LC_{50} values can be described by a statistical distribution. The parameters of the distribution have to be estimated by parametric (Wagner and Løkke, 1991; Aldenberg and Slob, 1993) or nonparametric methods (Jagoe and Newman, 1997; Chapter 7), and these estimates are uncertain (Aldenberg and Jaworska, 2000). To deal with this uncertainty, a safety or uncertainty factor may be used, or uncertainty may be shown as confidence intervals. Methodological questions about the choice of toxicological endpoint, data set, distribution type, protection criterion (cutoff value), and method for incorporating uncertainty, have been addressed and are presented in Section II of this volume (Chapters 5 through 8).

1.4.3 Issues Related to Ecological Interpretation of SSD Output

1.4.3.1 Environmental Quality Criteria

Ecological issues focus on the level of protection of natural community structure or function afforded by EQCs derived from SSDs. SSDs do not use any ecological information on communities. It may be assumed that by protecting most of the species with a conservative cutoff value, the associated percentile (concentration) is also protective of ecosystem properties (Van Straalen and Denneman, 1989; Van Leeuwen, 1990), but that assumption remains to be validated. SSDs, which are empirically derived, do not account for ecological interactions, habitat factors, or the specific importance of keystone species and functional groups.

Simultaneously with the recommended use of SSDs by the OECD (1992) for EQC derivation, attempts were undertaken to address the validity of the protection argument of the HC_5. This was done by comparing the HC_5 with effects observed in multispecies tests or experiments (Okkerman et al., 1993; Emans et al., 1993), and such efforts have continued (Versteeg et al., 1999). In general the HC_5 appeared to be lower than the mean model ecosystem NOEC and is thus protective of such systems on average. However, replication of the model ecosystems may be limited, so that large variance masks the response patterns, or sensitive species in model ecosystems may still be affected at the HC_5 indicating that the HC_5 is not overly conservative in that respect. A logical extension of this reasoning is analysis of the relationship between the percentiles of SSDs and the nature and levels of effects seen in contaminated communities in the field. That type of analysis is particularly important to the use of SSDs in ERA (Chapter 9).

SSDs were first applied for the derivation of environmental quality criteria (U.S. EPA, 1985a; Kooijman, 1987). These regulatory uses of SSDs vary depending on statistical, ecological, and ecotoxicological choices and assumptions. Methodologies have evolved partly due to scientific developments as reported in Section II of this book, partly due to pragmatic (nonscientific) choice that are specific to the regulatory context of the respective countries. The reasons for methodological differences and the relative contributions of science and policy are explored in Chapters 11 through 14. Stakeholder discussions regarding acceptance or rejection of SSDs in risk management frameworks are illustrated in Chapter 10, using a historical case that has been crucial for acceptance of SSDs in regulatory contexts.

1.4.3.2 Ecological Risk Assessment

Soon after SSDs were used to derive EQCs, SSDs began to be used to estimate exposure risks by the forward use (Van Straalen and Denneman, 1989; Suter, 1993; Cardwell et al., 1993; Baker et al., 1994). Two interpretations of risk in the forward use are common. The statistical interpretation is that at a given environmental concentration, the probability that a random species from a community is exposed to concentrations above its ecotoxicological endpoint, such as LC_{50} or NOEC, can be calculated from the CDF. Another interpretation is that the proportions on the Y-axis are a measure of the fraction of species in a community at risk. The units of the Y-axis have been defined as the potentially affected fraction (PAF) (Chapters 4 and 16). The use of the word *potential* indicates that PAF refers to a risk, the fraction of species estimated to be exposed beyond an effective concentration, and not an empirically observed fraction of the species in a community that are affected. Recent ecological risk assessments use the distributions of both exposure concentrations and species sensitivity to calculate a measure of ecological risk (Chapter 15; Parkhurst et al., 1996; Solomon et al., 1996; Manz et al., 1999; Suter et al., 1999). Central in these assessments is the translation of the calculated risk to real-world phenomena: is the calculated risk a probability of the occurrence of a natural event, and can this be the subject of validation, or is it a risk index based on statistical reasoning (Chapter 21)?

SSDs may be used to estimate risks from multiple contaminants by addition of probabilities rather than addition of toxic units (Hamers et al., 1996a; Klepper and Van de Meent, 1997; Steen et al., 1999; Chapters 16 and 20). The derivation of SSDs and risks for mixtures may be a useful addition to the ecological risk assessment process.

Software has been developed by a number of groups for the practical application of SSDs in the contexts of EQC derivation and ERA. Various applications are publicly available (e.g., Aldenberg, 1993; Parkhurst et al., 1996; Twining et al., 2000). Information on how to obtain the software is given in Appendix B.

1.5 AIMS OF THE BOOK

Results of SSDs are currently used in different environmental policy settings, and for different risk management purposes. The use of SSDs originated independently in different organizations, and as a consequence methods, assumptions, and terminology differ. At present, an overview of this field is lacking, and various developments are hidden in the gray literature, because of their local applications in environmental policy. Given the divergence and partly hidden coevolution of a variety of SSD applications, this book was written to:

1. Present the historical origins and developments of the SSD concept. The use of SSDs is introduced in Section I with an overview of the historical

developments, independently in North America and Europe, which have led to the present array of methods and applications presented in this volume.

2. Bring together, describe, and evaluate existing general and basic principles and assumptions of SSDs, whether for deriving environmental quality criteria for different environmental compartments or for risk estimates, by employing statistical theory (Section II).

3. Describe specific practical applications and differences among countries in the use of SSDs to set EQCs and evaluate site pollution (Section III).

4. Evaluate the SSD concept and the practical methods in view of their principles and usage and to look forward to possible future applications and necessary developments (Section IV).

This book should serve to improve the use of SSDs in practice and to advance the scientific bases for environmental protection by bringing the diverse SSD methods together and by discussing the common issues.

2 North American History of Species Sensitivity Distributions

Glenn W. Suter II

CONTENTS

Abstract — This chapter presents a brief historical review of the derivation and use of species sensitivity distributions (SSDs) in the United States and Canada. It does not address differences in techniques that distinguish North American from European practice; those issues are addressed in Chapter 21. The chapter is organized in terms of three uses of SSDs: the derivation of regulatory criteria, the derivation of benchmarks for screening assessments, and the estimation of ecological risks.

2.1 REGULATORY CRITERIA

The first use of SSDs was in the derivation of National Ambient Water Quality Criteria (NAWQCs) by the U.S. Environmental Protection Agency (EPA). As discussed in Chapter 11, EPA staff members decided in 1978 to replace the use of expert judgment to derive criteria with a formal method based on protection of a percentage of species. The new method was based on the insight: "We can see that the species sensitivity (LC_{50} or LD_{50}) distributes itself in a rather consistent way for most chemicals. The distribution resembles a lognormal one. Thus, each species we test is not representative of any other species but is one estimate of the general species sensitivity" (Mount, 1982). The method for calculating criteria based on 5th percentiles of SSDs (HC_5) was repeatedly revised until the U.S. EPA (1985a) method, which is still in use (Chapter 11). This method calculates two criteria for each chemical, a final acute value (FAV) and a final chronic value (FCV). The FAV is the HC_5 of acute LC_{50} and EC_{50} values for at least eight fish and invertebrates, divided by 2 to correspond to a lethality rate much lower than 50%. The FCV is derived as the HC_5 of chronic values if sufficient data are available; otherwise, it is derived by multiplying the acute HC_5 by a chronic–acute ratio. The EPA has continued to derive NAWQC values for additional chemicals and to update old NAWQCs using this method.

Independently, Klapow and Lewis (1979) proposed a method for deriving marine water quality standards in California using the 10th percentile of empirical SSDs of LC_{50} values. However, California, like other state regulatory agencies in the United States, now follows the 1985 EPA method.

The period of independent development of SSD models and criteria ended in 1990 with an OECD workshop on extrapolation of laboratory aquatic toxicity data to the field (OECD, 1992). The workshop brought North American assessors together with their counterparts from Europe and Australia. It originated the term *species sensitivity distribution* and recognized for the first time that SSDs are a class of ecological models and not simply a set of regulatory techniques. The workshop endorsed the EPA log-triangular method along with the log-logistic and lognormal methods of the Netherlands and Denmark, respectively. This result served to reinforce the confidence of the U.S. EPA in its method. The workshop also raised issues for research and consensus development concerning SSDs, which are still being considered (see Chapter 21).

The method for deriving water quality criteria based on HC_5 values was subsequently used to derive proposed sediment quality criteria for nonionic organic chemicals by the U.S. EPA (1993). This method used aquatic test data, supplemented with aqueous tests of benthic species to estimate aqueous-phase criteria using the method of the U.S. EPA (1985a). This value could then be converted to a sediment criterion using equilibrium partitioning models. The proposed sediment criteria have not been officially adopted, largely because of controversy concerning the assumptions of equilibrium partitioning and aqueous-phase toxicity. That is, there is not a sufficient consensus that toxicity is associated with the aqueous phase, that aqueous phase concentrations can be adequately predicted by equilibrium partitioning with the sediment organic matter, or that aqueous tests of fish and plankton should be used to derive sediment criteria.

Because the data requirements for calculating NAWQCs are relatively demanding (eight toxicity values from different families of fish and invertebrates), the number of chemicals for which NAWQCs have been derived is relatively small. As a result there has been a demand for regulatory benchmarks for less-tested chemicals. Rather than reducing the data requirements for criteria, the EPA has developed factors to be applied to the lowest value in small data sets to generate conservative estimates of the NAWQC (U.S. EPA, 1995c). The factors were derived by resampling data sets that had been used for NAWQCs to simulate data sets of one to seven species, creating distributions across chemicals of ratios of the lowest value in each sample to the actual NAWQC for that chemical, and then deriving the 80th percentile of that distribution. The method is presented in Host et al. (1995). Benchmarks derived by this method are termed Tier II values (the NAWQCs are Tier I). One result of this method was that chemicals that had been relatively poorly tested have lower values. This was considered a desirable trait since it encourages testing.

Although single-chemical criteria continue to be derived and revised, the methods have not been updated since 1985 and few new criteria have been developed in the United States during the 1990s relative to the 1980s and late 1970s. This is in part because of increased emphasis on alternative methods. First, during the 1980s, subchronic tests were developed to determine the toxicity of mixtures of chemicals (Mount and Norberg, 1985; Norberg and Mount, 1985; Norberg-King and Mount, 1986; Weber et al., 1989). Because these tests addressed the concern that single-chemical criteria did not adequately address effects of mixtures, they have become important alternative tools for regulation of water quality (Grothe et al., 1996). Second, bioassessments based on surveys of aquatic communities have been an important area of research and show promise for protecting aquatic communities from changes in physical conditions as well as chemical pollutants. The development of these methods absorbed funds that might otherwise have been devoted to updating the methods of the U.S. EPA (1985a). In addition, although there have been objections to some individual criteria values, there has been little pressure from regulators or industry to change the methods for calculating aquatic criteria.

An area of difference from the European history is the absence of soil quality criteria in the United States, where there has been no legal mandate to develop such criteria. This is in part attributable to the lack of concern by the public and environmental organizations for ecological effects of soil contamination relative to water contamination. Also, soil contamination has been largely addressed by site-specific risk assessments for Superfund sites, rather than setting national criteria. However, this leaves unresolved some issues such as acceptable levels of heavy metals and other hazardous materials in fertilizers, sludges, and other soil amendments.

In Canada, water quality guidelines have not been based on SSDs, but SSD-like distributions are used to derive soil and sediment guidelines (Chapter 13). One method for deriving soil guidelines uses the 25th percentile of the combined effects and no-effects values from tests of plants and invertebrates in spiked soils. The sediment threshold effects level (TEL) is the geometric mean of the 15th percentile of the effects data distribution and the 50th percentile of the no-effects distribution. This method is derived from the National Oceanic and Atmospheric Administration (NOAA) method for deriving screening benchmarks, discussed below. The data used to derive TELs come from tests of spiked or contaminated sediment or observed effects in contaminated field sediments. For both soil and sediments, an uncertainty factor may be applied based on professional judgment.

2.2 SCREENING BENCHMARKS

SSD-like distributions have also been used to establish benchmark values for screening sediments and soils in the United States. Screening differs from criteria setting in that screening values are not enforceable standards. Rather, they are used to determine whether potentially toxic concentrations exist at a site so that more definitive studies can be done, focusing on the chemicals of concern. Because of this difference, it has been relatively easy to develop screening values for sediment and soil, whereas criteria for these media have not been achievable in the United States.

The NOAA established effects range–low (ER-L) and effects range–median (ER-M) values based on the 10th and 50th percentiles of concentrations associated with effects of individual chemicals on sediment organisms or communities (Long and Morgan, 1990). These values are used to screen sediment contaminant concentrations measured in the NOAA National Status and Trends Program. The NOAA distributions were not strictly SSDs because some of the effects were on communities rather than species and because much of the variance in the distributions is due to differences in endpoints and sediment properties. They were developed independently of the EPA's water quality criteria.

Distributions of toxic concentrations for soil invertebrates, soil heterotrophic processes, and terrestrial plants in soil and solution culture were derived for screening chemicals of potential concern at contaminated sites (Efroymson et al., 1997a,b). The tenth percentile was used as the benchmark, following the precedent of the NOAA ER-L. The benchmarks for soil invertebrates and plants used tests of individual species. Because of the importance of variance in soil properties, the distributions were described as distributions of species–soil pairs (e.g., tomatoes in Yolo silt-loam), rather than SSDs. The distributions for heterotrophic processes included tests with microbial communities as well as individual microbial species, so they were further from being considered simple SSDs.

2.3 ECOLOGICAL RISK ASSESSMENT

Ecological risk assessment began in the United States in the early 1980s simultaneously with the development of the current U.S. water quality criteria methods (Barnthouse et al., 1982; 1987; O'Neill et al., 1982; Suter et al., 1983; Barnthouse and Suter, 1986). The early methods did not include SSDs, but they did provide a conceptual groundwork by emphasizing the development of probabilistic models of effects based on extrapolation from conventional laboratory toxicity data to effects in the field on individuals, populations, and ecosystems. The early methods also incorporated the idea that ecological risk could be estimated from the joint probability of an exposure distribution based on uncertainty analysis of a chemical fate model and an effects distribution based on uncertain extrapolation from laboratory test data to an assessment endpoint. The extrapolation methods included regression models, population demographic models, and ecosystem simulation models. Ecological risk assessment (ERA) did not become widely practiced in the United States until it was mandated for remedial investigations of contaminated sites (U.S. EPA, 1989). It became established with the publication of an EPA framework (U.S. EPA, 1992).

While probabilistic ecological risk assessment was developing without SSDs, a use of SSDs in environmental assessments was developed independently of the risk assessment formalism. McLaughlin and Taylor (1985) represented the percent of plant species visibly injured by SO_2 as empirical SSDs (Figure 2.1). The data were obtained by examination of plants in the field exposed to emissions from a power plant burning high-sulfur coal. This analysis allowed the authors to estimate acute effects for different averaging times. The authors did not associate this approach with other SSD-related work; they considered it to be simply another type of exposure-response model.

SSDs were first proposed as ecological risk assessment models in the early 1990s. Suter (1993) included them in an ecological risk text, as an extrapolation

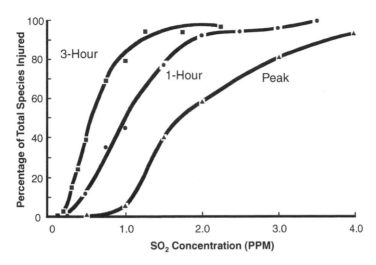

FIGURE 2.1 Species sensitivity distributions for plants exposed in the field to sulfur dioxide from a coal-fired power plant. Results are expressed as peak, 1-h, and 3-h averaged concentrations. (Redrawn from McLaughlin, S. and Taylor, G., in *Sulfur Dioxide and Vegetation,* Winner, W. E. et al., Eds., Stanford University Press, Stanford, CA; © 1985 by the Board of Trustees of the Leland Stanford, Jr. University. With permission.)

model and as an illustration of the treatment of uncertainty in probabilistic assessment. He pointed out that SSDs might be treated as models of the proportion of a community or taxon that is affected or of the probability of effects on a species. He adapted the idea from its use in standard setting in the United States, the Netherlands, and Denmark (U.S. EPA, 1985a; Van Straalen and Denneman, 1989; Van Leeuwen, 1990; Wagner and Løkke, 1991) and included McLaughlin and Taylor's (1985) field-derived SSD.

A method for assessment of aquatic ecological risks was developed for the Water Environment Research Foundation (WERF) consisting of three tiers of increasingly complex methods (Cardwell et al., 1993; Parkhurst et al., 1996). The second tier was based on SSDs including the characterization of risks in terms of either a graphical or statistical comparison of a distribution of exposure concentrations (either measured or estimated) and an SSD. The authors provided software and a manual for fitting SSDs and calculating risks (Parkhurst et al., 1996). They adapted SSDs from their use by the U.S. EPA in calculating national ambient water quality criteria. They argued that the use of SSDs in criterion-setting made their use in ecological risk assessment more acceptable than other potential methods. However, they used the logistic distribution rather than the EPA log-triangular distribution and they did not follow the EPA definition of adequate data sets. They interpreted SSDs as models of communities of species. This WERF method has been used in risk assessments by its authors and others for effluents, contaminated sites, and individual chemicals (Parkhurst et al., 1996; Cardwell et al., 1999). The method is discussed in Chapter 17.

Suter's proposed use of SSDs in ecological risk assessment and that of the WERF team (Rick Cardwell, Ben Parkhurst, and Bill Warren-Hicks) originated independently; both were based on the use of SSDs for standard setting in the United States and Europe. However, each was aware of the other's interest in the use of SSDs, and, at a review workshop for the WERF methods, they disagreed concerning the proper use of the models. In particular, Suter wanted more explicit identification of the endpoints that were to be estimated using SSDs, while the WERF team relied on the regulatory status of SSDs rather than their ability to estimate effects.

A set of methods was developed for risk assessment of pesticides by a team of scientists, the Aquatic Risk Assessment and Mitigation Dialog Group (ARAMDG), assembled by SETAC for the National Agricultural Chemical Association and the U.S. EPA (Baker et al., 1994). The ARAMDG's use of SSDs was derived from the WERF method and the methods for setting standards in the Netherlands. Their use of SSDs for risk assessment resembled standard setting in that they used an HC_p (the HC_{10}) as a threshold value. If the 90th percentile of the expected environmental concentrations exceeds the 10th percentile of the SSD, effects were considered to be potentially significant. This method has been applied to a cooling system biocide (Klaine et al., 1996a), an herbicide (Solomon et al., 1996), two metals (Hall et al., 1998), an algaecidal antifoulant (Hall et al., 1999), a broad-spectrum antifoulant (Hall et al., 2000), and an insecticide (Giesy et al., 1999).

In 1998, the U.S. EPA published guidelines for ecological risk assessment (U.S. EPA, 1998c). This guidance document presents SSDs in an example of risk characterization based on the comparison of exposure and effects distributions. The EPA authors cited the Health Council of the Netherlands, the WERF methods, and the ARAMDG methods as sources of this approach (Cardwell et al., 1993; Health Council of the Netherlands, 1993a; Baker et al., 1994; Solomon et al., 1996). Although the EPA endorsed this use of SSDs as one alternative for ecological risk characterization, it did not endorse any particular SSD technique.

The U.S. EPA commissioned an Ecological Committee on FIFRA Risk Assessment Methods (ECOFRAM) to develop probabilistic methods for assessment of the ecological risks of pesticides (ECOFRAM, 1999a,b). ECOFRAM terrestrial and aquatic workgroups developed draft reports that have been reviewed and may be revised and ultimately adopted in some form by the U.S. EPA. The use of SSDs in the aquatic ECOFRAM is essentially the same as in ARAMDG including the use of 10th percentiles as standards for significance. This similarity is not surprising given the overlapping memberships of the two groups. The ARAMDG and aquatic ECOFRAM method is discussed in Chapter 15. The terrestrial ECOFRAM focused on effects on birds, and their SSDs were distributions of avian LD_{50} values. This is reasonable, given the concern for mass mortalities of birds acutely exposed to pesticides.

Another set of methods for assessing pesticide risks to aquatic ecosystems was recently published (Campbell et al., 2000). The third tier of that method is based on visual inspection of the overlap of exposure distributions from simulation modeling with empirical SSDs. This tier follows the ARAMDG in using the HC_{10} as an endpoint and in comparing it to the 90% exposure concentration. The fourth tier uses

Monte Carlo simulation to calculate the probability that the exposure/effects quotient exceeds 1.

Although SSD-based risk assessment methods were developed to address individual chemicals, they are now commonly used in ecological risk assessments for contaminated sites. However, few of these assessments find their way into the open literature. An exception is the assessment of risks to fish from contaminants in the Clinch River, which used SSDs as the primary model of effects of individual contaminants (Suter et al., 1999). In these assessments, SSDs are commonly one of multiple lines of evidence used to estimate risks by weighing the evidence.

The Canadian government has not endorsed or mandated the use of SSDs in ecological risk assessment. However, individual Canadians have been active in the development and use of SSD in ecological risk assessment, including the ARAMDG and ECOFRAM methods discussed above (see also Chapter 15).

2.4 SUMMARY

SSDs were first developed and used by the U.S. EPA and the State of California in the late 1970s to derive water quality criteria. Subsequently, they were used by the EPA to derive proposed sediment quality criteria. Those uses and the Dutch and Danish use of SSDs inspired the use of SSDs, beginning in the early 1990s, for ecological risk assessment of chemicals and contaminated sites. Subsequently, the EPA endorsed the use of SSDs in ecological risk assessment. In the 1990s, Canadian soil and sediment standards were developed using distributions of species sensitivities, and individual Canadians participated in the development of SSDs as ecological risk models.

3 European History of Species Sensitivity Distributions

Nico M. van Straalen and
Cornelis J. van Leeuwen

CONTENTS

Abstract — The notion that species show considerable variability in their susceptibility to toxicants developed in the beginning of the 1980s when results of systematic testing programs became available. Challenged by policy makers, scientists began to use these data in the course of the 1980s and developed statistical approaches that allowed the derivation of safety factors from the mean and standard deviation of sensitivity distributions. In this chapter a brief account is given of the developments in Europe; we summarize the successive publications that contributed to the development of the theory and its application in environmental management.

A seminal paper by Kooijman (1987) introduced the concept of HCS, hazardous concentration for sensitive species. This was defined as a concentration of toxicant in the environment such that the probability that the LC_{50} of the most sensitive species in a finite community is below this concentration equals an arbitrary small value. The model developed by Kooijman included a statistical argument that allowed for uncertainty due to the fact that parametric species sensitivity distributions are estimated from a limited sample of data. Although the concept of HCS already included all the elements of probabilistic risk assessment, it was not accepted readily because HCS estimates came out extremely low and policy makers doubted the assumptions in the methodology.

The assumption underlying HCS that species could have arbitrarily low LC_{50} values was questioned by Van Straalen and Denneman (1989), who proposed to define a cutoff point in the distribution of sensitivities. Consequently, the concept of HC_p was introduced, defined as a concentration of toxicant in the environment such that the probability of finding a species with a no-effect level below this concentration equals an arbitrary small value (p). The approach taken by Van Straalen and Denneman was discussed intensively in various European countries and expert groups of the OECD and was adopted in some countries as the basis for derivation of environmental quality criteria. Statistical improvements to the methodology were made by Wagner and Løkke (1991) and Aldenberg and Slob (1993). The idea of HC_p fitted well in the risk assessment methodology developed around the same time, which acknowledged that environmental policy cannot eliminate undesired events completely, but should aim to reduce their occurrence to an acceptable low level.

When the framework of species sensitivity distributions was adopted by policy makers in the beginning of the 1990s, it triggered a variety of criticisms. The most frequently raised objection was that if $p\%$ of the species were left "unprotected," and these $p\%$ included species in a crucial trophic position or with "ecosystem engineering" properties, the ecological effects would be much larger than expected. Other objections regarded the shape of the distribution, the representativeness of test species, and the applicability to biologically essential elements. Despite these objections a variety of multispecies tests, mesocosm tests, and field studies provided good support for a numerical similarity between HC_5 and concentrations in the environment at which the first adverse effects start to become apparent.

The development of species sensitivity distributions is still ongoing. For ecotoxicology as a scientific discipline, three fundamental issues deserve attention for the future: (1) what are the reasons why some species are more sensitive than others, (2) which factors control differences between the laboratory and the field, and (3) how do distributed sensitivities relate to higher levels of ecological organization.

3.1 INTRODUCTION

The use of statistical methods to support risk assessment of chemicals and the derivation of environmental quality criteria in Europe followed a course that, in hindsight, was quite independent from the developments in North America. This was because the earlier publications in North America were reports of the Environmental Protection Agency, which at that time were not widely distributed in Europe (U.S. EPA, 1984b; U.S. EPA, 1985a). Conversely, the first papers published by European authors were in journals that were not well known by North American scientists involved in environmental policy. The first common element to the continents was that policy makers involved in environmental management, i.e., chemicals management and water quality management, drove these developments. In those days, the protection of the aquatic environment was a clear common goal of people working in the two fields. In the early 1990s the attention shifted toward other environmental compartments such as aquatic sediments and soils. The second common element was the question: What assessment factor should be used if many chronic aquatic effect concentrations are available? Normally, an assessment factor of 10 was applied to the lowest of a few no-observed-effect concentrations (NOECs) or LC_{50} values to arrive at a concern level for the aquatic environment (U.S. EPA, 1984b). But what should one do if many NOECs or LC_{50} values are available? It is quite evident that with increasing information uncertainty reduces; that is, a more reliable estimate can be made when more acute or chronic single-species toxicity data are available. In such situations, applying an assessment factor of 10 to the lowest value can be seen as a punishment for large sample sizes.

In the Netherlands, the science-policy question about how to assess the risk of toxic chemicals for ecosystems was addressed by the Minister of Housing, Physical Planning and Environment to the Health Council. A scientific advisory body was established and in 1988 the Health Council published its advice, followed by an English translation in 1989 (Health Council of the Netherlands, 1989). This was implemented in a policy paper on risk management (VROM, 1989b) and discussed in Parliament. Similar science-policy discussions and methodological developments took place in Denmark, Germany, and Spain. Several informal meetings were held in Europe, one of which was a meeting organized by the Commission of the European Communities in October 1990 (CEC, 1990).

It was also in 1990 that on both sides of the Atlantic Ocean people recognized that there was a common science-policy question and that there was a need for a transatlantic dialogue. A workshop was held in Arlington, initiated by the Hazard Assessment Advisory Board of the OECD. The proceedings of this OECD workshop were published in 1992 (OECD, 1992). After this workshop the methodology was implemented on a wider scale. In particular, risk assessment of the compound DTDMAC, a cationic surfactant used as fabric softener, contributed intensely to further discussion about the strength and weaknesses of the statistical extrapolation tools then implemented (Van Leeuwen et al., 1992a). This stressed the need for further validation of the extrapolation methodology (Emans et al., 1993; Okkerman et al., 1993; Van Leeuwen et al., 1994), which at the time was already applied in risk assessment and the derivation of environmental quality guidelines in the Netherlands.

In this chapter a brief account is given of the European developments, complementing the North American history given by Suter in Chapter 2. We summarize the successive publications that contributed to the development of the theory and its application in environmental management; we add some comments from the perspective of the present-day situation.

3.2 SPECIES SENSITIVITY DISTRIBUTIONS

The use of species sensitivity distributions is based on the recognition that not all species are equally susceptible to toxicants. This trivial observation must have been commonplace knowledge to even the earliest toxicologists; however, a statistical treatment of toxicological data became possible only after systematic investigations were made with a large-scale comparison of species. One of the earliest scientists systematically reporting on interspecies variability in sensitivity to toxicants was W. Slooff (Slooff and Canton, 1983; Slooff et al., 1983). The aim of Slooff's investigations was not so much to erect species sensitivity distributions, but to compare the relative sensitivity of species considered as indicators of water quality. This question was addressed in a systematic research program in which acute tests were conducted with 22 different aquatic species, including bacteria, algae, crustaceans, insects, fish, and amphibia, all subjected to 15 different substances. The data illustrated that sensitivity differences between species can be considerable, depending on the substance. The largest interspecies variation in Slooff's data was a factor of 8970 between the LC_{50} values for allylamine, and the smallest difference was a factor of 30 for n-heptanol (Slooff and De Zwart, 1984). The message was that the interspecies variability had a large, unpredictable component and that there was hardly any pattern in the data. As to Slooff's aims, the conclusion was that indicator species with a universal susceptibility to toxicants did not exist and that aquatic toxicity testing should involve a taxonomically diverse array of species.

In retrospect, it is remarkable that little attention was paid to the reasons species would differ in their sensitivity to toxicants. Interspecies variability can be broken down into a series of factors:

- Differences in uptake–elimination kinetics
- Differences in internal sequestering mechanisms
- Differences in biotransformation rates
- Differences in the nature or presence of the biochemical receptor
- Differences in the rate of receptor regeneration
- Differences in the efficiency of repair mechanisms

Examples in which species sensitivity was explained in terms of physiological or biochemical factors can be found in the field of pesticides. For these compounds the toxic mode of action is usually well known and the selectivity of the compound to the target species is a main issue of research. Long before the concept of species sensitivity distributions was invented, several authors had already noted the large differences between species in their metabolic potential toward pesticides and other organic compounds. Some authors were discussing these issues in an ecological context and explained species-specific metabolic capacities in terms of adaptations to certain environments or diets (Walker, 1978; Brattsten, 1979). For example, systematic differences were noted in arylhydrocarbon hydroxylase activities between aquatic and terrestrial animals and between animals with a broad diet range, and food specialists.

It is interesting to note that almost simultaneously with Slooff's research, a paper appeared in which it was proposed to assess pesticides with respect to their "structure–selectivity relationships" (Smissaert and Janssen, 1984). These authors considered a hypothetical distribution of species over tolerances and argued that for an ideal pesticide, the target species would be located in the utmost left tail of the distribution. They compared seven test species with respect to their mean tolerance toward seven insecticides and calculated a normalized tolerance per species, averaged over the insecticides. The mean and variation of these tolerances were correlated with physicochemical parameters of the insecticides in so-called "structure–selectivity relations." Their concept of selectivity came very close to the idea of a species sensitivity distribution.

At about the same time, interspecies variability in toxicant sensitivity was explored by Blanck (1984) and Blanck et al. (1984). These authors were concerned with the question of how to derive the EC_{50} for a sensitive species from EC_{50} values obtained for other species. If a new chemical is tested on a series of, say, six species, the toxicity to a seventh (sensitive) species may be predicted by applying a "predictivity factor" to the already existing data. Again, this method came very close to the idea of a species sensitivity distribution, but it was more a way of treating data than a theoretical concept.

3.3 THE CONCEPT OF HCS

The (sub)acute toxicity data collected by Slooff et al. (1983) were used by Kooijman (1987) to illustrate the concept of a hazardous concentration for sensitive species (HCS). Kooijman was concerned with the derivation of a safety factor that could be applied to LC_{50} data and that would allow for differences in sensitivity between species. Such safety factors are often used by policy makers to extrapolate experimental data obtained for a limited set of test species, to sensitive species in a community. Kooijman remarks that he was repeatedly approached by civil servants (specifically, H. Könemann) who asked him to develop an algorithm by which a safety factor could be derived in an objective way. He also noted, "All honest scientific research workers will feel rather uncomfortable with such a task, and the author is no exception." Nevertheless, one of the reasons for him to write his paper was that by providing an algorithm based on explicit assumptions, the procedure of deriving environmental quality standards would become more transparent and accessible to regular update in the light of new findings. This is very much in accordance with the analyis made a decade later by Chapman et al. (1998), who argued that safety factors (or application factors) should not be taken as fixed values with an "eternal" validity, but should be subject to continuous scientific scrutiny. Chapman et al. (1998) also argued that application factors for different components of a risk assessment procedure should be kept separate from each other, to allow their scientific investigation. Kooijman's paper actually deals only with one aspect of the derivation, extrapolation to sensitive species.

The basic assumption in Kooijman's derivation of HCS was that the LC_{50} values of test species can be conceived of as random trials from a log-logistic distribution. The data of Slooff et al. (1983) on LC_{50} values for 22 species and 14 chemical compounds were examined and although there were some deviations, the fits in general were good; the conclusion was that there were no indications of large deviations from the log-logistic distribution for most of the data.

The probability distribution of the log-logistic is a symmetric, bell-shaped function on a logarthmic concentration axis that very much resembles the better-known lognormal (Gaussian) distribution. Compared to a lognormal with the same mean and standard deviation, it is slightly leptokurtotic; that is, beyond a certain point the tails of the distribution fall down to zero less quickly. The main reason for using the logistic was that its integral (the cumulative distribution) can be written as an explicit mathematical formula, whereas for the normal distribution, the integral can only be approximated numerically.

The second step in Kooijman's argument was that a local community comprising n species (e.g., all invertebrates living in a certain pond) can be conceived as a random sample from the log-logistic distribution describing all species in the world (Figure 3.1). The HCS is then defined as: A concentration such that the probability

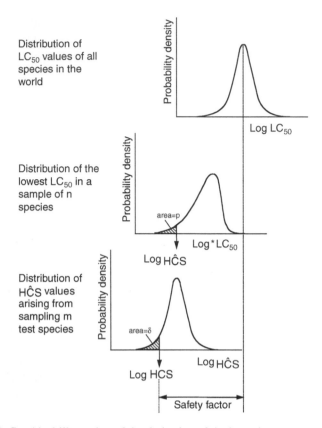

FIGURE 3.1 Graphical illustration of the derivation of the hazardous concentration for sensitive species (HCS), according to Kooijman (1987). $*LC_{50}$ = lowest LC_{50}. $\hat{HCS}$ = estimate for HCS derived from m test species.

that the LC_{50} of the most sensitive of n species is below this concentration equals an arbitrary small value, p.

HCS obviously depends not only on the mean and the standard deviation of the distribution from which the local community is sampled, but also on the number of species in it. The larger the community, the more likely it will be that it includes very sensitive species. In fact, when community size approaches infinity, the distribution for the local community becomes equal to the distribution of all species in the world, log HCS approaches minus infinity, and HCS becomes zero. Obviously, to estimate the sensitivity of the most sensitive species in a community, not only p must be specified (called δ_1 by Kooijman), but also the size (n) of the community. This brought an arbitrary argument in the derivation; however, it turned out that beyond a certain value, the effect of n becomes smaller and smaller.

The effect of community size actually has a good ecological rationale, because it expresses that large communities are more likely to comprise sensitive species. Such large communities (e.g., the ocean) should therefore be treated with more care than small communities (e.g., a pond), and this effect is thus incorporated in the derivation by HCS decreasing with n. Kooijman (1987) proposed that n could be estimated from the surface area of the ecosystem involved, using the species–area relationships developed in island biogeography (MacArthur and Wilson, 1967). In the absence of any knowledge a default value of 1000 could be used for n. This aspect of the method is a bit dubious, since the number of species in a community depends on more factors than the surface area.

The third step in Kooijman's algorithm was concerned with the estimation of HCS from real data. Obviously, the sensitivity distribution of all species in the world remains unknown because for any given substance, only a few species have actually been tested. Instead, it was assumed that the test data, obtained for m species, are a sample from the unknown distribution. The mean and the standard deviation of the m data are considered as estimators of the parametric mean and standard deviation. The method of moment estimation was used to derive the estimators. The estimation introduces uncertainty about how well the sample mean and sample standard deviation estimate the corresponding parametric values. The desired degree of confidence has to be specified by means of another small number, δ (called δ_2 by Kooijman), which represents the probability that the actual HCS is larger than the HCS estimated from the sample mean and standard deviation.

The statistical argument has the effect that HCS decreases with decreasing sample size, m. In fact HCS becomes very low when the sample size is below five and it increases to a fixed value when m approaches infinity. This acts as a built-in punishment for small sample sizes. If there are only few toxicity data for a given substance, any estimate of a safe concentration in the environment that includes uncertainty should be very low. In other words, the advice of the scientist will be very conservative when there are few data, a situation that can be relieved only by providing the scientist with the means to collect more data!

In summary, the algorithm developed by Kooijman can be written as the following equation:

$$HCS = \exp\left\{ x_m - f(p, \delta, m, n) s_m \right\} \qquad (3.1)$$

where x_m is the mean of the ln-transformed LC_{50} values, s_m the sample standard deviation of the ln-transformed LC_{50} values, and f is a function depending on p, δ, m (number of species tested) and n (assumed size of the community). The function f can be looked up in a table in Kooijman's paper. The equation expresses that HCS is located f standard deviations to the left of the mean sensitivity of the sample. A graphical overview of Kooijman's argument is given in Figure 3.1.

Although the concept of HCS included already all the elements of the probabilistic risk assessment methodology elaborated in this book, it was never actually applied to derive real environmental quality criteria. The main reason for this was that HCS estimates invariably came out very low, usually below any detection limit or background concentration. Despite the logic of the argument, policy makers suspected that there was a flaw in Kooijman's presentation and did not accept his argument that HCS could be so low.

3.4 THE CONCEPT OF HC_p

Kooijman's method explained in the previous section extrapolates sensitivities into the utmost left tail of the distribution, an area where no data are obtained. The reason HCS values could be so low is that the species sensitivity distribution assumed does not have a lower threshold; it extends its tail to minus infinity on a logarithmic scale. This was the main argument that triggered Van Straalen and Denneman (1989) to develop the concept of HC_p. This concept, first proposed at a Dutch conference in 1986 (Van Straalen, 1987) was developed as part of ministerial advice prepared by the Soil Protection Technical Committee in the Netherlands (VTCB, later TCB). The question was how ecotoxicological data for soil organisms could be used to evaluate so-called reference values for contaminant concentrations in soil. These reference values had been derived as the upper limits of background concentrations in soils that were considered clean and of good quality (De Haan, 1996). The reference values were reviewed by the VTCB from the perspective of human health using the concept of TDI; there was hardly any framework for an evaluation from an ecotoxicological perspective.

The basic idea of Van Straalen and Denneman (1989) was to get rid of the tail of the distribution of sensitivities by choosing an arbitrary cutoff point (Figure 3.2). The most sensitive species was not the target of interest, but a certain percentile of the distribution. Consequently, it was not necessary to assume that a local community was a sample from all the species in the world, the distributions were considered to be the same. This had another advantage: it removed the influence of community size, n, which gave an arbitrary element in the method of Kooijman (1987).

Van Straalen and Denneman (1989), like Kooijman (1987), assumed a log-logistic distribution of species sensitivities and developed an analytical expression for the p-percentile of the distribution, which was called HC_p, hazardous concentration for p of the species. They then adopted the statistical argument of Kooijman; that is, the mean and standard deviation of the distribution were estimated from the sample mean and standard deviation using moment estimation and the uncertainty involved produced an extra safety margin. The formula for HC_p became:

$$HC_p = \exp\left\{x_m - g(p, \delta, m)\, s_m\right\} \qquad (3.2)$$

where g is a function similar to f in Equation 3.1, that specifies how many standard deviations HC_p is located to the left of the sample mean. This function depends only on p, δ (the uncertainty in estimation), and m (sample size), not on n. The model has the same property as Kooijman's model: it produces lower estimates with decreasing sample size. When m approaches infinity, HC_p approaches the true p-percentile of the distribution. The difference between HC_p derived from a certain set of data and the parametric HC_p of the distribution can be seen as the research effort needed to increase information and decrease the necessary safety factor. Figure 3.2 provides a graphical illustration of the method.

Van Straalen and Denneman (1989) also presented an inverse version of their model. By rewriting the equation, an expression was derived for the percentage of

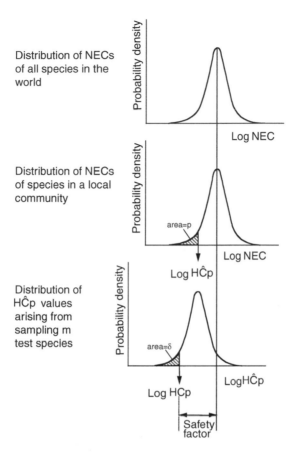

FIGURE 3.2 Graphical representation of the derivation of the hazardous concentration for $p\%$ of the species (HC_p), according to Van Straalen and Denneman (1989). NEC = no-effect concentration. $H\hat{C}_p$ = estimate for HC_p derived from m test species.

protected species as a function of the concentration. In this way, the choice of p could be left to the policy maker; at any value for p the corresponding concentration could be read from a graph. However, the only p-value actually used was 5, and the corresponding concentration was named HC_5. There were no clear reasons why a value of 5 should be chosen; it seemed to be the most logical choice in the light of statistical conventions. The inverse approach was later elaborated by Aldenberg and Jaworska (2000) and it became the basis for the concept of potentially affected fraction (PAF, Klepper et al., 1998; Chapter 16).

Van Straalen and Denneman (1989) applied their method to NOECs, rather than LC_{50} values. NOECs were considered to be more representative for the field situation, because effects on growth and reproduction will usually dominate the ecological effects of toxicants in the environment, rather than effects on mortality, which occur at higher concentrations. Most of the consequent discussion on species sensitivity distributions is therefore done in terms of NOECs, despite the fact that serious objections have been raised against the use of this endpoint (Hoekstra and van Ewijk,

1993; Laskowski, 1995; Kooijman, 1996; Van der Hoeven et al., 1997). The concept of species sensitivity distributions can be applied equally to LC_{50} values, NOECs, and EC_{10} values, so the aspect of the endpoints is not elaborated here.

The numerical values derived using the algorithm of Van Straalen were considered more "reasonable" than Kooijman's; they were usually low, but above background concentrations. This fact contributed greatly to the acceptance of the method. A committee of the Dutch Health Council (Health Council of the Netherlands, 1989) reviewed the methods of Blanck (1984), Slooff et al. (1986), Kooijman (1987), and Van Straalen and Denneman (1989), along with two methods developed in the United States (U.S. EPA, 1984b; U.S. EPA, 1985a; Chapter 11) and advised a combined procedure that placed the concept of HC_p in a central position. Consequently, the method of Van Straalen and Denneman (1989) became the official method by which environmental quality criteria for existing substances were derived in the Netherlands (VROM, 1989b). This was supported by a study of Okkerman et al. (1991), who applied the various methods to aquatic ecotoxicity data for eight different chemicals and 11 species.

The idea of HC_p fitted very well in the risk philosophy that was developed in several countries around the same time. This philosophy acknowledged that environmental policy cannot completely prevent undesired events, but has to develop strategies that limit their occurrence to an acceptable low level. For toxicants, the parameter p in HC_p was considered equivalent to risks estimated for industrial accidents, cancer risks from radiation, etc. Consequently, HC_5 could be considered as a concentration with an ecological risk of 5%, with risk in this case being the probability of finding a species exposed to a concentration higher than its NOEC (Van Leeuwen, 1990; Chapter 4). However, this was not the original intention of the HC_5. The HC_5 is an approach to arrive at a predicted no-effect level (PNEC) for ecosystems (see Section 3.5).

The developments depicted were also discussed in other European countries, first in Denmark (Løkke, 1989; Scott-Fordsmand et al., 1996), later also in Germany, Switzerland, and Spain. Wagner and Løkke (1991) questioned the assumption of the log-logistic distribution in Kooijman's and Van Straalen's methods and proposed a normal distribution instead. This distribution holds a central position in statistics and a huge body of statistical inference is built around it. By using a normal distribution, better use can be made of the already existing statistical theory. The model developed by Wagner and Løkke (1991) is similar to Van Straalen and Denneman (1989), except that the function specifying how far HC_p is located to the left of the mean is now different:

$$HC_p = \exp\left\{x_m - k(p, \delta, m)\, s_m\right\} \qquad (3.3)$$

Wagner and Løkke (1991) provided a table in which the values of k can be read as a function of the parameters p, δ, and m. The procedure is equivalent to finding the lower confidence limit of a fractile of the normal distribution, which is called the tolerance limit in statistical quality control. A similar, slightly different, table was later given by Smith and Cairns (1993).

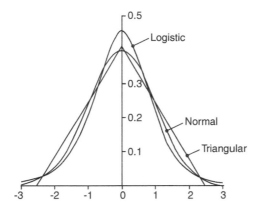

FIGURE 3.3 Illustrating the similarity of normal, logistic, and triangular distributions. All three distributions plotted have mean 0 and standard deviation 1. The triangular distribution has zero points at $\pm \sqrt{6}$, the other two distributions have tails extending to $\pm \infty$. The height of the top is $\pi/4/\sqrt{3}$ for the logistic, $1/\sqrt{6}$ for the triangular, and $1/\sqrt{2\pi}$ for the normal distribution.

Numerical exercises performed by OECD (1992) using a set of aquatic ecotoxicity data for eight chemicals showed that HC_p values calculated by the method of Wagner and Løkke (1991) were very similar to HC_p values obtained when the method of Aldenberg and Slob (1993) was applied to the same data. The latter method (which was available as a prepublication during the study of Wagner and Løkke, 1991) was based on the log-logistic distribution, like Van Straalen's. Hence, the conclusion can be that the distribution itself (normal or logistic) is not crucial for the derivation of HC_p, as long as p is not too small. For practical purposes the two distributions can hardly be distinguished from each other. This is illustrated in Figure 3.3, in which normal, logistic, and triangular distributions with the same mean and standard deviation are compared with each other. By using the mathematical expressions for the distributions, it can be shown that the 5th percentile is at −1.645 for the standard normal distribution, at −1.623 for the standard logistic, and at −1.675 for the standard triangular. So, if the mean and standard deviations are equal, the triangular distribution produces the lowest HC_5 and the logistic the highest; however, the differences are small.

Despite the similarity of the distributions, both Wagner and Løkke (1991) and Aldenberg and Slob (1993) noted that the method of Van Straalen and Denneman (1989) tended to overestimate HC_p. Aldenberg and Slob (1993) illustrated this by conducting Monte Carlo simulations, showing that the frequency of estimated HC_p values that were higher than the true HC_p was considerably greater than the value of δ suggested in the formula. Consequently, the authors developed new (tabulated) values for the function g in Equation 3.2. In retrospect, it appears that Van Straalen and Denneman (1989) overlooked an approximation made by Kooijman (1987), which is valid in the case of HCS, but not in the case of HC_5. In his derivation Kooijman (1987) at a certain point neglects the influence of sampling error in the mean and continues by taking only sampling error in the standard deviation into

account. This is justified when deriving the expression for HCS, because HCS is far
from the mean and is affected mainly by errors in the standard deviation. The same
approximation does not hold, however, in the derivation of HC_5, because at the 5%
level errors in the mean still have an effect on the estimation of percentiles.

In all the papers mentioned thus far, the value of δ was usually choosen to be
0.05; that is, HC_p was considered as the lower (one-sided) 95% confidence limit of
the p-percentile. Aldenberg and Slob (1993), however, also tabulated the function g
for $\delta = 0.5$. Estimates derived in this way represent a 50% confidence level for HC_p;
that is, the value will be too high in 50% of the cases. Clearly, this new element
removed the built-in safety effect that allowed for very low values of HC_p when the
data set was very small. The impact of sample size, m, in the method of Aldenberg
and Slob (1993) became much smaller than it was in the methods of Kooijman
(1987), Van Straalen and Denneman (1989), and Wagner and Løkke (1991). The
reasons for choosing the 50% confidence level for HC_p, rather than the 95% confidence level, were mainly political. It was not acceptable to policy makers that
environmental quality criteria reflected so strongly the influence of dearth of data.
For many substances, the number of test data was much lower than the number of
data on which the methods had been tried. The method of Aldenberg and Slob (1993)
with the 50% confidence level quickly became the national standard for deriving
environmental quality criteria in the Netherlands (Slooff, 1992).

3.5 CONCERNS AND VALIDATION

When the framework of species sensitivity distributions was adopted by policy
makers to derive environmental quality criteria, it triggered a variety of activities
and criticisms. First, several authors expressed their concern about the idea of HC_5
and its ecological implications. Second, studies were started that aimed to compare
HC_5 values derived from laboratory ecotoxicity tests with field observations and
multispecies tests. Third, it was felt that in certain areas (specifically soil) too little
data were available to apply the methodology and programs were started to develop
new test methods. This section discusses some of the main criticisms, as far as they
are part of European history. Other papers (e.g., Smith and Cairns, 1993) are discussed elsewhere in this book.

Many authors have raised objections against the idea of HC_p on the ground that
the $p\%$ of species in a community left "unprotected" could comprise species of great
ecological relevance (e.g., ecological engineers). In that case an environmental
concentration equal to HC_5 would not only affect the sensitive 5% of the species,
but also, through indirect consequences from the loss of an engineer, other parts of
the ecosystem. For example, Hopkin (1993) discussed the ecological implication of
the "95% protection level" for metals in soil and argued that if a species of large
earthworm was among the 5%, its absence could result in a decrease in the decomposition rate of dead plant material and accumulation of undecomposed leaf litter
on the soil surface, which could lead to a substantial reduction in the rate of nutrient
release to the soil. Van Straalen (1993a) argued that if evidence were available for
such effects to occur, environmental quality criteria should be set below HC_5, to
protect these species. The same argument is valid for species that have a special

value for conservation, recreation, or fisheries. This specification was in fact already adopted in policy documents (Health Council of the Netherlands, 1989; Van Leeuwen, 1990); however, concrete examples demonstrating toxicity to an engineering or commercial species below HC_5 have never been published. Maybe the concern should not be with the ecological engineers, as these often represent robust species that will not likely have NOECs below HC_5 (Eijsackers, 1997), but with rare species that escape testing and may be sensitive due to specific habitat requirements.

As a result of the concerns discussed above, the new risk assessment methodology triggered several studies that aimed to compare HC_p values derived from laboratory tests with critical concentrations derived from multispecies tests or field tests. In the Netherlands, the issue was primarily driven by policy makers, who were confronted with questions asked in Parliament. Van Leeuwen (see VROM, 1991) argued that the theoretical cutoff value of 5% coincided with no-effect concentrations from field tests. This was supported by several desk studies. Emans et al. (1993) and Okkerman et al. (1993) analyzed 300 literature references reporting effect levels of toxicants in multispecies test systems. After applying basic criteria of quality control, 29 different NOEC values could be derived from these studies. For 18 of these chemicals an estimate for HC_5 was derived from single species toxicity data. A good correlation was obtained between the multispecies NOEC and the HC_5 values estimated by the methods of Aldenberg and Slob (1993) and Wagner and Løkke (1991), both with 50% confidence. On the average, the HC_5 values derived from these methods were a factor of 3.4 and 4.3, respectively, lower than the average multispecies NOECs. The (cautiously formulated) conclusion was that the single-species test results, extrapolated to sensitive species using one of the distribution-based approaches, are indeed predictive of effects in multispecies systems.

For soil such extensive comparisons have not been made because of the lack of a good database. On a limited scale, Posthuma (1997) reviewed data obtained by Hågvar and Abrahamsen (1990) on the abundance of soil microarthropods in a naturally occurring gradient of lead pollution. The data were used to estimate the concentration in soil that corresponded to 5% loss of the number of species. This was estimated as 219 mg/kg, which was higher, but not significantly different from the HC_5 value of 77 mg/kg estimated from laboratory toxicity data. More extensive studies comparing field surveys with semifield experiments and laboratory toxicity tests allowed a similar conclusion: at concentrations below the laboratory-based HC_5 no effects are seen under field conditions (Posthuma et al., 1998a).

The validation studies reported in the literature seem to agree on the fact that the distribution-based approach produces estimates that are a factor of 2 to 5 below the values established in field tests or multispecies enclosures. Despite the ecological concerns discussed above, HC_5 values actually seem to be more conservative than field tests. For soil, the single largest factor dominating differences between laboratory and field is the question of bioavailability. Because of the lack of equilibration of freshly prepared media, toxicants in laboratory tests are usually more available to the test organisms, which causes an overestimation of toxicity. If the bioavailability differences between laboratory and field can be diminished, for example, by aging of the media, laboratory tests should be able to predict toxicity under field conditions quite well (Smit and Van Gestel, 1998).

In conclusion, probably one of the biggest mistakes in communicating the HC_5 approach to the scientific and regulatory community has been the percentage of "unprotected" species or the implication of the 95% protection level. Many people interpreted this as if 5% of the species were sacrificed with each chemical that came on the market. In retrospect, it would have been better to state that the HC_5 approach is one of the methods to derive PNECs for ecosystems and that HC_5 values are derived to protect ecosystems against the adverse effects of chemicals. As a consequence, a "statistical cutoff value" of 5% is needed to obtain the PNEC. This value of 5% is a pragmatic choice but has been validated on the basis of field studies.

Another point of discussion was the fact that the species sensitivity distribution approach places an emphasis on species, not on trophic structures or ecological processes. This was explicitly recognized by Health Council of the Netherlands (1989) and defended by stating that the protection of species is not a goal in itself, but an instrument to protect ecosystems as a whole. Consequently, the assumption was made that protection of all the species in an ecosystem (its biotic structure) is a sufficient condition for protecting ecosystem functions. Forbes and Forbes (1993) questioned the validity of this assumption and argued that as long as it remained untested, the HC_5 approach should not be applied. Forbes and Forbes (1994), Fawell and Hedgecott (1996), and Calow (1996) considered the statistical approach as too complicated and argued that the classical approach using assessment factors (e.g., dividing the lowest NOEC by 10) should be retained.

Unfortunately, ecology does not have a good answer yet to the question of how species are related to ecosystem function. Lawton (1994) distinguished three different hypotheses about the relationship between biodiversity and ecosystem function, but none of these can be proved or disproved at the moment. In experiments with microcosms containing algae, bacteria, and protozoans, Naeem and Li (1997) showed that more species per functional group made the systems more stable. The loss of one species in a functional group could be compensated for by growth of other species in the same group. In an elegant modeling study, Klepper et al. (1999) showed that measures of overall ecosystem functioning are affected by toxic stress only at high levels of pollution, especially in communities of high biodiversity. This study and other papers (see also Levin et al., 1989) seem to support the idea of functional redundancy, that is, that not all species contribute equally to the function of an ecosystem and functions are sustained even when some species are lost. Consequently, the assumption that species protection is more conservative than protection of functions still stands, despite the objections raised.

A more serious concern relates to the representativeness of test species. Even in the first publications, it was realized that test species do not represent a random sample from a local community. Test species are usually selected for their manageability under laboratory conditions, their sensitivity to toxicants, and their ecological relevance. If a true random sample from the environment were taken, at least 50% of all test species would be insects, as these represent about one half of the world's biodiversity (ignoring bacteria). Instead, test species are usually chosen to include representatives from different taxonomic groups and different trophic levels (Van Leeuwen and Hermens, 1995). It is evident that such a biased selection of species can have a serious effect on the estimation of HC_p, especially when the number of species is small. The species sensitivity approach derives its strength from large numbers. Empirical calculations show that HC_5 estimates stabilize if the number of input data becomes higher than ten (Vega et al., 1999).

The issue of representativeness of test species cannot be solved without further knowledge about patterns in sensitivity to toxicants among taxonomic, trophic, or ecological groups (Van Straalen, 1994). The impression of many ecotoxicologists is that there is relatively little correlation among the sensitivities of species to toxicants. The issue of patterns in ecotoxicological data sets was explored by Hoekstra et al. (1994), Calabrese and Baldwin (1994), and Vaal et al. (1997a,b). It appeared that most of the variation in ecotoxicity data for the aquatic environment was due to differences between chemicals, rather than intrinsic differences between species. Almost every species could be used to represent the mean toxicity of chemicals and to rank them. Some group-specific patterns were revealed; for example, fish and amphibians were relatively sensitive to dieldrin, lindane, and pentachlorophenol, in comparison with invertebrates. It also appeared that the largest interspecies variation (amounting to a factor of 10^5 to 10^6) was present among toxicants with a specific toxic mode of action that were very toxic to some species.

The adoption of the species sensitivity distribution approach by environmental policy makers also revealed a lack of data for certain categories of substances and certain environmental compartments. This was often due to the absence of internationally accepted guidelines for laboratory tests. For soil, in particular, significant efforts were made to increase the number of test guidelines. This resulted in research projects conducted during the 1990s in various European countries commissioned by national authorities as well as the European Union. One of the results of these activities was the *Handbook of Soil Invertebrate Toxicity Tests* (Løkke and van Gestel, 1998).

Van Leeuwen et al. (1991) have derived quantitative structure–activity relationships (QSARs) in which HC_5 values for certain categories of chemicals were related to log K_{ow}. It appeared that for nonreactive aquatic pollutants, good relationships could be derived; this offered a realistic alternative to testing every chemical on a large number of species, as would be required for the derivation of new HC_5 values. Notenboom et al. (1995) coined the term *quantitative species sensitivity relationships* (QSSRs) to denote possible predictive functions for species sensitivity. The idea was that some attributes of species such as body size, lipid content, or phylogenetic position might be used as predictive variables for tolerance to toxicants, similar to the way the octanol–water coefficient is used in QSARs to predict the action of a toxicant. The idea of QSSR, despite its attractiveness, has, however, remained a rather theoretical concept up to now.

3.6 CONCLUSIONS

In evolutionary terms, the idea of species sensitivity distributions originated as a single event (Kooijman, 1987), but soon radiated into a variety of approaches. The evolution was quite independent of what happened in North America, but at the same time very similar, including even the idea of a 5% cutoff value. The whole situation is strongly reminiscent of a case of convergent evolution, driven by similar science-policy discussions on risk assessment.

As illustrated by other chapters in this book, the developments in the use of species sensitivity distributions are still under way. Some recently added new developments include the incorporation of food chain modeling (Traas et al., 1998a; Klepper et al., 1999), the assessment of transient effects including ecological recovery (Van Straalen and Van Rijn, 1998), and the inverse assessment approach leading to the concept of "potentially affected fraction" (Klepper et al., 1998; Knoben et al., 1998; Aldenberg and Jaworska, 2000). Debate has arisen about the question whether the approach is applicable to essential metals, which have a natural background level and a lower limit related to deficiency phenomena (Hopkin, 1993). Undoubtedly, new additions will be made to the theory in the coming years. For ecotoxicology as a scientific discipline, some fundamental issues deserve attention:

- What are the reasons some species are more susceptible than others?
- Which factors control the differences between species sensitivity in laboratory and field?
- How do distributed sensitivities relate to higher levels of ecological organization?

It remains to be emphasized that the distribution-based approach is only one aspect of extrapolation in ecotoxicology; it does not deal with issues such as mixture toxicity, genetic adaptation, bioavailability, etc. Still it remains a lively area of activity where ecotoxicologists and policy makers meet and sometimes join.

ACKNOWLEDGMENTS

The authors are grateful to Bas Kooijman, two reviewers, and three editors for various comments on an early version of the manuscript.

Section II

Scientific Principles and Characteristics of SSDs

This section deals with theoretical and technical issues in the derivation of species sensitivity distributions (SSDs). The issues include the assumptions underlying the use of SSDs, the selection and treatment of data sets, methods of deriving distributions and estimating associated uncertainties, and the relationship of SSDs to distributions of exposure. A unifying statistical principle is presented, linking all current definitions of risk that have been proposed in the framework of SSDs. Further, methods for using SSDs for data-poor substances are proposed, based on effect patterns related to the toxic mode of action of compounds. Finally, the issue of SSD validation is addressed in this section. These issues are the basis for the use of SSDs in setting environmental criteria and assessing ecological risks (Sections IIIA and B).

4 Theory of Ecological Risk Assessment Based on Species Sensitivity Distributions

Nico M. van Straalen

CONTENTS

Abstract — The risk assessment approach to environmental protection can be considered as a quantitative framework in which undesired consequences of potentially toxic chemicals in the environment are identified and their occurrence is expressed as a relative frequency, or a probability. Risk assessment using species sensitivity distributions (SSDs) focuses on one possible undesired event, the exposure of an arbitrarily chosen species to an environmental concentration greater than its no-effect level. There are two directions in which the problem can be addressed: the forward approach considers the concentration as given and aims to estimate the risk; the inverse approach considers the risk as given and aims to estimate the concentration. If both the environmental concentration (PEC) and the no-effect concentration (NEC) are distributed variables, the expected value of risk can be obtained by integrating the product of the probability density function of PEC and the cumulative distribution of NEC over all concentrations. Analytical expressions for the expected value of risk soon become rather formidable, but numerical integration is still possible. An application of the theory is provided, focusing on the interaction between soil acidification and toxicity of metals in soil. The analysis shows that the concentration of lead in soil that is at present considered a safe reference value in the Dutch soil protection regulation may cause an unacceptably high risk to soil invertebrate communities when soils are acidified from pH 6.0 to pH 3.5. This is due to a strong nonlinear effect of lead on the expected value of risk with decreasing pH. For cadmium the nonlinear component is less pronounced. The example illustrates the power of quantitative risk assessment using SSDs in scenario analysis.

4.1 INTRODUCTION

The idea of risk has proved to be an attractive concept for dealing with a variety of environmental policy problems such as regulation of discharges, cleanup of contaminated land, and registration of new chemicals. The reason for its attractiveness seems to be because of the basically quantitative nature of the concept and because it allows a variety of problems associated with human activities to be expressed in a common currency. This chapter addresses the scientific approach to ecological risk assessment, with emphasis on an axiomatic and theoretically consistent framework.

The most straightforward definition of risk is *the probability of occurrence of an undesired event*. As a probability, risk is always a number between 0 and 1, sometimes multiplied by 100 to achieve a percentage. An actual risk will usually lie closer to 0 than to 1, because undesired events by their nature are relatively rare. For the moment, it is easiest to think of probabilities as relative frequencies, being measured by the ratio of actual occurrences to the total number of occurrences possible. The mechanisms generating probabilities will be discussed later in this chapter.

In the scientific literature and in policy documents, many different definitions of risk have been given. Usually, not only the occurrence of undesired events is considered as part of the concept of risk, but also the magnitude of the effect. This is often expressed as risk = (probability of adverse effect) × (magnitude of effect). This is a misleading definition, because, as pointed out by Kaplan and Garrick (1981), it equates a low probability/high damage event with a high probability/low damage event, which are obviously two different types of events, not the same thing at all. This is not to negate the importance of the magnitude of the effect; rather, magnitude of effect should be considered *alongside* risk. This was pictured by Kaplan and Garrick (1981) in their concept of the "risk curve": a function that defines the relative frequency (occurrence) of a series of events ordered by increasing severity. So risk itself should be conceptually separated from the magnitude of the effect; however, the *maximum acceptable risk* for each event will depend on its severity.

When the above definition of risk is accepted, the problem of risk assessment reduces to (1) specifying the undesired event and (2) establishing its relative incidence. The undesired event that I consider the basis for the species sensitivity framework is *a species chosen randomly out of a large assemblage is exposed to an environmental concentration greater than its no-effect level*.

It must be emphasized that this endpoint is only one of several possible. Suter (1993) has extensively discussed the various endpoints that are possible in ecological risk assessments. Undesired events can be indicated on the level of ecosystems, communities, populations, species, or individuals. Specification of undesired events requires an answer to the question: what is it that we want to protect in the environment? The species sensitivity distribution (SSD) approach is only one, narrowly defined, segment of ecological risk assessment. Owing to its precise definition of the problem, however, the distribution-based approach allows a theoretical and quantitative treatment, which adds greatly to its practical usefulness.

It is illustrative to break down the definition of our undesired event given the above and to consider each part in detail:

- *"A species chosen randomly"* — This implies that species, not individuals, are the entities of concern. Rare species are treated with the same weight as abundant species. Vertebrates are considered equal to invertebrates. It also implies that species-rich groups, e.g., insects, are likely to dominate the sample taken.
- *"Out of a large assemblage"* — There is no assumed structure among the assemblage of species, and they do not depend on each other. The fact that some species are prey or food to other species is not taken into account.
- *"Is exposed to an environmental concentration"* — This phrase assumes that there is a concentration that can be specified, and that it is a constant. If the environmental concentration varies with time, risk will also vary with time and the problem becomes more complicated.
- *"Greater than its no-effect level"* — This presupposes that each species has a fixed no-effect level, that is, an environmental concentration above which it will suffer adverse effects.

Considered in this very narrowly defined way, the distribution-based assessment may well seem ridiculous because it ignores all ecological relations between the species. Still the framework derived on the basis of the assumptions specified is interesting and powerful, as is illustrated by the various examples in this book that demonstrate applications in management problems and decision making.

4.2 DERIVATION OF ECOLOGICAL RISK

If it is accepted that the undesired event specified in the previous section provides a useful starting point for risk assessment, there are two ways to proceed, which I call the forward and the inverse approach (Van Straalen, 1990). In the *forward problem*, the exposure concentration is considered as given and the risk associated with that exposure concentration has to be estimated. This situation applies when chemicals are already present in the environment and decisions have to be made regarding the acceptability of their presence. Risk assessment can be used here to decide on remediation measures or to choose among management alternatives. Experiments that fall under the forward approach are bioassays conducted in the field to estimate *in situ* risks. In the *inverse problem*, the risk is considered as given (set, for example to a maximum acceptable value) and the concentration associated with that risk has to be estimated. This is the traditional approach used for deriving environmental quality standards. The experimental counterpart of this consists of ecotoxicological testing, the results of which are used for deriving maximum acceptable concentrations for chemicals that are not yet in the environment.

Both in the forward and the inverse approach, the no-effect concentration (NEC) of a species is considered a distributed variable c. For various reasons, including asymmetry of the data, it is more convenient to consider the logarithm of the concentration than the concentration itself. Consequently, I will use the symbol c to denote the logarithm of the concentration (to the base e). Although the concentration itself can vary, in principle, from 0 to ∞, c can vary from $-\infty$ to $+\infty$, although, in practice a limited range may be applicable. Denote the probability density distribution for c by $n(c)$, with the interpretation that

$$\int_{c_1}^{c_2} n(c)dc \tag{4.1}$$

equals the probability that a species in the assemblage has a log NEC between c_1 and c_2. Consequently, if only c is a distributed variable and the concentration in the environment is constant, ecological risk, δ, may be defined as:

$$\delta = \int_{-\infty}^{h} n(c)dc \tag{4.2}$$

where h is the log concentration in the environment. In the forward problem, h is given and δ is estimated; in the inverse problem, δ is given and h is estimated.

There are various possible distribution functions that could be taken to represent $n(c)$, for example, the normal distribution, the logistic distribution, etc. These distributions have parameters representing the mean and the standard deviation. Consequently, Equation 4.2 defines a mathematical relationship among δ, h, and the mean and standard deviation of the distribution. This relationship forms the basis for the estimation procedure.

The assumption of a constant concentration in the environment can be relaxed relatively easily within the framework outlined above. Denote the probability density function for the log concentration in the environment by $p(c)$, with the interpretation that:

$$\int_{c_1}^{c_2} p(c)dc \tag{4.3}$$

equals the probability that the log concentration falls between c_1 and c_2. The ecological risk δ, as defined above, can now be expressed in terms of the two distributions, $n(c)$ and $p(c)$ as follows:

$$\delta = \int_{-\infty}^{\infty} (1 - P(c))n(c)dc = 1 - \int_{-\infty}^{\infty} P(c)n(c)dc \tag{4.4}$$

where $P(c)$ is the cumulative distribution of $p(c)$, defined as follows:

$$P(c) = \int_{-\infty}^{c} p(u)du \tag{4.5}$$

where u is a dummy variable of integration (Van Straalen, 1990). Again, if n and p are parameterized by choosing specific distributions, δ may be expressed in terms of the means and standard deviations of these distributions. The actual calculations can become quite complicated, however, and it will in general not be possible to derive simple analytical expressions. For example, if n and p are both represented by logistic distributions, with means μ_n and μ_p, and shape parameters β_n and β_p, Equation 4.4 becomes

$$\delta = 1 - \int_{-\infty}^{\infty} \frac{\exp\left(\dfrac{\mu_n - c}{\beta_n}\right)}{\beta_n \left[1 + \exp\left(\dfrac{\mu_n - c}{\beta_n}\right)\right]^2 \left[1 + \exp\left(\dfrac{\mu_p - c}{\beta_p}\right)\right]} dc \qquad (4.6)$$

Application of this equation would be equivalent to estimating PEC/PNEC ratios (predicted environmental concentrations over predicted NECs). Normally in a PNEC/PEC comparison only the means are compared and their quotient is taken as a measure of risk (Van Leeuwen and Hermens, 1995). However, even if the mean PEC is below the mean PNEC, there still may be a risk if there is variability in PEC and PNEC, because low extremes of PNEC may concur with high extremes of PEC. In Equation 4.6 both the mean and the variability of PNEC and PEC are taken into account.

Equation 4.4 is graphically visualized in Figure 4.1a. This shows that δ can be considered an area under the curve of $n(c)$, after this is multiplied by a fraction that becomes smaller and smaller with increasing concentration. If there is no variability in PEC, $P(c)$ reduces to a step function, and δ becomes equivalent to a percentile of the $n(c)$ distribution (Figure 4.1b). So the derivation of HC_p (hazardous concentration for $p\%$ of the species) by the methods explained in Chapters 2 and 3 of this book, is a special case, which ignores the variability in environmental exposure concentrations, of a more general theory.

Equation 4.4 can also be written in another way; applying integration by parts and recognizing that $N(-\infty) = P(-\infty) = 0$ and $N(\infty) = P(\infty) = 1$, the equation can be rewritten as

$$\delta = \int_{-\infty}^{\infty} p(c)N(c)dc \qquad (4.7)$$

where $N(c)$ is the cumulative distribution of $n(c)$, defined by

$$N(c) = \int_{-\infty}^{c} n(u)du \qquad (4.8)$$

where u is a dummy variable of integration. A graphical visualization of Equation 4.7 is given in Figure 4.1c. Again, δ can be seen as an area under the curve, now the

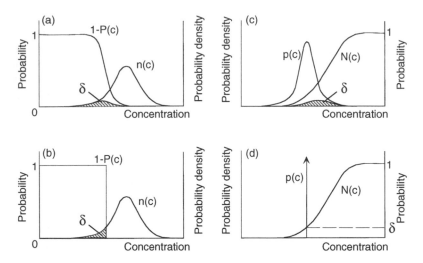

FIGURE 4.1 Graphical representation of the calculation of ecological risk, δ, defined as the probability that environmental concentrations are greater than NECs. The probability density distribution of environmental concentrations is denoted $p(c)$, the distribution of NECs is denoted $n(c)$. $P(c)$ and $N(c)$ are the corresponding cumulative distributions. In a and c, both variables are subject to error; in b and d, the environmental concentration is assumed to be constant. Parts a and b illustrate the calculation of δ according to Equation 4.4; parts c and d illustrate the (mathematically equivalent) calculation according to Equation 4.7. Part b illustrates the derivation of HC_p (see Chapter 3), and part d is equivalent to the graphical representation in Solomon (1996).

curve of $p(c)$, after it is multiplied by a fraction that becomes larger and larger with increasing c. In the case of no variability in $p(c)$, it reduces to an impulse (Dirac) function at $c = h$. In that case δ becomes equal to the value of $N(c)$ at the intersection point (Figure 4.1d). This graphical representation was chosen by Solomon et al. (1996) in their assessment of triazine residues in surface water.

The theory summarized above, originally formulated in Van Straalen (1990), is essentially the same as the methodology described by Parkhurst et al. (1996). These authors argue from basic probability theory, derive an equation equivalent to Equation 4.7, and also provide a simple discrete approximation. This can be seen as follows. Suppose that the concentrations of a chemical in the environment can be grouped in a series of discrete classes, each class with a certain frequency of occurrence. Let p_i be the density of concentrations in class i, with width Δc_i, and N_i the fraction of species with a NEC below the median of class i, then

$$\delta = \sum_{i=1}^{m} p_i N_i \Delta c_i \tag{4.9}$$

if there are m classes of concentration covering the whole range of occurrence. The calculation is illustrated here using a fictitious numerical example with equal class widths (Table 4.1). The example shows that, given the values of p_i and N_i provided,

TABLE 4.1
Numerical (Fictitious) Example Illustrating the Calculation of Expected Value of Risk (δ)

Class No., i	Concentration Interval, Δc	Probability of Concentration in the Environment, $p_i \Delta c$	Cumulative Probability of Affected Species at Median of Class, N_i	Risk per Interval[a]
1	0–10	0	0.01	0
2	10–20	0.10	0.05	0.005
3	20–30	0.48	0.10	0.048
4	30–40	0.36	0.30	0.108
5	40–50	0.06	0.60	0.036
6	>50	0	1	0
Total		1		**0.197** (= δ)

[a] From the probability density of concentrations in the environment (p_i) and the cumulative probability of affected species (N_i), according to Equation 4.9 in the text.

the expected value of risk is 19.7%. In the example, the greatest component of the risk is associated with the fourth class of concentrations, although the third class has a higher frequency (Table 4.1).

In summary, this section has shown that the risk assessment approaches developed from SSDs, as documented in this book, can all be derived from the same basic concept of risk as the probability that a species is exposed to an environmental concentration greater than its no-effect level. Both the sensitivities of species and the environmental concentrations can be viewed as distributed variables, and once their distributions are specified, risk can be estimated (the forward approach) or maximum acceptable concentrations can be derived (the inverse approach).

4.3 PROBABILITY GENERATING MECHANISMS

The previous section avoided the question of the actual reason that species sensitivity is a distributed variable. Suter (1998a) has rightly pointed out that the interpretation of sensitivity distributions as probabilistic may not be quite correct. The point is that probability distributions are often postulated without specifying the mechanism generating variability.

One possible line of reasoning is: "Basically the sensitivity of all species are the same, however, our measurement of sensitivity includes errors. That is why species sensitivity comes as a distribution."

Another line of reasoning is: "There are differences in sensitivity among species. The sensitivity of each species is measured without error, but some species appear to be inherently more sensitive than others. That is why species sensitivity comes as a distribution."

In the first view, the reason one species happens to be more sensitive than another is not due to species-specific factors, but to errors associated with testing, medium preparation, or exposure. A given test species can be anywhere in the distribution, not at a specific place. The choice of test species is not critical, because each species can be selected to represent the mean sensitivity of the community. The distribution could also be called "community sensitivity distribution." According to this view, ecological risk is a true probability, namely, the probability that the community is exposed to a concentration greater than its no-effect level.

In the second view, the distribution has named species that have a specified position. When a species is tested again, it will produce the same NEC. There are patterns of sensitivity among the species, due to biological similarities. The choice of test species is important, because an overrepresentation of some taxonomic groups may introduce bias in the mean sensitivity estimated.

Suter (1998a) pointed out that the second view is not to be considered probabilistic. The mechanism generating the distribution in this case is entirely deterministic. The cumulative sensitivity distribution represents a gradual increase of effect, rather than a cumulative probability. When the concept of HC_5 (see Chapters 2 and 3) is considered as a concentration that leaves 5% of the species unprotected, this is a nonprobabilistic view of the distribution. The problem is similar to the difference between LC_{50}, as a stochastic variable measured with error, and EC_{50}, as a deterministic 50% effect point in a concentration–response relationship. According to the second view, the SSD represents variability, rather than uncertainty.

Although the SSD may not be considered a true probability density distribution, there is an element of probability in the estimation of its parameters. Parameters such as μ and β in Equation 4.6 are unknown constants whose values must be estimated from a sample taken from the community. Since the sampling procedure will introduce error, there is an element of uncertainty in the risk estimation. This, then, is the probabilistic element. Ecological risk itself can be considered a deterministic quantity (a measure of relative effect, i.e., the fraction of species affected), which is estimated with an uncertainty margin due to sampling error. The probability generating mechanism is the uncertainty about how well the sample represents the community of interest. This approach was taken when establishing confidence intervals for HCS and HC_p (see Chapter 3). It is also similar to the view expressed by Kaplan and Garrick (1981), who considered the relative frequency of events as separate from the uncertainty associated with estimating these frequencies. Their concept of risk curve includes both types of probabilities.

It is difficult to say whether the probability generating mechanism should be restricted to sampling only. In practice, the determination of sensitivity of one species is already associated with error and so the SSD does not represent pure biological differences. In the extreme case, differences between species could be as large as differences between replicated tests on one species (or tests conducted under different conditions). A sharp distinction between variance due to specified factors (species) and variance due to unknown (random) error is difficult to make. This shows that the discussion about the interpretation of distributions is partly semantic. Considering the general acceptance of the word *risk* and its association with probabilities, there does not seem to be a need for a drastic change in terminology, as long as it is understood what is analyzed.

4.4 SPECIES SENSITIVITY DISTRIBUTIONS IN SCENARIO ANALYSIS

Most of the practical applications of SSDs in ecotoxicology have focused on the derivation of environmental quality criteria. However, a perhaps more powerful use of the concept is the estimation of risk (δ) for different options associated with a management problem. The lowest value for δ would then indicate the preferable management option. Different scenarios for environmental management or emission of chemicals could be assessed, based on minimization of δ or on an optimization of risk reduction vs. costs. An example of this approach is given in a report by Kater and Lefèvre (1996). These authors were concerned with management options for a contaminated estuary, the Westerschelde, in the Netherlands. Different scenarios were considered, dredging of contaminated sediments and emisson reductions. Risk estimations showed that zinc and copper were the most problematic components.

Another interesting aspect of the species sensitivity framework is that the concept of ecological risk (δ) can integrate different types of effects and can express their joint risk in a single number. If we consider two independent events, for example, exposure to two different chemicals, the joint risk, δ_T can be expressed in terms of the individual risks, δ_1 and δ_2, as follows:

$$\delta_T = 1 - (1 - \delta_1)(1 - \delta_2) \tag{4.10}$$

or in general:

$$\delta_T = 1 - \prod_{i=1}^{n} (1 - \delta_i) \tag{4.11}$$

if there are n independent events. The concept of δ lends itself very well to use in maps, where it can serve as an indicator of toxic effects if the concentrations are given in a geographic information system. In this approach, δ can be considered as the fraction of a community that is potentially affected by a certain concentration of chemical, abbreviated PAF. The PAF concept was applied by Klepper et al. (1998) to compare the risks of heavy metals with those of pesticides, in different areas of the Netherlands, and by Knoben et al. (1998) to measure water quality in monitoring programs. In general, PAF can be considered an indicator for "toxic pressure" on ecosystems (Van de Meent, 1999; Chapter 16).

To illustrate further the idea of scenario analysis based on SSDs, I will review an example concerning interaction between soil acidification and ecological risk (Van Straalen and Bergema, 1995). In this analysis, data on ecotoxicity of cadmium and lead to soil invertebrates were used to estimate ecological risk (δ) for the so-called reference values of these metals in soil. Reference values are concentrations equivalent to the upper boundary of the present "natural" background in agricultural and forest soils in the Netherlands. Literature data were collected about metal concentrations in earthworms as a function of soil pH. These data very clearly showed that internal metal concentrations increased in a nonlinear fashion with decreasing pH. The increase of bioconcentration with pH was modeled by means of regression lines (see Van Gestel et al., 1995). Van Straalen and Bergema (1995) subsequently assumed that the NECs of the individual invertebrates (expressed in mg per kg of soil) were proportionally lower at lower pH, because a higher internal concentration implies a higher risk, even at a constant external concentration. To quantify the increase of effect, the regressions for the bioconcentration factors were applied to the NECs of the individual invertebrates. In this way, it became possible to estimate δ as a function of soil pH.

The analysis is summarized in Figure 4.2. The scenario is a decrease of pH from 6.0 (pertaining to most of the toxicity experiments) to pH 3.5, an extreme case of acidification that would arise, for example, from soil being taken out of agriculture, planted with trees, and left to acidify when a forest would develop on it. The total concentration of metal in soil is assumed to remain the same under acidification. Because of the rise of concentrations of metals in invertebrates, the community would become more "sensitive" when sensitivity remains expressed in terms of the total concentration. Consequently, a larger part of community is exposed to a concentration above the no-effect level. The effect is stronger for lead than for cadmium. For Cd, δ would increase from 0.051 to 0.137, whereas for Pb it would increase from 0.015 to 0.767. The strong increase in the case of Pb is due to the nonlinear effect of pH on Pb bioavailability and the fact that the SSD for Pb is narrower than the one for Cd. Interestingly, the present reference value for Pb (85 mg/kg) is less of a problem than the reference value for Cd (0.8 mg/kg); however, when soils are acidified, Pb becomes a greater problem than Cd.

Of course, this example is a rather theoretical exercise and should not be judged on its numerical details. It nevertheless shows that quantitative risk estimations can be very well combined with scenario analysis and that even though the absolute values of expected risks may not be realistic, a comparison of alternatives can be helpful. In such calculations, estimations of δ may be considered indicators rather than actual risks pertaining to a concrete situation.

4.5 CONCLUSIONS

There are many different kinds of ecological risks associated with environmental contamination. Each corresponds to an undesired event whose incidence we want to minimize. The theory of SSDs can be seen as a framework that elaborates on one of these events, the exposure of a species above its no-effect level. There are two approaches in the theory, one arguing forward from concentrations to risks, the other arguing inversely from risks to concentrations. The theory is now well developed and parts of it are beginning to be accepted by environmental policy makers. Risk estimates derived from SSDs can be taken as indicators in scenario analysis.

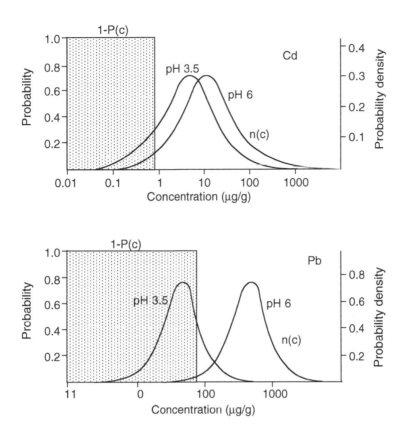

FIGURE 4.2 SSDs for lead and cadmium effects on soil invertebrates, estimated from the literature. For both metals, a scenario of soil acidification is illustrated, in which no-effect levels expressed as total concentrations in soil are assumed to decrease with decreasing pH in proportion to documented increases of bioconcentration of metal in the body. The result is that the distributions are shifted to the left when pH is lowered from 6.0 to 3.5. A larger part of the community is then exposed to a concentration greater than the reference value (0.8 mg/kg for Cd, 85 mg/kg for Pb). (From Van Straalen, N. M. and Bergema, W. F., *Pedobiologica*, 39, 1, 1995. With permission from Gustav Fischer Verlag, Jena.)

Some authors have dismissed the whole idea of risk assessment as too technocratic and not applicable to complicated systems (Lackey, 1997). I do not agree with this view. I believe that a precise dissection of risk into its various aspects will actually help to better define management questions, since a quantitative risk assessment requires the identification of undesired events (endpoints). When doing a risk assessment, we are forced to answer the question, "What is it that we want to protect in the environment?" Clearly, the SSD considers just one type of event. It does not deal with extrapolations other than the protection of sensitive species. The challenge for risk assessment now is to define other endpoints and develop quantitative approaches as strong as the SSD approach for them. Risk assessment then becomes a multidimensional problem, in which a suite of criteria has to be considered at the same time.

ACKNOWLEDGMENTS

I am grateful to three anonymous reviewers and the editors of this book for comments on an earlier version of the manuscript. In particular I thank Theo Traas for drawing my attention to The Cadmus Group report and Tom Aldenberg for pointing out some inconsistencies in the formulas.

5 Normal Species Sensitivity Distributions and Probabilistic Ecological Risk Assessment

Tom Aldenberg, Joanna S. Jaworska, and Theo P. Traas

CONTENTS

1-56670-578-9/02/$0.00+$1.50
© 2002 by CRC Press LLC

Abstract — This chapter brings together several statistical methods employed when identifying and evaluating species sensitivity distributions (SSDs). The focus is primarily on normal distributions and it is shown how to obtain a simple "best" fit, and how to assess the fit. Then, advanced Bayesian techniques are reviewed that can be employed to evaluate the uncertainty of the SSD and derived quantities. Finally, an integrative account of methods of risk characterization by combining exposure concentration distributions with SSDs is presented. Several measures of ecological risk are compared and found to be numerically identical. New plots such as joint probability curves are analyzed for the normal case. A table is presented for calculating ecological risk of a toxicant when both exposure concentration distributions and SSDs are normal.

5.1 INTRODUCTION

A species sensitivity distribution (SSD) is a probabilistic model for the variation of the sensitivity of biological species for one particular toxicant or a set of toxicants. The toxicity endpoint considered may be acute or chronic in nature. The model is probabilistic in that — in its basic form — the species sensitivity data are only analyzed with regard to their statistical variability. One way of applying SSDs is to protect laboratory or field species assemblages by estimating reasonable toxicant concentrations that are safe, and to assess the risk in situations where these concentrations do not conform to these objectives.

The Achilles' heel of "SSDeology" is the question: From what (statistical) population are the data considered a sample? We want to protect communities, and the — on many occasions — implicit assumption is that the sample is *representative* for some target community, e.g., freshwater species, or freshwater species in some type of aquatic habitat. One may develop SSDs for species, genera, or other levels of taxonomic, target, or chemical-specific organization. Hall and Giddings (2000) make clear that evaluating single-species toxicity tests alone is not sufficient to obtain a complete picture of (site-specific) ecological effects. However, this chapter investigates what statistical methods can be brought to bear on a set of single-species toxicity data, when that is the only information available.

The SSD model may be used in a *forward* or *inverse* sense (Van Straalen, Chapter 4). The focus in forward usage is the estimation of the proportion or fraction of species (potentially) affected at given concentration(s). Mathematically, forward usage employs some estimate of the *cumulative distribution function* (CDF) describing the toxicity data set. The fraction of species (potentially) affected (FA or PAF), or "risk," is defined as the (estimated) proportion of species for which their sensitivity is exceeded. Inverse usage of the model amounts to the inverse application of the CDF to estimate *quantiles* (percentiles) of species sensitivities for some (usually low) given fraction of species not protected, e.g., 5%. In applications, these percentiles may be used to set *ecological quality criteria*, such as the *hazardous concentration* for 5% of the species (HC_5).*

Toxicity data sets are usually quite small, however, especially for new chemicals. Samples below ten are not exceptional at all. Only for well-known substances, there may be tens of data points, but almost never more than, say, 120 sensitivity measures (see Newman et al., Chapter 7; Warren-Hicks et al., Chapter 17; De Zwart, Chapter 8). For relatively large data sets, one may work with empirically determined quantiles and proportions, neglecting the error of the individual measurements. In the unusual case that the data set covers all species of the target community, that is all there is to it. Almost always, however, the data set has to be regarded as a (representative) sample from a (hypothetical) larger set of species. If the data set is relatively large, statistical *resampling* techniques, e.g., *bootstrapping* (Newman et al., Chapter 7), yield a picture of the uncertainty of quantile and proportion estimates.

If the data set is small (fewer than 20), we have to resort to *parametric* techniques, and must assume that the selection of species is unbiased. If the species selection is biased, then parameters estimated from the sample species will also be biased.

The usual parametric approach is to assume the underlying SSD model to be *continuous*, that is, the target community is considered as basically "infinite." The ubiquitous line of attack is to assume some continuous statistical *probability density function* (PDF) over log concentration, e.g., the *normal* (*Gaussian*) PDF, in order to yield a mathematical description of the variation in sensitivity for the target community.

These statistical distributions are on many occasions *unimodal*, i.e., have one peak. Then, the usual number of parameters to specify the distribution is not more than two or three. Data sets may not be homogeneous for various reasons, in which

---

* Aldenberg and Jaworska (2000) have used the term *inverse Van Straalen method* for what in this volume is called the forward application of the model; this was because the inverse method, e.g., HC_5, was first, historically.

case the data may be subdivided into different (target) groups. Separate SSDs could be developed for each subgroup. One may also apply *bi-* or *multimodal* statistical distributions, called mixtures, to model heterogeneity in data sets (Aldenberg and Jaworska, 1999). A mixture of two normal distributions employs five parameters.

Section 5.2 explains the identification of a single-fit normal distribution SSD from a small data set ($n = 7$) that has been the running example in previous papers (Aldenberg and Slob, 1993; Aldenberg and Jaworska, 2000). This single-fit approach is easier and more elementary than the one in Aldenberg and Jaworska (2000) by *not* taking the uncertainty of the SSD model into account initially. However, in the latter paper, we treated the exposure concentration (EC) as given, or fixed, with no indication how to account for its uncertainty. This assumption will be relaxed in Section 5.4.

The single-fit SSD estimation allows reconciliation of *parameter estimation* with the assessment of the fit through *probability plotting* and *goodness-of-fit testing*. There are several ways to estimate parameters of an assumed distribution. Moreover, the *estimation* of the parameters of the probabilistic model may not be the same thing as *assessing* the fit. We will use the ordinary sample statistics (*mean* and *standard deviation*) to estimate the parameters.

Graphical assessment of the fit often involves probability plotting, but the statistical literature on how to define the axes, and what *plotting positions* to use, is confusing, to say the least. There are two major kinds of probability plot: *CDF* plots and *quantile–quantile* (Q-Q) plots (D'Agostino and Stephens, 1986).

The CDF plot is the most straightforward, intuitively, with the data on the horizontal axis and an estimate of the CDF on the vertical axis (the ECDF, empirical CDF, is the familiar staircase-shaped function). It turns out that the sensitive, and easily calculable Anderson–Darling goodness-of-fit test is consistent with ordinary CDF plots.

In Q-Q plots, the data are placed on the *vertical* axis. One may employ plotting positions, now on the horizontal axis, that are tailored to a particular probability distribution, e.g., the normal distribution. The relationships between Q-Q plots and goodness-of-fit tests are quite complicated. We employ some well-known goodness-of-fit tests based on regression or correlation in a Q-Q plot.

After having studied the single-fit normal SSD model, we review the extension of the single-fit SSD model to Bayesian and sampling statistical confidence limits as studied by Aldenberg and Jaworska (2000) in Section 5.3. We compare the single-fit median posterior CDF with the classical single fit obtained earlier and observe that the former SSD is a little wider.

All this still assumes the data to be without error. We know from preparatory calculations leading to species sensitivity data, e.g., from *dose–response curve* fitting, that their uncertainty should be taken into account as well. We will only touch upon one aspect: the *sensitivity* of the lower SSD quantiles (percentiles) for individual data points. We learn that these sensitivities are *asymmetric*, putting more emphasis on the accuracy of the lower data points. Given the apparent symmetry in the parameter estimation of a symmetric probability model, this is surprising, but fortunate, when compared to risk assessment methodologies involving the most sensitive species (data point).

When both species sensitivities and exposure concentrations are uncertain, we enter the realm of *risk characterization*. In Section 5.4, we put together a mathematical theory of risk characterization. We review early approaches in reliability engineering in the 1980s called the *probability of failure*. Certain integrals describe the risk of ECs to exceed species sensitivities, and formally match Van Straalen's ecological risk, δ (Chapter 4). Another interpretation of these integrals is indicated by the term *expected total risk*, as developed in the Water Environment Research Foundation (WERF) methodology (Warren-Hicks et al., Chapter 17).

In environmental toxicology, risk characterization employs plots of exposure concentration exceedence probability against fraction of species affected for a number of exposure concentrations. These so-called *joint probability curves* (JPCs) (Solomon and Takacs, Chapter 15), graphically depict the risk of a substance to the species in the SSD. We will demonstrate that the area under the curve (AUC) of these JPCs is mathematically equal to the probability of failure, Van Straalen's ecological risk δ, and expected total risk from the WERF approach.

Finally, we discuss the nature and interpretation of probabilistic SSD model statistics and predictions.

To smooth the text, we have deferred some technical details to a section called Notes (Section 5.6) at the end of the chapter. The appendix to this chapter contains two tables that expand on those in Aldenberg and Jaworska (2000).

5.2 THE NORMAL SPECIES SENSITIVITY DISTRIBUTION MODEL

5.2.1 Normal Distribution Parameter Estimates and Log Sensitivity Distribution Units

Since we are focusing on small species sensitivity data sets (below 20, often below 10), we use parametric estimates to determine the SSD, as well as quantiles and proportions derivable from it.

To set the stage, Table 5.1 reproduces our running example of the Van Straalen and Denneman (1989) $n = 7$ data set for no-observed-effect concentration (NOEC) cadmium sensitivity of soil organisms, also analyzed in two previous papers (Aldenberg and Slob, 1993; Aldenberg and Jaworska, 2000). We add a column of standardized values by subtracting the mean and dividing by the sample standard deviation, and give a special name to standardized (species) sensitivity units: log SDU (log sensitivity distribution units). The mean and sample standard deviation of log species sensitivities in log SDU are 0 and 1, respectively.

Mean and sample standard deviation are simple descriptive statistics. However, when we hypothesize that the sample may derive from a normal (Gaussian) distribution, they are also reasonable estimates of the parameters of a normal SSD. This is just one of many ways to estimate parameters of a (normal) distribution (more on this in Section 5.6.1).

Figure 5.1 shows the normal PDF over standardized log cadmium concentration with the data displayed as a dot diagram. The area under the curve to the left of a particular value gives the proportion of species that are affected at a particular

TABLE 5.1
Seven Soil Organism Species Sensitivities (NOECs)
for Cadmium (Van Straalen and Denneman, 1989),
Common Logarithms and Standardized log Values
(log SDU = log Sensitivity Distribution Unit)

Species	NOEC (mg Cd/kg)	$\log_{10}$	Standardized
1	0.97	−0.01323	−1.40086
2	3.33	0.52244	−0.63862
3	3.63	0.55991	−0.58531
4	13.50	1.13033	0.22638
5	13.80	1.13988	0.23996
6	18.70	1.27184	0.42774
7	154.00	2.18752	1.73071
Mean ($\bar{x}$)		0.97124	0.00000
Std. Dev. (s)		0.70276	1.00000
(5th Percentile) 0.65		−0.18469	−1.64485
(EC) 0.80		−0.09691	−1.51994

Note: The 5th percentile is estimated from mean − 1.64485 standard deviation. EC (a former quality objective) is used to demonstrate estimation of the fraction of species affected.

concentration. The shaded region indicates the 5% probability of selecting a species from the fitted distribution below the standardized 5th percentile, or log hazardous concentration (log HC_5) for 5% of the species.

We call the fit in Figure 5.1 a *single*-fit normal SSD, in contrast to the Bayesian version developed in Section 5.3, where we estimate the uncertainty of PDF values (see Figure 5.5). Now we confine ourselves to a PDF "point" estimate.

With the aid of the fitted distribution, we can estimate any quantile (inverse SSD application) or proportion (forward SSD application). As an inverse example, the 5th percentile (log HC_5) on the standardized scale is at $\Phi^{-1}(0.05) = -1.64$ [log SDU]. Φ^{-1} is the inverse normal CDF, available in Excel™ through the function NORMSINV(p). The z-value of −1.64 corresponds to −0.185 [$\log_{10}$ mg Cd/kg] in the unstandardized logs, amounting to 0.65 [mg Cd/kg] in the original units (see Table 5.1).

As a forward example: at a given EC of 0.8 [mg Cd/kg], the z-value is −1.52 [log SDU]. The fraction (proportion) of species affected would be $\Phi(-1.52) = 6.4\%$, with Φ the standard normal CDF. This function is available in Excel as NORMSDIST(x).

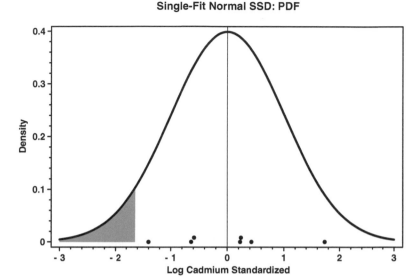

Single-Fit Normal SSD: PDF

FIGURE 5.1 Single-fit normal SSD to the cadmium data of Table 5.1 on the standardized log scale. Parameters are estimated through mean and sample standard deviation. Shaded area under the curve (5%) is the probability of selecting a species below log HC_5 (5th percentile) equal to z-value −1.64.

5.2.2 PROBABILITY PLOTS AND GOODNESS-OF-FIT

To assess the fit of a distribution one can make use of probability plots, and/or apply goodness-of-fit tests. In the previous section, we estimated the distribution parameters directly from the data through the sample statistics, mean and standard deviation, without plotting first, but one may also fit a distribution from a probability plot. To further complicate matters, a goodness-of-fit test may involve a certain type of fit implicit in its mathematics. The statistical literature on this is not easy to digest.

Probability plots make use of so-called plotting positions. These are empirical or theoretical probability estimates at the ordered data points, such as: i/n, $(i − 0.5)/n$, $i/(n + 1)$, and so on. However, plotting positions may depend on the purpose of the analysis (e.g., type of plot or test), on the distribution hypothesized, on sample size, or the availability of tables or algorithms. Tradition plays a role, too (Berthouex and Brown, 1994: p. 44). The monograph of D'Agostino and Stephens (1986) contains detailed expositions of different approaches toward probability plotting and goodness-of-fit.

Most methods make use of ordered samples. In Table 5.1 the data are already sorted, so the first column contains the ranks, and both original and standardized data columns contain estimates of the order statistics.

There are two major types of probability plots: CDF plots (D'Agostino, 1986a) with the data on the horizontal axis and Q-Q plots (Michael and Schucany, 1986) with the data on the vertical axis.

5.2.2.1 CDF Probability Plot and CDF-Based Goodness-of-Fit (Anderson-Darling Test)

The classical sample or ECDF (E for empirical), $F_n(x)$, defined as the number of points *less than or equal to* value x divided by the number n of points, would in our case attribute 1/7 to the first data point, 2/7 to the second, and so on. But just below the first data point, $F_n(x)$ equals 0/7, and just below the second data point, $F_n(x)$ equals 1/7. So, exactly *at* the data points the ECDF makes jumps of 1/7, giving it a staircase-like appearance (Figure 5.2, thin line).

The ECDF jumps are given by the plotting positions $p_i = (i - 0)/n$ and $p_i = (i - 1)/n$ at the data points. As a compromise, one may plot at the midpoints halfway ECDF jumps: $p_i = (i - 0.5)/n$, named *Hazen* plotting positions (Cunnane, 1978) (Figure 5.2, dots). We used Hazen plotting positions in Aldenberg and Jaworska (2000: figure 5). With Hazen plotting positions, the estimated CDF value at the first data point is taken as 1/14 = 0.0714, instead of 1/7 = 0.1429.

As is evident from Figure 5.2, one can make quick *nonparametric* forward and inverse SSD estimates by linearly interpolating the Hazen plotting positions. For example, with $x_{(1)}, x_{(2)}, ..., x_{(7)}$ denoting the ordered data (standardized logs), the first quartile can be estimated as $0.75 \cdot x_{(2)} + 0.25 \cdot x_{(3)} = -0.625$, which compares nicely with the fitted value: $\Phi^{-1} (0.25) = -0.674$. The nonparametric forward and inverse algorithms are given in Section 5.6.2.

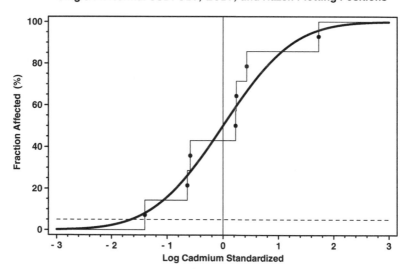

Single-Fit Normal SSD: CDF, ECDF, and Hazen Plotting Positions

FIGURE 5.2 Single-fit normal SSD with CDF estimated through mean and sample standard deviation. ECDF displays a staircase-like shape. Halfway the jumps of 1/7th, dots are plotted: the so-called Hazen plotting positions $p_i = (i - 0.5)/n$. The plot is compatible with the Anderson–Darling goodness-of-fit test and other quadratic CDF-based statistics (see text).

This will not work as easily for probabilities below $0.5/n$, or above $(n - 0.5)/n$, since we then have to extrapolate. In particular, the 5th percentile is out of reach at a sample size of 7.

A remarkable consequence of interpolated quantile estimation is that for $n = 10$ (*not* 20), the 5th percentile can be estimated by taking the lowest point. Selecting the most sensitive species is part of many risk assessment methodologies. This estimate may not be very accurate, however.

The Hazen plotting positions can be considered as a special case ($c = 0.5$) of a more general formula:

$$p_i = \frac{i - c}{n + 1 - 2c}, \text{ with } 0 \le c \le 1 \tag{5.1}$$

At least ten different values of c are proposed in the statistical literature (Cunnane, 1978; Mage, 1982; Millard and Neerchal, 2001: p. 96). In Section 5.6.4, we review some rationale behind different choices.

The curved line in Figure 5.2 displays the fitted normal CDF over standardized log concentration, as given by the standard normal CDF: $\Phi(x)$. We note that it nicely interpolates the Hazen plotting positions.

A formal way to judge the fit, consistent with the curvilinear CDF plot, is the Anderson–Darling goodness-of-fit test statistic, which measures vertical quadratic discrepancy between $F_n(x)$ and the CDF where it may have come from (Stephens, 1982; 1986a: p. 100). The Anderson–Darling test is comparable in performance to the Shapiro and Wilk test (Stephens, 1974). Both are considered powerful *omnibus* tests, which means they are strong at indicating departures from normality for a wide range of alternative distributions (D'Agostino, 1998).

The Anderson–Darling test uses the Hazen plotting positions and mean and sample standard deviation for the fit, thus being fully consistent with the plot in Figure 5.2. In Section 5.6.5.1, we show how it builds upon the data in Table 5.1. The modified A^2 test statistic equals 0.315, which is way below the 5% critical value of 0.752 (Stephens, 1986a: p. 123). Hence, on the basis of this small data set, there is no reason to reject to normal distribution as an adequate description. In D'Agostino (1986b: p. 373), the procedure is said to be valid for $n \ge 8$, which is not mentioned by Stephens (1986a: p. 123).

Figure 5.3 results from transforming the vertical CDF axis through the inverse normal CDF, which causes the fitted normal to become a straight line. This used to be done on specially graded graphing paper (normal probability paper). Analytically, one only needs the standard normal inverse CDF: $\Phi^{-1}(p)$. *Any* normal CDF then plots as a straight line (Section 5.6.3), with slope and intercept depending on the mean and standard deviation. We also transformed the Hazen plotting positions and ECDF staircase function. Note that the first and last ECDF jumps extend indefinitely to minus and plus infinity after employing the transformation. The Hazen plotting positions do not transform to midpoints of the transformed jumps, except for the middle observation for n odd.

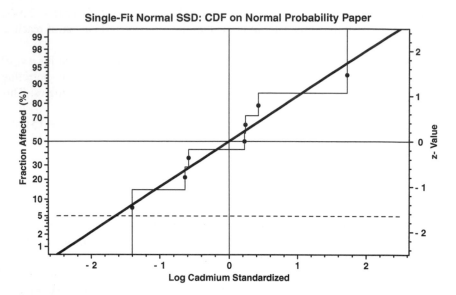

FIGURE 5.3 Straightened CDF/FA plot on computer-generated normal probability paper by applying the standard normal inverse CDF, Φ^{-1}, to CDF values on the vertical axis in Figure 5.2. Dots are transformed Hazen plotting positions. Because of the transformation, they are generally not midpoints of ECDF jumps. The straight line CDF z-values correspond 1:1 with standardized log concentrations.

We only compare the fitted line with the data points to judge the fit. The line is not fitted by regression, but by parameter estimation through mean and standard deviation. The fit can be judged to be quite satisfactory. Regression- or correlation-based goodness-of-fit is treated in the next section.

5.2.2.2 Quantile-Quantile Probability Plot and Correlation/Regression-Based Goodness-of-Fit (Filliben, Shapiro and Wilk Tests)

CDF plots such as the ones in Figures 5.2 and 5.3 are attractive when we try to estimate the CDF from the data to use as SSD, both in forward and inverse modes. From the point of view of sampling statistics, the data will vary over samples, which leads authors (e.g., D'Agostino, 1986a) to emphasize that *horizontal* deviations of the points in Figure 5.3 are important. (Bayesian statisticians should appreciate CDF plots, however, because they assume the data to be fixed, and the model to be uncertain.)

In Q-Q plots (Wilk and Gnanadesikan, 1968, and generally available in statistical packages), the data (order statistics) is put on the *vertical* axis against inverse CDF transformed plotting positions on the *horizontal* axis, tailored to the probability distribution hypothesized. This is not only more natural from the sampling point of view, but also allows one to formally judge the goodness-of-fit on the basis of regression or correlation.

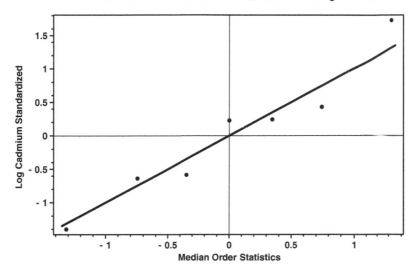

FIGURE 5.4 Q-Q plot of log standardized cadmium against exact median order statistics for the normal distribution. These are found by applying the inverse standard normal CDF to median plotting positions for the uniform distribution (see text). Straight line is the identity line indicating where the theoretical medians are to be found.

The transformed plotting positions on the horizontal axis can be based on means, or medians of order statistics, at any rate on chosen measures of location (see Section 5.6.4). Means ("expected" values) are used often, but they are difficult to calculate, and, further, means before transformation do not transform to means after transformation. Medians, however, transform to medians and this leads to what is called a property of invariance (Filliben, 1975; Michael and Schucany, 1986: p. 464). Hence, we only need medians of order statistics for the uniform distribution. Medians for other distributions are found by applying the inverse CDF for that distribution.

Median plotting positions for the uniform distribution are not easy to calculate either. *Approximate* median plotting positions for the uniform distribution can be calculated from the general plotting position formula with the Filliben constant $c = 0.3175$, as compared with the Hazen constant $c = 0.5$.

Figure 5.4 gives the Q-Q plot with the standardized Table 5.1 data on the vertical axis against normal median plotting positions on the horizontal axis at ± 1.31, ± 0.74, ± 0.35, and 0.00. (In Section 5.6.4, normal order statistics distributions are plotted, of which these numbers are the medians.) The straight 1:1 line indicates where the median of the data is located, if the model is justified.

Filliben (1975) developed a correlation test for normality based on median plotting positions; see Stephens (1986b). Since correlation is symmetric, it holds for both types of probability plot. The test is carried out easily, by calculating the correlation coefficient, either for the raw data, or the standardized data (Section 5.6.5.2). The correlation coefficient is 0.968, which is within the 90% critical region: 0.897 to 0.990. Hence, there is no reason to reject the normal distribution as a useful description of the data.

The power of the Filliben test is virtually identical to that of the Shapiro and Francia test based on *mean* plotting positions, and compares favorably with the high-valued Shapiro and Wilk test, treated next, for longer-tailed and skewed alternative distributions. (It is entertaining to see the Filliben test outperform the Shapiro and Wilk for the *logistic* alternative.) For short-tailed departures of normality, like the uniform or triangular distributions, the Shapiro and Wilk does better.

The Shapiro and Wilk test for normality is based on regression of the sorted data on *mean* normal plotting positions in the Q-Q plot. Since the order statistics are correlated, the mathematics is based on *generalized* linear regression. The mathematics is quite involved, so we have not plotted the regression line. However, as explained in Section 5.6.5.3, the test statistic W is easy to calculate, especially for standardized data. One needs some tabulated, and generally available, coefficients, however.

The test statistic $W = 0.953$ is within the 5 to 95% region (0.803 to 0.979), and since departures from normality usually result into *low* values of W, there is no reason to reject the normal distribution as an adequate description of the data in Table 5.1.

For larger samples sizes, W becomes equal to W', the Shapiro and Francia correlation test statistic (Stephens, 1986b: p. 214), which, as we have seen, relates to the Filliben correlation test.

We may conclude that the Filliben test, the Shapiro and Francia test, and the Shapiro and Wilk test are largely equivalent to each other, and all relate to the Q-Q plot. The first two are correlation tests that are also compatible with the normal probability paper CDF plot (Figure 5.3), if normal median or mean, respectively, plotting positions were used instead of Hazen plotting positions. The Anderson–Darling test is a strong alternative for ordinary-scale CDF plots based on the intuitive midpoint Hazen plotting positions. This test is also very powerful (D'Agostino, 1998).

Neither test indicates that the normal distribution would not be an adequate description of the cadmium data in Table 5.1.

5.3 BAYESIAN AND CONFIDENCE LIMIT–DIRECTED NORMAL SSD UNCERTAINTY

The preceding analysis of a single normal SSD fit and judgments of goodness-of-fit will by itself not be sufficient, given the small amount of data. We need to assess the uncertainty of the SSD and its derived quantities. Except for questions concerning species selection bias, which are very difficult to assess, this can be done with confidence limits, either in the spirit of classical sampling statistics theory, or derived from the principles of Bayesian statistical inference.

Aldenberg and Jaworska (2000) analyzed the uncertainty of the normal SSD model assuming unbiased species selection, and showed that both theories lead to numerically identical confidence limits. A salient feature of this extrapolation method is that, in the Bayesian interpretation, PDFs and corresponding CDFs are not determined as single curves, but as distributed curves. It acknowledges the fact that density estimates are uncertain. The methodology is related to second-order Monte Carlo methods in which uncertainty is separated from variation (references in Aldenberg and Jaworska, 2000). In the normal PDF SSD model, the variation is the variation of species sensitivities for a particular toxicant. Testing more species will not reduce this variation (on the contrary, it is more likely to increase), but it will reduce our uncertainty in the density estimates and therefore percentile estimates.

We should always keep in mind that this reduction in uncertainty may be misguided if the species selection process is biased. We will show in Section 5.3.4 how 5th percentile estimates depend on the precision of individual data points in an asymmetric fashion, which illustrates that bias toward sensitive or insensitive species does matter.

5.3.1 PERCENTILE CURVES OF NORMAL PDF VALUES AND UNCERTAINTY OF THE 5TH PERCENTILE

In the Bayesian point of view, the working hypothesis is that the data are fixed, while the model is uncertain. That means that contrary to the single PDF SSD fit in Figure 5.1, we now have a collection of PDF curves that may fit the data. The intuitively attractive graphical illustration is the *spaghetti plot* (Aldenberg and Jaworska, 2000: figure 2). The uncertainty of PDF values at each given concentration can be summarized through percentiles of PDF values. In this way, PDF percentile curves are obtained, e.g., for the 5th, 50th (median), and 95th percentiles. The same can be done for uncertain CDF values. The single-fit PDF and CDF estimates are now replaced by three curves giving percentiles of PDF and CDF estimates at given log concentration: a median curve in the middle and two confidence limits* to PDF or CDF values. The technology, e.g., the noninformative prior employed, is further explained in Aldenberg and Jaworska (2000).

Figure 5.5 shows Bayesian confidence limits of normal PDF SSD values for the standardized data of Table 5.1. It is the secondary uncertainty analogue of Figure 5.1. The three line confidence (credibility) limits of the PDF are curves joining individual pointwise confidence limits of the posterior distribution of PDF values at each given (standardized) concentration. They are not PDFs themselves.

The 5th percentile, −1.64 before, i.e., a fixed number, now becomes *distributed*, representing its uncertainty given such a small sample. The PDF of log HC_5 is displayed as a gray line in Figure 5.5. It can be shown that the posterior distribution of log HC_5 has a *noncentral t distribution* (Section 5.6.6).

The 5th, median, and 95th percentiles of the 5th percentile distribution of the SSD can be calculated as −3.40, −1.73, and −0.92, respectively, on the standardized scale. These are exactly the minus extrapolation constants for a FA of 5% at $n = 7$ (Aldenberg and Jaworska, 2000: table 1, third set), and replace −1.64 in the single-fit normal SSD. Table 5.A1 (see Appendix 5) contains an extended set of extrapolation

---

* Bayesian confidence limits are also called "credibility" limits, to indicate that they are not ordinary (sampling statistics) confidence limits. We use confidence limits also for the Bayesian approach. Moreover, in the case of the normal SSD both limits are numerically identical.

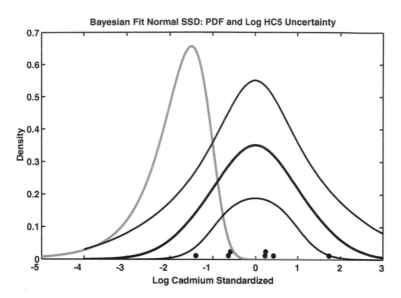

FIGURE 5.5 Bayesian fit of normal SSD to standardized log cadmium NOEC data of Table 5.1. Black lines are 5th, median, and 95th percentile curves of PDF values at given concentration. Gray line is PDF of the 5th percentile (log HC$_5$) and illustrates the uncertainty of the 5th percentile of the normal SSD for all $n = 7$ cases over standardized log concentration.

constants for the 5th percentile for sample sizes of n from 2 to 150. That will suffice for almost any toxicity data set.

Note that the median estimate of the 5th percentile is somewhat lower than the point estimate derived from the single fit. Note also the confidence limits indicate that even the first digit (–2, rounded from –1.73) is uncertain. Most (90%) of the uncertainty is confined within the interval (–3.40, –0.92).

It can be shown mathematically that, if the normal model can be assumed, these numbers hold for all $n = 7$ cases, *independent* of the data (Aldenberg and Jaworska, 2000: pp. 15, 17). On the standardized scale, all PDF (and CDF) posterior percentile curves are identical for the same sample size and confidence levels. Accordingly, Figure 5.5 is generic for $n = 7$. Hence, provided that the species selected are representative for the target community, the goodness-of-fit tests are important to reveal whether the hypothesis of normality has to be rejected or not.

5.3.2 PERCENTILE CURVES OF NORMAL CDF VALUES AND THE LAW OF EXTRAPOLATION

Figure 5.6 displays the three-line Bayesian CDF uncertainty for the cadmium NOEC data of Table 5.1, to be compared to Figure 5.2. In Figure 5.6, we have used the same Hazen plotting positions as in Figure 5.2. We have not tried a Bayesian generalization of the Anderson–Darling statistic. One could perhaps determine or simulate its posterior distribution from the posterior distribution of CDF fits.

Figure 5.7 displays an enlarged portion of the CDF uncertainty plot of Figure 5.6, with both a horizontal cut at a FA of 5% and a vertical cut at the median estimate

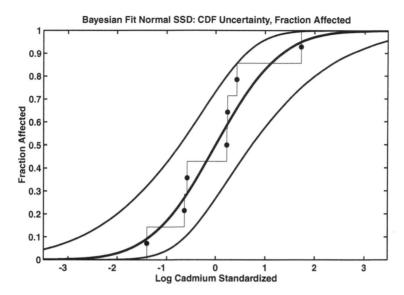

FIGURE 5.6 Bayesian fit of normal SSD over log standardized concentration; ECDF (staircase line), Hazen plotting positions $(i - 0.5)/n$ (dots), and 5th (thin), median (thick), and 95th (thin) percentile curves of posterior CDF values (FA at EC).

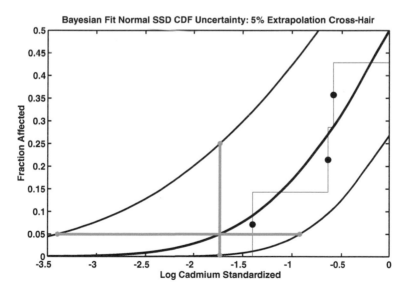

FIGURE 5.7 Enlarged lower portion of Figure 5.6 with 5% extrapolation cross-hair: horizontal cut at FA of 5%, and vertical cut at median log HC_5 added (gray lines). Horizontal line illustrates law of extrapolation (see text); vertical line demonstrates uncertainty of FA at median log HC_5 estimate: 0.34% (5th percentile) up to 25.0% (95th percentile). Lines cross at 5% FA. Percentile curve intersection points with horizontal line are minus the extrapolation constants $(n = 7)$.

of log HC$_5$ (–1.73) added (gray lines): the *extrapolation cross-hair* at 5%. The horizontal and vertical slicing (extrapolation cross-hair) is easy to interpret from the Bayesian point of view, as the horizontal cut determines the posterior distribution of the log HC$_5$, while the vertical cut governs the posterior distribution of the FA at median log HC$_5$. One may graphically derive the consequence of the horizontal cut, as expressed in the *law of extrapolation* (Aldenberg and Jaworska, 2000: p. 13):

> The upper (median, lower) confidence limit of the fraction affected at the lower (median, upper) confidence limit of log HC$_p$ is equal to p% fraction affected used to define the hazardous concentration.

With FA$^\gamma$, the γth percentile of the posterior PDF of the FA, and HC$_p^{1-\gamma}$ the $(1 - \gamma)^{th}$ percentile of the hazardous concentration for p% of the species, the Bayesian version of the law of extrapolation reads

$$\mathrm{FA}^\gamma\left(\log\!\left(\mathrm{HC}_p^{1-\gamma}\right)\right) = p\% \tag{5.2}$$

This is true for any confidence level as well as any protection level.

The classical sampling statistics version requires a mind-boggling inversion of confidence limit statements over repeated samples (Aldenberg and Jaworska, 2000: p. 4). The Bayesian version is definitely the easier one to interpret, since we may talk about distributed percentiles and proportions, given a (one) sample, without referring to a theoretically infinite number of possible samples from a (one) true model, which is the classical view in sampling statistics. The results are numerically similar.

5.3.3 FRACTION AFFECTED AT GIVEN EXPOSURE CONCENTRATION

The vertical cut defines the FA distribution at a given exposure concentration. The FA is the probability that a randomly selected species is affected, and we observe that this probability has a probability distribution in the present Bayesian analysis. It is rather skewed for standardized concentrations in the tails of the SSD. This is exemplified by the FA percentiles (5%, median, and 95%) at an exposure concentration equal to the median estimate of log HC$_5$ = –1.73: 0.34% (5th percentile), 5.0% (median), and 25.0% (95th percentile). Note that the median FA equals exactly 5%. The upper limit seems to be unacceptably high, when the objective is to protect 95% of the species. Unfortunately, it comes down very slowly as a function of sample size: at a sample size of 30 it is almost 12%, while at 100 it still amounts to 8% (Aldenberg and Jaworska, 2000: table 5).

Figure 5.8 is the Bayesian equivalent of Figure 5.3 with percentile curves to indicate the credibility limits. It results from plotting Figure 5.7 on computer-generated normal probability paper (Section 5.6.3). The vertical linear scale is equivalent to standard normal z-values. The unequally spaced ticks refer to the FA. The Hazen plotting positions are transformed in the same way. The 5% extrapolation cross-hair lines (gray) are indicated too. The horizontal cut defines the same distribution for

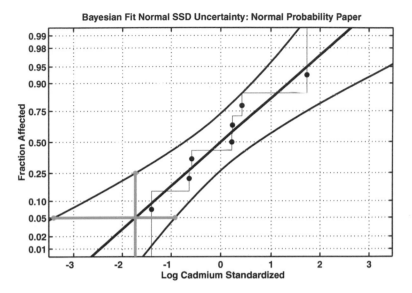

FIGURE 5.8 Bayesian normal SSD CDF percentile curves (5%, median, and 95%) on computer-generated normal probability paper. This is just Figure 5.6 with the inverse standard normal CDF $z = \Phi^{-1}$ (FA) applied to transform the vertical axis. Gray lines show the 5% extrapolation cross-hair. Dots are transformed Hazen plotting positions. Staircase line is the ECDF.

log HC_5 as in Figure 5.7; the vertical cut yields a transformed FA distribution over z-values. The transformed ECDF (staircase line) is also plotted.

Percentiles of the FA at given standardized log exposure concentration are tabulated in Aldenberg and Jaworska, 2000: table 2). In particular, column 0 of this table gives the fundamental uncertainty of FA for a median estimate of the log HC_{50}, i.e., for 50% of the species (26.7, 50.0, and 73.3%, respectively; see Figure 5.8).

However, that table is somewhat difficult to interpolate. Moreover, because of symmetry, half is redundant. In the appendix Table 5.A2, we present a better FA table with a finer spacing for the left portion of the standardized log concentration axis. Entries are given as $z = -K_p = \Phi^{-1}$ (FA), with Φ^{-1} the inverse standard normal CDF, as depicted in Figure 5.8. To convert to FA, one has to apply the standard normal CDF: $\Phi(x)$, available in Excel™ as `NORMSDIST(x)`.

The median FA curve in Figure 5.8 seems to be linear. Table 5.A2 reveals that median z-values do not exactly scale linearly with standardized log concentration for the smaller sample sizes. An approximate normal distribution that fits the median log HC_5 and median log HC_{50} exactly is derived in Section 5.6.7.

5.3.4 SENSITIVITY OF LOG HC_p TO INDIVIDUAL DATA POINTS

Up to now, we have neglected any possible error in the data points. Moreover, a commonly expressed fear about SSD-based extrapolation is that high points (insensitive species) have an unduly, and negative, that is, lowering, influence on log HC_p.

TABLE 5.2

Sensitivity Quotients of the Lower, Median, and Upper Estimates of log HC_5 for Individual Cadmium Data Points from Table 5.1

Rank (Species)	Data (NOEC)	$\log_{10}$	Standardized	Sensitivity Quotients log HC_5		
				Lower	Median	Upper
1	0.97	−0.01323	−1.40086	0.94	0.55	0.36
2	3.33	0.52244	−0.63862	0.50	0.33	0.24
3	3.63	0.55991	−0.58531	0.47	0.31	0.23
4	13.50	1.13033	0.22638	0.01	0.08	0.11
5	13.80	1.13988	0.23996	0.01	0.07	0.11
6	18.70	1.27184	0.42774	−0.10	0.02	0.08
7	154.00	2.18752	1.73071	−0.84	−0.36	−0.12

Note: Sensitivity depends on the extrapolation constant involved, as well as the standardized log concentration. The highest points (insensitive species) have negative (i.e., lowering) influence on the 5th percentile estimates, but not as large in absolute value as the low points (sensitive species).

In Section 5.6.8, we derive an expression for the sensitivity of $\bar{x} - k_s \cdot s$ to an individual data point x_i as the differential quotient:

$$\frac{\partial(\bar{x} - k_s \cdot s)}{\partial x_i} = \frac{1}{n} - \frac{k_s}{n-1}\left(\frac{x_i - \bar{x}}{s}\right) \tag{5.3}$$

This sensitivity quotient is dimensionless, and a function of the standardized log concentration of the point, as well as the extrapolation constant. With k_s positive, and focusing on estimating the 5th percentile, we observe that points below the mean have positive influence on the 5th percentile estimate. However, at *positive* value $(n-1)/(n \cdot k_s)$, sensitivity changes sign, which indicates that the influence of individual data points on lower SSD percentiles is *asymmetric*.

Table 5.2 shows the sensitivity quotients of log HC_5 confidence limits for individual cadmium data points (Table 5.1). The sensitivity pattern indeed depends on the extrapolation constant: the lower confidence limits express greater sensitivity than the median and upper confidence limits of log HC_5. Sensitivity is highest in absolute value for the *lower* data points. Apparently, the fourth- and fifth-order statistics (data points) have nil influence, and need not be as precise as the other estimates. The influence of the lowest data point on the lower confidence limit of log HC_5 is almost unity (0.94). This means that the precision of this data point is reflected in the precision of the lower log HC_5.

We note that in parametric extrapolation to low quantiles, e.g., to log HC_5, the lower data values have highest influence on the estimates. This is in line with nonparametric alternatives and approaches based on the lowest data value (most sensitive species) exclusively.

5.4   THE MATHEMATICS OF RISK CHARACTERIZATION

The characterization of the risk of toxicants to species, when *both* EC and SS are uncertain is the central issue in probabilistic ecological risk assessment (PERA). The methodology centers on CDF-type probability plots of both the *exposure concentration distribution* (ECD) and the SSD, and is well developed (Cardwell et al., 1993; 1999; Parkhurst et al., 1996; The Cadmus Group, Inc., 1996a,b; Solomon, 1996; Solomon et al., 1996; 2000; Solomon and Chappel, 1998; Giesy et al., 1999; Giddings et al., 2000; Solomon and Takacs, Chapter 15; Warren-Hicks et al., Chapter 17).

The basic problem characterizing the risk of a toxicant is the overlap between CDFs, or PDFs, of ECD and SSD. At first, both CDFs over log concentration were plotted on normal probability paper and the risk characterization was confined to comparing high ECD percentiles to low SSD percentiles. Later, EC exceedence probabilities were plotted against FA for all kinds of concentrations to construct JPCs, or exceedence profile plots (EPPs). Then researchers started to determine the AUC of these plots as a (numerical) measure of the risk of the toxicant to the species represented by the SSD (Solomon and Takacs, Chapter 15).

Here, we will further develop the probabilistic approach of risk characterization and relate several risk measures by putting the probability plots and equations into a general mathematical perspective. We will do so in two steps.

First, we review the probability of failure in reliability engineering that preceded the identical integrals for calculating ecological risk, δ, due to Van Straalen (1990; see Chapter 4). Another version of these integrals is expected total risk (ETR) as calculated by Cardwell et al. (1993).

Second, we show how JPCs, and the AUC in particular, relate to the probability of failure, ecological risk, and ETR. It can be shown mathematically that the AUC is identical to both the probability of failure, as well as the ETR, irrespective of the particular distributions involved. For normally distributed ECD and SSD, we provide a comprehensive lookup table.

5.4.1   PROBABILITY OF FAILURE AND ECOLOGICAL RISK: THE RISK OF EXPOSURE CONCENTRATIONS EXCEEDING SPECIES SENSITIVITIES

Species sensitivities are usually determined in laboratory toxicity tests. Exposure concentrations, however, relate to field data. To assess overlap, both sets of values must be compatible. One cannot compare 96-h toxicity tests to hourly fluctuating concentrations at a discharge point, or an instantaneous concentration profile from a geographic information system (GIS), without any change.

Suter (1998a,b) pointed out that distributions must be compatible with regard to what is distributed, to make sense comparing them in a risk assessment. One example is averaging a time-series to make the data compatible to the toxicity endpoints, as through the device of time-weighted mean concentrations (Solomon et al., 1996). Other corrections might employ adjustments for bioavailability. If spatial exposure distributions are assessed, one may treat the data to express territory-sized weighting adapted to the ecological endpoint studied.

In this section (Section 5.4), we assume that this data preparation has been carried out. Hence, by EC, we mean any measured, predicted, proposed, or assumed concentration of a toxicant that has been suitably adapted to match the toxicity endpoint of concern. After this data preparation, the remaining variation of the EC is considered to be a sample of a random variable (RV) to model its uncertainty. We also consider SS as an RV with probability model given by the SSD. Analogous data preparations may have been applied to the toxicity data.

We now regard the probability of some randomly selected EC exceeding some randomly selected SS as a measure of risk to concentrate on. If $X1$ is the RV of the logarithm of EC, and $X2$ is the RV of the logarithm of SS, then the problem is to evaluate the probability, or risk, that one exceeds the other, i.e.,

$$Pr(\log EC > \log SS) = Pr(X1 > X2) \tag{5.4}$$

This probability is expressible in either of two integrals (Ang and Tang, 1984: p. 335):

$$Pr(X1 > X2) = \int_{-\infty}^{\infty} PDF_{X1}(x) \cdot CDF_{X2}(x) dx \tag{5.5}$$

or, alternatively,

$$Pr(X1 > X2) = \int_{-\infty}^{\infty} \left(1 - CDF_{X1}(x)\right) \cdot PDF_{X2}(x) dx \tag{5.6}$$

Here, $PDF_X(x)$, respectively, $CDF_X(x)$, stands for the PDF (CDF) of random variable X taking values, i.e., log concentrations, x. An analytical derivation of these integrals is given in Section 5.6.9. An important proviso is that $X1$ and $X2$ are independent. This will usually be the case, as EC values do not depend on SS values, and vice versa.

The term $1 - CDF_{X1}(x)$ in Equation 5.6 is the probability of $X1$ to *exceed* value (log concentration) x. In classical probability theory this function is known as the survival function, apparently motivated by things or organisms surviving certain periods of *time* in survival analysis. In environmental toxicology, we use *exceedence function* or *exceedence* for short. Proposing the mnemonic EXF, the second integral (Equation 5.6) can be concisely written as

$$Pr(X1 > X2) = \int_{-\infty}^{\infty} EXF_{X1}(x) \cdot PDF_{X2}(x) dx \tag{5.7}$$

These integrals are known in the reliability engineering literature as the probability of failure (Ang and Tang, 1984: p. 333; EPRI, 1987; Jacobs, 1992), and can be used for *any* risk of $X1$ exceeding $X2$, e.g., load exceeding strength, demand exceeding supply, etc.

Van Straalen (1990; Chapter 4) developed the same expressions for the ecological risk δ of concentrations exceeding no-effect levels. He shows, in a graphical way, how the integrals essentially quantify the *overlap* between the two distributions, as AUCs of reduced PDFs. The validity of either representation is further substantiated in the special case of a fixed log(EC): $X1 = x_1$, since then either integral reduces to

$$\Pr(X1 > X2) = \Pr(X2 < x_1) = \mathrm{CDF}_{X2}(x_1) \tag{5.8}$$

which is the probability of selecting a random log(SS) below this fixed log(EC) (δ_1 in Van Straalen and Denneman, 1989). Compare the figures in Van Straalen (Chapter 4).

5.4.2 The Case of Normal Exposure Concentration Distribution and Normal SSD

An important special case arises when both $X1$ and $X2$ are normally distributed. In the preceding paragraphs, we have considered normally distributed SSDs. ECs are often analyzed as lognormal distributions, implying that log(EC) is normally distributed. We abbreviate the log(EC) distribution as ECD.

For example, with respect to the standard normal SSD fitted to the cadmium data (Table 5.1), we take the standardized log(EC) value of −1.52 rounded to −1.5 as the mean of the normal ECD and consider various standard deviations: 0.0, 0.2, 0.5, 1.0, and 2.0 (Figure 5.9). An ECD standard deviation of 0.5 means that the ECD is half as variable as the SSD.

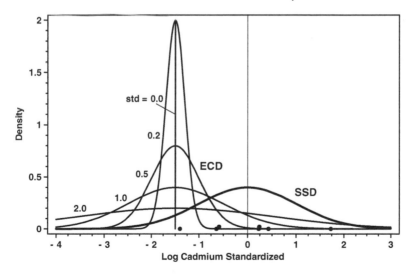

FIGURE 5.9 Standard normal SSD fitted to soil–organism cadmium data (Table 5.1), and five hypothetical normal ECDs at standardized log(EC) = −1.5 (Table 5.1) with increasing standard deviation: σ = 0.0, 0.2, 0.5, 1.0, and 2.0. The location and variability of an ECD is considered relative to the SSD.

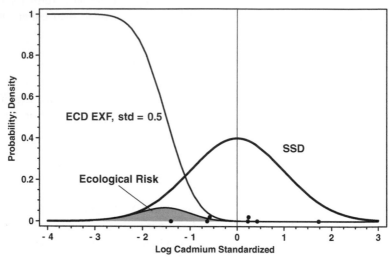

FIGURE 5.10 Normal ECD EXF at $\mu_{\mathrm{ECD}} = -1.5$ and $\sigma_{\mathrm{ECD}} = 0.5$, standard normal SSD for cadmium (Table 5.1), and ecological risk/probability of failure integral (Equation 5.6) of the probability of a random log(EC) to exceed a random log(SS). The ecological risk is 9.0% (exact value 8.9856% in Table 5.3).

Figure 5.10 displays a Van Straalen ecological risk/probability of failure plot of Equation 5.6 or 5.7 for normally distributed ECD and SSD with $\mu_{\mathrm{ECD}} = -1.5$ and $\sigma_{\mathrm{ECD}} = 0.5$ on the standardized SSD scale.

Figure 5.11 is the analogous ecological risk/probability of failure plot for Equation 5.5. Note that the positions and shapes of the ecological risk curves differ, but that their integral is identical.

To calculate the ecological risk/probability of failure, we numerically integrate integral (Equation 5.6) for two normal distributions:

$$\Pr(\log \mathrm{EC} > \log \mathrm{SS}) = \int_{-\infty}^{\infty} \left(1 - \Phi_{\mathrm{ECD}}(x)\right) \cdot \phi_{\mathrm{SSD}}(x)\,dx \tag{5.9}$$

with $\Phi_{\mathrm{ECD}} = \Phi_{\tilde{\mu}1,\tilde{\sigma}1}(x)$ the normal ECD CDF with mean $\tilde{\mu}_1$ and standard deviation $\tilde{\sigma}_1$ on the standardized SSD scale and $\phi_{\mathrm{SSD}} = \phi(x)$ the standard normal SSD PDF. The result is found to be 9.0%.

Alternatively, one may numerically integrate integral (Equation 5.5) after substituting normal distributions:

$$\Pr(\log \mathrm{EC} > \log \mathrm{SS}) = \int_{-\infty}^{\infty} \phi_{\mathrm{ECD}}(x) \cdot \Phi_{\mathrm{SSD}}(x)\,dx \tag{5.10}$$

with $\phi_{\mathrm{ECD}} = \phi_{\tilde{\mu}1,\tilde{\sigma}1}(x)$ the normal ECD PDF and $\Phi_{\mathrm{SSD}} = \Phi(x)$ the standard normal SSD CDF. The result is again found to be 9.0%.

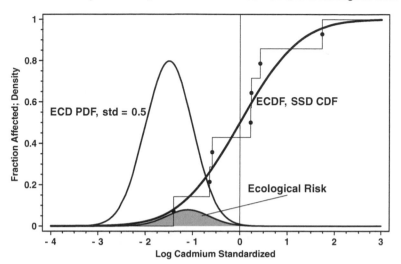

Normal Exposure PDF, Standard Normal SSD CDF, and Ecological Risk

FIGURE 5.11 Normal ECD PDF of log exposure at $\mu_{ECD} = -1.5$ and $\sigma_{ECD} = 0.5$, standard normal SSD CDF for cadmium (Table 5.1), and ecological risk/probability of failure integral (Equation 5.5) expressing the probability of a random log(EC) to exceed a random log(SS). The AUC (ecological risk) is identical to the one in Figure 5.10: 9.0% (exact value 8.9856% in Table 5.3), although the curves differ.

However, one may wonder whether some analytical shortcut is possible to evaluate these integrals. When switching back to the $X1$, $X2$ notation, the risk of $X1$ exceeding $X2$ can be written as

$$\Pr(X1 > X2) = \Pr(X1 - X2 > 0) \tag{5.11}$$

So, in the special case of normal distributions for both ECD and SSD, we are essentially asking for the probability of the difference of two normal RVs to exceed zero. A well-known result in probability theory is that the difference of two independent normal RVs $X1$ and $X2$ is *also normal* with mean $\mu = (\mu_1 - \mu_2)$ and standard deviation $\sigma = \sqrt{\sigma_1^2 + \sigma_2^2}$ (e.g., Mood et al., 1974: p. 194; Ang and Tang, 1984: p. 338), where the indices refer to the respective RVs.

In the case of Figures 5.10 and 5.11, the difference of log(EC) and log(SS) is normally distributed with mean

$$\mu = (\tilde{\mu}_1 - 0.0) = -1.5 \tag{5.12}$$

and standard deviation

$$\sigma = \sqrt{\tilde{\sigma}_1^2 + 1.0^2} = \sqrt{0.5^2 + 1.0^2} = \sqrt{1.25} = 1.11803 \tag{5.13}$$

Since we know from normal CDF lookup that

$$\Pr(X1 - X2 > 0) = 1 - \Phi_{\mu,\sigma}(0) = 1 - \Phi\left(\frac{0-\mu}{\sigma}\right) = \Phi\left(\frac{\mu-0}{\sigma}\right) \tag{5.14}$$

the ecological risk/probability of failure of the difference to exceed 0 equals

$$\Pr(X1 - X2 > 0) = \Phi\left(\frac{-1.5-0}{1.11803}\right) = \Phi(-1.34165) = 8.9856\% \tag{5.15}$$

which matches the 9.0% found above through numerical integration.

It follows that the ecological risk can be calculated as

$$\Pr(X1 > X2) = \Phi\left(\frac{\mu_1 - \mu_2}{\sqrt{\sigma_1^2 + \sigma_2^2}}\right) \tag{5.16}$$

Here, μ_1 and σ_1 are the mean and standard deviation of the normal PDF of $X1$, log(EC), and μ_2 and σ_2 are the parameters of the normal PDF of $X2$, log(SS). $\Phi(x)$ denotes the standard normal CDF as a function of log concentration. Hence, all we need is means and standard deviations of ECD and SSD and a normal CDF lookup.

It remains to be shown that the standardization to the SSD leaves the result the same. One standardizes on the SSD by subtracting μ_2, the mean of the SSD, from *both* ECD and SSD means and then divides each by σ_2. Also, both standard deviations are divided by the standard deviation of the SSD to obtain:

$$\log(\text{EC}): \quad \tilde{\mu}_1 = \frac{\mu_1 - \mu_2}{\sigma_2} \quad \text{and} \quad \tilde{\sigma}_1 = \frac{\sigma_1}{\sigma_2}$$

$$\log(\text{SS}): \quad \tilde{\mu}_2 = 0 \qquad \text{and} \quad \tilde{\sigma}_2 = 1$$

It follows that

$$\Phi\left(\frac{\tilde{\mu}_1 - \tilde{\mu}_2}{\sqrt{\tilde{\sigma}_1^2 + \tilde{\sigma}_2^2}}\right) = \Phi\left(\frac{\mu_1 - \mu_2}{\sqrt{\sigma_1^2 + \sigma_2^2}}\right) \tag{5.17}$$

and the risk of log(EC), scaled to the SSD, to exceed a scaled log(SS) is identical:

$$\Pr(X1 > X2) = \Pr\left(\frac{X1 - \mu_2}{\sigma_2} > \frac{X2 - \mu_2}{\sigma_2}\right) = \Phi\left(\frac{\tilde{\mu}_1}{\sqrt{\tilde{\sigma}_1^2 + 1}}\right) \tag{5.18}$$

TABLE 5.3

Ecological Risk/Probability of Failure (%) of log EC Exceeding log SS for Two Independent Normal Distributions[a]

$\tilde{\mu}_1$ / $\tilde{\sigma}_1$	-5.0	-4.5	-4.0	-3.5	-3.0	-2.5	-2.0	-1.5	-1.0	-0.5	0.0
0.00	0.0000	0.0003	0.0032	0.0233	0.1350	0.6210	2.2750	6.6807	15.8655	30.8538	50.0000
0.10	0.0000	0.0004	0.0034	0.0248	0.1417	0.6431	2.3291	6.7777	15.9859	30.9412	50.0000
0.20	0.0000	0.0005	0.0044	0.0300	0.1632	0.7114	2.4930	7.0663	16.3400	31.1964	50.0000
0.30	0.0001	0.0008	0.0064	0.0401	0.2030	0.8320	2.7705	7.5396	16.9075	31.6000	50.0000
0.40	0.0002	0.0015	0.0102	0.0578	0.2673	1.0138	3.1659	8.1853	17.6580	32.1238	50.0000
0.50	0.0004	0.0028	0.0173	0.0873	0.3645	1.2674	3.6819	8.9856	18.5547	32.7360	50.0000
0.60	0.0009	0.0057	0.0302	0.1344	0.5049	1.6027	4.3174	9.9180	19.5586	33.4054	50.0000
0.70	0.0021	0.0114	0.0525	0.2070	0.6992	2.0276	5.0662	10.9564	20.6327	34.1044	50.0000
0.80	0.0047	0.0221	0.0894	0.3138	0.9575	2.5459	5.9175	12.0738	21.7440	34.8108	50.0000
0.90	0.0101	0.0412	0.1474	0.4640	1.2878	3.1568	6.8562	13.2438	22.8652	35.5078	50.0000
1.00	0.0203	0.0731	0.2339	0.6664	1.6947	3.8550	7.8650	14.4422	23.9750	36.1837	50.0000
1.10	0.0385	0.1235	0.3565	0.9277	2.1795	4.6315	8.9257	15.6485	25.0578	36.8309	50.0000
1.20	0.0685	0.1983	0.5223	1.2525	2.7394	5.4748	10.0208	16.8458	26.1026	37.4449	50.0000
1.30	0.1150	0.3038	0.7367	1.6422	3.3690	6.3720	11.1342	18.0210	27.1027	38.0238	50.0000
1.40	0.1829	0.4454	1.0037	2.0959	4.0604	7.3099	12.2521	19.1643	28.0540	38.5671	50.0000
1.50	0.2773	0.6277	1.3250	2.6102	4.8046	8.2759	13.3629	20.2690	28.9550	39.0756	50.0000
1.60	0.4025	0.8540	1.7003	3.1798	5.5918	9.2586	14.4573	21.3307	29.8056	39.5505	50.0000
1.70	0.5621	1.1257	2.1276	3.7984	6.4122	10.2479	15.5282	22.3469	30.6070	39.9937	50.0000
1.80	0.7587	1.4430	2.6034	4.4589	7.2568	11.2353	16.5703	23.3165	31.3610	40.4072	50.0000
1.90	0.9937	1.8047	3.1232	5.1539	8.1171	12.2138	17.5799	24.2395	32.0699	40.7930	50.0000
2.00	1.2674	2.2086	3.6819	5.8762	8.9856	13.1776	18.5547	25.1167	32.7360	41.1532	50.0000

[a] As a function of $\tilde{\mu}_1$ and $\tilde{\sigma}_1$: mean and standard deviation of log(EC) scaled to mean and standard deviation of the SSD.

Hence, when tabulating ecological risk/probability of failure, we need not vary all four parameters. By scaling to the SSD, that is, expressing everything in log SDU (sensitivity distribution units, see Table 5.1), we obtain a two-parameter dependent ecological risk, by only varying the mean $\tilde{\mu}_1$ and standard deviation $\tilde{\sigma}_1$ of the ECD (log ECD) relative to the SSD (log SSD).

In Table 5.3, the ecological risk/probability of failure (%) of log(EC) to exceed log(SS) is tabulated as a function of $\tilde{\mu}_1$ for -5.0(0.5)0.0, and $\tilde{\sigma}_1$ for 0.0(0.1)2.0. Entries for positive $\tilde{\mu}_1$ are obtained by subtracting the ecological risk at $-\tilde{\mu}_1$ from 100%. The first line in Table 5.3 ($\tilde{\sigma}_1 = 0$) consists of CDF values of the standard normal distribution. When log(EC) is a fixed number, the risk of exceeding some log(SS) is equal to the CDF value of the standardized SSD at that point, as observed by Van Straalen (Chapter 4). Note that when both means are identical ($\tilde{\mu}_1 = 0$) the risk is 50%, independent of the scaled standard deviation $\tilde{\sigma}_1$ of log(EC).

To explain the use of Table 5.3, let us continue the example with $\tilde{\mu}_1 = -1.5$, which is near the standardized value of log(EC) for cadmium in Table 5.1, and

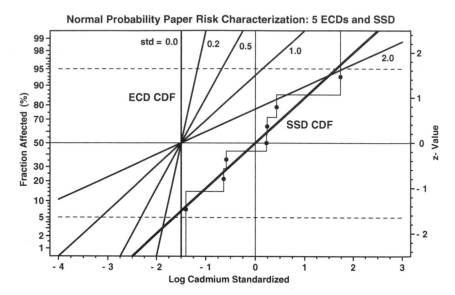

FIGURE 5.12 Normal probability paper risk characterization plot of five hypothetical ECD CDFs centered at standardized log concentration −1.5 (cadmium EC, Table 5.1) and increasing standard deviations (0.0, 0.2, 0.5, 1.0, and 2.0) compared to the standardized normal SSD for cadmium (data from Table 5.1).

evaluate the ecological risk/probability of failure at an increasing range of $\tilde{\sigma}_1$ values: 0, 0.2, 0.5, 1.0, and 2.0 (see Figure 5.9), that is, for increasing uncertainty of log(EC) relative to the SSD. From Table 5.3, the risks of a log(EC) to exceed a log(SS), respectively, are 6.7*, 7.1, 9.0, 14.4, and 25.1%. We observe how the risk increases dramatically when the uncertainty of a random log(EC) is of the same order of magnitude, or higher, as the uncertainty of a random log(SS) from the SSD.

5.4.3 JOINT PROBABILITY CURVES AND AREA UNDER THE CURVE

Cardwell et al. (1993; see also Warren-Hicks et al., Chapter 17) plotted CDFs of the ECD and of SSDs for chronic and acute toxicity over log concentration. CDF values of the SSD express the percentage of species affected, and CDF values of the ECD are converted to probabilities (%) of exceeding certain log concentrations. These are CDF type probability plots with the data on the horizontal axis.

In Solomon (1996), Solomon et al. (1996), Klaine et al. (1996a), and Solomon and Chappel (1998), the CDF plots were linearized by plotting on the normal probability scale (vertical axis). Figure 5.12 displays these linearized CDFs by applying the inverse *standard* normal CDF to CDF values. The left vertical axis shows the nonlinear probability scale (%) and the right vertical axis has ticks at standard normal z-values.

---

* The result of 6.7 vs. 6.4% in Section 5.2.1 is due to rounding −1.51994 to −1.5.

In assessing the risk of single toxicants to aquatic species, Cardwell et al. (1993) developed a discrete approximation to the integral:

$$\int_{-\infty}^{\infty} PDF_{X1}(x) \cdot CDF_{X2}(x)dx \qquad (5.19)$$

with, as before, $X1$ standing for log(EC) and $X2$ shorthand for log(SS), which they call expected total risk (ETR); see the WERF report and software (Parkhurst et al., 1996; The Cadmus Group, Inc., 1996a,b; Cardwell et al., 1999; Warren-Hicks et al., Chapter 17). The integral is the same as the one that originated as the probability of failure; see previous section and Section 5.6.9. Hence, it can also be interpreted as the probability of a random log(EC) exceeding a random log(SS), i.e., $Pr(X1 > X2)$.

The term *expected total risk* can be understood, when it is realized that probabilities of occurrence of concentrations, $PDF_{X1}(x)dx$, are multiplied by probabilities that a randomly selected species will be exceeded by these concentrations as given by $CDF_{X2}(x)$. The latter is regarded as the risk. Therefore, the integral is the statistical *expectation* of this risk.

The discrete approximation to the integral can also be seen as a sum of *joint probabilities*, since $X1$ and $X2$ are assumed independent, so that probabilities multiply (see Section 5.6.9).

The plotting of joint probabilities came to the fore in graphs called *risk distributions* or *risk distribution functions* (Parkhurst et al., 1996; The Cadmus Group, Inc., 1996a,b; Warren-Hicks et al., Chapter 17), and joint probability curves or exceedence profile plots (ECOFRAM, 1999a,b; Giesy et al., 1999; Solomon et al., 2000; Giddings et al., 2000; Solomon and Takacs, Chapter 15). These curves amount to plotting exceedence probabilities of ECs against FAs of species associated with these concentrations.

In the present notation, one plots exceedence (EXF) values, $EXF_{X1}(x) = 1 - CDF_{X1}(x)$, of the ECD on the vertical axis against CDF values of the SSD, $CDF_{X2}(x)$, that is, fraction of species affected, on the horizontal axis for relevant log concentrations x. The procedure is illustrated in Figures 5.13 and 5.14 for the case $\tilde{\mu}_1 = -1.5$ and $\tilde{\sigma}_1 = 0.5$, which are the normal distribution parameters of the ECD relative to the standard normal SSD.

The values read off in Figure 5.13 define a curve that is *parameterized* by log concentration x. (A parametric plot results when the variables on both axes are defined as functions of a third variable.) The curve is plotted as an EPP (Figure 5.14). Since products of probabilities on the individual axes are joint probabilities for the same log(EC) = x, the term JPC is justified. JPCs as EPPs are decreasing curves apart from possible plateaus.

Figure 5.14 shows the joint probability that a random log(SS) is below -1.5 *and* that a random log(EC) is above -1.5 as a shaded rectangle. Because of the assumption of independence, this joint probability is the product of the individual probabilities: $6.7\% * 50\% = 3.3\%$, and therefore equal to the area of the shaded rectangle.

Figure 5.15 displays EPP JPC curves for our running cadmium example (Table 5.1) with $\tilde{\mu}_1 = -1.5$, and increasing $\tilde{\sigma}_1$ values: 0.0, 0.2, 0.5, 1.0, and 2.0.

ECD Exceedence Function and SSD CDF Joint Probability Curve Construction

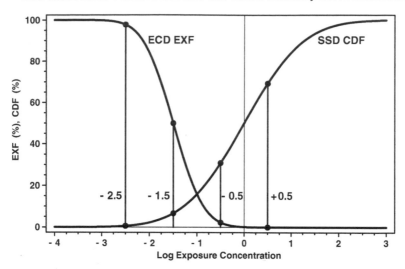

FIGURE 5.13 Construction of JPC by reading off ECD exceedence values and SSD cumulative distribution values at different values of log(EC). SSD is standard normal distribution. ECD is normal distribution with $\tilde{\mu}_1 = -1.5$ and $\tilde{\sigma}_1 = 0.5$ on the standardized SSD scale.

ECD EXF against SSD CDF Joint Probability Curve

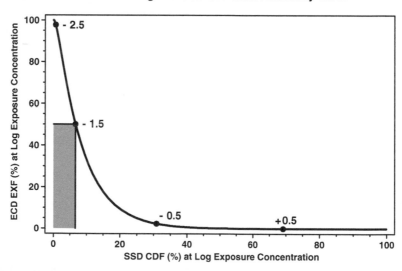

FIGURE 5.14 EPP JPC constructed from ECD EXF values and SSD CDF values read off in Figure 5.13. Numbers (–2.5, …, +0.5) refer to log EC. The shaded rectangle indicates the joint probability that a random species has log sensitivity below –1.5 *and* that a random log EC is above –1.5. This joint probability equals 6.7 $*$ 50% = 3.3%. This follows from the assumption of independence of the two distributions. Each point on the curve defines a joint probability. The AUC equals 9.0% and is the (total) ecological risk.

Exceedence Profile Plot with 5 Joint Probability Curves, Normal Distributions

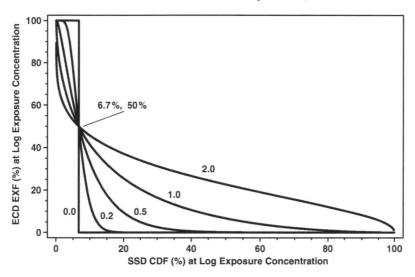

FIGURE 5.15 EPP with five JPCs relating the probability of exceedence of log(EC), vertical axis, to CDF values (FA) of the SSD for normal log(EC) distributions with $\tilde{\mu}_1 = -1.5$, and $\tilde{\sigma}_1 = 0.0, 0.2, 0.5, 1.0,$ and 2.0, relative to the standard normal SSD. A JPC going down slowly indicates a relatively high risk of some log(EC) to exceed some log(SS).

The AUC of an EPP JPC is considered as a numerical measure of the risk of the toxicant to species (Solomon et al., 2000; Solomon and Takacs, Chapter 15), which a risk manager may want to minimize. Analytically, the EPP JPC is a parametric plot of $1 - \text{CDF}_{X1}(x)$ on the vertical axis against $\text{CDF}_{X2}(x)$ on the horizontal axis, with the curve parameterized by x (log concentration); see Figure 5.14. Writing $d\text{CDF}_{X2}(x) = \text{PDF}_{X2}(x) \cdot dx$ as the differential of the FA, the AUC is equal to:

$$\text{AUC}_{\text{EPP}} = \int_{-\infty}^{\infty} \left(1 - \text{CDF}_{X1}(x)\right) \cdot d\text{CDF}_{X2}(x) = \int_{-\infty}^{\infty} \left(1 - \text{CDF}_{X1}(x)\right) \cdot \text{PDF}_{X2}(x) dx \quad (5.20)$$

which is identical to the integral expression in Equation 5.6 for the risk of log(EC) exceeding log(SS), or $\Pr(X1 > X2)$, derived as the probability of failure. Consequently, the AUC of an exceedence profile plot JPC *is equal to* the ecological risk of $X1$ exceeding $X2$, and hence constitutes a sensible measure of risk.

A similar type of plot arises, when plotting SSD CDF values (FA) on the *vertical* axis, against ECD CDF (not EXF) values on the horizontal axis at corresponding log(EC) x (Figure 5.16).

We call these plots *cumulative profile plots* (CPPs) and they display increasing JPCs. CPP JPCs start at (0%, 0%), and end at (100%, 100%). They are probably easier to draw and interpret than EPPs, since they involve CDFs only. Moreover, *cause* (EC) is plotted horizontally, while *effect* (FA) is plotted vertically. The interpretation of a CPP JPC going up near the lower right corner of the display means

Cumulative Profile Plot with 5 Joint Probability Curves, Normal Distributions

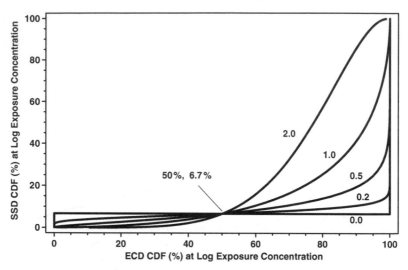

FIGURE 5.16 CPP with five JPCs of SSD CDF values against ECD CDF values at identical log(EC): same normal distributions as in Figure 5.15. The AUC also equals ecological (total) risk. CPPs involve only CDFs, and hence are simpler to construct and interpret. A risk manager may want to push the curve as much as possible to the lower right corner to reduce the ecological risk of a substance.

that most exposure concentrations are below values where species sensitivities are occurring. A CPP JPC going up early means a relatively high risk to the species.

Writing $dCDF_{X1}(x) = PDF_{X1} \cdot dx$ for the differential of the exposure distribution CDF, the AUC of a CPP JPC is equal to:

$$AUC_{CPP} = \int_{-\infty}^{\infty} CDF_{X2}(x)dCDF_{X1}(x) = \int_{-\infty}^{\infty} PDF_{X1}(x) \cdot CDF_{X2}(x)dx \quad (5.21)$$

which is identical to the integral expression in Equation 5.5 for the ecological risk/probability of failure. Hence, CPP JPCs describe the same risk as EPP JPCs.

The ETR calculation of Cardwell et al. (1993) involves tabulating probabilities $PDF_{X1}(x) \cdot \Delta x$ and $CDF_{X2}(x)$ over a range of x (log concentration) values. Then, entries are multiplied and summed to obtain ETR. This can be interpreted as a discrete approximation to the integral.

Consequently, the probability of failure, ecological risk, ETR, and the AUC of a JPC are all equal to the same number. The AUC values of the different JPCs in Figures 5.15 and 5.16 for increasing imprecision of the log EC relative to the SSD, therefore, equal the ecological risk/probability of failure values obtained earlier: 6.7, 7.1, 9.0, 14.4, and 25.1% from Table 5.3.

Concluding, the discrete summation of Cardwell et al., Van Straalen's ecological risk δ, the numerical integration of risk distribution curves in the WERF methodology, as well as the AUC of JPCs in EPRs and CPPs are all numerically equal to, and may be interpreted as, the risk of some log(EC) to exceed some log(SS), as originally implemented by the probability of failure in reliability engineering. It follows that, in the special case of *normally distributed* ECD and SSD, these risk measures can be obtained from Table 5.3, or the underlying equation. The quickest way to estimate the risk in the normal case is to calculate the respective two means and standard deviations, then standardize to the SSD, and look up the appropriate normal CDF value.

5.5 DISCUSSION: INTERPRETATION OF SPECIES SENSITIVITY DISTRIBUTION MODEL STATISTICS

5.5.1 SSD MODEL FITTING AND UNCERTAINTY

In this chapter, the emphasis of the analysis has been on small samples of species sensitivity data, i.e., sets of data points usually below 20, as frequently encountered in daily practice when SSDs are applied. By hypothesis, the set of species for which data are present in the sample is thought to be representative of some target community. This may or may not be an ecological community in the classical sense. The community may consist of species from selected taxa in an area or a spatially defined compartment, or it may be a set of laboratory species, either distributed over taxa or chosen with respect to a certain mode of action of a toxicant.

Toxicant sensitivity data sets may be compiled in a specific way to cover a wide range of taxa, or directed toward taxa sharing a common feature that makes them more sensitive to a toxicant than other taxa. This difference shows up, for example, when common practice for derivation of environmental quality criteria (all data in one SSD, at least a certain number of different taxa) is compared to practice in ecological risk assessment for contaminated ecosystems. The "community" or statistical "population" that the SSD, or a set of SSDs, addresses clearly depends on the philosophy that is used when collecting the data. Without having defined what the SSD describes, one cannot reasonably assume that the species tested constitute a random sample. Most of the statistical methods available to infer properties of the statistical population from the sample do assume random sampling.

In this chapter, we do not investigate whether this key assumption is satisfied or not, and how one would proceed to assess that. Instead, we bring together some statistical approaches and their interpretation, when one can reasonably assume the sample to be representative for the target species or taxa.

The SSD is usually modeled as a probability distribution. A certain distribution may fit or fail to fit the data. How can one determine whether the fit is adequate in a statistical sense? An assessment of whether the model applies is often lacking in current practices. It is obviously preferable to develop the model and assess its adequacy, rather than to employ a model blindly. Various methods, pertaining to probability plotting and analyses of goodness-of-fit, are available and have been described in this chapter. These methods are not commonly used in present SSD applications. It is recommended that they be incorporated in SSD-based assessments, although we discuss some potential shortcomings, or possible misinterpretations.

We have discussed CDF-based plots and so-called Q-Q plots, and some good-ness-of-fit tests related to them. With the exception of the familiar ECDF plots and the distribution-free Hazen plotting positions, most plots and tests depend on the model employed. The ubiquitous choice of the normal distribution is treated at length. Even the best tests, called omnibus tests, which have highest power for a wide range of alternative distributions, are not too informative in the case of small data sets. In our opinion, this means that there is no reason to reject a certain model as an adequate description of a small data set. However, other models are likely to pass the test as well, and therefore are equally reasonable descriptions. This implies that the determination of a particular model from a small data set may be undecidable.

We think that there is another reason probability plotting and goodness-of-fit testing should be employed, and interpreted, with care. In their classical form, they refer to single "best" fits, as discussed in Section 5.2, but the single fit probability curve should be regarded as a first try only. The Bayesian fit, Section 5.3, does more justice to the inherent uncertainty with regard to the model, by essentially employing a multitude of possible probability curves. For small data sets, the Bayesian statistical paradigm is more appropriate than the repeated sampling viewpoint in classical statistics, because the parameters in the probability model are explicitly allowed to be uncertain. The Bayesian method permits determination of our *uncertainty* with regard to the *variation* of species sensitivities for a toxicant.

We extended CDF probability plotting for the Bayesian method through percentile curves (Figures 5.6 through 5.8). Even when we apply one specific probability model, such as the normal distribution, the Bayesian approach shows how uncertain we apparently are about its fit given the small data set. The acceptance or rejection of a razor-sharp single point estimate of an SSD fit is at variance with this inherent imprecision. We did not study Bayesian versions of measures of goodness-of-fit, however.

The classical methods of goodness-of-fit testing may not be appropriate for another reason: that is, when one is not interested in the whole SSD. Suppose one is concerned with the determination of low percentiles only. Why reject a model when it fails in the upper tail? We saw in Section 5.3.4 that the sensitivity of low percentile estimates to individual data points does depend more on lower data points than on higher data points. Hence, the results may not depend very much on the higher points. Yet, the classical goodness-of-fit tests studied here are symmetric with respect to the data. We have to better understand aspects of local vs. global goodness-of-fit in relation to the purpose of the model.

The fact that the single-fit normal SSD model is rejected in roughly half the cases (Newman et al., Chapter 7) should not be interpreted as the normal SSD model to be invalid as a first try. In practice, one should always make an ordinary CDF plot (see Figures 5.2 and 5.3) to see where deviations are most severe, in cases of trouble followed by inspection of individual entries of a spreadsheet version of the Anderson–Darling test (Section 5.6.5.1). Then, Bayesian CDF plots (see Figures 5.6 or 5.8) may help in judging the departures of the empirical points with regard to the inherent uncertainty of CDF values. Correlated departures from the straight-line CDF on normal probability paper may indicate bimodality or multimodality of the SSD (see figure 2 in Aldenberg and Jaworska, 1999). If that is clearly the case, one may divide the data into two or more groups and develop separate SSDs, or one SSD for the more sensitive group only. This sequence of steps yields much more information than a yes–no decision resulting from a goodness-of-fit test when treated as a black-box device. One should *always* plot the data and the SSD together to see what is happening. Even when the model does *not* hold, this makes sense; for example, a plot on normal probability paper may help in detecting *non*-normality and how to model it.

5.5.2 RISK CHARACTERIZATION AND UNCERTAINTY

In principle, the approaches toward fitting the SSD could be largely applied to fitting ECDs. In particular, the secondary normal distribution model (Bayesian approach) could be applied to ECDs. We have not done so, yet. Section 5.4.1 touches upon the issue, put forward by Suter (1998a,b), that the ECD and SSD should be commensurate, in the sense that the type of exposure (duration, spatial average) should match the nature of the toxicity test of the SSD. More statistical effort should focus on methods to accomplish this. In Section 5.4, we assume ECs and SSs to have been preprocessed to be compatible.

Sections 5.2 and 5.3 focus on the SSD as primary and secondary distributions, respectively, while treating the EC as fixed. The approach followed in Section 5.4 is that both ECD and SSD are primary single-fit (normal) distributions that only capture the variation, not the uncertainty of this variation. The normality is not an essential ingredient in the theory developed. We have not considered an uncertainty-variability treatment of the ECD, or combined a primary uncertain ECD with a secondary SSD, or even thought of risk characterization of two secondary distributions: both ECD and SSD. (We will come back to this in the next section.)

What has been said about the assumption of the SSD sample to be representative and random now obviously holds for the EC sample as well. Much more study is needed on how the assumption of randomness relates to the sample deriving from time-series, GIS type of data, or model predictions. As stated above for the SSD, decisions about what data to collect and to combine in one distribution have to be made, or have been made, *prior* to analyzing the data. Statistical methods may be of little help in deciding what data to collect. These decisions strongly influence both the fit of a model to the data (unimodal vs. bimodal, for example) and its further interpretation in the risk assessment.

Assuming in Section 5.4 the ECD and SSD to be compatible, and assuming their single-fit probability curves to describe (summarize) their variation adequately, we analyzed the mathematical aspects of risk characterization. We showed how the method of risk curves, or JPCs, relates to the elementary exercise of calculating the probability of a random EC to exceed a random SS. This method has been known as the probability of failure in reliability engineering for decades, and yields an *unequivocal* interpretation of the AUC in JPCs. Van Straalen's ecological risk is identical to this measure, and hence can be interpreted in the same way. We, moreover, present a simplified diagram to draw JPCs: the CPP.

Another (probabilistic) interpretation of the ecological risk to species, due to Cardwell et al. (1993), is called ETR. To understand this wording, one may reason as follows. When the SSD is fixed and the EC is a fixed number, then the FA, or risk, is a fixed number. Hence, for distributed EC, the FA has to be distributed as well.* The interpretation of the ecological risk integral (probability of failure integral, AUC of a JPC, and so forth) is that we do calculate the *mean* or *expected value* of this FA distribution. A full assessment of this distribution would reveal the *uncertainty* of the ecological risk, which is now considered a fixed number only. (A simple numerical example will be given in the next section.)

We observe from Sections 5.2 through 5.4 that we have quite some options in specifying the variability and uncertainty of the quantities of interest in a probabilistic ecological risk assessment. We will expand on this theme in the next section.

5.5.3 PROBABILISTIC ESTIMATES AND LEVELS OF UNCERTAINTY

SSDs have been used in practice to determine EC objectives, to estimate the fraction of species affected at certain exposure concentrations, to extrapolate to other systems or conditions, and so on. This has been done by either considering the SSD alone, or combined with the ECD. SSDs and ECDs are very simple models, structurally. In fact, their variance structure is the only structure present. There are no independent variables or process-oriented concepts involved. Historically, mathematical modeling shows a gradual shift from fixed geometrical thinking, developed in ancient times, via deterministic analytical models toward probabilistic reasoning. These categories do not exclude each other: one may have a process-oriented deterministic model and apply Monte Carlo analysis to it, to study its probabilistic behavior.

An SSD model is a probabilistic model to describe the variation of species sensitivities for a toxicant. There are many ways to specify the model and to address problems with it. The same is true for the ECD. A "deterministic" version might go like this: the sensitivity of soil species to cadmium (Table 5.1) is (on the average) $10^{0.97124} = 9.36$ [mgCd/kg]. That may be enough in some risk assessments, and totally inappropriate in others. For example, we may infer that an EC of 0.80 [mgCd/kg] (Table 5.1) is more than one order of magnitude below the mean SS. This may suffice for the first round ("tier").

When we account for the variation of SSs, while taking the EC to be a fixed number, we find the probability of the given EC to exceed a random SS to be 6.7% (6.4% without rounding the standardized log EC to −1.5). Hence, this fixed EC may affect more than 5% of the species. This is based on a "point" estimate of the mean and standard deviation of the normal SSD, i.e., a single-fit probability curve.

However, when we account for the uncertainty of the SSD, with the EC still fixed, Table 5.A2 (column −1.5 and rows $n = 7$) yields (after applying the standard normal CDF) a median FA of 7.7% with 90% confidence limits of 0.1 and 29.9%. So, what first seemed an order of magnitude difference now results into a FA distribution with values up to almost 30% of the species.

Now, we fix the SSD again, and account for the variability (or uncertainty) of the EC. We then find an ecological risk of 9.0% when the ECD is half as variable as

---

* We have derived its distribution, but it will not be presented here.

the SSD, 14.4% when the ECD is as variable as the SSD, and 25.1% when it is twice as variable as the SSD. These are *means* (expected values) of the FA distribution.

Percentiles of this FA distribution can be regarded as confidence limits of this ecological risk. They are easily calculated by applying the standard normal CDF (FA of a standardized SSD) to $-1.5 \pm 1.64 \cdot \sigma_{ECD}$ on the standardized log(EC) scale, since ECD percentiles map to FA distribution percentiles. If the ECD is half as variable as the SSD, we obtain 1.0 and 24.9% lower and upper confidence limits. An ECD as variable as the SSD yields FA confidence limits of 0.1 and 55.8% (!). In the case where the ECD is twice as variable as the SSD, the 90% confidence limits are 0.0% and a devastating FA of 96.3%. (In all three cases, the *median* FA is 6.7%, of course.) Hence, in the latter case, although the average FA is 25.1%, FA values seem to scatter near 0% *and* near 100%, apparently in the ratio 3:1.

The different options of incorporating uncertainty illustrate that we have a cascade, or hierarchy, of uncertainty levels. Each time we account for the uncertainty or variability of some quantity initially fixed in the assessment, output quantities (a percentile, a fraction) that were fixed become distributed, primary distributions become secondary, and so on.

There has been a debate in the literature whether to call quantities derivable from a fixed probability model "probabilistic" (Solomon, 1996; Suter, 1998a,b). Indeed, the concept of a fixed probability model has some self-contradictory flavor. By recognizing the hierarchy of uncertainty levels in probabilistic modeling, this issue can be resolved. Let us consider four steps in climbing the uncertainty ladder.

First, in a fixed SSD probability model, the primary uncertainty is the variation of possible species sensitivities, assuming that there is no uncertainty for each data point, and no model uncertainty. In this model, a CDF value stands for a probability, since that is the way a CDF is defined. It can be interpreted as the probability that the sensitivity of a random species is exceeded by the given concentration. A plot over relevant concentrations presents these probabilities as a fixed increasing function. Now, we ask the question: What will the number x of species sensitivities exceeded by a given concentration be in a (new) randomly drawn *finite* set n of species? Then the probability, as given by the fixed CDF value, would be the binomial parameter p, and equal the *expected fraction*, or *expected proportion* of the sample. The observed number x exceeded may substantially deviate from the expected value np which is not necessarily an integer. It is a discrete random variable with possible values x from 0 to n.

The binomial probability mass function

$$\binom{n}{x} p^x (1-p)^{n-x}$$

models the probability of occurrence of these values on the basis of random sampling. Hence, the fixed probability, coming from a distribution of variability, may trigger an uncertain event in a new (conceptual) experiment.

A satisfying consistency appears, when considering a "sample" of *one*, i.e., $n = 1$. Then the binomial formula degenerates to p in case of exceeded ($x = 1$), and $(1 - p)$

in case of not exceeded ($x = 0$). Apparently, the CDF value *is* the probability of a random species to be exceeded by the given concentration.

The theoretical (fixed) population distribution may be seen as an "infinite" sample. Perceived in this way, the "observed" fraction exceeded (affected) stabilizes to p, and becomes "deterministic." The FA (Aldenberg and Jaworska, 2000) of a theoretically infinite population of species, as specified by a fixed continuous probability model, at a given concentration is equal to the probability of one randomly selected species sensitivity to be exceeded. Moreover, it stands for the *expected* fraction of species affected in a *finite* sample of tested species.

In this view, there is no conflict between a fixed probability and the fact that it models something uncertain in a (new) experiment. That experiment may be that we have a species at hand with unknown sensitivity and a given concentration. Hence, there is no self-contradictory aspect in the interpretation of quantities derived from a fixed SSD. The methodology is definitely probabilistic, and some quantities derivable are fixed numbers, of which the imprecision cannot be given, unless we further climb the ladder of uncertainty levels. We also observe that a clear separation between variability and uncertainty is difficult to maintain. The SSD may be fitted to variability data. Then the model may be interpreted as the uncertainty regarding the outcome of a random drawing.

Second, in assessing the uncertainty of the SSD model through Bayesian statistics (while still assuming that the data points are not uncertain), the step is made from primary uncertainty to secondary uncertainty, by explicitly allowing for uncertainty of the probability model itself. The PDFs and CDFs, as well as quantities derived from them, become distributed, as if we say that different communities may have generated these kind of data. Percentiles become distributed, and so are fractions (proportions) at given concentration. Since the FA is the probability of a randomly selected SS to be at the wrong side of a given EC, the philosophical consequence of this second step in the hierarchy of uncertainty is that we indeed may consider probabilities of probabilities (see Suter, 1998b). And so on, of course. However, analytical theories of imprecise probabilities are still under development.

As a third option of combining uncertainties, the EC can be considered distributed. In this case, the above two options can both be hypothesized again. In this case, the risk of a concentration to exceed some SS is a fixed number that has to be interpreted as the *expected* FA. Apparently, the FA is distributed again, but now due to the variation in the exposure concentration. In the first case, we obtain secondary uncertainty by combining a fixed number with an uncertain distribution; now, in the third case, we get secondary uncertainty by feeding an uncertain input into a fixed distribution.

One can speculate further on a fourth case, namely, what would happen with the two cases combined: both the SSD (secondary) uncertain and the EC (primary) uncertain. This will increase the order of uncertainty by 1, leading to tertiary uncertainties. Quaternary uncertainty could follow when the effect of accounting for the uncertainty of the data would be taken into account, which we have neglected all together.

One may ask whether this fractal organization of uncertainty will end somewhere. In practice, it may be cut artificially at some level, guided by the problem at hand. Just as we may confine ourselves with a point estimate of something uncertain, without explicitly denying its uncertain nature, we may do the same thing in the hierarchy of uncertainty levels. In communicating results of probabilistic ecological risk assessments to the public, or to risk managers, one may present summarizing statistics by taking appropriate means, medians, or other percentiles of quantities and distributions at some uncertainty level. Probabilistic ecological risk assessment is an exciting field to explore such a methodology systematically.

5.6 NOTES

This section contains technical details and expansions on issues in the main text. A neat introduction to probability basics and an overview of the most common distributions is Evans et al. (2000). For developing manipulative skills in handling random variables and probability distributions, Hsu (1997) in the Schaum's Outline Series is recommended. More classical textbook-style introductions to mathematical statistics are Mood et al. (1974) and Hogg and Craig (1995). A thorough overview of environmental statistics dedicated to S-PLUS is Millard and Neerchal (2001).

5.6.1 Normal Distribution Parameter Estimators

In parametric point estimation, as well as interval estimation, one may use different estimators; see Mood et al. (1974: p. 273) and Millard and Neerchal (2001: p. 201). Let x_i be n sample values, with sample mean:

$$\bar{x} = \frac{1}{n} \sum_{i=1}^{n} x_i$$

We have two versions of the sample variance:

$$s_m^2 = \frac{1}{n} \sum_{i=1}^{n} (x_i - \bar{x})^2$$

the moment version, and

$$s^2 = \frac{1}{n-1} \sum_{i=1}^{n} (x_i - \bar{x})^2$$

the unbiased version. The latter is often called "the" sample variance.

The minimum variance unbiased estimators for the normal distribution parameters are

$$\hat{\mu}_{MVUE} = \bar{x}$$

$$\hat{\sigma}^2_{MVUE} = s^2$$

By using $\bar{x}$ and s as normal distribution parameter estimators of μ and σ, we apparently employ the MVUE estimators. The square root of the $(n-1)$ sample variance, as an estimator for σ, is not itself unbiased. We prefer it above the maximum likelihood estimator (MLE): $\hat{\sigma}_{MLE} = s_m$, for which the variance estimate is biased.

5.6.2 FORWARD AND INVERSE LINEARLY INTERPOLATED HAZEN PLOT SSD ESTIMATES

Let $x_{(i)}$, $i = 1,2,\ldots,n$ be the order statistics.

The inverse algorithm is $j = n \cdot \text{fraction} + 0.5$; $i = \text{IntegerPart}(j)$; $f = j - i$; quantile $= (1-f) \cdot x_{(i)} + f \cdot x_{(i+1)}$. This form is available as $\texttt{InterpolatedQuantile}$ in the Mathematica™ Standard Add-on Packages (Martin and Novak, 1999: p. 426).

The forward algorithm is: select largest $x_{(i)} \leq x$; if $x_{(i+1)} > x_{(i)}$, then calculate

$$\text{fraction} = \frac{i - 0.5}{n} + \frac{x - x_{(i)}}{n\left(x_{(i+1)} - x_{(i)}\right)}$$

if $x_{(i+1)} = x_{(i)}$, and $x = x_{(i)}$, then

$$\text{fraction} = \frac{i}{n}$$

if there are more ties, take the median of the respective $(i - 0.5)/n$.

5.6.3 NORMAL PROBABILITY PAPER AND THE INVERSE STANDARD NORMAL CDF

With $\Phi_{\mu,\sigma}(x)$ the normal CDF for a normal distribution with parameters μ and σ, $\Phi(x)$ the standard normal CDF, and $\Phi^{-1}(p)$ the standard normal inverse CDF, we may plot an arbitrary normal CDF by transforming through the standard normal inverse CDF to obtain z-values. In this way, any normal CDF plots as a straight line:

$$z = \Phi^{-1}\left(\Phi_{\mu,\sigma}(x)\right) = \Phi^{-1}\left(\Phi\left((x-\mu)/\sigma\right)\right) = (x-\mu)/\sigma$$

The standard normal inverse CDF is available in Excel™ through the function: $\texttt{NORMSINV(x)}$.

5.6.4 ORDER STATISTICS AND PLOTTING POSITIONS

A general formula covering several different plotting positions is

$$p_i = \frac{i - c}{n + 1 - 2c}, \text{ with } 0 \leq c \leq 1$$

To appreciate different suggestions for c, one has to understand the theoretical uncertainty of points in sorted samples. Suppose n numbers are drawn from a *uniform* (rectangular) distribution. From the point of view of sampling statistics, realizations of $x_{(1)}$, the lowest sample point, would vary over different samples. So, the first order statistic, denoted $X_{(1)}$, is a random variable with a distribution. Since it only addresses $X_{(1)}$, and not the joint probability of all order statistics, it is called a *marginal* distribution. Similarly, the other order statistics $X_{(i)}$ for $i = 2,\ldots,n$ are random variables. All, except the middle one for n odd, are skewed (asymmetric) distributions. Their PDFs are beta distributions:

$$\text{PDF}_{X(i)}(x) = \frac{n!}{(i-1)!(n-i)!} x^{i-1}(1-x)^{n-i} \quad (i=1,2,\ldots,n)$$

with mean:

$$E\left(X_{(i)}\right) = \frac{i}{n+1}$$

One often-stipulated mathematical *desideratum* is that plotting positions match expected (mean) values of order statistics. This motivates the Weibull plotting positions ($c = 0$) for *uniformly* distributed data. (These have no relationship with the Weibull distribution.)

If one would go for the order statistics *modes* (most probable values) in the uniform case, then differentiation would reveal:

$$\text{Mode}\left(X_{(i)}\right) = \frac{i-1}{n-1}$$

This is the general formula plotting position with $c = 1$. Note, that for $i = 1$ and $i = n$, the modes are at 0 and 1, respectively (Figure 5.17), which may cause trouble on probability paper.

For skewed distributions, however, *medians* make a lot of sense, too. Only the first and last order statistics have explicit expressions:

$$\text{Median}\left(X_{(1)}\right) = 1 - 0.5^{1/n}$$

$$\text{Median}\left(X_{(n)}\right) = 0.5^{1/n}$$

Fortunately, by calibrating c in the general plotting formula, the inner medians ($i = 2,3,\ldots, n-1$) can be approximated by means of $c = 0.3175$ with a maximum error of 0.0003 for all i and all n (Filliben, 1975; Michael and Schucany, 1986). Exact order statistic medians can be found by numerical integration. The next table shows

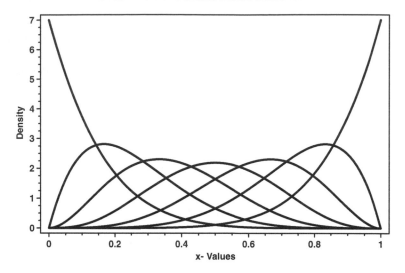

FIGURE 5.17 PDFs of order statistic distributions for samples of size 7 drawn from the [0,1] uniform distribution.

the exact median uniform order statistics for $n = 7$, and the Filliben plotting position approximation.

Rank	Exact Medians	Filliben Approx.
1	0.094276	0.094276
2	0.228490	0.228445
3	0.364116	0.364223
4	0.500000	0.500000
5	0.635884	0.635777
6	0.771510	0.771555
7	0.905724	0.905724

By equating $0.5^{1/n}$ to $(n - c)/(n + 1 - 2c)$, we derive an expression c_n that equals 0.29 at $n = 2$, becomes 0.30 for $n = 3,\ldots,14$, and stabilizes at 0.31 for all higher n. This may explain early plotting positions with $c = 0.30$ (Benard and Bos-Levenbach, 1953), and $c = 0.31$ due to Beard in 1943 (Cunnane, 1978), although the Filliben value works better for the inner medians.

Other percentiles of the order statistics can be calculated by numerically integrating the PDF. For example, for $n = 7$, the 5th, 50th (median), and 95th percentiles of $X_{(7)}$ are 0.651836, 0.905724, and 0.992699, respectively. Note that this range is more than twice the ECDF jumps (1/7). The Benard and Bos–Levenbach plotting position for the median would be 6.7/7.4 = 0.905405.

Apparently, much of the variation results from choosing different measures of *location* (median, mean, mode) for a skewed distribution. There is no unique answer. However, this all relates to the uniform distribution only.

For other distributions specified by PDF and CDF, the general expression of the order statistics reads

$$PDF_{X(i)}(x) = \frac{n!}{(i-1)!(n-i)!} \cdot \left[CDF_X(x)\right]^{i-1} \cdot \left[1 - CDF_X(x)\right]^{n-i} \cdot PDF_X(x)$$

Unfortunately, means of these order statistics do *not* generally lead to the above general plotting position formula. Not all is lost, however. To draw random numbers for a specific distribution, a well-known (Monte Carlo) method is to draw uniform random numbers and transform the sample according to the inverse CDF, the so-called probability–integral transformation. Similarly, the uniform distribution plotting positions can be transformed through the inverse CDF to distribution-specific plotting positions. This may involve a recalibration of c in the general plotting position formula.

Thus, for the standard normal distribution:

$$E\left(X_{(i)}\right) \approx \Phi^{-1}\left(\frac{i - 0.375}{n + 0.25}\right)$$

with Φ^{-1} the inverse standard normal CDF, and $c = 0.375$ (3/8), well-known as the Blom (1958) plotting position formula for means of normal-order statistics. These numbers are also called *normal scores* or *rankits* (Davison, 1998).

Clearly, the uniform order statistic means do *not* transform in a similar fashion with the inverse CDF to yield order statistic means of the specific distribution. For the normal case, we then would evaluate

$$\Phi^{-1}\left(\frac{i}{n+1}\right)$$

which *can* be defended, however, as the first-order term in a series approximation (Gibbons, 1971: p. 36). The Blom formula turns out to be superior.

Percentiles, however, *do transform faithfully with the probability–integral transformation*. Thus, with the above calculated 0.651836, 0.905724, and 0.992699 for the 5th, 50th, and 95th percentiles of $X_{(7)}$ for samples of size 7 from a uniform distribution, we can readily calculate the same percentiles of $X_{(7)}$ for the standard normal distribution as the z-values: Φ^{-1} (0.651836) = 0.390282, Φ^{-1} (0.905724) = 1.31487, and Φ^{-1} (0.992699) = 2.44211, which is slightly skewed to the right (Figure 5.18).

To try the same for the means, the mean of $X_{(7)}$ for the uniform distribution equals 7/8 = 0.875, which is somewhat below the median 0.905724, since $X_{(7)}$ is

Normal Order Statistic Distributions for n = 7

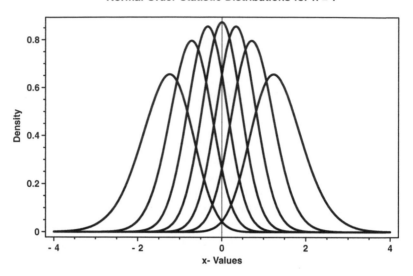

FIGURE 5.18 PDFs of order statistic distributions for samples of size 7 drawn from a standard normal distribution.

skewed to the left in the uniform case. To find the mean of $X_{(7)}$ for the normal case, we note that

$$\text{PDF}_{X(i)}(x) = \frac{n!}{(i-1)!(n-i)!} \cdot \left[\Phi(x)\right]^{i-1} \cdot \left[1-\Phi(x)\right]^{n-i} \cdot \phi(x)$$

with $\phi(x)$ the standard normal PDF. Figure 5.18 displays the seven order statistics PDFs for the standard normal distribution.

The exact value of the mean of $X_{(7)}$ is found by numerically integrating $7 \cdot x \cdot [\Phi(x)]^6 \cdot \phi(x)$ over the whole axis, yielding 1.35218. The Blom formula for the mean of $X_{(7)}$ yields $\Phi^{-1}\left((7 - 0.375)/7.25\right) = 1.36449$, quite close to the true value, and somewhat above the median 1.31487. Median standard normal order statistics for $n = 7$, exact mean normal order statistics, and the Blom approximation plotting positions are given by the next table:

Rank	Normal Medians	Normal Means	Blom Approx.
1	−1.31487	−1.35218	−1.36449
2	−0.74383	−0.75737	−0.75829
3	−0.34748	−0.35271	−0.35293
4	0.00000	0.00000	0.00000
5	+0.34748	+0.35271	+0.35293
6	+0.74383	+0.75737	+0.75829
7	+1.31487	+1.35218	+1.36449

However, the first-order approximation of the mean of $X_{(7)}$ for the standard normal distribution yields a poor $\Phi^{-1}(0.875) = 1.15035$. Application of the inverse normal CDF to the Hazen plotting position $\Phi^{-1}(6.5/7) = 1.46523$ is not satisfactory either, but not as bad.

With an eye to hydrologists interested in the extreme value distribution (0.44 due to Gringorten), and close to the Blom value (0.375), Cunnane (1978) proposed $c = 0.4$ as a compromise between the transformation invariant median plotting systems (0.3/0.31/0.3175) and the intuitive Hazen ECDF midpoints (0.5). Helsel and Hirsch (1992: p. 24) make use of Cunnane plotting positions throughout.

Blom (1958: p. 71) and Tukey (1962) proposed the easy-to-remember plotting positions $(i - 0.33)/(n + 0.33)$ for exploratory work.

5.6.5 GOODNESS-OF-FIT TESTS

5.6.5.1 Anderson–Darling CDF Test

The Anderson–Darling test belongs to a wide class of quadratic statistics measuring vertical discrepancy in a CDF-type probability plot (Stephens, 1986a: p. 100). It is designed to be sensitive to departures of the distribution in the "tails," i.e., close to CDF values 0 *and* 1. The final formula reads

$$A^2 = -n - 2\sum_{i=1}^{n}\left[\left(\frac{i-0.5}{n}\right)\ln\left(z_{(i)}\right)+\left(1-\left(\frac{i-0.5}{n}\right)\right)\ln\left(1-z_{(i)}\right)\right]$$

with $z_{(i)} = \Phi((x_{(i)} - \bar{x})/s)$, i.e., the normal CDF estimates for the standardized order statistics based on mean and sample standard deviation. The $(i - 0.5)/n$ terms are exactly the Hazen plotting positions. Hence all terms are easily calculated from the ranks and standardized log SDU values in Table 5.1 to yield $A^2 = 0.274$:

i	log SDU	$z(i)$	$(i - 0.5)/n$	
1	−1.40086	0.08063	0.07143	−0.25791
2	−0.63862	0.26153	0.21429	−0.52561
3	−0.58531	0.27917	0.35714	−0.66613
4	0.22638	0.58955	0.50000	−0.70945
5	0.23996	0.59482	0.64286	−0.65661
6	0.42774	0.66558	0.78571	−0.55458
7	1.73071	0.95825	0.92857	−0.26646
			Sum:	−3.63676
			$A^2 = -n - 2 *$ Sum:	0.273512

Note that $\Phi(x)$ is available in Excel as NORMSDIST(x). The modified statistic $A^2(1.0 + 0.75/n + 2.25/n^2) = 0.315$ accounts for the distribution being derived from the sample itself. The 5% critical value equals 0.752 and is not exceeded.

5.6.5.2   Filliben Correlation Test

The Filliben (1975) Correlation Test for Normality is calculated by correlating the normal median plotting positions with the data (whether standardized or not). Since correlation is symmetric, the test works in both probability plots (data either horizontally or vertically). The data of Table 5.1 yield correlation coefficient 0.968:

(*i*)	log SDU	Norm. Medians
1	−1.40086	−1.31487
2	−0.63862	−0.74383
3	−0.58531	−0.34748
4	0.22638	0.00000
5	0.23996	0.34748
6	0.42774	0.74383
7	1.73071	1.31487
	Correlation *R*:	0.968

The 5th and 95th percentile critical values are at 0.899 (0.897 in our simulations) and 0.990 for $n = 7$ (Filliben, 1975: table 1). So, normality need not be rejected.

If *mean* normal plotting positions are used instead of median plotting positions, the correlation test is known as the Shapiro and Francia test (Shapiro and Francia, 1972; Stephens, 1986b: p. 202), with test statistic W'. Filliben (1975) noted that the power of both tests is virtually identical for a wide range of alternative distributions. This means that nothing has been lost by using median plotting positions, rather than mean plotting positions. Median plotting positions are attractive since they transform consistently and are easily calculated in general. Mean plotting positions are difficult to calculate for distributions other than uniform and normal.

It would be worthwhile to recalculate the Filliben critical values in the form $n \cdot (1 - R^2)$, as tabulated by Stephens (1986b: p. 203) for the Shapiro and Francia test.

5.6.5.3   Shapiro and Wilk Regression Test

The Shapiro and Wilk Regression/Analysis of Variance Test for Normality (Shapiro and Wilk, 1965; see Hahn and Shapiro, 1969: p. 295; Stephens, 1986b: p. 208) is based on regression of the Q-Q plot of ordered sample data against mean standard normal order statistics. However, since order statistics are correlated (the second is greater than the first and so on), generalized least squares with given covariance matrix is to be preferred over ordinary regression based on independence. The regression line is cumbersome to calculate, and not shown, but the resulting test statistic W is easily calculated given some tabulated coefficients (italic in the next table). We have:

$$W = \frac{b^2}{S^2}$$

with

$$b = \sum_{i=1}^{k} b_{(i)} = \sum_{i=1}^{k} a_{(i)} \cdot \left(x_{(n+1-i)} - x_{(i)} \right)$$

Here $k = n/2$ for k even, and $k = (n-1)/2$ for k odd. The $a_{(i)}$ are tabulated in Shapiro and Wilk (1965: p. 603), Hahn and Shapiro (1969: p. 330), and Stephens (1986b: p. 209). For $n = 7$, we have: $b = a_{(1)} \cdot (x_{(7)} - x_{(1)}) + a_{(2)} \cdot (x_{(6)} - x_{(2)}) + a_{(3)} \cdot (x_{(5)} - x_{(3)})$.

The denominator is $S^2 = (n-1) \cdot s^2$, with s the sample standard deviation. When using standardized data, for which the test is invariant, we have: $S^2 = (n-1)$, since the sample standard deviation equals 1.0. Note, that the individual entries building up S^2, then, although not required, are the squares of the standardized order statistics, because the mean is 0.0. The calculation is given in the next table:

(i)	log SDU	Squares	A(i)	b(i)
1	−1.40086	1.96241	0.6233	1.95191
2	−0.63862	0.40784	0.3031	0.32321
3	−0.58531	0.34259	0.1401	0.11562
4	0.22638	0.05125		
5	0.23996	0.05758		
6	0.42774	0.18296		
7	1.73071	2.99537	Sum (b):	2.39075
		$S^2 = 6.00000$	B^2:	5.71566
			$W = b^2/S^2$:	0.95261

The calculated $W = 0.953$ is within the 5 to 95% region (0.803 to 0.979), and since departures from normality usually result into *low* values of W, there is no reason to reject the normal distribution as an adequate description of the data in Table 5.1.

For sample sizes over 50, W becomes equivalent to W', the Shapiro and Francia correlation statistic, since the correlation structure of the order statistics becomes irrelevant. Above we have seen that the latter statistic is similar to the Filliben correlation statistic that was also useful for small sample sizes.

5.6.6 PROBABILITY DISTRIBUTION OF STANDARDIZED LOG HC$_5$

To derive the probability distribution of $\log(\mathrm{HC}_5^{\mathrm{SDU}}) = \mu - K_p \cdot \sigma$ on the standardized scale, we take equation (A4*), the Bayesian version, in Aldenberg and Jaworska (2000; p. 15), then substitute $\bar{x} = 0$, $s = 1$ in view of the standardized log SDU scale, to obtain $\Pr(\mathrm{NCT} \leq k_s \cdot \sqrt{n}) = \gamma$. Here, random variable

$$\mathrm{NCT} = \frac{\left[\dfrac{-\mu}{\sigma/\sqrt{n}} + K_p \cdot \sqrt{n} \right]}{[1/\sigma]}$$

has a noncentral t distribution with noncentrality parameter $\lambda = K_p \cdot \sqrt{n}$ and $(n - 1)$ degrees of freedom.

The stated probability can be written as $\Pr(\mu - K_p \cdot \sigma \geq -k_s) = \gamma$, which is the probability of the 5th percentile exceeding constant $-k_s$ on the standardized scale. Hence, the uncertainty of $X = \log(\mathrm{HC}_5^{\mathrm{SDU}}) = \mu - K_p \cdot \sigma$ is the same as the uncertainty of $-\mathrm{NCT}/\sqrt{n}$.

If NCT has density function: $\mathrm{PDF}_{\mathrm{NoncentralT}(n-1,\lambda)}(\mathrm{nct})$, then $X = -\mathrm{NCT}/\sqrt{n}$, which is a linear transformation (Mood et al., 1974: p. 200; Hsu, 1997: p. 131), has density function: $\sqrt{n} \cdot \mathrm{PDF}_{\mathrm{NoncentralT}(n-1,\lambda)}(-\sqrt{n} \cdot x)$, with x taking negative standardized values for the 5th percentile. This PDF plots as the Bayesian simulation in Figure 5.5 (gray line).

5.6.7 Approximate Normal Distribution Fitted to Median Bayesian FA Curve

To fit median log HC_{50} and median log HC_5 exactly, one may employ

$$\hat{\mu} = \bar{x}$$

and

$$\hat{\sigma} = \frac{k_s^{\mathrm{median}}}{1.64485} \cdot s$$

This somewhat *inflated* standard deviation (see Aldenberg, 1993; Van Beelen et al., 2001) does the job, since

$$\Phi_{\hat{\mu},\hat{\sigma}}(\bar{x}) = \Phi\left(\frac{\bar{x} - \hat{\mu}}{\hat{\sigma}}\right) = \Phi(0) = 0.50$$

and

$$\Phi_{\hat{\mu},\hat{\sigma}}\left(\bar{x} - k_s^{\mathrm{median}} \cdot s\right) = \Phi\left(\frac{\bar{x} - k_s^{\mathrm{median}} \cdot s - \hat{\mu}}{\hat{\sigma}}\right) = \Phi(-1.64485) = 0.05$$

At $n = 7$, the inflated standard deviation becomes

$$\hat{\sigma} = \frac{1.73179}{1.64485} \cdot s = 1.05286 \cdot s$$

5.6.8 Sensitivity of log (HC_p) for Individual Data Points

The sensitivity of $\log(\mathrm{HC}_p)$ with respect to small changes in individual species sensitivity data is obtained by differentiating $\log(\mathrm{HC}_p) = \bar{x} - k_s \cdot s$ with respect to x_i (the ith data point). With mean

$$\bar{x} = \frac{1}{n}\sum_{i=1}^{n}x_i$$

and sample standard deviation

$$s = \left[\frac{1}{n-1}\sum_{i=1}^{n}(x_i - \bar{x})^2\right]^{1/2}$$

we have

$$\frac{\partial \bar{x}}{\partial x_i} = \frac{1}{n}$$

and

$$\frac{\partial s}{\partial x_i} = \frac{1}{2}(s^2)^{-1/2}\frac{2}{n-1}\left[(x_i - \bar{x})(1-1/n) + \sum_{j \neq i}^{n}(x_j - \bar{x})(-1/n)\right] = \frac{1}{n-1}\left(\frac{x_i - \bar{x}}{s}\right)$$

so that

$$\frac{\partial(\bar{x} - k_s \cdot s)}{\partial x_i} = \frac{1}{n} - \frac{k_s}{n-1}\left(\frac{x_i - \bar{x}}{s}\right)$$

Hence, sensitivity of log HC_p to a toxicity data point is a dimensionless function of the standardized log concentration of the data point and the extrapolation constant. This sensitivity changes sign at the positive standardized value: $(n-1)/(n \cdot k_s)$, and hence is asymmetric.

5.6.9 DERIVATION OF PROBABILITY OF FAILURE

Let the probability that RV X takes values smaller than x be given by the CDF:

$$\Pr(X \leq x) = CDF_X(x) = \int_{-\infty}^{x} PDF_X(t)dt$$

The exceedence function (or survival probability), with new mnemonic EXF, is the complementary probability

$$\Pr(X > x) = EXF_X(x) = 1 - CDF_X(x) = \int_{x}^{\infty} PDF_X(t)dt$$

Suppose RVs $X1$ and $X2$ both take values x. We are interested in the probability of $X1$ exceeding $X2$, that is, $\Pr(X1 > X2)$, or equivalently $\Pr(X1 - X2 > 0)$. Thus, we consider a new RV: $Z = X1 - X2$ for the difference, taking values z, and require: $EXF_{X1-X2}(0)$.

The analytical derivation of the probability of failure integrals closely follows Papoulis (1965: p. 189, see his figure 7-2), Mood et al. (1974: pp. 185–186), or Hsu (1997: p. 137). The difference of a pair of values x_1 and x_2 exceeds value z, if $x_2 < x_1 - z$. Consequently, we have to sum (integrate) the *joint probability* of x_1 and x_2 over all values satisfying this inequality:

$$EXF_{X1-X2}(z) = \iint_{x1-x2>z} PDF_{X1,X2}(x_1, x_2) dx_1 dx_2 = \int_{-\infty}^{\infty} \left[\int_{-\infty}^{x1-z} PDF_{X1,X2}(x_1, x_2) dx_2 \right] dx_1$$

We now assume that the RVs of $X1$ and $X2$ are *independent*:

$$PDF_{X1,X2}(x_1, x_2) = PDF_{X1}(x_1) \cdot PDF_{X2}(x_2)$$

that is, the joint probability density factorizes into the univariate PDFs.
It follows that

$$EXF_{X1-X2}(z) = \int_{-\infty}^{\infty} PDF_{X1}(x_1) \cdot \left[\int_{-\infty}^{x1-z} PDF_{X2}(x_2) dx_2 \right] dx_1$$

$$= \int_{-\infty}^{\infty} PDF_{X1}(x) \cdot CDF_{X2}(x - z) dx$$

The required exceedence of $Z = X1 - X2$ at $z = 0$ equals

$$\Pr(X1 > X2) = EXF_{X1-X2}(0) = \int_{-\infty}^{\infty} PDF_{X1}(x) \cdot CDF_{X2}(x) dx \qquad (5.5)$$

An alternative expression for this exceedence can be derived as follows. Instead of integrating $PDF_{X1,X2}(x_1, x_2)$ over the region $x_2 < x_1 - z$ for pairs of values x_1 and x_2 given z, one could have done the double integration over the region $x_1 > x_2 + z$:

$$EXF_{X1-X2}(z) = \int_{-\infty}^{\infty} \left[\int_{x2+z}^{\infty} PDF_{X1,X2}(x_1, x_2) dx_1 \right] dx_2 = \int_{-\infty}^{\infty} EXF_{X1}(x + z) \cdot PDF_{X2}(x) dx$$

The exceedence of $Z = X1 - X2$ at 0 equals:

$$\Pr(X1 > X2) = EXF_{X1-X2}(0) = \int_{-\infty}^{\infty} (1 - CDF_{X1}(x)) \cdot PDF_{X2}(x) dx \qquad (5.6)$$

ACKNOWLEDGMENTS

We thank Leo Posthuma and Glenn Suter for their coordinative efforts, sensitive reviews, and inspiring discussions. Dik van de Meent is thanked for his drive to apply and further develop the PAF concept. John Warmerdam explained the probability of failure integral that was on a SETAC poster and pointed to relevant reliability engineering literature. Eric Smith and Tim Springer reviewed two substantially different versions of the chapter, making a lot of detailed and constructive remarks. Especially Tim's careful reading triggered a completely rewritten, and substantially enhanced, version. Keith Solomon read the second version and suggested several improvements.

APPENDIX

TABLE 5.A1
Extrapolation Constants k_s (lower, median, and upper percentiles: 5%, 50%, 95%, Bayesian) to Estimate the 5th Percentile of a Normal SSD, through $\bar{x} - k_s \cdot s$ in Original Log Concentration Units, or $-k_s$ in log SDU[a]

n	Lower	Median	Upper	n	Lower	Median	Upper	n	Lower	Median	Upper
				27	2.26005	1.66398	1.23135	53	2.05071	1.65435	1.33749
2	26.25968	2.33873	0.47479	28	2.24578	1.66326	1.23780	54	2.04625	1.65417	1.34005
3	7.65590	1.93842	0.63914	29	2.23241	1.66259	1.24395	55	2.04193	1.65399	1.34254
4	5.14388	1.82951	0.74330	30	2.21984	1.66197	1.24981	56	2.03774	1.65383	1.34497
5	4.20268	1.77928	0.81778	31	2.20800	1.66140	1.25540	57	2.03367	1.65366	1.34734
6	3.70768	1.75046	0.87477	32	2.19682	1.66086	1.26075	58	2.02972	1.65351	1.34965
7	3.39947	1.73179	0.92037	33	2.18625	1.66035	1.26588	59	2.02589	1.65336	1.35191
8	3.18729	1.71872	0.95803	34	2.17623	1.65988	1.27079	60	2.02216	1.65321	1.35412
9	3.03124	1.70906	0.98987	35	2.16672	1.65943	1.27551	61	2.01853	1.65307	1.35627
10	2.91096	1.70163	1.01730	36	2.15768	1.65901	1.28004	62	2.01501	1.65294	1.35837
11	2.81499	1.69574	1.04127	37	2.14906	1.65861	1.28441	63	2.01157	1.65281	1.36043
12	2.73634	1.69096	1.06247	38	2.14085	1.65823	1.28861	64	2.00823	1.65268	1.36244
13	2.67050	1.68700	1.08141	39	2.13300	1.65788	1.29266	65	2.00498	1.65256	1.36441
14	2.61443	1.68366	1.09848	40	2.12549	1.65754	1.29657	66	2.00180	1.65244	1.36633
15	2.56600	1.68082	1.11397	41	2.11831	1.65722	1.30035	67	1.99871	1.65232	1.36822
16	2.52366	1.67836	1.12812	42	2.11142	1.65692	1.30399	68	1.99569	1.65221	1.37006
17	2.48626	1.67621	1.14112	43	2.10481	1.65663	1.30752	69	1.99274	1.65210	1.37187
18	2.45295	1.67433	1.15311	44	2.09846	1.65635	1.31094	70	1.98987	1.65199	1.37364
19	2.42304	1.67266	1.16423	45	2.09235	1.65609	1.31425	71	1.98706	1.65189	1.37538
20	2.39600	1.67116	1.17458	46	2.08648	1.65584	1.31746	72	1.98431	1.65179	1.37708
21	2.37142	1.66982	1.18425	47	2.08081	1.65559	1.32058	73	1.98163	1.65170	1.37875
22	2.34896	1.66861	1.19330	48	2.07535	1.65536	1.32360	74	1.97901	1.65160	1.38038
23	2.32832	1.66752	1.20181	49	2.07008	1.65514	1.32653	75	1.97645	1.65151	1.38199
24	2.30929	1.66651	1.20982	50	2.06499	1.65493	1.32939	76	1.97394	1.65142	1.38357
25	2.29167	1.66560	1.21739	51	2.06007	1.65473	1.33216	77	1.97149	1.65133	1.38511
26	2.27530	1.66475	1.22455	52	2.05532	1.65453	1.33486	78	1.96909	1.65125	1.38663

TABLE 5.A1 (continued)
Extrapolation Constants k_s (lower, median, and upper percentiles: 5%, 50%, 95%, Bayesian) to Estimate the 5th Percentile of a Normal SSD, through $\bar{x} - k_s \cdot s$ in Original Log Concentration Units, or $-k_s$ in log SDU[a]

n	Lower	Median	Upper	n	Lower	Median	Upper	n	Lower	Median	Upper
79	1.96674	1.65117	1.38812	103	1.92191	1.64968	1.41744	127	1.89143	1.64875	1.43838
80	1.96444	1.65109	1.38959	104	1.92041	1.64963	1.41845	128	1.89036	1.64872	1.43912
81	1.96218	1.65101	1.39103	105	1.91894	1.64958	1.41944	129	1.88931	1.64869	1.43986
82	1.95997	1.65093	1.39244	106	1.91749	1.64954	1.42042	130	1.88827	1.64866	1.44060
83	1.95781	1.65086	1.39383	107	1.91607	1.64949	1.42139	131	1.88724	1.64863	1.44132
84	1.95568	1.65078	1.39520	108	1.91466	1.64945	1.42234	132	1.88622	1.64861	1.44204
85	1.95360	1.65071	1.39654	109	1.91328	1.64941	1.42328	133	1.88522	1.64858	1.44274
86	1.95156	1.65064	1.39786	110	1.91191	1.64937	1.42421	134	1.88423	1.64855	1.44344
87	1.94955	1.65058	1.39916	111	1.91057	1.64932	1.42513	135	1.88325	1.64852	1.44414
88	1.94759	1.65051	1.40044	112	1.90924	1.64928	1.42604	136	1.88228	1.64849	1.44482
89	1.94566	1.65045	1.40170	113	1.90794	1.64924	1.42693	137	1.88133	1.64847	1.44550
90	1.94376	1.65038	1.40294	114	1.90665	1.64920	1.42781	138	1.88038	1.64844	1.44617
91	1.94190	1.65032	1.40415	115	1.90538	1.64917	1.42869	139	1.87945	1.64841	1.44684
92	1.94007	1.65026	1.40535	116	1.90413	1.64913	1.42955	140	1.87852	1.64839	1.44750
93	1.93828	1.65020	1.40654	117	1.90289	1.64909	1.43040	141	1.87761	1.64836	1.44815
94	1.93651	1.65014	1.40770	118	1.90168	1.64906	1.43124	142	1.87671	1.64834	1.44879
95	1.93478	1.65009	1.40885	119	1.90048	1.64902	1.43207	143	1.87582	1.64831	1.44943
96	1.93308	1.65003	1.40998	120	1.89929	1.64898	1.43289	144	1.87493	1.64829	1.45006
97	1.93140	1.64998	1.41109	121	1.89812	1.64895	1.43370	145	1.87406	1.64827	1.45069
98	1.92975	1.64993	1.41219	122	1.89697	1.64892	1.43450	146	1.87320	1.64824	1.45131
99	1.92813	1.64987	1.41327	123	1.89583	1.64888	1.43530	147	1.87234	1.64822	1.45192
100	1.92654	1.64982	1.41433	124	1.89471	1.64885	1.43608	148	1.87150	1.64820	1.45253
101	1.92497	1.64977	1.41538	125	1.89360	1.64882	1.43685	149	1.87067	1.64817	1.45313
102	1.92343	1.64972	1.41642	126	1.89251	1.64879	1.43762	150	1.86984	1.64815	1.45372
Inf	1.64485	1.64485	1.64485	Inf	1.64485	1.64485	1.64485	Inf	1.64485	1.64485	1.64485

[a] The latter are scaled to the toxicity data sample mean ($\bar{x}$) and sample standard deviation (s).

TABLE 5.A2
Normal Probability Paper z-Values of Upper, Median, and Lower Percentiles (95%, 50%, and 5%, Bayesian) of the FA

$-k_s$: n	-5.0	-4.5	-4.0	-3.5	-3.0	-2.5	-2.0	-1.5	-1.0	-0.5	0.0
2	-0.06452	-0.00442	0.05998	0.12973	0.20631	0.29191	0.38998	0.50638	0.65224	0.85188	1.16309
	-3.40601	-3.07245	-2.73978	-2.40833	-2.07849	-1.75039	-1.42319	-1.09352	-0.75369	-0.39120	0.00000
	-9.89733	-8.92806	-7.96141	-6.99847	-6.04101	-5.09214	-4.15771	-3.25023	-2.39995	-1.67626	-1.16309
3	-0.99165	-0.86167	-0.72773	-0.58855	-0.44232	-0.28637	-0.11658	0.07373	0.29644	0.57422	0.94966
	-4.17768	-3.76292	-3.34847	-2.93443	-2.52092	-2.10802	-1.69564	-1.28281	-0.86628	-0.43969	0.00000
	-8.74944	-7.89443	-7.04192	-6.19295	-5.34915	-4.51331	-3.69058	-2.89121	-2.13713	-1.47213	-0.94966
4	-1.59226	-1.40790	-1.22044	-1.02875	-0.83115	-0.62507	-0.40650	-0.16885	0.09922	0.41816	0.82243
	-4.45014	-4.00708	-3.56421	-3.12158	-2.67925	-2.23725	-1.79550	-1.35341	-0.90908	-0.45857	0.00000
	-8.15904	-7.36174	-6.56675	-5.77502	-4.98802	-4.20827	-3.44028	-2.69282	-1.98371	-1.34660	-0.82243
5	-2.00046	-1.77822	-1.55331	-1.32475	-1.09103	-0.84985	-0.59744	-0.32763	-0.02977	0.31479	0.73560
	-4.58770	-4.13038	-3.67320	-3.21620	-2.75940	-2.30282	-1.84638	-1.38966	-0.93130	-0.46852	0.00000
	-7.78385	-7.02282	-6.26394	-5.50806	-4.75655	-4.01169	-3.27754	-2.56184	-1.87990	-1.25981	-0.73560
6	-2.29552	-2.04581	-1.79369	-1.53830	-1.27826	-1.01146	-0.73443	-0.44139	-0.12236	0.23986	0.67151
	-4.67039	-4.20452	-3.73875	-3.27311	-2.80763	-2.34231	-1.87709	-1.41161	-0.94484	-0.47463	0.00000
	-7.51817	-6.78264	-6.04911	-5.31838	-4.59172	-3.87122	-3.16061	-2.46686	-1.80358	-1.19526	-0.67151
7	-2.52007	-2.24945	-1.97662	-1.70078	-1.42067	-1.13433	-0.83853	-0.52785	-0.19290	0.18234	0.62170
	-4.72553	-4.25395	-3.78245	-3.31106	-2.83980	-2.36867	-1.89760	-1.42631	-0.95394	-0.47877	0.00000
	-7.31719	-6.60084	-5.88638	-5.17455	-4.46653	-3.76428	-3.07124	-3.07124	-1.74438	-1.14484	-0.62170
8	-2.69780	-2.41064	-2.12143	-1.82941	-1.53341	-1.23160	-0.92096	-0.59635	-0.24892	0.13639	0.58154
	-4.76489	-4.28924	-3.81365	-3.33816	-2.86277	-2.38749	-1.91226	-1.43683	-0.96047	-0.48174	0.00000
	-7.15821	-6.45697	-5.75752	-5.06057	-4.36721	-3.67929	-3.00003	-2.33538	-1.69671	-1.10405	-0.58154
9	-2.84277	-2.54213	-2.23957	-1.93436	-1.62541	-1.31100	-0.98825	-0.65232	-0.29479	0.09859	0.54828
	-4.79440	-4.31568	-3.83704	-3.35847	-2.87999	-2.40160	-1.92325	-1.44472	-0.96538	-0.48399	0.00000
	-7.02832	-6.33940	-5.65218	-4.96732	-4.28589	-3.60961	-2.94151	-2.28720	-1.65724	-1.07016	-0.54828

TABLE 5.A2 (continued)
Normal Probability Paper z-Values of Upper, Median, and Lower Percentiles (95%, 50%, and 5%, Bayesian) of the FA

$-k$; n	-5.0	-4.5	-4.0	-3.5	-3.0	-2.5	-2.0	-1.5	-1.0	-0.5	0.0
10	-2.96383	-2.65195	-2.33824	-2.02203	-1.70228	-1.37734	-1.04450	-0.69914	-0.33323	0.06678	0.52015
	-4.81733	-4.33624	-3.85522	-3.37426	-2.89338	-2.41258	-1.93180	-1.45087	-0.96921	-0.48575	0.00000
	-6.91960	-6.24095	-5.56393	-4.88918	-4.21769	-3.55110	-2.89231	-2.24658	-1.62386	-1.04143	-0.52015
11	-3.06685	-2.74540	-2.42221	-2.09665	-1.76771	-1.43383	-1.09242	-0.73906	-0.36605	0.03954	0.49594
	-4.83566	-4.35268	-3.86975	-3.38688	-2.90408	-2.42135	-1.93864	-1.45578	-0.97228	-0.48716	0.00000
	-6.82682	-6.15693	-5.48860	-4.82244	-4.15941	-3.50107	-2.85017	-2.21173	-1.59514	-1.01666	-0.49594
12	-3.15586	-2.82615	-2.49478	-2.16114	-1.82427	-1.48267	-1.13386	-0.77360	-0.39449	0.01587	0.47483
	-4.85065	-4.36612	-3.88163	-3.39720	-2.91283	-2.42852	-1.94423	-1.45981	-0.97479	-0.48831	0.00000
	-6.74644	-6.08412	-5.42330	-4.76457	-4.10884	-3.45763	-2.81355	-2.18139	-1.57008	-0.99502	-0.47483
13	-3.23375	-2.89682	-2.55829	-2.21758	-1.87378	-1.52543	-1.17016	-0.80387	-0.41945	-0.00495	0.45620
	-4.86314	-4.37731	-3.89153	-3.40580	-2.92012	-2.43450	-1.94889	-1.46316	-0.97689	-0.48928	0.00000
	-6.67589	-6.02021	-5.36598	-4.71375	-4.06443	-3.41945	-2.78133	-2.15466	-1.54796	-0.97591	-0.45620
14	-3.30265	-2.95933	-2.61448	-2.26752	-1.91759	-1.56328	-1.20229	-0.83068	-0.44157	-0.02344	0.43961
	-4.87370	-4.38678	-3.89990	-3.41307	-2.92628	-2.43955	-1.95283	-1.46600	-0.97866	-0.49010	0.00000
	-6.61333	-5.96353	-5.31512	-4.66866	-4.02500	-3.38554	-2.75269	-2.13088	-1.52826	-0.95885	-0.43961
15	-3.36415	-3.01513	-2.66464	-2.31211	-1.95670	-1.59707	-1.23099	-0.85464	-0.46137	-0.04002	0.42470
	-4.88275	-4.39489	-3.90707	-3.41929	-2.93157	-2.44388	-1.95621	-1.46843	-0.98018	-0.49080	0.00000
	-6.55735	-5.91279	-5.26960	-4.62829	-3.98969	-3.35515	-2.72702	-2.10953	-1.51055	-0.94352	-0.42470
16	-3.41948	-3.06534	-2.70977	-2.35223	-1.99190	-1.62749	-1.25683	-0.87623	-0.47921	-0.05498	0.41121
	-4.89059	-4.40192	-3.91329	-3.42469	-2.93614	-2.44763	-1.95914	-1.47054	-0.98150	-0.49141	0.00000
	-6.50686	-5.86704	-5.22854	-4.59187	-3.95783	-3.32772	-2.70383	-2.09024	-1.49453	-0.92964	-0.41121
17	-3.46960	-3.11082	-2.75065	-2.38858	-2.02380	-1.65506	-1.28026	-0.89579	-0.49540	-0.06858	0.39894
	-4.89744	-4.40807	-3.91872	-3.42941	-2.94014	-2.45092	-1.96170	-1.47238	-0.98266	-0.49194	0.00000
	-6.46102	-5.82549	-5.19125	-4.55879	-3.92888	-3.30280	-2.68274	-2.07268	-1.47993	-0.91698	-0.39894

n											
18	-3.51528	-3.15228	-2.78792	-2.42171	-2.05287	-1.68019	-1.30161	-0.91365	-0.51018	-0.08101	0.38770
	-4.90349	-4.41349	-3.92351	-3.43357	-2.94367	-2.45381	-1.96396	-1.47401	-0.98367	-0.49241	0.00000
	-6.41916	-5.78755	-5.15719	-4.52856	-3.90243	-3.28001	-2.66346	-2.05661	-1.46657	-0.90539	-0.38770
19	-3.55714	-3.19026	-2.82207	-2.45207	-2.07952	-1.70322	-1.32119	-0.93001	-0.52374	-0.09243	0.37736
	-4.90887	-4.41831	-3.92777	-3.43727	-2.94681	-2.45638	-1.96596	-1.47545	-0.98458	-0.49283	0.00000
	-6.38072	-5.75271	-5.12591	-4.50081	-3.87813	-3.25908	-2.64574	-2.04184	-1.45427	-0.89472	-0.37736
20	-3.59567	-3.22523	-2.85350	-2.48002	-2.10405	-1.72443	-1.33922	-0.94509	-0.53624	-0.10297	0.36780
	-4.91368	-4.42262	-3.93159	-3.44058	-2.94962	-2.45868	-1.96776	-1.47674	-0.98539	-0.49321	0.00000
	-6.34527	-5.72057	-5.09706	-4.47520	-3.85571	-3.23976	-2.62937	-2.02819	-1.44290	-0.88485	-0.36780
21	-3.63130	-3.25756	-2.88257	-2.50587	-2.12674	-1.74405	-1.35590	-0.95905	-0.54781	-0.11273	0.35894
	-4.91800	-4.42650	-3.93501	-3.44356	-2.95214	-2.46076	-1.96937	-1.47791	-0.98612	-0.49355	0.00000
	-6.31244	-5.69081	-5.07034	-4.45148	-3.83494	-3.22186	-2.61420	-2.01553	-1.43235	-0.87569	-0.35894
22	-3.66437	-3.28757	-2.90955	-2.52986	-2.14780	-1.76226	-1.37139	-0.97200	-0.55856	-0.12181	0.35068
	-4.92192	-4.43000	-3.93812	-3.44626	-2.95443	-2.46263	-1.97083	-1.47896	-0.98678	-0.49385	0.00000
	-6.28191	-5.66313	-5.04549	-4.42942	-3.81562	-3.20520	-2.60009	-2.00375	-1.42252	-0.86716	-0.35068
23	-3.69517	-3.31552	-2.93469	-2.55221	-2.16742	-1.77922	-1.38581	-0.98408	-0.56859	-0.13028	0.34298
	-4.92547	-4.43319	-3.94094	-3.44870	-2.95650	-2.46433	-1.97216	-1.47992	-0.98738	-0.49413	0.00000
	-6.25344	-5.63731	-5.02230	-4.40884	-3.79759	-3.18965	-2.58691	-1.99274	-1.41334	-0.85918	-0.34298
24	-3.72394	-3.34164	-2.95817	-2.57309	-2.18575	-1.79508	-1.39930	-0.99537	-0.57796	-0.13821	0.33575
	-4.92872	-4.43610	-3.94351	-3.45094	-2.95840	-2.46588	-1.97337	-1.48079	-0.98793	-0.49438	0.00000
	-6.22679	-5.61315	-5.00061	-4.38958	-3.78072	-3.17510	-2.57457	-1.98243	-1.40474	-0.85171	-0.33575
25	-3.75091	-3.36612	-2.98017	-2.59267	-2.20293	-1.80994	-1.41194	-1.00595	-0.58675	-0.14565	0.32897
	-4.93169	-4.43877	-3.94587	-3.45299	-2.96014	-2.46731	-1.97449	-1.48159	-0.98843	-0.49462	0.00000
	-6.20178	-5.59048	-4.98025	-4.37150	-3.76488	-3.16144	-2.56298	-1.97275	-1.39665	-0.84468	-0.32897
30	-3.86418	-3.46891	-3.07261	-2.67488	-2.27512	-1.87238	-1.46507	-1.05046	-0.62376	-0.17702	0.30031
	-4.94349	-4.44935	-3.95522	-3.46111	-2.96702	-2.47296	-1.97889	-1.48477	-0.99042	-0.49554	0.00000
	-6.09638	-5.49490	-4.89440	-4.29526	-3.69807	-3.10379	-2.51406	-1.93184	-1.36246	-0.81497	-0.30031
35	-3.95147	-3.54815	-3.14386	-2.73825	-2.33077	-1.92053	-1.50606	-1.08482	-0.65237	-0.20133	0.27803
	-4.95181	-4.45681	-3.96181	-3.46684	-2.97188	-2.47694	-1.98200	-1.48701	-0.99183	-0.49620	0.00000
	-6.01471	-5.42083	-4.82785	-4.23615	-3.64625	-3.05905	-2.47608	-1.90003	-1.33586	-0.79183	-0.27803

TABLE 5.A2 (continued)
Normal Probability Paper z-Values of Upper, Median, and Lower Percentiles (95%, 50%, and 5%, Bayesian) of the FA

$-k_s$; n	-5.0	-4.5	-4.0	-3.5	-3.0	-2.5	-2.0	-1.5	-1.0	-0.5	0.0
40	-4.02139	-3.61162	-3.20094	-2.78903	-2.37536	-1.95911	-1.53891	-1.11237	-0.67533	-0.22087	0.26007
	-4.95800	-4.46235	-3.96671	-3.47109	-2.97549	-2.47990	-1.98431	-1.48868	-0.99288	-0.49668	0.00000
	NA	-5.36123	-4.77431	-4.18857	-3.60454	-3.02303	-2.44547	-1.87439	-1.31439	-0.77316	-0.26007
50	-4.12758	-3.70801	-3.28763	-2.86614	-2.44309	-2.01774	-1.58884	-1.15427	-0.71029	-0.25069	0.23262
	-4.96658	-4.47004	-3.97351	-3.47700	-2.98050	-2.48401	-1.98752	-1.49099	-0.99433	-0.49736	0.00000
	NA	NA	-4.69257	-4.11594	-3.54083	-2.96798	-2.39868	-1.83516	-1.28151	-0.74456	-0.23262
75	-4.29105	-3.85640	-3.42110	-2.98489	-2.54740	-2.10804	-1.66578	-1.21888	-0.76427	-0.29686	0.18993
	NA	NA	-3.98247	-3.48478	-2.98709	-2.48941	-1.99174	-1.49404	-0.99625	-0.49825	0.00000
	NA	NA	NA	-4.00317	-3.44188	-2.88245	-2.32591	-1.77409	-1.23028	-0.69999	-0.18993
Inf	-5.00000	-4.50000	-4.00000	-3.50000	-3.00000	-2.50000	-2.00000	-1.50000	-1.00000	-0.50000	0.00000

Columns:Standardized log(EC): $-k_s$. Rows: At each sample size (n), confidence limits and median of z-values: $-K_p = \Phi^{-1}(FA)$, with Φ^{-1} the inverse standard normal CDF and FA the fraction of species affected. Apply Φ, the standard normal CDF, to get FA. Positive exposure concentrations follow from symmetry.

Source: Extended version of table 2 in Aldenberg and Jaworska (2000); see also Figures 5.3 and 5.8.

6 Extrapolation Factors for Tiny Toxicity Data Sets from Species Sensitivity Distributions with Known Standard Deviation

Tom Aldenberg and Robert Luttik

CONTENTS

Abstract — Extrapolation techniques based on species sensitivity distributions are usually applied to data sets with four or more toxicity data. In this chapter, extrapolation factors are developed for SSDs with known standard deviation that also apply to very small ("tiny") data sets (n = 1, 2, 3, …). The concept of a standard extrapolation factor (SEF) is introduced, which is an extrapolation factor (EF) for estimating some percentile of a unit standard deviation SSD of common log-transformed species sensitivities. EFs result from raising SEFs to the power of the SSD standard deviation. The determination of EFs for tiny LD_{50} data sets of birds and mammals is demonstrated, both for an "average" SSD standard deviation derived from a collection of pesticide data sets, as well as for a more conservative (high) estimate of the SSD standard deviation. In the discussion the following aspects are reviewed:

1-56670-578-9/02/$0.00+$1.50
© 2002 by CRC Press LLC

- Whether LD_{50} data sets of bird and mammals can be modeled with normal distributions (one of the premises made in this chapter);
- Relationships between acute SSD extrapolation based on LD_{50} or LC_{50} values and chronic SSD extrapolation based on NOEC values (including a method for attacking this problem); and
- Limitations of extrapolation based on tiny samples and possible improvements.

6.1 INTRODUCTION

The evaluation of the possible environmental hazard for birds and mammals arising from the use of agricultural pesticides is usually based on the lowest median lethal dose (LD_{50}) available. The LD_{50} is a statistically derived single oral dose of a compound that will cause 50% mortality of the test population of a species. Generally there are only *one* or *two* LD_{50} values for birds available (mallard and quail) and usually *one* LD_{50} for mammals (rat) (Anonymous, 1991; U.S. EPA, 1994d; EEC, 1996). We call such sample sizes *tiny*. With tiny toxicity data sets, underestimation of the potential hazard is a real possibility.

A general approach toward extrapolating laboratory toxicity data to estimate levels of protection or fraction of species exceeded is to model the species sensitivity distribution (SSD) for the toxicity endpoint and to estimate percentiles and cumulative fractions statistically. This SSD extrapolation methodology in principle applies to both chronic and acute toxicity data. One way of extrapolating no-observed-effect concentrations (NOECs) to acceptable concentrations in the field is to estimate a low, typically the 5th, percentile of the NOEC SSD. This percentile is called the HC_5, the hazardous concentration affecting 5% of the species (Kooijman, 1987; Van Straalen and Denneman, 1989; Wagner and Løkke, 1991; Aldenberg and Slob, 1993; Aldenberg and Jaworska, 2000). The percentage of species affected is called the fraction affected (FA).

To estimate a percentile of a SSD on the basis of a (small) sample of toxicity data for different species, at a certain level of confidence, one may apply extrapolation *constants** to sample mean and sample standard deviation of log-transformed toxicity data. Extrapolation constants k_s have been tabulated for the logistic distribution (Aldenberg and Slob, 1993) and the normal distribution (Wagner and Løkke, 1991; Aldenberg and Jaworska, 2000: table 1). The latter contains extrapolation constants for lower, median, and upper estimates of the HC_p (95, 50, and 5% confidence level respectively, as one-sided underestimates), at six different levels of FA ($p\%$): 1, 2, 5, 10, 25, and 50%. An extended table for the 5th percentile only, covering sample sizes up to 150, can be found in Aldenberg et al., Chapter 5: Table 5.A1, this volume).

---

* To clean up terminology: we use extrapolation *constants* for the number of sample standard deviation units (log concentration) to be subtracted from the sample mean (log concentration), and extrapolation *factors* for numbers to be applied multiplicatively to the geometric sample mean in original concentration units.

Extrapolation constants only depend on the FA employed, sample size, and confidence level. When the log toxicity data set is *standardized* by subtracting the sample mean and dividing by the sample standard deviation, logarithmic percentile estimates become equal to minus the extrapolation constants (Aldenberg et al., Chapter 5).

Extrapolation *factors* (see Mineau et al., 1996; 2001; Luttik and Aldenberg, 1997) are factors to be applied multiplicatively to the geometric mean of the original (not log-transformed) toxicity data. Extrapolation factors depend on the FA, sample size, confidence level, *and* on the (log) standard deviation of the SSD. Wider SSDs need larger extrapolation factors.

Distribution-based extrapolation is applicable to *small* sample sizes, e.g., below 20. Although the procedure can be applied to samples even from $n = 2$ onward, in practice the method is of limited use in the case of tiny samples (say, $n = 2$, or 3). In these cases, the extrapolation constants are large, especially for the lower confidence limit, because of the uncertainty in estimating the SSD mean and standard deviation from the sample. If only one NOEC datum is available ($n = 1$), the method cannot be applied, since it is impossible to estimate the SSD standard deviation from the sample.

In this chapter, we develop a distribution-based SSD extrapolation methodology analogous to the one developed for small samples, for tiny samples of chronic or acute toxicity data. The basic idea is that the standard deviation is *not* derived from the sample, but comes from somewhere else. For example, one may use information on the standard deviation of SSDs for similar substances as the one under study. We show how distribution-based extrapolation theory simplifies if the standard deviation of the SSD is given, whatever its source.

We develop so-called *standard extrapolation factors* (SEF) to calculate estimates of the FA (e.g., 5%) that are defined for a SSD with unit standard deviation. To obtain *actual* extrapolation factors (EF), one has to raise the SEFs to the power of the standard deviation of the SSD.

As an illustration, we derive EFs for tiny LD_{50} toxicity data sets of birds and mammals assuming *normal* distribution SSDs, similar to those developed for the logistic distribution by Luttik and Aldenberg (1995; 1997). The device of SEFs allows employing of *any* "average" or conservative estimate of the SSD standard deviation derived from other data sets.

EFs for SSDs on the basis of *acute* toxicity data, e.g., median lethal dose in birds, are usually based on the 5th percentile of the SSD (Luttik and Aldenberg, 1997; Mineau et al., 2001). The interpretation of a 5th percentile estimate of acute data differs dramatically from a 5th percentile estimate from chronic data (see the discussion in Mineau et al., 2001: pp. 20, 21). To shed some light on this, we address the question: What percentile of acute data would be compatible with the 5th percentile of chronic data, given an acute-to-chronic ratio for the respective SSD *means*, and *equal* SSD standard deviations?

When the SSD standard deviation is known (given), acute EFs can be defined that are equal to the chronic EFs multiplied by the acute-to-chronic ratio of the means of the respective SSDs.

We would like to acknowledge the fact that for birds in particular the *species dependent* EFs of Mineau et al. (2001), as derived for the 5th percentile of median lethal dose SSDs, are more informative than the generic EFs derived here. We use birds, and mammals for that matter, only as examples for illustrating a general methodology, when such information is not available and, moreover, when other acute percentages may be considered appropriate.

6.2 EXTRAPOLATION WITH KNOWN SSD STANDARD DEVIATION

Let the SSD for a toxicant under consideration after log transformation of the toxicity data be normally distributed (see Aldenberg and Jaworska, 2000; Aldenberg et al., Chapter 5). The hypothesis of normality is examined in Section 6.5.1 for the example data sets used.

Suppose we had full knowledge of the normal distribution parameters μ and σ. Then, the logarithmic HC for some FA ($p\%$) would be

$$\log\left(\mathrm{HC}_p\right) = \mu - K_p \cdot \sigma \qquad (6.1)$$

The z-value constant K_p depends on the fraction affected. The ubiquitous choice of 5% yields: $K_p = 1.64.$*

Let log concentration be denoted as x. Sample-based extrapolation would employ estimates of (HC_p) on the basis of mean $\bar{x}$ and sample standard deviation s:

$$\log\left(\hat{\mathrm{HC}}_p\right) = \bar{x} - k_s \cdot s \qquad (6.2)$$

The circumflex indicates that we have an estimate based on a sample. The coefficients k_s are the extrapolation constants (Aldenberg and Jaworska, 2000: table 1; Aldenberg et al., Chapter 5: Table 5.A1).

Suppose the sample is too small to yield reliable information about the SSD standard deviation. We are interested in the case where we have information on the standard deviation σ, e.g., from similar substances, and assuming it to be a fixed number, we estimate $\log(\mathrm{HC}_p)$ through

$$\log\left(\hat{\mathrm{HC}}_p\right) = \bar{x} - K_p \cdot \sigma \qquad (6.3)$$

Confidence levels follow from the uncertainty of the sample mean only, since the standard deviation is taken as a fixed number.

Luttik and Aldenberg (1997) developed tiny sample extrapolation ("safety") factors along similar lines for the logistic distribution. It will be shown in Section 6.5.1 that the normal distribution is an equally valid, even slightly better, description of the majority of acute data sets used by Luttik and Aldenberg (1995;

---

* The six-digit representation of the 5th percentile of the standard normal distribution is −1.64485.

1997). Moreover, for the normal distribution, the equations simplify, because the sample mean $\bar{x}$ has an exact normal distribution with mean μ and standard deviation $\sigma/\sqrt{n}$. In the logistic case, we had to assume *approximate* normality for the mean. Note that we do *not* assume σ to be necessarily derived as a pooled variance estimate, as in Luttik and Aldenberg (1997): σ may derive from historic averages, conservative estimates, what-if assumptions, etc.

Clearly, for a "sample" of size 1, the standard deviation of $\bar{x}$ (i.e., the single data point itself) is σ. The uncertainty of a single data point is given by the SSD. For $n = 2$, the sample mean $\bar{x}$ has standard deviation $\sigma/\sqrt{2}$, and so on.

The sample mean $\bar{x}$ is a median estimate of μ, because it has a 50% chance of being higher than μ, and a 50% chance of being lower than μ. Hence, a median estimate of $\log(HC_p)$ is given by

$$\log\left(HC_p\right)^{\text{median}} = \bar{x} - K_p \cdot \sigma \tag{6.4}$$

The 90% two-sided confidence limits for the mean of the SSD μ are $\bar{x} \pm 1.64 \cdot \sigma/\sqrt{n}$. Consequently, lower and upper confidence limits of $\log(HC_p)$ can be calculated as:

$$\log\left(HC_p\right)^{\text{lower,upper}} = x \pm 1.64 \cdot \sigma/\sqrt{n} - K_p \cdot \sigma \tag{6.5}$$

It follows that in the case of given σ, the sampling distribution of $\log(HC_p)^{\text{lower}}$ is a normal distribution with mean $\mu - 1.64 \cdot \sigma/\sqrt{n} - K_p \cdot \sigma$, and standard deviation $\sigma/\sqrt{n}$; the sampling distribution of $\log(HC_p)^{\text{median}}$ is a normal distribution with mean $\mu - K_p \cdot \sigma$, and standard deviation $\sigma/\sqrt{n}$ while $\log(HC_p)^{\text{upper}}$ has a normal sampling distribution with mean $\mu + 1.64 \cdot \sigma/\sqrt{n} - K_p \cdot \sigma$, and standard deviation $\sigma/\sqrt{n}$ as well.

Figure 6.1 illustrates these three normal sampling distributions of $\log(HC_5)$ for the 5th percentile of the SSD ($K_p = 1.64$), when sample size n equals 1, i.e., in case of only one data point. The SSD is taken to be standard normal ($\mu = 0$, $\sigma = 1$). Clearly, the lower, median, and upper confidence limit sampling distributions of $\log(HC_5)$, in the case of one sample point, have the same spread as the SSD, and the means, except for the upper confidence limit, are shifted to the left. The lower confidence limit sampling distribution has mean: $-2(1.64) = -3.29$; the median estimate of $\log(HC_5)$ has mean: -1.64.

The upper confidence limit sampling distribution of the 5th percentile of the SSD for a single data point has the same PDF (probability density function) as the SSD. This follows from Equation 6.5 with $K_p = 1.64$ and $n = 1$. The explanation is that the SSD PDF is the distribution of a single value drawn at random from the SSD (see Aldenberg et al., Chapter 5: Discussion). Samples of a single value have a chance of 5% to be lower than $\log(HC_5)$, and 95% to be higher. A 95% one-sided upper confidence limit of $\log(HC_5)$, therefore, is given by the value sampled. Hence, the SSD PDF can also be interpreted as the sampling distribution of the 95% one-sided upper confidence limit of the 5th percentile of the SSD on the basis of samples of size one. Consequently, a single toxicity data value can be utilized as an upper confidence limit of the 5th percentile of the SSD.

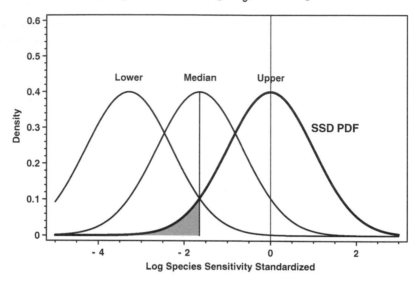

FIGURE 6.1 Sampling distributions of log(HC$_5$) for toxicity data samples of size 1, when the standard deviation of the SSD is a given number. Lower and upper confidence limit distributions and median estimate distribution (Equations 6.4 and 6.5) are plotted over standardized log concentration. The FA (5%) is shaded. Estimates overpredict the true 5th percentile in 5, 50, and 95% of the cases, respectively. For samples of size 1, the sampling distribution of the upper confidence limit of log(HC$_5$) is identical to the PDF of the SSD.

Figure 6.2 shows sampling distributions of lower and upper confidence limits, and median estimate of log(HC$_5$) for samples of size 2. Note that now the upper confidence limit distribution is distinct from the SSD. The standard deviations of the three sampling distributions are equal to $1/\sqrt{2} = 0.71$.

The sampling distributions (Equations 6.4 and 6.5) displayed in Figures 6.1 and 6.2 imply that for known σ, one can estimate log(HC$_5$) from tiny samples as follows. For one toxicity data point x, i.e., $n = 1$, it follows from Equation 6.5 that the statistics $x - 3.29\sigma$, $x - 1.64\sigma$, and x are lower, median, and upper estimates, respectively, of log(HC$_5$). (Note that for a single datum, and estimating the 5th percentile HC, the upper 95% one-sided confidence limit is taken to be the point itself, as just inferred.)

For $n = 2$, and estimating the 5th percentile of the SSD again, the respective estimates are $\bar{x} - 2.81\sigma$, $\bar{x} - 1.64\sigma$, and $\bar{x} - 0.48\sigma$. Here, $\bar{x}$ is the arithmetic average of the two data points. Similarly, for $n = 3$, we have $\bar{x} - 2.59\sigma$, $\bar{x} - 1.64\sigma$, and $\bar{x} - 0.70\sigma$. Estimates for sample sizes of 4 and higher can be derived likewise.

Obviously, estimates based on a known standard deviation shift to lower values for smaller FA percentages that might be considered reasonable for acute SSDs. For a single datum and estimating the 0.5th percentile ($K_p = 2.58$), we may employ $x - 4.22\sigma$, $x - 2.58\sigma$, and $x - 0.93\sigma$, as lower, median, and upper estimates, respectively. An FA of 0.01% ($K_p = 3.72$), and a single toxicity value x, one would use $x - 5.36\sigma$, $x - 3.72\sigma$, and $x - 2.07\sigma$. These acute FAs are motivated in Section 6.5.2.

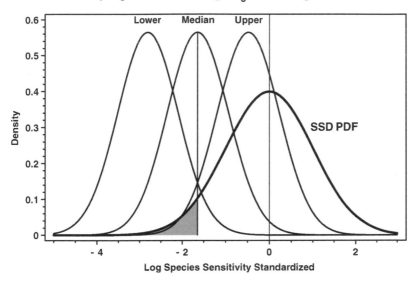

Sampling Distributions Log HC₅, Known Sigma: n = 2

FIGURE 6.2 Sampling distributions of log(HC$_5$) for toxicity data samples of size 2, when the standard deviation of the SSD is a given number. Lower and upper confidence limit distributions and median estimate distribution (Equations 6.4 and 6.5), are plotted over standardized log concentration. The FA (5%) is shaded. Estimates overpredict the true 5th percentile in 5, 50, and 95%, of the cases, respectively.

6.3 STANDARD EXTRAPOLATION FACTORS

EFs are defined as the ratio of the geometric mean of the untransformed SSD data, to some estimate of the HC in original units (Kooijman, 1987). Let us assume that common logarithms (base 10) have been used for transforming the toxicity concentration data. Then, the geometric mean of the data in original units is $10^{\bar{x}}$.

Let us consider the median estimate of $\log_{10}(HC_p)$, Equation 6.4. In original data units, the median estimate becomes

$$HC_p^{\text{median}} = 10^{\left(\bar{x} - K_p \cdot \sigma\right)} \tag{6.6}$$

The EF to apply to the geometric mean needed to arrive at this estimate, therefore, is

$$EF^{\text{median}} = \frac{10^{(\bar{x})}}{10^{\left(\bar{x} - K_p \cdot \sigma\right)}} = 10^{\left(K_p \cdot \sigma\right)} = \left(10^{K_p}\right)^{\sigma} \tag{6.7}$$

We call 10^{K_p} in Equation 6.7 a (base 10) SEF. It is the EF for an SSD with $\sigma = 1$ in common logarithms. Thus, we have

$$SEF^{\text{median}} = 10^{K_p} \tag{6.8}$$

To find the real median EF, one has to raise the median SEF to the power of the SSD standard deviation of $\log_{10}$-transformed data according to Equation 6.7.

High and low EFs follow from the lower and upper confidence limits of $\log_{10}(HC_p)$, respectively, as given by Equation 6.5:

$$\mathrm{EF}^{\mathrm{high,low}} = \frac{10^{(\bar{x})}}{10^{(\bar{x} \pm 1.64 \cdot \sigma/\sqrt{n} - K_p \cdot \sigma)}} = \left(10^{K_p \pm 1.64/\sqrt{n}}\right)^{\sigma} \tag{6.9}$$

Thus, a high EF leads to a lower confidence limit, while a low EF yields an upper confidence limit.

The (base 10) standard high and low EFs can be defined as

$$\mathrm{SEF}^{\mathrm{high,low}} = 10^{K_p \pm 1.64/\sqrt{n}} \tag{6.10}$$

Again, to find the real high and low EFs, one has to raise the respective high and low SEFs to the power of the standard deviation of the $\log_{10}$-transformed data, Equation 6.9.

Another way of writing Equations 6.7 and 6.9 is

$$\mathrm{EF}^{\mathrm{median}} = \left(10^{\sigma}\right)^{K_p} \tag{6.11}$$

and

$$\mathrm{EF}^{\mathrm{high,low}} = \left(10^{\sigma}\right)^{K_p \pm 1.64/\sqrt{n}} \tag{6.12}$$

So, to obtain an EF, one may take the anti-log of the SSD standard deviation of log-transformed data, then raise the result to the power of the appropriate z-value.

As a mathematical aside, note that when estimating the 5th percentile HC_5, i.e., when $K_p = 1.64$, and in the case of one point ($n = 1$), the high EF is the *square* of the median EF irrespective of the standard deviation: $10^{2 \times 1.64\sigma} = (10^{1.64\sigma})^2$.

Table 6.1 gives base 10 SEFs according to Equations 6.8 and 6.10 for estimates of the 5th percentile HC ($K_p = 1.64$).

6.4 EXTRAPOLATION FACTORS FOR TINY LD$_{50}$ DATA SETS OF BIRDS AND MAMMALS

As an example of how to apply the SEFs presented in Table 6.1, we will calculate EFs based on SSD standard deviations derived from LD$_{50}$ values for birds and mammals (Luttik and Aldenberg, 1995; 1997).

Figure 6.3 displays a scatter diagram of sample means and sample standard deviations of common log-transformed LD$_{50}$ bird species sensitivities for 55 pesticides. The horizontal line at 0.465 is the pooled variance estimate of the standard

TABLE 6.1
Base 10 SEFs (Equations 6.8 and 6.10), to Calculate Lower Confidence Limit, Median Estimate, or Upper Confidence Limit of the HC_5

Sample size	High SEF	Median SEF	Low SEF
1	1948.5	44.14	1.000
2	642.6	44.14	3.032
3	393.1	44.14	4.957
4	293.3	44.14	6.644
5	240.1	44.14	8.114

Note: To obtain EFs raise to the power of the SSD standard deviation of $\log_{10}$-transformed data, and use as divisor of the geometric mean of the untransformed toxicity data.

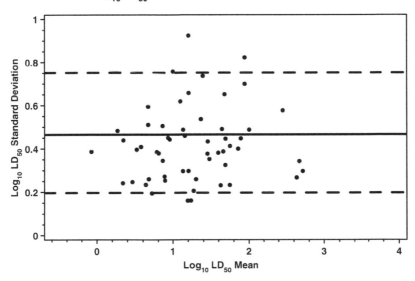

Log_{10} LD$_{50}$ Statistics of 55 Pesticides for Birds

FIGURE 6.3 Means and standard deviations of $\log_{10}$-transformed LD_{50} values (mg/kg body weight) of birds for 55 pesticides. Horizontal line: pooled variance estimate (0.465). Dashed lines: 5th and 95th percentile estimates (0.197 and 0.752). (Modified from Luttik and Aldenberg, 1997.)

deviation (Luttik and Aldenberg, 1997). The dashed lines are 5th and 95th percentile estimates, 0.197 and 0.752, respectively, of the standard deviation.

Figure 6.4 is a scatter diagram similar to Figure 6.3 for 69 LD_{50} pesticide data sets referring to mammals. The pooled variance estimate of the standard deviation

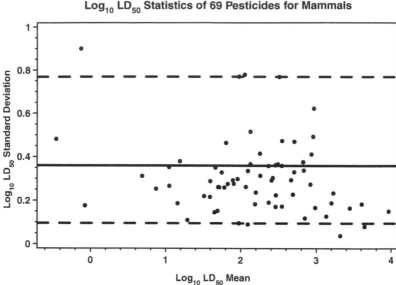

FIGURE 6.4 Means and standard deviations of $\log_{10}$ transformed LD_{50} values (mg/kg body weight) of mammals for 69 pesticides. Horizontal line: pooled variance estimate (0.360). Dashed lines: 5th and 95th percentile estimates (0.095 and 0.768). (Modified from Luttik and Aldenberg, 1997.)

(horizontal line) for mammals is 0.360. The dashed lines are 5th and 95th percentile estimates: 0.095 and 0.768, respectively.

Suppose we use the pooled variance estimates from multiple LD_{50} data sets as an "average" SSD standard deviation (common logarithms) for pesticides: 0.465 for birds, and 0.360 for mammals.

Confidence limit and median estimate EFs for the 5th percentile of bird LD_{50} values, in the case of one toxicity test for some toxic compound, follow from raising the SEFs in Table 6.1, first line, to the power of the standard deviation (0.465):

$$EF^{high} = 1948.5^{0.465} = 33.9,$$
$$EF^{median} = 44.14^{0.465} = 5.8, \text{ and}$$
$$EF^{low} = 1.000^{0.465} = 1.0.$$

Analogous results for mammals (one data point) result by applying the pooled variance estimate standard deviation for mammals (0.360):

$$EF^{high} = 1948.5^{0.360} = 15.3,$$
$$EF^{median} = 44.14^{0.360} = 3.9, \text{ and}$$
$$EF^{low} = 1.000^{0.360} = 1.0.$$

Note that the high EFs are the median EFs squared. These EFs are nearly identical to those reported by Luttik and Aldenberg (1997: table 1, there called safety

---

TABLE 6.2
High, Median, and Low EFs (Equations 6.7 and 6.9),
to Calculate Lower Confidence Limit, Median
Estimate, or Upper Confidence Limit of the HC_5
of Tiny LD_{50} Data Sets of Birds or Mammals

Sample Size	High EF	Median EF	Low EF
1	316.4	17.8	1.0
2	136.2	17.8	2.3
3	93.7	17.8	3.4
4	75.0	17.8	4.2
5	64.4	17.8	4.9

Note: Based on conservative estimate of the SSD $\log_{10}$ standard deviation: 0.760.

---

factors). The slight numerical differences are due to a change of distribution: logistic vs. normal.

Other, more conservative estimates of a generic SSD standard deviation, in comparison to a pooled variance estimate over different data sets, may be readily applied to the SEFs of Table 6.1. For example, both scatter diagrams (Figures 6.3 and 6.4) reveal comparable 95th percentiles of the standard deviation of LD_{50} samples of pesticides for birds and mammals: 0.752 and 0.768, respectively. If we consider using 0.760 as an average, EFs based on this conservative estimate of the standard deviation of LD_{50} SSDs are suitable for birds *and* mammals exposed to pesticides (Table 6.2).

The distributions of the standard deviations in Figures 6.3 and 6.4 could be used to derive *distributed* EFs. We will not do so here.

6.5 DISCUSSION

In this section, we will put the methodology of SSD-based extrapolation for a known standard deviation into perspective. First, we discuss the relevance of the normal distribution used for pesticide LD_{50} SSDs of birds and mammals. Second, we discuss a possible approach toward deriving acute FAs and acute EFs. In Section 6.5.3, we will discuss some limitations and possible improvements.

6.5.1 GOODNESS-OF-FIT TEST FOR THE NORMAL DISTRIBUTION

One of the premises when carrying out an SSD extrapolation method is the type of distribution assumed. In deriving the above generic standard deviations, we have only made use of sample means and standard deviations, without any presupposition with regard to the distributions involved. However, SSD extrapolation based on these generic standard deviations in this chapter assumes bird and mammal LD_{50} SSDs to be normally (Gaussian) distributed. Therefore, it makes sense to investigate how the normal PDF fits the bird and mammal LD_{50} data sets used by Luttik and Aldenberg (1997).

The Anderson–Darling goodness-of-fit test for normality (see Aldenberg et al., Chapter 5, Section 5.6.5.1) was applied to test the compliance of each of the log-transformed bird and mammal LD_{50} data sets to the normal distribution. The results are that, for birds, 48 out of 55 toxicant data sets of size 4 or larger, that is, 87%, are not rejected as deriving from a normal PDF (5% significance). For mammals, 67 of 69 toxicants of size 4 or larger, i.e., 97%, pass the test likewise. These figures are very similar to, and slightly better than, those for the logistic distribution: 84 and 93%, respectively (Luttik and Aldenberg, 1997), based on the Kolmogorov–Smirnov test.

These results indicate that in the majority of cases, and given the small sample sizes, the assumption that these acute toxicity data for birds and mammals derive from a normal distribution cannot be rejected on statistical grounds. This implies that the normal distribution can be taken as a reasonable, *however, not unique*, description of these types of acute toxicity data sets. The resulting EFs are not sensitive to whether the normal or the logistic distribution is used.

6.5.2   TOWARD A FRACTION AFFECTED AND EXTRAPOLATION FACTORS FOR ACUTE TOXICITY DATA COMPARABLE TO CHRONIC EXTRAPOLATION

Suppose both the acute (index *A*) and chronic (index *C*) SSD for a specific toxicant to be normally distributed on the log concentration scale. Let us consider common logarithms (base 10) again. We assume the "average" acute-to-chronic ratio (ACR) to hold for the SSD *means* in original concentration units (see De Zwart, Chapter 8, this volume):

$$\text{ACR} = \frac{10^{\mu_A}}{10^{\mu_C}} = 10^{\mu_A - \mu_C} \tag{6.13}$$

Hence, the difference between the logarithmic SSD means is given by

$$\mu_C = \mu_A - \log_{10}(\text{ACR}) \tag{6.14}$$

For example, with an ACR of 10, μ_C is one log unit below μ_A.

For simplicity, we assume the SSD standard deviations ($\log_{10}$) to be *equal* (see De Zwart, Chapter 8):

$$\sigma = \sigma_C = \sigma_A \tag{6.15}$$

Figure 6.5 displays a standard normal chronic SSD and a unit standard deviation normal acute SSD shifted one log unit to the right. The lightly shaded FA of the chronic SSD equals 5%.

Let a log HC for the chronic SSD be defined by

$$\mu_C - K_p^C \cdot \sigma \tag{6.16}$$

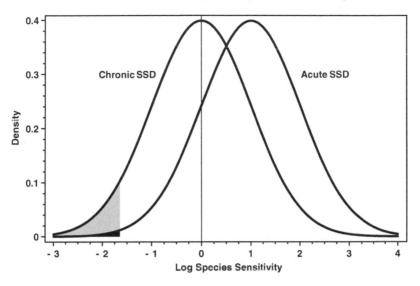

FIGURE 6.5 Standard normal chronic SSD and a unit standard deviation normal acute SSD shifted one log unit to the right. The lightly shaded FA of the chronic SSD equals 5%. The darkly shaded acute FA corresponding to the chronic 5th percentile equals 0.41% (0.0041).

For example, when considering log HC_5, the 5th percentile of the chronic SSD, as a chronic level to calibrate to, we take: $K_5^C = 1.64$. We search for a similar K_q^A value to be applied to the acute SSD that leads to comparable safety levels:

$$\mu_A - K_q^A \cdot \sigma = \mu_C - K_p^C \cdot \sigma \qquad (6.17)$$

It follows that

$$K_q^A = K_p^C + \frac{\log_{10}(\text{ACR})}{\sigma} \qquad (6.18)$$

The percentage q of the acute SSD that corresponds to an FA (%) p of chronic data becomes:

$$q = 100 \cdot \Phi\left(-K_q^A\right) \qquad (6.19)$$

In this equation, Φ is the standard normal cumulative distribution function (CDF).

For example, for the case of the hypothetical SSDs plotted in Figure 6.5, if common logarithms were used, we apparently have an ACR of 10, $\sigma = 1$, and $p = 5\%$. The "acute FA," (Equation 6.19), yields $q = 100 \cdot \Phi(-2.64) = 0.41\%$ (darkly shaded in Figure 6.5).

Section 6.4 demonstrated that "average" acute SSD standard deviations for pesticides with regard to birds and mammals do not seem to be larger than 0.5. We have at present little information available on a generic ACR for birds or mammals, with a value of 10 typical for *aquatic* toxicity (De Zwart, Chapter 8). However, with this value and $\sigma = 0.5$ (common logarithms), one would arrive at an acute FA of $q = 100 \cdot \Phi(-3.64) = 0.013\%$, i.e., 0.00013. (We illustrated tiny sample extrapolation for these acute FAs at the end of Section 6.2.)

Whatever acute FA one would employ, it is informative to see how "acute EFs" can be defined that relate to chronic EFs based on 5% of the species. From Equations 6.7 and 6.18, we derive

$$EF_A^{median} = \left(10^{K_p^A}\right)^{\sigma} = \left(10^{K_p^C + \log_{10}(ACR)/\sigma}\right)^{\sigma} = ACR \cdot 10^{K_p^C \cdot \sigma} = ACR \cdot EF_C^{median} \quad (6.20)$$

Clearly, given the current assumptions, one may multiply the "chronic" EFs by the ACR to obtain "acute" EFs.

Another way of looking at acute extrapolation and getting the same result is that we consider a pseudo-chronic sample mean:

$$\bar{x}_A - \log_{10}(ACR) \quad (6.21)$$

and take a median log percentile estimate for "chronic" extrapolation:

$$\bar{x}_A - \log_{10}(ACR) - K_p^C \cdot \sigma \quad (6.22)$$

Then, the acute EF becomes:

$$EF_A^{median} = \frac{10^{(\bar{x}_A)}}{10^{\left(\bar{x}_A - \log_{10}(ACR) - K_p^C \cdot \sigma\right)}} = ACR \cdot 10^{\left(K_p^C \cdot \sigma\right)} = ACR \cdot EF_C^{median} \quad (6.23)$$

which is identical to Equation 6.20. This illustrates that for known σ, it is justified to create pseudo-chronic data from acute data through the ACR for the means (given equal standard deviations), and employ chronic EFs to these pseudo-chronic data. The same answer results when applying acute EFs to the acute data themselves that are *products* of the ACR and chronic EFs.

Similar relationships may be derived for acute upper and lower confidence limit EFs from Equations 6.9 and 6.18:

$$EF_A^{high,low} = ACR \cdot EF_C^{high,low} \quad (6.24)$$

Apparently, in order to convert the 5th percentile acute EFs of Table 6.2 for birds and mammals to 5th percentile pseudo-chronic EFs, the numbers have to be multiplied by some (assumed, measured, or statistically derived) ACR.

The assumption that the SSD standard deviation is known, and identical for both distributions, is important here. If the standard deviation is to be derived from the sample as well, we know from sample-based extrapolation (see Aldenberg and Jaworska, 2000: table 1) that smaller FAs involve larger differences between upper and lower extrapolation constants at the same sample size and confidence level. That would imply that with acute data extrapolation in small samples, and considering smaller FAs, one generally needs bigger samples to obtain equivalent precision (difference between upper and lower extrapolation constant).

6.5.3 LIMITATIONS OF TINY DATA SET EXTRAPOLATION AND POSSIBLE IMPROVEMENTS

Although tiny data set extrapolation is illustrated in calculating EFs for pesticide LD_{50} values of birds and mammals, the methodology is generally applicable to both acute and chronic toxicity data, when information on the SSD standard deviation can be obtained from other sources. Essentially, we developed and explored the theory of SSD-based extrapolation when the standard deviation is known and taken as a fixed number.

By taking a point estimate, we neglect its possible uncertainty (see Aldenberg et al., Chapter 5: Discussion). For bird and mammal LD_{50} values, we dealt with this uncertainty in a crude way: by deriving a conservative (i.e., relatively high) upper percentile point of the SSD standard deviation from other (small sample) pesticide data sets.

In the transition zone from tiny, e.g., fewer than 4, to small, fewer than 20, sample sizes, whether it can be treated as given depends on the strength of the information regarding the SSD standard deviation. The Bayesian small sample method (see Aldenberg et al., Chapter 5) is completely opposite to the approach taken here. It takes the prior information on the SSD standard deviation as *noninformative*, which is Bayesian idiom for the mean of the SSD and the logarithm of the standard deviation of the SSD to be distributed uniformly with no prior bounds. This is one way of modeling that we know very little about μ and σ, except what the sample tells us.

If prior information on σ *is* available, we are in an intermediate situation. One could use this information in Bayesian estimation as prior distribution. The Bayesian method allows combining this information with the actual sample standard deviation in a systematic way. An alternative approach is to assess the uncertainty of σ and neglect the information in the sample if sample size is too small, which means evaluating the distribution of σ in Figures 6.3 and 6.4 and taking the variability into account in estimating percentiles.

Another route is the incorporation of explanatory covariates. Bird species may differ significantly in (average) body weight. Mineau et al. (1996; see also ECOFRAM, 1999b: pp. 4–56 and further; Mineau et al., 2001) have shown that correcting LD_{50} values for body weight may lead to a substantial reduction of variance of the SSD. The best example is Methiocarb, where the correction amounts to

$$LD_{50}^{*} = LD_{50}/W^{0.4079} \quad \left[mg/(kg \text{ body weight})^{1.4079} \right] \qquad (6.25)$$

These corrected LD_{50} values lead to a considerable reduction in the sensitivity ratio*
of the SSD: 30 down to 14 (ECOFRAM, 1999b). It is interesting to note that in the
data used in this study, Methiocarb ($n = 31$) is one of the data sets failing the
Anderson–Darling test ($A^2 = 0.852 > 0.752$ at 5% significance). This turns out to
be largely due to the two highest data points. When we leave them out, the sensitivity
ratio is reduced from 30 to 11, whereas the Anderson–Darling test value reduces to
a nonsignificant 0.303. At least in the case of Methiocarb, statistical truncation of a
small fraction of outlying insensitive species is an alternative possibility to adjust
the sample and use the data not additionally corrected for body weight.**

A limitation of the method of incorporating explanatory factors that possibly
reduce the SSD standard deviation is that they may not be easy to find for SSDs
involving multiple taxonomic groups. It is difficult to imagine explanatory factors
that work generically for multiple taxonomic groups, because of the large differences
between organisms representing a diversity of animal and plant kingdoms. Separating the SSD into different SSDs on the basis of taxonomic groups, toxic mode of
action, or statistical techniques may be the best procedure for such SSDs.

---

* Sensitivity ratio: 95th percentile of the SSD divided by the 5th percentile of the SSD.
** Lethal dose is already expressed per unit of body weight.

7 Species Sensitivity Distributions in Ecological Risk Assessment: Distributional Assumptions, Alternate Bootstrap Techniques, and Estimation of Adequate Number of Species

*Michael C. Newman, David R. Ownby,
Laurent C. A. Mézin, David C. Powell,
Tyler R. L. Christensen, Scott B. Lerberg,
Britt-Anne Anderson, and Tiruponithura V. Padma*

CONTENTS

119

Abstract — Predicting effects to ecological communities with species sensitivity distributions (SSDs) involves several poorly tested assumptions including (1) data fit the model distribution, (2) sample size is sufficient for precise prediction, and (3) the species included in the data set adequately represent the community, species assemblage, or taxocene being protected. Previously, we found that 15 of 30 data sets failed to fit the frequently assumed lognormal distribution and no single model provided adequate fit for all data sets. A bootstrap-based alternative was proposed that did not require a specific distribution. This chapter further assesses a bootstrap approach that yields a distribution-free measure of species sensitivity with associated confidence limits, and allows estimation of adequate sample size. It improves the use of the bootstrap method by developing an explicit method for estimating minimal sample size and by applying these methods to subsets of species data; specifically, species subsets that are focused on sensitive, taxonomically similar species. Inadequacy of the lognormal model was further indicated by testing an additional 21 NOEC and EC_{50}/LC_{50} data sets in the present study: 27 of 51 data sets failed tests of lognormality. Sample sizes recommended for various applications (e.g., five to seven species sensitivity values for pesticide registration; Baker et al., 1994) were determined to be insufficient for precise estimation. Application of methods to data subsets based on taxonomic similarity lessened the impact of, but did not eliminate, these shortcomings.

7.1 INTRODUCTION

Species sensitivity distribution (SSD) methods combine effects data from single-species tests to predict a concentration below which only an acceptably small percentile (p) of species would be affected. This hazardous concentration (HC_p) is notionally protective of all but $p\%$ of species, i.e., the test species in the data set adequately represent the community or taxocene to be protected and the effect metric adequately reflects exposure consequences at the population level.

Several distribution types are applied to species sensitivity data including the log-logistic (Kooijman, 1987; Aldenberg and Slob, 1993), Gompertz (Newman et al., 2000), and lognormal (Wagner and Løkke, 1991) distributions; however, method formalization in North America increasingly favors the lognormal distribution (e.g., Baker et al., 1994; U.S. EPA, 1998c). The predicted mean value for 5% (HC_5) or 10% (HC_{10}) is frequently the inferred protective concentration. Safety factors have been included in some cases because application of the HC_p without such adjustment assures the protection of $100 - p$ percentile of species in only 50% of all cases. As an early example, Van Straalen and Denneman (1989) described conservative adjustments to predictions from a log-logistic model. The lower 95% confidence or tolerance limits for the HC_p were also promoted as conservative estimators of a protective concentration (Wagner and Løkke, 1991; Jagoe and Newman, 1997; Newman et al., 2000).

To be conservative and to avoid the assumption of a specific distribution, Jagoe and Newman (1997) proposed using the lower 95% confidence limit for HC_5 values produced with a resampling (bootstrap) method. In addition to being distribution independent and conservative, this approach has the advantage that it rewards the inclusion of more information because the concentration at the lower 95% confidence limit tends to increase as more data are included. The bootstrap approach compared favorably to distribution-dependent methods (Newman et al., 2000). The assumption of a lognormal distribution was not supported in this study: 15 of 30 data sets failed tests of lognormality. Also, comparison of the log-logistic, lognormal, and Gompertz models demonstrated that the lognormal model was frequently not the best for fitting these data. Newman et al. (2000) extended the resampling approach to qualitatively assess sample size (number of species observations) adequacy for HC_p estimation. Large sample sizes of 15 to 55 observations were needed to achieve minimal 95% confidence intervals for the HC_5 with the resampling approach.

The bootstrap approach is extended here to (1) further assess the general use of the lognormal distribution for SSD data, (2) develop an explicit estimation method of adequate sample size via resampling, (3) apply these techniques to subsets of species data, and (4) provide a FORTRAN program for convenient implementation of the method.

7.2 MATERIALS AND METHODS

7.2.1 SOURCES OF SSD DATA

Toxicant data sets from Newman et al. (2000) were augmented using the AQUIRE database (U.S. EPA, 1997) (Table 7.1). The compiled data sets were typical of those currently used in SSD applications. Acute effect metrics expressed as EC_{50} or LC_{50} values were combined as done in assessments of atrazine (Solomon et al., 1996), diazinon (Giddings et al., 1996), tributyltin (Hall et al., 2000), and two metals (copper and cadmium) (Hall et al., 1998). U.S. Ambient Water Quality Criteria are also derived by combining EC_{50} and LC_{50} values. Of the 51 data sets, 12 were compilations of no-observed-effect concentration (NOEC) information.

7.2.2 ESTIMATION METHODS

7.2.2.1 Lognormal-Based Estimation

The lognormal-based approach involved ranking the n species in a data set from the most ($i = 1$) to the least ($i = n$) sensitive. Rank was calculated for each species with the approximation, $i/(n + 1)$ for purposes of plotting, lognormality testing, and estimating prediction error for parametric models. Application of the preferred approximation ($i - 0.5)/n$ to randomly selected data sets resulted in minimal improvement. The $i/(n + 1)$ transformation was used in this study because it is the most commonly used approximation in SSD studies and minimal improvement resulted from use of ($i - 0.5)/n$. The probit of rank of each species and the logarithm of effect concentration were paired for all species and fit to a linear model with a maximum likelihood procedure. The HC_5 and its confidence interval were calculated by the inverse prediction methods of conventional probit analysis.

The lognormal-based approach was scrutinized further because Newman et al. (2000) found that approximately half of data sets failed tests of lognormality. This implied that error could result from inverse prediction done with an inappropriate lognormal model. The error or residual sum of squares (SS_{error}) was calculated as

TABLE 7.1
Summary of Toxicant Data Used in Analyses and the Results of SSD Calculations

Chemical	CAS	Endpoint	Medium[a]	Passed Log-Normality Test?	n	SS_{error}	PRESS	Parametric HC_5 (mg/l)	Parametric 95% C.I.	Resampling HC_5 (mg/l)	Resampling 95% C.I.
Acetone	64641	LC/EC$_{50}$	FW	No	35	9.86	13.52	1415	—	51	—
Aldrin	309002	LC/EC$_{50}$	FW	No	35	5.24	6.29	0.7	0.1–1.6	1.3	0.1–4.5
Atrazine	1912249	LC/EC$_{50}$	FW	No	38	3.63	4.02	39	16–77	42	21–60
	1912249	LC/EC$_{50}$	SW	Yes	22	0.79	1.14	14	5–28	72	20–85
	1912249	NOEC	FW	Yes	20	0.86	1.06	0.266	0.076–0.641	3.0	1.0–4.0
Benzene	71432	LC/EC$_{50}$	FW	Yes	32	0.86	1.18	6461	4522–8622	10,000	1745–15,356
Cadmium chloride	10108642	NOEC	FW	No	38	1.35	1.78	0.341	0.201–0.515	1.000	1.000–1.000
	10108642	NOEC	SW	No	23	1.47	1.81	0.463	0.125–1.218	10.0	7.0–10.0
Carbofuran	1563662	LC/EC$_{50}$	FW	Yes	54	1.60	1.77	3.20	2.08–4.67	1.53	0.23–5.69
Chlordane	57749	LC/EC$_{50}$	FW	Yes	26	1.63	2.42	2.9	1.7–4.3	3.0	3.0–6.3
Chlorphyrifos	2921882	LC/EC$_{50}$	FW	Yes	35	1.28	1.48	0.016	0.008–0.027	0.070	0.057–0.110
Chromium (III)	1354384	LC/EC$_{50}$	FW	Yes	21	1.31	1.65	952	494–1506	772	397–2221
Chromium (VI)	10588019	LC/EC$_{50}$	FW	No	33	4.95	5.56	22.3	3.5–77.8	32.3	23.1–40.9
Chromium (VI)	10588019	LC/EC$_{50}$	SW	Yes	23	0.90	1.11	1047	665–1471	2030	2000–3100
Copper	7440508	LC/EC$_{50}$	FW	No	48	2.48	2.83	9.4	7.4–11.5	13.0	13.0–23.8
	7440508	LC/EC$_{50}$	SW	Yes	25	1.33	2.31	9.7	6.1–13.7	7.8	5.8–13.9
DDVP	62737	LC/EC$_{50}$	FW	Yes	54	1.46	1.62	0.35	0.21–0.57	0.19	0.09–2.14
	62737	LC/EC$_{50}$	SW	Yes	26	1.25	1.51	13.8	6.4–24.9	7.0	4.0–15.0
Diazinon	100155473	LC/EC$_{50}$	FW	No	55	2.09	2.38	0.99	0.63–1.48	0.80	0.03–2.00
2,4-Dichloroanilin	554007	NOEC	FW	No	21	0.75	1.16	72.0	1.0–246.0	32.0	32.0–320.0
Dieldrin	60571	LC/EC$_{50}$	FW	Yes	39	0.93	1.17	0.70	0.40–0.90	0.80	0.50–2.30
	60571	LC/EC$_{50}$	SW	Yes	27	0.22	0.25	1.0	0.7–1.4	0.9	0.7–1.5
Dimethoate	60515	NOEC	FW	No	21	1.02	1.26	19	1–89	66	32–100
2,4-Dinitro-o-cresol	534521	NOEC	FW	Yes	21	0.15	0.18	44.0	22.0–73.0	66.0	32.0–100.0

Compound	CAS	Endpoint	Water		n						
Endosulfan	115297	LC/EC$_{50}$	FW	No	47	6.31	7.45	0.09	0.03–0.21	0.24	0.18–0.44
	115297	LC/EC$_{50}$	SW	No	21	1.28	1.58	0.007	0.002–0.018	0.07	0.040–0.100
Endrin	72208	LC/EC$_{50}$	FW	No	66	2.78	3.25	0.056	0.042–0.072	0.150	0.060–0.315
	72208	LC/EC$_{50}$	SW	No	23	2.39	3.71	0.045	0.023–0.727	0.050	0.037–0.100
Fenitrothion	122145	LC/EC$_{50}$	FW	No	107	10.82	11.27	0.258	0.148–0.420	1.600	0.64–3.100
Guthion (azinphos-methyl)	86500	LC/EC$_{50}$	FW	Yes	31	0.83	1.00	0.050	0.023–0.091	0.180	0.130–0.250
Heptachlor	76448	LC/EC$_{50}$	FW	Yes	29	0.29	0.34	1.4	0.8–2.2	1.1	0.9–2.3
	76448	LC/EC$_{50}$	SW	Yes	20	0.47	0.68	0.16	0.06–0.32	0.43	0.06–0.86
Hexachloro-cyclohexane	58899	LC/EC$_{50}$	FW	No	113	1.53	1.61	2.9	2.5–3.4	4.0	2.5–8.6
	58899	LC/EC$_{50}$	SW	No	33	2.35	2.73	0.38	0.19–0.66	3.8	1.7–4.4
	58899	NOEC	FW	Yes	20	0.30	0.35	1.0	0.4–2.2	2.0	1.0–5.0
Malathion	121755	LC/EC$_{50}$	FW	Yes	91	0.94	1.01	4.7	3.8–5.8	7.2	0.9–19.5
	121755	LC/EC$_{50}$	SW	Yes	22	0.38	0.51	3.7	1.9–6.0	4.4	1.3–12.0
Mercury	7439976	LC/EC$_{50}$	FW	Yes	20	0.65	0.79	8.8	3.9–15.4	7.4	4.8–14.5
Methyl parathion	298000	LC/EC$_{50}$	FW	No	42	11.33	12.31	10.6	1.1–39.5	3.4	2.9–5.0
p-Nitrotoluene	99990	NOEC	FW	No	21	0.28	0.42	555	312–832	660	320–1000
Parathion	56382	LC/EC$_{50}$	FW	No	78	5.69	6.04	0.060	0.034–0.101	0.535	0.280–1.100
Pentachlorophenol	87865	LC/EC$_{50}$	FW	No	95	9.67	10.96	19.5	11.5–29.3	44.0	32.5–77.7
	87865	LC/EC$_{50}$	SW	Yes	33	0.56	0.63	47.7	33.0–64.2	67.0	37.0–141.5
	87865	NOEC	FW	No	41	3.32	4.05	6.0	3.0–10.0	15.0	3.2–32.0
Potassium dichromate	7778509	NOEC	FW	Yes	22	0.23	0.28	88	41–138	100	32–100
Sodium bromide	767156	NOEC	FW	No	34	1.59	1.82	4692	2578–7692	7800	2800–10,800
TCE	79016	LC/EC$_{50}$	FW	No	27	4.87	6.04	8041	724–15,846	2300	1700–11,000
Tetrapropylene benzene sulfonate	11067815	NOEC	FW	No	25	0.72	0.89	599	280–964	940	320–1000
Toluene	108883	LC/EC$_{50}$	FW	Yes	26	0.75	0.92	3382	2107–4875	7474	7219–8756
Toxaphene	8001352	LC/EC$_{50}$	FW	No	52	2.67	3.47	1.1	0.8–1.3	2.0	1.3–3.0
Tributyltin oxide	56359	LC/EC$_{50}$	SW	No	22	5.03	7.69	0.2	—	1.0010	0.0006–1.030

[a] FW = fresh water; SW = salt water.

an overall measure of the deviation between observed and predicted probit values for all observations in the data set.

$$SS_{error} = \sum_{i=1}^{n} \left(Y_i - \hat{Y}_i \right)^2 \tag{7.1}$$

The prediction sum of squares (PRESS) was also calculated with SAS Procedure REG (SAS, 1988) because the SS_{error} can be an unacceptably biased indicator of prediction error (Neter et al., 1990; Montgomery, 1997). The PRESS is similar to the SS_{error} with one important difference. The ith observation is omitted from the data set and a regression model developed with the $n - 1$ remaining observations. A $\hat{Y}_{i(i)}$ is produced by placing the X_i for the omitted observation into the regression model and solving for Y_i. This procedure is repeated n times with a different Y_i omitted each time. A set of predictions for observations sequentially omitted from the models is used to produce the PRESS.

$$PRESS = \sum_{i=1}^{n} \left(Y_i - \hat{Y}_{i(i)} \right)^2 \tag{7.2}$$

The resulting PRESS will be equal to or larger than SS_{error} because it contains the additional error associated with prediction for observations not included in building the model. How much larger PRESS is than SS_{error} depends on the magnitude of prediction error for the model. If there were small differences between the SS_{error} and PRESS statistics, the SS_{error} accounts for most of the prediction error and can be used directly to gauge the level of prediction error. Otherwise, the PRESS statistic is required to reflect accurately prediction inaccuracy.

Estimation of the accuracy of the lognormal-based SSD methods required more complicated computations than simple prediction by regression because HC_p estimation involves inverse prediction of the logarithm of the effect concentration. A regression model is developed with the intent to estimate 10^X, not Y. Also, assessment focused on inverse predictions at or below $p = 10\%$ because this is the region of interest in most SSD studies. Consequently, assessments of inverse prediction accuracy included only inverse predictions of HC_i for each observation (i) at or below $p = 10\%$.

7.2.2.2   Bootstrap-Based Estimation

The bootstrap approach, as implemented with Resampling Stats Version 4 (Resampling Stats, 1995), produced estimates of HC_5 and its associated 95% confidence limits without assuming a specific distribution (Newman et al., 2000). Each data set was sampled randomly with replacement to create a resampling data set of 100 observations. The observations in the resampled data set were ranked from smallest to largest and the concentration for the observation ranked at the 5th percentile was taken as an estimate of HC_5. This resampling was repeated until 10,000 HC_5 estimates were produced. The 10,000 HC_5 estimates were ranked and those ranked at 2.5th,

50th, and 97.5th percentiles were the bootstrap estimates of the HC_5 and its 95% confidence interval. Interpolation between concentrations was linear if necessary.

Notice that this bootstrap approach will be biased upward if the number of observations is too small. For example, the smallest proportion associated with an observation would be 0.2 if only five observations were available. There would be a bias if one were trying to estimate HC_5 from such a data set because no number in the 10,000 bootstrapped data set will be less than the lowest value of the five observations, i.e., the concentration corresponding with 0.2 with no interpolation. For an HC_5, 20 or more observations would be needed. Linear interpolation between zero and the lowest value would reduce, but not remove, this bias. However, as demonstrated below, this is not a major shortcoming because sample sizes of fewer than 20 are often unacceptably imprecise.

7.2.3 SAMPLE SIZE ADEQUACY

The bootstrap estimates described above were produced for different resampling sizes to determine the change in the 95% confidence interval with increasing sample size (Newman et al., 2000). A pilot data set is obtained and resampled at increasing resampling sizes in increments of 5. The point at which the decrease in width of the confidence interval is slight was taken as the best sample size.

This approach does not assume a specific distribution as required by the conventional sample size techniques (i.e., Kooijman, 1987; Kupper and Hafner, 1989). It was similar to that of Bros and Cowell (1987); however, resampling was done with replacement. Sampling without replacement limits sample size estimation to sample sizes less than one half of the original sample size. Sampling with replacement allows estimation for sample sizes exceeding the number of observations in the original data set (Manly, 1992).

No formal rule for satisfactory convergence on a best sample size was presented in Newman et al. (2000). A formal, albeit arbitrary, stopping rule is described here such that convergence is satisfied when the decrease in the magnitude of the 95% confidence interval with an increase in observation number is less than 10% of the interval estimated for the previous sample size. (Increments of sample size used here were set at five species.) This approach is illustrated with four large data sets using smoothing. Alternate improvements include the use of moving averages or specific models. Smoothing was used because it is simple and results for small data sets tend to oscillate between a few points as sample size reaches the point of minimal improvement. Smoothing involved three adjacent observations but could have included more if necessary (Diggle, 1990).

$$s_n = \frac{y_{n-1} + y_n + y_{n+1}}{3} \tag{7.3}$$

where y_n = the observed width of the 95% confidence interval with n observations included. Plots of residuals ($y_n - s_n$) against the number of observations could be used to determine if more points should be involved in smoothing. A random pattern in such a plot suggests adequate smoothing. Alternatively, the moving average or regression fitting to specific models (e.g., $\sqrt{n}$ vs. 95% confidence interval width fits adequately for the fenitrithion data set) could be explored as a means of more easily identifying the point of minimal improvement. If there were significant increases in costs associated with accumulating more data, the cost–benefit approach of Bros and Cowell (1987) would be applied. However, sampling with replacement is recommended if their approach were used.

7.2.4 ESTIMATION FOR SPECIES SUBSETS

Some SSD applications (e.g., Solomon et al., 1996) use subsets of species based on knowledge of toxic mode of action and species group sensitivity. Four large data sets were divided according to such rules (all species, all species except plants, only arthropods) and results compared with those for the entire data set. The objective was to determine whether subsetting based on differences in subset sensitivity influences deviation from a lognormal distribution and precision of HC_p estimation.

7.3 RESULTS

7.3.1 LOGNORMAL-BASED AND BOOTSTRAP ESTIMATION OF HC_5

By using Shapiro–Wilk's test (SAS, 1988) with log transformed data, the null hypothesis of lognormality was rejected ($\alpha = 0.05$) for half (53%) of the data sets (Table 7.1) (e.g., Figure 7.1). A larger proportion of NOEC data sets (8 of 12) failed the test than the LC/EC_{50} data sets (19 of 39). Previous study of eight of these NOEC data sets indicated that the Gompertz or log-logistic models provided better fit than the lognormal (Newman et al., 2000). Estimates from the lognormal- and bootstrap-based approaches are provided in Table 7.1. Data for one chemical (acetone) failed to generate confidence intervals with either method. Another (tributyltin oxide) failed to generate confidence intervals with the lognormal-based method.

Linear regression of the log of HC_5 estimates for the two methods (Figure 7.2) produced the following model: $\log_{10} HC_5$ for the bootstrap method = $0.85(\log_{10} HC_5$ for lognormal-based method) + 0.32. The intercept was significantly different from 0 ($t = 5.10$, $P = 0.0001$). This intercept and the slope smaller than unity (0.85 with a standard error of 0.04) suggested a consistent difference in the two methods despite a high r^2 of 0.91. However, one point (acetone) showed poor agreement between lognormal- and resampling-based HC_5 estimates and may have had a disproportionate influence on the slope. Estimates of HC_5 from the resampling method were generally (37 out of 51 estimates), but not always, higher than those from the lognormal-based method. There was no obvious difference in deviation from the linear model for points from data sets passing or failing tests of lognormality.

Comparison of the SS_{error} and PRESS indicated that SS_{error} underestimated prediction error. Therefore, the following SS_{error}-based discussion of differences between predictions from models including all data and actual observation values understate the deviation from perfect prediction.

Figure 7.3 presents these data by dividing the inverse predicted concentration for all points at or below p = 10% by the observed concentration. A value of unity

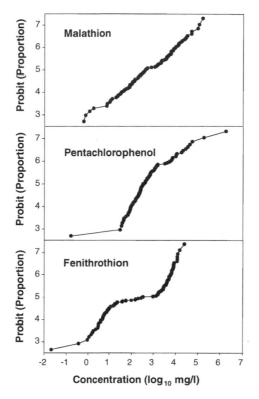

FIGURE 7.1 Plots for three example data sets (fresh water, FW) used in analyses. These data were also used in subsampling analysis. The pentachlorophenol and fenitrothion, but not the malathion, data sets failed tests of lognormality.

for the inverse predicted concentration/observed concentration would indicate perfect correspondence. Respectively, values of 10 and 0.1 would indicate a predicted value deviation that is ten times higher or lower than the observed concentration.

A total of 173 observations (26 from NOEC data sets and 147 from EC/LC$_{50}$ data sets) were included in Figure 7.3. Based on SS$_{error}$ (top panel), many estimated concentrations were more than tenfold too high (33 observations) or low (26). Only 40 quotients were between 0.5 and 2.0: 27% of observations had predicted concentrations within one half to two times the observed concentration. This suggested poor inverse prediction with the lognormal-based method.

Inverse prediction of each concentration below $p = 10\%$ was more accurately assessed by the cross-validation approach of removing one observation at a time, building a regression model without that observation, and then performing inverse prediction for the omitted point using the model. These results (bottom panel of Figure 7.2) more accurately reflect prediction error than those above. When this was done for the 173 observations, nearly one third of the predicted concentrations (36%) were more than tenfold higher (49 observations) or lower (14 observations) than the observed concentrations. Only one of every five of the 173 observations had inversely

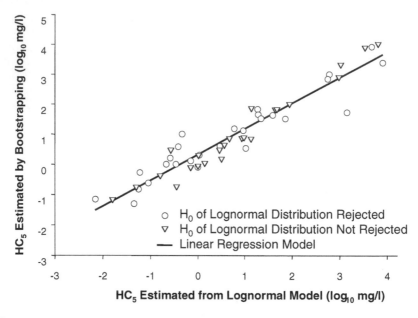

FIGURE 7.2 Linear relationship between $\log_{10}$ of HC_5 estimates from the lognormal- and bootstrap-based methods.

predicted concentrations between one half to two times the observed concentration. The results indicated poor inverse prediction with the lognormal model.

7.3.2 SAMPLE SIZE ADEQUACY

Sample sizes were determined for four large data sets with an explicit, although arbitrary, rule for determining adequate sample size (Figure 7.4). Resampling from each data set was done with resampling size steps of five species in each simulation, e.g., 5, 10, 15, 20, etc. samples. The results for each resampling size were used with a stopping rule of a smoothed improvement of no more than 10% in the 95% confidence interval. By applying this method, adequate sample sizes were estimated as 40 (fenitrothion) to 60 (hexachloro-cyclohexane). These sample sizes are much larger than at present recommended in several publications (e.g., Baker et al., 1994). If a lower level of precision were acceptable or the cost of accumulating more data became prohibitive, a different set of criteria could justifiably be applied. However, a clear statement of justification regarding the acceptable level of precision would be required. Therefore, although the rules established here are somewhat arbitrary, they are explicit. They oblige the researcher to consider sample size adequacy and to make a clear decision about acceptable precision.

7.3.3 ESTIMATION WITH SPECIES SUBSETS

Because three (fenitrothion, hexachloro-cyclohexane, and malathion) of the four selected chemicals were insecticides, data were split into the following subsets: all species, all species except plants, and only arthropod species (Table 7.2). With

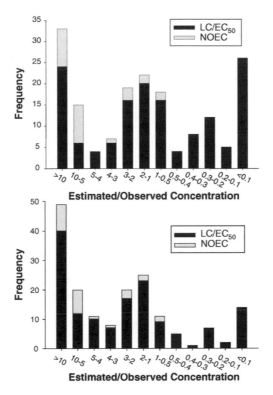

FIGURE 7.3 Frequency distribution for the regression estimated concentration/observed concentration for all observations at or below $p = 10\%$ in all 51 data sets. The top panel provides results from methods using all points to produce one model for prediction. The bottom panel involves inverse prediction using a method in which models are built leaving one point out at a time and performing inverse prediction to estimate the concentration corresponding to the omitted value.

pentachlorophenol (see Figure 7.1) and hexachloro-cyclohexane, subsetting the species data improved conformity to the lognormal model. By considering only nonplant species or arthropods, data sets for these two toxicants did not fail tests of lognormality. Subsetting did not improve conformity to the lognormal model for fenitrothion (see Figure 7.1) and the assumption of a lognormal distribution was rejected for subsets of arthropods only and all species except plants. The results suggested that restricting estimation to subsets of species may improve fit to the lognormal assumption but the assumption of lognormality was not generally valid for subsets of species. One distinct advantage of considering taxonomic subsets of species was the general decrease in HC_5 values with increased subsetting. The HC_5 dropped as the more sensitive subsets became the focus of estimation. This occurred for both the lognormal- and resampling-based estimates. With the exception of HC_5 for arthropods exposed to the general ADP phosphorylation disruptor, pentachlorophenol, the 95% confidence intervals became smaller as subsetting narrowed downward toward the arthropods. This seemed to be less the case with those for the lognormal-based

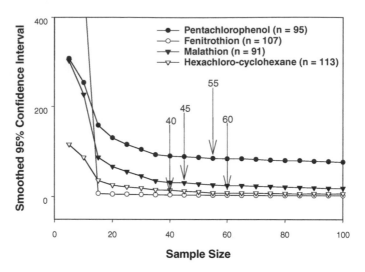

FIGURE 7.4 Sample size estimation illustrated for four large data sets. Arrows indicate the sample sizes at which the improvement in precision falls below 10% at each successive step.

estimate. A final advantage of considering species subsets was the decreased sample size needed to obtain an acceptable level of precision. This advantage must be balanced against the loss of species data as subsetting increases in order to generate the most meaningful and precise estimates of effect.

7.4 DISCUSSION

Approximately half of all data sets failed the test for conformity to a lognormal model, the model assumed in many North American SSD methods. These results do not imply that the remaining data sets come from lognormal distributions. The power of the tests might have been inadequate to detect a deviation from the null hypothesis. This may be one reason the scatter of observations for data sets failing the test for lognormality was not obviously broader than that for data sets that were not rejected. Regardless, these results clearly indicate that the general assumption of a lognormal model is invalid and can contribute to estimation error for the HC_5 and its confidence interval.

Prediction of effect concentration was judged unreliable for percentiles (p) in the range of 10% or lower based on analyses of simple inverse prediction and cross-validation inverse prediction. Estimated concentrations were often more than tenfold too high or low. An undefined portion of this error was associated with application of an inappropriate model.

A bootstrap method was demonstrated that eliminates error that can be introduced by assuming an inappropriate distribution. A bootstrap HC_p and its associated 95% confidence intervals can be estimated for EC/LC_{50} and NOEC data sets. Because convenient programs only exist for the conventional parametric method for HC_p (e.g., that listed in appendix 2 of Bacci, 1994), a FORTRAN program can be provided

TABLE 7.2
Lognormality Tests and Estimation for Taxonomically-Based Subsets of Four Data Sets (freshwater, LC/EC$_{50}$ data sets)

Toxicant	Subset	Passed Test for Lognormality?	n	Parametric HC$_5$ (mg/l)	Parametric 95% C.I.	Resampling HC$_5$ (mg/L)	Resampling 95% C.I.	Optimal Sample Size
Fenitrothion	All data	No ($p < 0.0001$)	107	0.258	0.148–0.420	1.600	0.64–3.100	40
	All but plants	No ($p < 0.0001$)	101	0.217	0.126–0.350	1.585	0.64–3.100	25
	Arthropods	No ($p = 0.0053$)	50	0.417	0.249–0.620	0.920	0.190–1.585	20
Hexachloro-cyclohexane	All data	No ($p = 0.0417$)	113	2.9	2.5–3.4	4.0	2.5–8.6	55
	All but plants	Yes ($p = 0.0616$)	110	2.8	2.4–3.2	4.9	2.4–8.3	60
	Arthropods	Yes ($p = 0.0754$)	35	1.6	1.0–2.2	3.3	3.2–4.0	40
Pentachlorophenol	All data	No ($p < 0.0001$)	95	19.5	11.5–29.3	44.0	32.5–77.7	50
	All but plants	No ($p < 0.0001$)	87	19.4	10.2–31.1	43.0	30.0–71.7	75
	Arthropods	Yes ($p = 0.3641$)	34	33.1	19.4–51.1	38.0	35.0–170.0	85
Malathion	All data	Yes ($p = 0.2200$)	91	4.7	3.8–5.8	7.2	0.9–19.5	45
	Arthropods	Yes ($p = 0.1784$)	23	0.23	0.07–0.56	1.10	0.71–1.68	15

by the authors to implement the bootstrap method (newman@vims.edu). However, commercial software such as Resampling Stat can also calculate these and additional metrics. Jagoe and Newman (1997) provide a program that implements a bootstrap procedure similar to that described here except their program focuses on the lower confidence limit of the HC_5 as the effect metric of choice.

Imprecision can be reduced by resampling estimation of adequate sample size. The programs mentioned above can be used for this purpose. A 10% stopping rule is described here, but other rules can be applied depending on the intended use of the estimates. Smoothing improves identification of the sample size above which there is minimal improvement in precision (see Diggle, 1990).

Neither the data presented here nor ecotoxicological theory supports the general application of a lognormal model to SSD data. Only recent convention and mathematical convenience provide impetus for the continued assumption of the lognormal model. An alternative method is described here that eliminates the need to assume any distribution. Simple code is provided for its implementation including a simple and intuitive way of determining an acceptable sample size. We conclude that the resampling or another nonparametric approach should be adopted for the general analysis of SSD data. More attention should be focused on adequate sample size and species subsetting.

ACKNOWLEDGMENTS

We gratefully acknowledge the insights of Bryan Manly and David Farrar regarding resampling size estimation. This is Contribution 2327 of the College of William and Mary, Virginia Institute of Marine Science and School of Marine Science.

8 Observed Regularities in Species Sensitivity Distributions for Aquatic Species

Dick de Zwart

CONTENTS

Abstract — This chapter presents an analysis of regularities observed in species sensitivity distributions (SSD) fitted on acute and chronic aquatic toxicity data. For a large number of both organic and inorganic toxicants, log-logistic species sensitivity distributions are fitted to acute ($L(E)C_{50}$) and chronic (NOEC) data obtained from several internationally available databases. The log-logistic sensitivity model is characterized by two parameters only: (i) α (alpha) which is the mean of the observed $\log_{10}$-transformed $L(E)C_{50}$ or NOEC values over a variety of test species; and (ii) β (beta), a scale parameter proportional to the standard deviation of the $\log_{10}$-transformed toxicity values.

A regression of acute and chronic α values for a large number of chemicals reveals that the average acute toxicity is approximately a factor of ten higher in concentration than the average chronic toxicity. Provided that sufficient species are tested, regression analysis on the acute and chronic β values indicates that the factorial difference between the sensitivities of the most and least sensitive species tested is about equal for both acute and chronic tests. The data suggest that the magnitude of β values may be related to the toxic mode of action of the compounds considered. The observed regularities may be used to assign surrogate SSD parameters in case an appropriate set of chronic toxicity data is lacking.

8.1 INTRODUCTION

This chapter deals with regularities that can be observed in statistical distributions of sensitivity toward chemical exposure over different species (species sensitivity distribution, SSD). These distributions can only be derived if the toxicity of a chemical is tested with a wide variety of species. To identify possible sources of the required toxicity data, it is essential to linger on the historical development of toxicity testing.

With only 30 to 40 years of experience, ecotoxicological testing is a fairly recent development. Testing the sensitivity of species to chemical exposure has been applied to obtain information on environmentally acceptable conduct with respect to the fabrication, use, and disposal of anthropogenic chemicals. From the start, ecotoxicology primarily focused on the exposure of aquatic species because it was soon realized that the world water resources play a prominent role in receiving and relocating chemicals. The use of aquatic toxicity tests was further promoted by the fact that the administration of toxicants dissolved in water is simple and highly controllable. The huge quantity of data available prompted limitation of this chapter to SSD for aquatic species.

By exposing different species to the same chemicals, it soon became evident that species differ in susceptibility. Together with the observation that individual water bodies display considerable differences in their species composition, this leads to the situation where the scientific community has conducted a multitude of tests with a large variety of species. This trend of diversification to study species indigenous to the ecosystem that may receive the chemical has been partly counteracted since the 1970s by an urge for standardization in procedures and test species (Davis, 1977). Standardization of test protocols with reference toxicants and uniform test species was considered essential for maximizing comparability, replicability, and reliability in the determination of relative toxicity of a chemical in a legally accepted framework (Buikema et al., 1982). For both fresh- and saltwater testing, officially approved acute and chronic test batteries were identified, mainly composed of tests with algae, fish, and invertebrates (e.g., ASTM, 1980; 1981; OECD, 1981).

The first steps in aquatic toxicity testing were taken by putting the proverbial goldfish in a jar and finding the aqueous concentration of chemical that caused acute mortality. This concept was soon extended to determine the median lethal concentration (LC_{50}) in a number of fish exposed for a prescribed number of hours (e.g., $LC_{50\text{-}96\,h}$). The application of acute lethality data in determining environmentally "safe" concentrations is obviously rather limited. In this respect, the need for conducting tests that were more appropriate was quickly identified. Chronic and subchronic test protocols (e.g., full life cycle tests or early life stage tests) were developed with a much longer exposure time. In these tests the magnitude (e.g., no-observed-effect concentration, NOEC) and type of effect (growth, development, reproduction) were defined to be ecologically more relevant. For acute toxicity tests, the test organisms can be collected from field populations in relatively unpolluted areas, purchased from commercial suppliers, or cultured in the laboratory. Acute tests are not very demanding in providing test species with near natural conditions, and even feeding is generally omitted during the test period. Therefore, results are available of acute tests with many species belonging to all major taxonomic groups. Chronic testing, however, requires that the organisms can be maintained successfully in the laboratory for a prolonged period of time. This restriction implies that chronic toxicity tests have only been conducted with a limited variety of species.

The relative shortage of chronic toxicity data invoked the first use of assumed regularities in the sensitivity of species. Mount and Stephan (1967) promoted the use of so-called application factors (AF), which were defined as the ratio of the chronically tolerated concentration and the acute LC_{50} for a given species. The AF was intended to provide an estimate of the relationship between chronic and acute toxicity as an inherent property of the chemical. Assuming uniformity over species, the AF could be applied to extrapolate from acute to chronic toxicity for those species producing difficulties in conducting chronic tests.

This chapter extends the work of Mount and Stephan (1967) by statistically analyzing the data currently available in the world's resources on aquatic toxicity. Based on the observed regularities, two main topics are addressed:

1. The possibility of predicting a chronic SSD from the more widely available data on acute toxicity is investigated.
2. Further investigated is whether a more appropriate prediction can be made if information on the mode of action of the toxicant is available.

8.2 SOURCES OF DATA AND DATA PREPARATION

8.2.1 DATA SOURCES

With the help of the U.S. Environmental Protection Agency, Mid-Continent Ecology Division (MED), Duluth, Minnesota, *all* data in the aquatic information retrieval toxicity database (AQUIRE) (U.S. EPA, 1984a) related to the test endpoints EC_{50}, LC_{50}, NOEC, LOEC, MATC, EC_0, EC_5, and EC_{10} have been retrieved (83,365 records). To enhance the coverage of toxicity data on pesticides, two other sources of data have been addressed:

1. The Centre for Substances and Risk Assessment belonging to the Netherlands National Institute of Public Health and the Environment (RIVM) contributed with a compilation of pesticide toxicity (7345 records) (Crommentuijn et al., 1997a; Tomlin, 1997).
2. The U.S. EPA Office of Pesticides Programs, Ecological Effects Branch, Washington, D.C., offered a set of toxicity data comprising 12,882 records.

TABLE 8.1
The 20 Most Recognized Toxic Modes of Action

TMoA	No. Chemicals
Nonpolar narcosis	169
Polar narcosis	97
Organophosphates	77
Multisite inhibitor	60
Uncoupler of oxidative phosphorylation	52
Photosynthesis inhibitor	50
Plant growth regulator	43
Carbamates	29
Plant growth inhibitor	29
Pyrethroids	28
Reactive dinitro group	26
Ergosterol synthesis inhibitor	24
Systemic fungicide	20
Neurotoxicant: cyclodiene-type	16
Alkylation or arylation reaction	15
Amino acid synthesis inhibitor	14
Diesters	12
Dithiocarbamates	11
Cell division inhibitor	11
Reactions with carbonyl compounds	10

For as many compounds as possible, information on their toxic mode of action (TMoA) was retrieved. Initially, the TMoA indication strongly relied upon the "assessment tools for evaluation of risk" (ASTER) of the MED. ASTER grossly distinguishes eight different modes of toxic action with the aid of a QSAR (quantitative structure–activity relationship) and effect oriented expert system (Russom et al., 1997). However, if none of the specific modes of action can be properly attributed, the expert system defaults to an indication of "nonpolar narcosis" (NP). The obvious presence of incorrect NP indication and the further diversification of pesticides in our data files on toxicity made it necessary to address other sources of data on toxic modes of action. The information contained in the *Agrochemicals Handbook* (Royal Society of Chemistry, 1994) and the *Pesticide Manual* (Tomlin, 1997) was manually attached to the records of toxicity data. This action extended the number of (pseudo) toxic modes of action recognized to 68. Based on the number of chemicals represented, the top 20 modes of action are given in Table 8.1.

8.2.2 DATA PREPARATION

All data were brought together into a single Microsoft Excel database with 103,592 records under the following field descriptors:

- CAS number
- Chemical name
- Chemical type (Organic: pesticide/nonpesticide/organo-heavy-metal; Inorganic: heavy metal/other)
- Toxic mode of action
- Species name (e.g., *Daphnia magna*)
- Major taxon (e.g., Crustaceans)
- Minor taxon (e.g., Cladocera)
- Water type (fresh water, salt water, mixed, and unknown)
- Endpoint (LC_{50}, EC_{50}, NOEC, etc.)
- Effect criterion (mortality, immobility, reproduction, growth, productivity, etc.)
- Test duration
- Effective concentration
- Reference number

Data preparation then followed a lengthy path where the following topics have been sequentially addressed:

- Unification of CAS number layout (no dashes).
- Unification of species name spelling.
- Removal of records with nonaquatic species.
- Addition of water type if necessary, depending on species.
- Unification of reported units in exposure duration (e.g., 96 h → 4 days).
- Unification of reported units in effective concentration (e.g., $mg \cdot l^{-1}$ → $\mu g \cdot l^{-1}$).
- Removal of records with deviating concentration units (e.g., ppm, $mg \cdot kg^{-1}$, $mM \cdot l^{-1}$, etc.).
- Designation of records to represent an acute or chronic toxicity criterion (A/C-criterion):
  1. Records with EC_{50} and LC_{50} are marked as "acute" when they have an appropriate test duration (Table 8.2) and effect criterion (e.g., mortality and immobility).
  2. Records with NOEC, LOEC, MATC, EC_0, EC_5, and EC_{10} are marked as "chronic" when they have an appropriate test duration (Table 8.2) and effect criterion (e.g., reproduction, growth, population growth, etc., next to mortality and immobility).
- Removal of records not fitting the above acute or chronic criteria.
- Removal of double entries originating from using multiple data sets by comparing references.
- Removal of entries with effective concentration indication "greater than" or "smaller than" unless they are, respectively, the highest or the lowest concentration reported for the particular chemical, species, and A/C-criterion combination. If not removed, modification of the effective concentration by leaving the numeric part only (e.g., $<200 \, \mu g \cdot l^{-1}$ → $200 \, \mu g \cdot l^{-1}$).

TABLE 8.2
Indication for Acute/Chronic Criterion

Species Group	Acute Test Duration	(Sub)chronic Test Duration
Algae	12 h	>Acute
Bacteria	12 h	>Acute
Unicellular animals	12–24 h	>Acute
Crustaceans	24–48 h	>72 h
Fish	4–7 days	>30 days
Mollusks, worms, etc.	2–7 days	>14 days

Source: Based on ECETOC, 1993b.

- Checking for outliers in effective concentration for multiple entries characterized by the same chemical, species, and A/C-criterion. Verification with original reference, followed by correction or removal. This check demonstrated that many of the multiple entries corresponding in chemical and species are derived from single references. These studies are in general concerned with the expression of toxicity in relation to other environmental factors, like temperature or pH, or they involve different life stages of the same species. To correct for this bias, the following action has been taken:
  - Removal of records with all but minimum effective concentration from multiple entries with corresponding chemical, species, A/C criterion, and reference number.

At this point, the working data set can be summarized as depicted in Figure 8.1. The 3462 chemicals comprise about 250 inorganic compounds of which about 180 contain heavy metals. Of the remaining 3212 organic substances, at least 750 compounds are used as pesticides, approximately 80 substances contain heavy metals, and 738 are organo-halogens.

Prior to analyzing SSD, two more steps were taken in the preparation of the working data set:

1. Log-transformation of the effective concentration expressed in $\mu g \cdot l^{-1}$. The $\log_{10}$ is taken because this enables easy interpretation of the concentration ranges involved.
2. Calculation of the average of the $\log_{10}$-transformed effective concentration over chemical, species, water type, and A/C criterion combinations to avoid multiple entries for the same endpoint.

The last step of calculating average toxicity has been done with and without the distinction of water type.

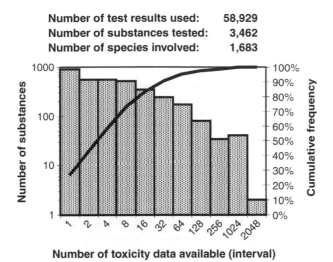

Number of test results used: 58,929
Number of substances tested: 3,462
Number of species involved: 1,683

Number of toxicity data available (interval)

FIGURE 8.1 Summary of the data set expressed as the number of tests performed per chemical after basic data preparation (bars) and cumulatively (line).

A summary of the reduced data set without the distinction of water type is given in Figure 8.2. The data set contains a total of 665 compounds on which both chronic and acute toxicity tests have been performed.

8.3 SSD CALCULATIONS

8.3.1 TYPE OF SPECIES SENSITIVITY DISTRIBUTION USED

Maximum permissible environmental concentration values for individual chemicals are generally derived from laboratory-measured NOEC for single species. In its simplest form, the associated risk for exposed ecosystems can be evaluated by regression analysis on effects observed in laboratory exposure tests and data from field or semifield experiments (e.g., Slooff et al., 1986). More sophisticated procedures may use SSDs to predict an environmental concentration below which only an acceptably small proportion of species would be affected. These methods assume that for every chemical the $L(E)C_{50}$ (Kooijman, 1987) or NOEC values (Van Straalen and Denneman, 1989; Van Straalen, 1990; Aldenberg and Slob, 1993; Wagner and Løkke, 1991) for single species in a community can be described as random variables that are characterized by a probability model for which the model parameters are unknown and must be estimated from scarcely available data. In the Netherlands, the SSD is generally taken to be logistic for log-transformed toxicity data, whereas in other parts of the world the normal or triangular distributions for log-transformed toxicity data are favored (Wagner and Løkke, 1991; Baker et al., 1994; U.S. EPA, 1998c; Aldenberg and Jaworska, 2000; Chapter 11). There are no theoretical grounds to select either of these distribution functions. The logistic distribution is very similar to the normal and the triangular distributions. According to Aldenberg and Slob

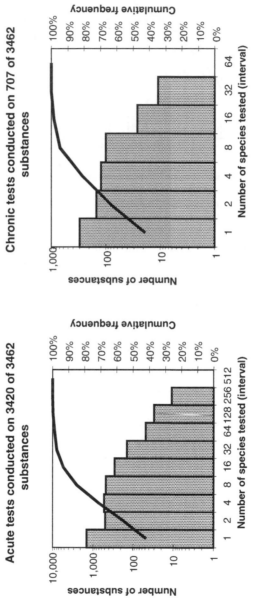

FIGURE 8.2 Frequency distributions of the number of species tested acutely and chronically per chemical.

(1993), the slightly extended tails of the logistic probability density function render marginally more conservative values in the estimation of hazard concentrations. Wagner and Løkke (1991) state that the logistic function has been designed to describe resource-limited population growth, whereas the normal distribution has already held a central position in general statistics for decades. For reasons of simplicity in calculus, the present chapter adheres to the logistic distribution based on $\log_{10}$-transformed toxicity data.

8.3.2 THE PARAMETERS OF THE LOG-LOGISTIC SSD

Aldenberg and Slob (1993) describe the logistic distribution function of toxicity values. The logistic function is totally determined by the two parameters α and β only.

The logistic distribution function is defined by

$$F(C) = \frac{1}{1 + e^{-\left(\frac{\log_{10} C - \alpha}{\beta}\right)}} \tag{8.1}$$

where C is the environmental concentration of the compound under consideration.

The toxicity data are log-transformed using the formula:

$$x = \log_{10}\left(\text{NOEC or L(E)C}_{50}\right) \tag{8.2}$$

By applying a log transformation to the effective concentrations, the distribution becomes log logistic.

The first parameter of the logistic distribution, α, is estimated by the sample mean of the $\log_{10}$-transformed toxicity values:

$$\hat{\alpha} = \bar{x} = \frac{1}{n}\sum_{i=1}^{n} x_i \tag{8.3}$$

The second parameter of the logistic distribution, β, is a scale parameter estimated from the standard deviation of the log-transformed toxicity values with the formula:

$$\hat{\beta} = \frac{\sqrt{3}}{\pi}\cdot s = 0.55 \cdot s = \frac{\sqrt{3}}{\pi}\cdot\sqrt{\frac{1}{n-1}\cdot\sum_{i=1}^{n}\left(x_i - \bar{x}\right)^2} \tag{8.4}$$

Figure 8.3 exemplifies the cumulative representation of an SSD as fitted by Equation 8.1. For derivation of environmental quality criteria (EQC), the cumulative distribution function is generally used for obtaining the HC_5, which is the concentration above which more than 5% of the species is exposed to the chemical exceeding its NOEC (Van Straalen and Denneman, 1989).

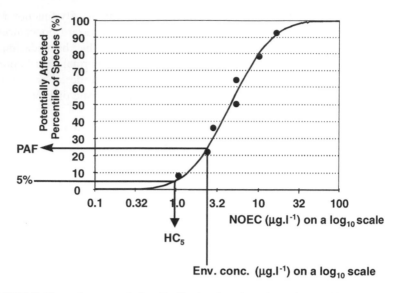

FIGURE 8.3 Exemplary cumulative distribution function of species sensitivity fitted (curve) to observed chronic toxicity values (NOEC; dots). The arrows indicate the inference of a PAF of species (PAF-value) and the HC_5.

By analogy, Hamers et al. (1996b) use the distribution curve to infer which fraction of species is exposed above the NOEC. This parameter, called the potentially affected fraction (PAF) of species, is used as a measure of the ecological risk of a chemical to the ecosystem at a given ambient concentration. Similar procedures can be used with $L(E)C_{50}$ data or other toxicity endpoints.

8.4 DATA ANALYSIS

8.4.1 FRESH- AND SALTWATER SPECIES

The data set contains toxicity values obtained with both freshwater and saltwater species. The SSD calculation would benefit from combining the data on freshwater and saltwater toxicity by producing the widest range of species tested. Combining fresh- and saltwater toxicity values for single chemicals is only justifiable if the average toxicities over species in both media are comparable. Since the average toxicity for fresh- and saltwater species both are subject to independent stochastic error, the relation between the two has been evaluated by orthogonal regression (Orthogonal Regression Analysis Software, version 4.0, Orthogonal Software, info@orthogonal.net). Orthogonal linear regression requires information about the relative errors in the x- and y-variables. One of the most common methods of providing this information is by using the ratio (λ) of the variance of the x-error divided by the variance of the y-error. Based on the lower number of species tested per single compound, the error in average saltwater toxicity is estimated to be about threefold the error in freshwater toxicity ($\lambda = 0.33$).

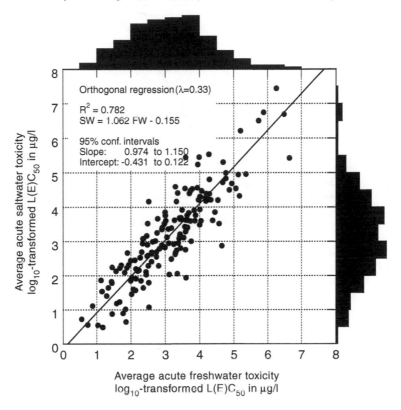

160 compounds - 4 or more species tested (28 heavy metals, 92 pesticides, 40 others)

FIGURE 8.4 Comparison of average freshwater and saltwater toxicity.

The orthogonal regression of acute average toxicity over fresh- and saltwater species, presented in Figure 8.4, demonstrates that there is no statistically significant difference between the two (the 95% confidence interval, or CI, for the slope comprises 1 and the 95% CI for the intercept comprises 0). Therefore, with the remaining calculations, fresh- and saltwater toxicity data have been lumped.

8.4.2 THE FIT OF THE LOG-LOGISTIC MODEL

Figures 8.5 through 8.8 give examples of the actual and modeled SSD for cadmium chloride, malathion, atrazine, and pentachlorophenol, respectively. In the quantile plots the n species in the data set were ordered from the most ($i = 1$) to the least ($i = n$) sensitive. The quantile (or PAF) for each species was calculated by the applying the approximation:

$$\frac{i - 0.5}{n} \tag{8.5}$$

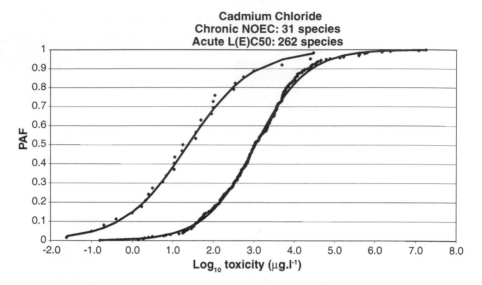

FIGURE 8.5 Observed $\log_{10}$-transformed data on chronic (left series of dots) and acute (right series of dots) toxicity for CaCl together with the respective fitted logistic distribution curves (Chronic: $\hat{\alpha} = 1.38$, $\hat{\beta} = 0.79$; Acute: $\hat{\alpha} = 3.05$, $\hat{\beta} = 0.63$).

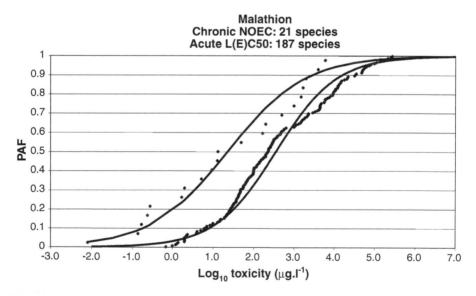

FIGURE 8.6 Observed $\log_{10}$-transformed data on chronic (left series of dots) and acute (right series of dots) toxicity for malathion together with the respective fitted logistic distribution curves (Chronic: $\hat{\alpha} = 1.36$, $\hat{\beta} = 0.95$; Acute: $\hat{\alpha} = 2.55$, $\hat{\beta} = 0.75$).

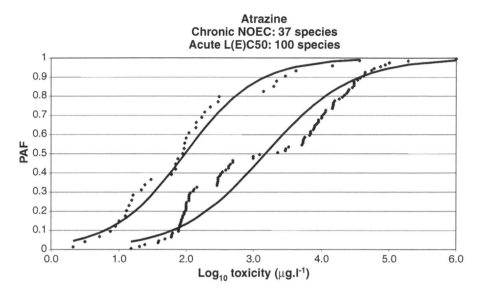

FIGURE 8.7 Observed $\log_{10}$-transformed data on chronic (left series of dots) and acute (right series of dots) toxicity for atrazine together with the respective fitted logistic distribution curves (Chronic: $\hat{\alpha} = 1.98$, $\hat{\beta} = 0.54$; Acute: $\hat{\alpha} = 3.18$, $\hat{\beta} = 0.63$).

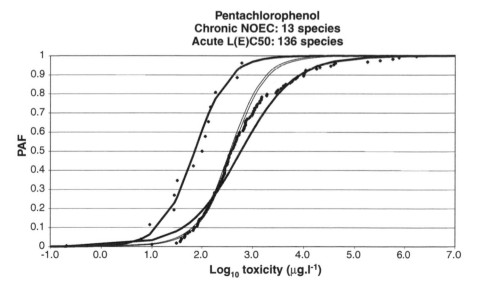

FIGURE 8.8 Observed $\log_{10}$-transformed data on chronic (left series of dots) and acute (right series of dots) toxicity for pentachlorophenol together with the respective fitted logistic distribution curves (Chronic: $\hat{\alpha} = 1.85$, $\hat{\beta} = 0.36$; Acute: $\hat{\alpha} = 2.77$, $\hat{\beta} = 0.52$; double line is first part fit by eye of acute data: $\hat{\alpha} = 2.57$, $\hat{\beta} = 0.36$).

TABLE 8.3
**Proportion (%) of Species Groups Tested for General
and Specific Toxicants**

Organism Group	Acute CdCl	Acute Malathion	Acute Atrazine	Acute PCP
Algae	4.6	0.0	**38.0**	11.0
Cyanobacteria	0.8	0.5	**3.0**	0.0
Water plants	0.4	0.0	**2.0**	1.5
Amphibia	1.9	1.6	3.0	2.2
Annelida	6.5	2.1	1.0	5.1
Crustacea	32.1	20.9	24.0	24.3
Insects	6.1	**28.9**	3.0	8.8
Mollusca	9.5	11.2	3.0	13.2
Fish	23.3	32.1	19.0	25.0
Protozoa	7.3	0.0	1.0	0.7

Note: The **bold** data indicate the groups of species that are supposed to be highly sensitive. PCP = pentachlorophenol.

For all compounds, the selection of test species is strongly influenced by the internationally accepted practice for a minimum test battery to comprise algae, crustaceans, and fish (ACF) (Table 8.3). In Figure 8.5, both modeled acute and chronic SSD for CdCl fit the available toxicity data nearly perfectly. The measured toxicities are evenly spread. Since Cd is not applied as a selective biocontrol agent, the selection of test species additional to ACF has obviously been quite random. With the testing of insecticides and herbicides, presented in Figures 8.6 and 8.7, the selection of test species tends to focus on the groups of organisms that are expected to be the most sensitive (bold in Table 8.3). Also with PCP (Figure 8.8), a more general biocide, the standard bias to test a wide variety of fish species, which happen to be very sensitive to PCP, leads to the observed high frequency of low EC_{50} values. It can be concluded that the overrepresentation of sensitive species causes a considerable misfit of the modeled SSD with the available data. This phenomenon is also observed by Newman et al. (Chapter 7) and Van de Brink et al. (Chapter 9). Newman et al. (Chapter 7) conclude that the lognormal model may not be a proper representation. The CdCl case (Figure 8.5) and also the lower part of the PCP graph (Figure 8.8) do indicate that the lognormal or log-logistic model may intrinsically be the most appropriate way to interpret SSD. The estimation of the model parameters, however, may be hampered by the bias in the available data.

8.4.3 REGRESSION OF ACUTE AND CHRONIC SSD PARAMETERS

To relate chronic toxicity to acute toxicity, the parameters of the SSD for the chemicals provided with both types of toxicity data have been subject to regression analysis. Since the SSD parameters (α and β) for chronic and acute toxicity tests are subject to independent stochastic error, the relation between the acute and chronic SSD parameters has been evaluated by orthogonal regression (Orthogonal Regression Analysis Software, version 4.0, Orthogonal Software, info@orthogonal.net). Based on the lower number of species tested per single compound, the error in the chronic SSD parameters is estimated to be about threefold the error in the acute SSD parameters ($\lambda = 0.33$). Chemicals with sufficient data for this type of analysis were selected by applying the rule that both acute and chronic tests are at least performed with one species each of the ACF. Application of this selection criterion resulted in 89 pairs of acute and chronic α and β values with numbers of species tested ranging from 3 to 262.

The maximum difference between acute and chronic average toxicity over species is a factor of 491. The mean of the average chronic toxicity of the 89 chemicals is about a factor of 18 (13 to 24) lower than the mean of the average acute toxicity. With a correlation coefficient of 0.768, a strong relationship between acute and chronic average toxicity over species is demonstrated. The 95% CI for the intercept of the regression line ranges from –1.973 to –0.888, which implies that average chronic toxicity is between a factor of 8 and 94 more sensitive than average acute toxicity. It should be realized that the intercept of the regression line is only indicative for the lower and most uncertain outskirts of the data range. Interpreting the regression by eye, and taking the rather large uncertainties into account, yields the overall impressions that average chronic toxicity is a factor of about ten lower than average acute toxicity. The confidence interval of the slope of the regression line encloses unity (0.889 to 1.217), which means that the difference between acute and chronic α values holds over the entire range (Figure 8.9).

The β values for acute toxicity range from about 0.2 to 1.3, whereas the β values for chronic toxicity are in the range of 0.02 to 1.65. At a β value of 1.25, the difference between the sensitivity of the least and most sensitive species amounts to a factor of about 10^{10}. This implies that if the effective concentration for the most sensitive species is about 1 ng/l, the least sensitive species demonstrates effects at a concentration of about 10 g/l. This spread in sensitivities is extremely unlikely for all known chemicals and should be regarded as an artifact. Extremely high or low β values can only occur if the number of species tested is too low to determine a reliable β value. With a rather low correlation coefficient ($R^2 = 0.314$), the chronic β values are not strongly related to the acute β values. In Figure 8.10, the number combination labels associated with the individual data points represent the numbers of species tested acutely (first) and chronically (second). Four of the five data points falling outside the 95% confidence ellipse are characterized by either an acute or chronic β exceeding the unlikely value of 1.25. Three of the five outlier β values are calculated with only three species tested either chronically or acutely.

For bivariate data that do have a stochastic component on both axes, orthogonal regression is the most appropriate technique to infer a relationship. When the outliers (the points outside the 95% confidence ellipse in Figure 8.10) are discarded, the orthogonal regression of the scatterplot in Figure 8.11 approximately reveals a one-to-one relationship.

As is illustrated in the box plot of the difference in acute and chronic β against the lower of the number of species tested in the acute and chronic tests (Figure 8.12), the outliers are obviously caused by the incidence of low numbers of species tested in both acute and chronic toxicity tests.

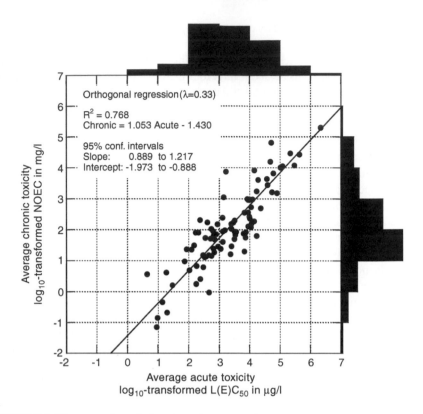

FIGURE 8.9 The regression of chronic and acute α values

8.4.4 RELATION OF β WITH TOXIC MODE OF ACTION

If, as a hypothesis, chemicals with the same toxic mode of action are considered to affect comparable species, there should be some resemblance of their log-logistic SSDs, which are only characterized by the α and the β parameter. Different chemicals with the same TMoA may differ considerably in their intrinsic toxicity. This implies that the average toxicity over species, or the SSD α parameter, will display considerable variance. Irrespective of the actual intrinsic toxicity, the SSD β parameter, or the slope of the distribution, should be equal for compounds with the same TMoA. As is indicated earlier in this chapter, the reliability of the estimated log-logistic SSD β value strongly relies on the number and variety of the species tested.

This is again demonstrated in Figures 8.13 through 8.15 where for three modes of action examples are given of acute β values plotted against the number of species tested. The gray areas in these graphs suggest that, when sufficient species are tested, the SSD β values level off to a value that is characteristic for the TMoA. For substances with an NP narcosis toxic mode of action (Figure 8.13), the intrinsic β value appears to narrow to about 0.5. For the more specifically acting compounds, such as organophosphates and photosynthesis inhibitors, the intrinsic β values appear to stabilize at values of around 0.8 and 0.6, respectively. The number of species

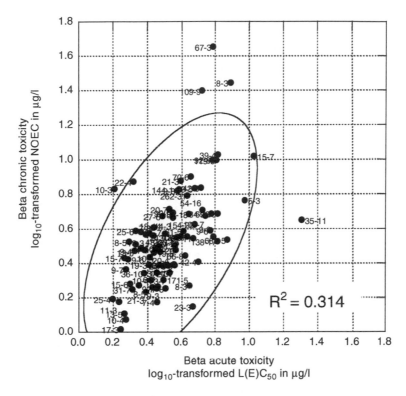

FIGURE 8.10 Acute and chronic β values plotted against each other. The ellipse is representing the 95% confidence region of the data. The number combination with each data point shows the number of species tested acutely and chronically, respectively.

tested, required to reach the constant level, is on the order of 25 to 50. It should be noted, however, that there are very few chemicals tested with these numbers of species. This finding corresponds nicely to the finding of Newman et al. (2000).

As a compromise, the compounds acutely tested with ten or more species and at least tested with ACF are selected for the estimation of the TMoA specific acute β value. Of these chemicals, the acute β values are averaged over the toxic modes of action (Table 8.4). The generally low standard errors of the means (SEM) indicate that the concept of TMoA specific β value may very well be valid.

8.5 DISCUSSION

The applicability of the concept of SSD for estimating acceptable contaminant levels or ecological risk strongly relies on proper parametrization. For adequate parametrization, the number and random variety of species-specific toxicity data required is of the order of 25 to 50, as has been demonstrated in Figures 8.13 to 8.15.

If the SSD must be based on chronic toxicity data, as is common practice, the cost and time required to generate reliable SSD information for the vast number of

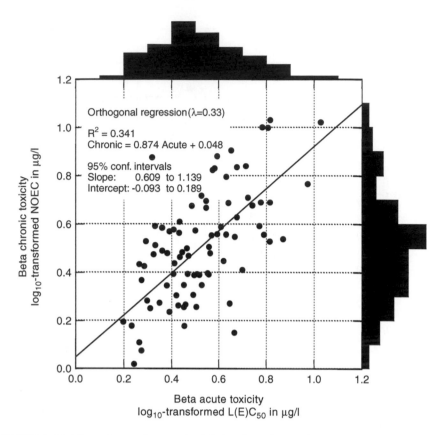

FIGURE 8.11 The orthogonal regression of chronic and acute β values.

chemicals that are potentially released to the environment are prohibitive. Actually, this chapter demonstrates that the required information on chronic toxicity is at present not available for any of the chemicals tested. The maximum number of species chronically tested on a single chemical by the combined efforts of the world's ecotoxicologists is 37 for atrazine (note the bias for primary producers in Table 8.3).

Restricting the input to acute toxicity data will render the concept of SSD far more applicable. Fortunately, the statistical analysis performed in this chapter indicates that, for many compounds, the chronic toxicity averaged over species is a fairly constant factor of approximately 10 lower than the average acute toxicity. This finding justifies the use of acute toxicity data for SSD parameterization. With the analysis of far fewer data, the same conclusion was drawn by Slooff and Canton (1983), who observed that the acute LC_{50} and chronic NOEC for the same species had a better correlation than both the acute and chronic endpoint concentrations over taxonomically different species. Therefore, they conclude that ... "it is not scientifically tenable that the margins of uncertainty in the predictive value of acute tests will be larger than those of chronic tests, nor that chronic toxicity data are indispensable for predicting environmental effects of chemicals."

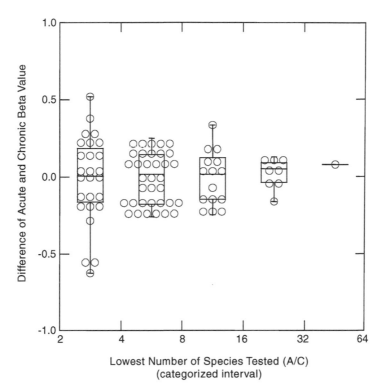

FIGURE 8.12 Box plot of the difference in acute and chronic β against the lower of the number of species tested in the acute and chronic tests (categorized bin). The dots represent the observations.

For the parameterization of the log-logistic and the lognormal SSD it is only necessary to estimate the true mean and the true standard deviation of the log-transformed toxicity data for the assembly of species to be modeled. In Figure 8.9, a rather high correlation can be observed between the acute and the chronic α for 89 chemicals. The high correlation is clearly not highly influenced by the sometimes low number of species tested for either of the α values. This can only mean that the average toxicity over species is not extremely sensitive to the number of species tested, provided that sufficient species diversity is guaranteed (minimum ACF). The standard deviation, or the slope of the curve, proved to be highly sensitive to the number of species tested (see Figures 8.12 through 8.15). The observation that the slope of the SSD (β value) is related to the toxic mode of action of the chemical under consideration may enable the introduction of surrogate β values as depicted in Table 8.4. This may reduce the need to collect vast numbers of toxicity data for the construction of reliable SSD curves.

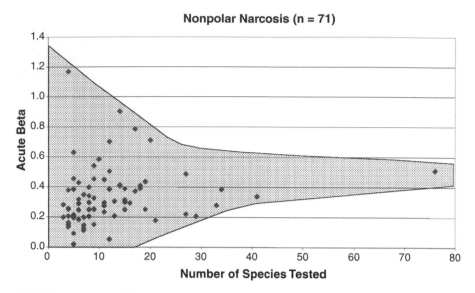

FIGURE 8.13 Acute β for 71 NP narcotics plotted against the number of species tested. The shaded area represents a subjective CI for the TMoA-dependent β value.

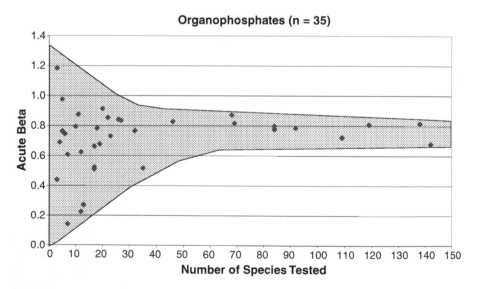

FIGURE 8.14 Acute β for 35 organophosphates plotted against the number of species tested. The shaded area represents a subjective CI for the TMoA-dependent β value.

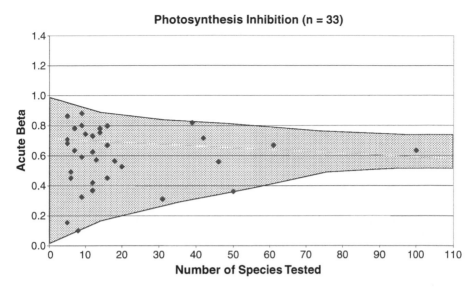

FIGURE 8.15 Acute β for 33 photosynthesis inhibitors plotted against the number of species tested. The shaded area represents a subjective CI for the TMoA-dependent β value.

TABLE 8.4
Acute β Values, Based on Ten or More Species, Averaged over Toxic Modes of Action

Toxic Mode of Action	n	Avg. β	SEM	Approx. 5–95% Factorial Sensitivity Interval
Nonpolar narcosis	34	0.39	0.03	1,300
Acetylcholinesterase inhibition: organophosphates	27	0.71	0.03	520,000
Inhibits photosynthesis	20	0.60	0.03	67,000
Polar narcosis	13	0.31	0.03	280
Acetylcholinesterase inhibition: carbamates	11	0.50	0.05	10,000
Uncoupler of oxidative phosphorylation	8	0.38	0.05	1,000
Multisite inhibition	6	0.62	0.07	91,000
Dithiocarbamates	6	0.57	0.05	38,000
Diesters	6	0.42	0.07	2,400
Systemic fungicide	5	0.46	0.04	4,500
Sporulation inhibition	5	0.37	0.05	950
Neurotoxicant: pyrethroids	4	0.65	0.03	160,000
Neurotoxicant: cyclodiene-type	4	0.61	0.01	75,000
Plant growth inhibition	4	0.52	0.06	15,000
Membrane damage by superoxide formation	3	0.69	0.01	350,000
Cell division inhibition	3	0.63	0.21	100,000
Systemic herbicide	3	0.52	0.12	14,000
Plant growth regulator	3	0.44	0.10	3,400
Neurotoxicant: DDT-type	2	0.50	0.13	9,800
Amino acid synthesis inhibition	2	0.47	0.03	5,600
Germination inhibition	2	0.40	0.02	1,600
Quinolines	2	0.28	0.02	180
Reactions with carbonyl compounds	2	0.28	0.07	170

n = number of compounds; SEM = standard error of the mean.

9 The Value of the Species Sensitivity Distribution Concept for Predicting Field Effects: (Non-)confirmation of the Concept Using Semifield Experiments

Paul J. van den Brink, Theo C. M. Brock, and Leo Posthuma

CONTENTS

1-56670-578-9/02/$0.00+$1.50
© 2002 by CRC Press LLC

Abstract — This chapter focuses on the field relevance of the output data of species sensitivity distribution (SSD) curves by seeking confirmation of laboratory-based SSD curves with population responses observed in semifield experiments. Two types of comparisons were made, namely, between full curve SSDs of laboratory and field toxicity data, and between SSDs based on laboratory data and the magnitude and nature of effects in exposed field communities. Attention is paid to the aquatic and terrestrial compartments and to chemicals with a specific and aspecific toxic mode of action (TMoA). This study aims (1) to substantiate the role of uncertainties and assumptions within the SSD concept, (2) to derive scientific conclusions on various aspects of the SSD concept, and (3) to express the limitations and points of concern regarding the use of the SSD-based extrapolation procedure in the derivation of environmental quality criteria (EQC) and ecological risk assessment (ERA).

The comparisons between the laboratory-based SSDs and the field-based SSDs for aquatic arthropods showed a high similarity for the readily bioavailable insecticide chlorpyrifos. This can be interpreted as a confirmation of the SSD concept per se, meaning that laboratory-based SSDs and field-based SSDs are similar when uncertainties are small. It should be noted that optimal predictions are obtained when a sensible division into taxonomic groups is made, one that reflects the knowledge on the TMoA of the chemical. The terrestrial example with zinc indicates that predictions for terrestrial assessments improve when additional attention is given to the difference in compound bioavailability among soils.

The comparisons between laboratory-based SSDs and the magnitudes and types of effects in the field suggest that the direct effects of long-term exposure generally start to emerge above the 5th percentile of the SSD curve based on chronic NOEC values. A similar percentile of the SSD curve based on acute EC_{50} values appeared to have predictive power where cases of acute exposure are concerned. The more numerous acute toxicity data may also be used to evaluate chronic exposure regimes when an appropriate safety factor is used. Indirect effects, however, do not correspond with a certain part of the SSD curve, but sometimes occur at low and sometimes at high percentage values. This indicates that ecological interactions are not explicitly modeled by SSD curves.

In the case of compounds with a specific TMoA, the use of TMoA-specific curves should be the preferred standard for EQC and ERA, to maximize the information content generated with the SSD concept and to allow the distinction of effects as observed in semifield communities in direct and indirect effects.

9.1 INTRODUCTION

"Numerical models are increasingly being used in the public arena, in some cases to justify highly controversial decisions. Therefore, the implication of 'truth' is a serious matter" (Oreskes et al., 1994). This citation comes from a paper on the level of truth in numerical earth science models that are used to predict the safety level of nuclear waste repositories in the Earth's crust. Public interest is obviously related to long-term public protection against radioactive contamination. Ecological risk assessment (ERA) models are often numerical models as well (Suter, 1993), with the species sensitivity distribution (SSD) concept just one example. This chapter concerns the level of truth that should be envisaged when the results obtained from the application of the SSD concept are used for environmental decision making.

Following the SSD concept, toxicity data collected from single-species laboratory toxicity studies is used to construct a statistical distribution of sensitivities. The SSD curve is thought to reflect some property of toxicant-exposed biological assemblages under field conditions. This property is usually not operationally defined, although it refers to a certain proportion of biological species within a certain environmental compartment that may, to a certain degree, be adversely affected at a given environmental concentration of a toxicant. In short, it most often links to something that relates to toxicant effects on the community or ecosystem structure, although function-related SSDs are also used (see Chapter 12).

The "serious matter" at stake is that SSD results are thought to be useful in environmental decision making, namely, both in setting environmental quality criteria (EQC) by national governments, as well as when assessing ecotoxicological risks (ERA) at contaminated sites by local authorities (see previous chapters for an overview). If the SSD concept is wrong, then any decisions taken on the basis of the concept may be unjustified. If the EQCs are too strict, for example, this may lead to costly preventive emission reduction measures, or to local cleanup activities with little environmental profits. On the other hand, if the criteria are too flexible, this may lead to unexpected and publicly undesired damage levels in the form of a change in biodiversity or nonsustainable ecosystem functioning.

This chapter seeks answers to questions on the level of truth regarding SSD results. Questions concerning truth have been put forward in the early stages of SSD concept development, when the sole use of SSDs was still aimed at the derivation of EQCs. For example, Van Straalen and Denneman (1989) have listed the major uncertainties in the laboratory-to-field extrapolation of toxicity data for terrestrial organisms. Questions relate to any of the following three key issues in the SSD concept: (1) the quality and field relevance of the input (single-species toxicity) data, (2) the statistical methods used to describe the SSD, and (3) the field relevance of the output data for the protection endpoint. One can, for example, question whether laboratory SSDs will approximate the field SSDs because the mechanisms governing the species composition in field communities are different from those used with the selection of an appropriate biotest battery. The currently prescribed biotests have been selected mainly according to practical reasons or to the trophic position of a species rather than to taxonomic differences in sensitivity. There have, however, been extensive studies on the relationship between toxicity parameters based on laboratory single-species tests and mesocosm-based community endpoints (e.g., Okkerman et al., 1993; Van den Brink et al., 1996; ECETOC, 1997; Van Wijngaarden and Brock, 1999; Brock et al., 2000a,b). These indicate that at least an empirical relation between the results of laboratory tests and mesocosm experiments might be present. Statistical aspects are still being debated (e.g., Chapter 7). The level of truth regarding the output data has, however, received limited attention so far (Campbell et al., 1999; Versteeg et al., 1999).

In an attempt to obtain confirmation of SSD, we have analyzed a limited series of case studies of various kinds, both for the aquatic and the terrestrial compartment. The studies were chosen to highlight cases of matching and mismatching, and to identify restrictions or "rules of thumb" in the application of SSDs.

Various SSD-related confirmation attempts have been made in the recent past (e.g., Posthuma et al., 1998a; 2001; Versteeg et al., 1999; Van Wijngaarden and Brock, 1999) with limited problem definitions (e.g., confirmation related only to point estimates for EQCs). In this chapter, we try to broaden the problem definition by showing (non-)confirmation in case studies for both the aquatic and terrestrial compartment, for compounds with specific and aspecific toxic modes of action (TMoA), and for the whole SSD curve where applicable. Two types of confirmation are sought: (1) between full curve SSDs derived from laboratory and field toxicity studies and (2) between the magnitude and nature of the effects in exposed field communities and laboratory data–based SSDs.

Based on the comparisons, this study aims to:

1. Substantiate the role of uncertainties and assumptions within the SSD concept
2. Derive scientific conclusions on various aspects of the SSD concept, and
3. Express the limitations and points of concern regarding the use of the SSD-based extrapolation procedure in the derivation of EQCs and ERA.

In view of space limitations, only the crucial aspects of SSD derivation and the field case studies are introduced; the reader is referred to the original papers for further information.

9.2 DERIVATION OF SSDS AND CASE STUDIES ON FIELD EFFECTS: GENERAL ASPECTS

To determine whether SSD-based predictions are confirmed by true field population/community effects, one requires two data sources when investigating a compound, namely, laboratory-based toxicity data and effect data collected under (semi-)field conditions. The former is ubiquitous and easily accessible from electronic databases (e.g., U.S. EPA, 1997) for many compounds, and it is common practice to construct and use SSDs based on this type of information. Appropriate sets of data regarding the latter, hereafter referred to as case studies, are scarcer. Case studies are scattered throughout various sources in "open" and "gray" literature. Moreover, the studies often lack information of sufficient quality to reconstruct quantitatively the relationship between cause (exposure to the compound) and observed effects (Posthuma, 1997), which is needed for a proper confirmation attempt. We selected compounds for which both appropriate laboratory and field data were available. The field studies were taken from the open literature and from our own recent work. We selected six examples from the available studies that together demonstrate most clearly the relevant issues regarding (non-)confirmation.

9.2.1 LABORATORY-BASED SSDs

Laboratory SSDs were derived from data series on various test endpoints, e.g., NOECs, EC_{50} or LC_{50} values, as indicated. Although the laboratory-based SSDs that are currently used are often based on chronic NOEC data, there is no theoretical limitation regarding the construction of separate SSDs from acute EC_{50} or LC_{50} data (see, e.g., Chapter 8), and we extended the range of test endpoints in this way for various reasons. Evidently, the distinction of exposure effects from control variation in the case studies is often only possible given very distinct population responses, which can most profitably be described with response endpoints (e.g., EC_{50} on abundance or biomass) rather than with nonobserved response endpoints (like NOEC).

For the compounds under investigation, the descriptions of the SSD curves derived either from laboratory data or from data on the semifield effects were obtained using a log-logistic model. The SSD curves were generated using the logistic model:

$$y = \frac{100}{1 + \exp^{-b*(\ln(x)-a)}} \tag{9.1}$$

where
  y = risk percentage
  a = ln(50th percentile)
  b = slope parameter
  x = exposure concentration

The model was programmed in GENSTAT, version 5.3.1 (Payne and Lane, 1987).

9.2.2 CASE STUDY INTERPRETATION

The community-level toxic effects in the case studies were calculated in various ways using original data (usually on species densities) collected by us or by colleagues. The effects on field populations were analyzed either by ANOVA or log-logistic regression. In the latter case, the model given above was used, with the exception that an extra parameter c was introduced in the numerator to quantify the control performance. Multivariate techniques were used for the evaluation of the treatment effects on a community level (Van Wijngaarden et al., 1995; Kedwards et al., 1999; Van den Brink and Ter Braak, 1999). The methods used, along with a justification, are summarized for each case study.

9.2.3 TWO TYPES OF COMPARISONS

In view of data limitations in the case studies, two types of comparisons were made.

First, we analyzed two examples in which a comparison could be made between the SSD curves derived from laboratory toxicity data and the field SSD curve derived from population specific responses in semifield conditions. The construction of a field-based SSD requires field data that meets the highest quality standards, with a level of detail seldom encountered in literature.

Second, we analyzed four examples in which a comparison is made between SSD curves derived from laboratory toxicity data and the occurrence and magnitude of direct and indirect effects at explicit exposure concentrations under semifield conditions. In these cases, the occurrence and magnitude of overall field responses could be determined with the original source, or it could be reconstructed, but it was not possible to construct a field SSD. This type of comparison allows one to deduce whether or not there is a general SSD percentage at which direct and indirect effects start to emerge.

9.3 CASE STUDIES I: LABORATORY DATA SSD VS. FIELD DATA SSD

The level of truth in SSDs can be established by comparing SSDs that are based on laboratory toxicity data (the common approach) with those based on field toxicity data. There are many reasons these curves may show no similarity, among which are aspects relating to compound behavior and the species incorporated in the two curves. What is the extent of the (dis)similarity if we compare SSDs based on these two sources, and what can we learn from these comparisons regarding the applicability of SSDs in solving practical problems? Two examples are shown. Both concern a comparison between a laboratory SSD constructed from single-species laboratory toxicity tests and a field SSD derived from multispecies field tests.

9.3.1 CHLORPYRIFOS IN EXPERIMENTAL DITCHES

9.3.1.1 Compound

Chlorpyrifos is an organophosphorus compound that displays broad-spectrum insecticidal activity against a number of important arthropod pests. Because its molecule is nonpolar, it has a low water solubility (2 mg/l) and a relatively high lipophilicity (log K_{ow} = 4.7 to 5.3). Chlorpyrifos is a degradable compound, with hydrolysis as the most important process (Racke, 1993). Because chlorpyrifos shows a field half-life of <0.08 to 2.4 days (Racke, 1993), the SSDs were compared using acute toxicity data.

9.3.1.2 Laboratory SSDs

An overview of the ecotoxicological profile of the test substance was obtained using the database AQUIRE (U.S. EPA, 1997). Acute EC_{50} values were used to construct the laboratory SSD. Compared with chronic NOEC values, these data were available in a much larger quantity. The EC_{50} values that we selected were all based on tests that lasted between 2 and 4 days. When more than one EC_{50} was available for the same species, the geometric mean was calculated.

As could be expected from the specific TMoA of the compound, the laboratory SSD curves given in Figure 9.1 show that arthropods and, to a lesser extent, fish are the most sensitive groups, followed by algae and other invertebrates. The SSD curves for the various groups differ by two or three orders of magnitude. This suggests that comparisons between laboratory SSDs and field SSDs should be based on the taxon-specific curves, since the contributions of the species groups to the local community may differ depending upon the circumstances (laboratory or field). Moreover, this analysis tells us that it can be profitable in practical SSD applications to distinguish between species groups prior to constructing an SSD curve.

9.3.1.3 Field SSDs

The experiment consisted of eight experimental ditches that were sprayed with the insecticide chlorpyrifos, and of four other ditches that served as controls. The treatment levels of 0.1, 0.9, 6, and 44 µg/l were each applied to two experimental ditches. The ditches were 40 m long, 3.4 m wide, and 0.5 m deep, resulting in a water volume of 60 m³. The ditches were macrophyte dominated. Detailed results of this experiment are described by Van Wijngaarden et al. (1996) and Van den Brink et al. (1996). The field EC_{50} values were calculated using the data obtained from samples taken 1 week after the application. A total of 59 different invertebrate taxa

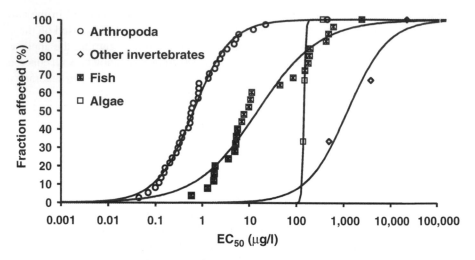

FIGURE 9.1 Laboratory-based SSD curves for acute toxicity (individual points are LC_{50}/EC_{50} in µg/l) for the insecticide chlorpyrifos, following the distinction of four groups of water organisms. Lines represent the logistic regressions on these data. For reasons of clarity, Figure 9.2 displays the SSD that has been constructed for all species simultaneously. (Data obtained from AQUIRE, U.S. EPA, 1997.)

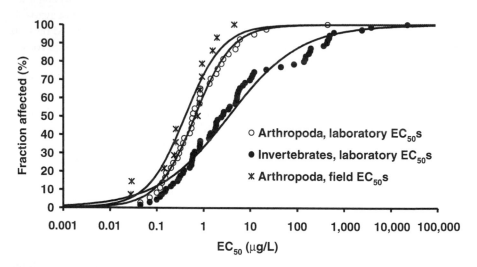

FIGURE 9.2 SSD curves for chlorpyrifos based on arthropod sensitivity expressed by acute LC_{50}/EC_{50} values collected from both laboratory tests and a semifield test. The figure also shows the laboratory SSD curve based on all available EC_{50} values of aquatic invertebrates.

were identified at that time, 36 of which were arthropod taxa. It was possible to construct an EC_{50} for 14 arthropod taxa. These were subsequently used to construct a field-based SSD (Figure 9.2). It was not possible to calculate EC_{50} values for the remaining arthropod taxa. In most cases, this was due to low abundance values and/or a high variability between replicates.

9.3.1.4 Comparisons and Discussion

Prior to interpreting the (dis)similarities between laboratory-based and field-based SSDs, one must rule out the possibility that the field-based SSD is biased by a lack of insensitive species, i.e., species for which the EC_{50} value is lacking because the experimental exposure is too low ($EC_{50} > 44$ μg/l). Such a bias seems, however, unlikely for the following reasons. First, the concentration of 44 μg/l in the laboratory SSD curve corresponds with the 99th percentile, indicating severe effects on all tested arthropod taxa. Second, 16 of the 30 most abundant taxa in the case study were arthropods. Only for 2 of the abundant arthropod taxa was it not possible to calculate an EC_{50} value due to convergence problems because of high control variation. The 16 arthropod data sets contained no indication of chlorpyrifos insensitivity ($EC_{50} > 44$ μg/l). This corresponds with the findings of the literature reviews by Brock et al. (2000b) and Giesy et al. (1999), who showed that, under field circumstances, chlorpyrifos concentrations of 5 μg/l and higher usually result in severe, long-term effects on arthropods.

Moreover, it should be noted that the measured toxicity endpoint that was quantified to determine the field EC_{50} is not necessarily the same as the test endpoints used in laboratory tests. The acute laboratory test endpoint normally represents death or immobility, whereas the measured effect in the field is "density at time t," a net effect of death, birth, ecological interactions, etc. The data that were used for the derivation of the field EC_{50} values was obtained from samples taken 1 week after the application of the insecticide, so the abundance data mainly reflect the carrying capacity of the test system (in the case of the controls) and their sensitivity (in the case of the treatment levels).

Figure 9.2 allows for a comparison between the laboratory-based and field-based SSD curves thus derived for the arthropods. Arthropod taxa are almost equally susceptible to chlorpyrifos in the laboratory and in the field, notwithstanding (among others) different exposure conditions and the species contributing to the SSD curves. The 50th percentiles and the slopes are almost the same (0.65 vs. 0.40, and 1.09 vs. 1.08, respectively). The latter indicates an equal width of the distribution of the data. The difference of a factor 1.6 between the 50th percentiles is well within the normal interexperimental range (Rand and Petrocelli, 1985). Apparently, laboratory-based and field-based SSDs can show a high similarity, one that may be unexpected in view of the large number of unknown and uncontrolled variables. Note that this pertains to an SSD comparison on laboratory and field data for a readily bioavailable compound in water and a specific group of sensitive organisms.

Figure 9.2 also shows the SSD curve based on all available laboratory data of invertebrates. This curve is clearly distinct from both arthropod-based curves. The composite curve, based on data from both sensitive and insensitive invertebrate groups, fits to a reasonable extent to the data (statistically) in the lower tail of the curve, but it would clearly yield an underestimation of the field effects for expected sensitive endpoints beyond concentrations of approximately 0.1 to 0.5 μg/l.

In conclusion, this case study can be interpreted as a confirmation of the SSD concept per se, meaning that under more or less ideal conditions (i.e., minimal uncertainties) laboratory SSDs and field SSDs are similar. It should be noted that optimal predictions are obtained when a sensible division into taxonomic groups is made, one that reflects the knowledge on the TMoA of the chemical.

9.3.2 ZINC IN EXPERIMENTAL FIELD PLOT SOILS

9.3.2.1 Compound

Zinc is a heavy metal with many biological functions (Bettger and O'Dell, 1981). The availability of zinc largely depends upon the characteristics of the substrate. In soil, the physicochemical sorption of zinc is strongly influenced by soil acidity (e.g., Janssen et al., 1997a), and this affects the uptake of zinc by soil invertebrates (e.g., Janssen et al., 1997b). It should be noted that the regulation of the zinc level by organisms additionally influences the toxicological availability of zinc, i.e., the zinc concentration at the target site within the organism (e.g., Peijnenburg et al., 1999a,b). The case study data pertains to a multiyear study on the chronic effects of zinc on a natural nematode species assemblage in an outdoor terrestrial field plot. Hence, the SSD comparisons are based on chronic toxicity data. Because zinc does not, in contrast to chlorpyrifos, have a specific TMoA, our working hypothesis was that all species (invertebrates and plants) can be represented with a single sensitivity distribution. This assumption is also made by the derivation of the EQC for zinc in the Netherlands (Crommentuijn et al., 1997b).

9.3.2.2 Laboratory SSDs

Chronic laboratory NOECs for zinc were available from the compilation of the original data by Crommentuijn et al. (1997b) for only seven terrestrial species, namely, for three plant species, two snail species, a worm, and an isopod. More than one NOEC was available for some species, pertaining to (probably) different exposure conditions, i.e., substrate. Contrary to the often-prescribed procedures, the geometric mean of the test results was *not* calculated for these species. Since the tests with these species were executed in substrates with different sorption characteristics (at different pH values), the "sensitivity" in the SSD of soil was in fact interpreted as the "intrinsic species sensitivity" following modification by the sorption status of the soil.

The data suggest a subdivision of toxicity data into plants and invertebrates, with plants apparently more sensitive than invertebrates (see original data). However, one should not interpret this as intrinsic differences between plants and invertebrates, because it is likely a consequence of the fact that the zinc toxicity for plants has been investigated in experimental series at pH values between approximately 4 and 8, with proper NOEC estimation only for the acid series; the other NOECs were reported as "beyond highest tested concentration," which could not be used as input data here. As a consequence of this flaw in the input data, the laboratory SSDs differ as shown in Figure 9.3, with the plant curve shifted to the left, probably as a consequence of a low pH in the test systems. Both curves correspond to the data to

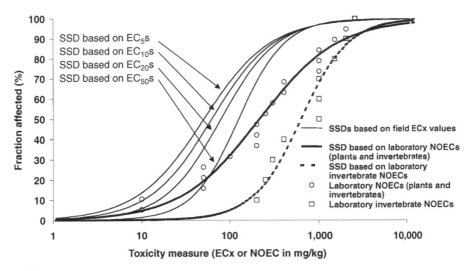

FIGURE 9.3 Comparison between SSD curves for zinc based on the sensitivity of various terrestrial taxa (no nematodes) in laboratory tests and on nematode taxa in the field. For the laboratory-based SSD, both a composite curve of "plants + invertebrates" and a separate curve for "invertebrates only" are shown.

an extent that is typical. For the comparisons, however, only the data of the soil invertebrates were used. In these tests, the pH was approximately 6.0 to 6.5 (like in the field test), or it was not reported.

9.3.2.3 Field SSDs

The experiment consisted of 50 outdoor soil mesocosms. The soil in the mesocosms originated from an uncontaminated field site. The soil mesocosms were spiked with a solution of $ZnCl_2$ at ten different zinc levels, yielding nominal concentrations ranging from 32 to 3200 mg/kg dry wt, with five mesocosms acting as water-treated control (with background zinc concentrations). The natural nematode assemblage of the soil was inoculated. Samples for nematode censuses were taken 3, 10, and 22 months after zinc spiking and inoculation. The pH of the field soil was between 6.5 and 7.1 on the last sampling date. Details on the experimental design and on nematode censuses are given in Smit et al. (in press) and Posthuma et al. (2001).

On the first sampling date, 37 different taxa were identified. No NOECs were derived from this data, however, since any NOECs would be overestimated by the control density variation for all species. Instead, attempts were made to determine full EC_x curves for density effects for these taxa. A low-effect endpoint (e.g., EC_5, EC_{10}) was extracted from the fitted curves for the purpose of laboratory/field SSD comparisons. For 13 taxa, corresponding with the 21 most abundant species, the fitting procedure converged. All the other taxa included in these 21 species were almost absent in the highest treatment level, indicating that the EC_x value could not be calculated because of high control variation rather than insensitivity. Field-based SSD curves were derived at the EC_{50}, EC_{20}, EC_{10}, and EC_5 level (see Figure 9.3). Obviously, the fitted SSD curves shifted to the left with decreasing x (effect level input data).

9.3.2.4 Comparisons and Discussion

Figure 9.3 shows that the SSD curve based on laboratory NOECs for invertebrates was positioned to the right of all of the SSD curves based on field EC_x values (x between 50 and 5% effect). The laboratory-based NOEC curve differed from the field EC_5-based SSD by approximately one order of magnitude, with a similar slope. Apparently, the nematode community in the soil mesocosms showed a more sensitive response to zinc than the invertebrates tested in the laboratory. As argued for the example with chlorpyrifos, this may be because different representatives of the soil invertebrate community are compared and because endpoints may differ between laboratory and field tests. In addition, the difference can also result from the fact that soil characteristics strongly influence the biological availability (and therefore the NOEC or EC_x) of zinc under both laboratory and field conditions, as substantiated by the laboratory tests with plants at different pH values. Disentangling the three possible (major) causes of the difference between SSDs required a further look at the laboratory data. This investigation showed that only two pH values were provided — one approximately 6.0 and the other approximately 6.5 — for tests with the worm *Eisenia andrei*, which yielded NOECs of 200 and 1500, respectively, for different endpoints. These two values are almost in both tails of the laboratory SSD. Since the pH of these tests is more or less similar to that of the test field after 22 months, it can be concluded that density reduction of the nematode species is a more sensitive endpoint than the sublethal toxicity of zinc for *E. andrei* under laboratory conditions at a similar physicochemically available zinc fraction. The pH was not provided for the other species. Concerning this case study, the laboratory SSD curves and field SSD curves likely differ as a result of "different representation of species groups" (nematodes could be more sensitive to zinc than *E. andrei*), but also due to differences in biological availability.

To reduce the uncertainties in the laboratory SSD derivation procedure, for example, it seems worthwhile to derive laboratory SSD curves on the basis of a biologically available zinc fraction. This has the advantage that the confirmation of the field SSD can take place when the field SSD is also based on this fraction measured in the field soil. However, an adequate correction method for the major issue of availability differences has not yet been formulated. Furthermore, in accordance with the chlorpyrifos example, it might be important to distinguish between the taxonomic groups of interest in relation to the TMoA of the compound.

9.4 CASE STUDIES II: SSD VS. ECOLOGICAL FIELD EFFECTS

A comparison between laboratory SSDs and field SSDs yields important information on similarity or divergence, and may help to identify likely causes of divergence, as shown in the first two examples. However, many case studies do not allow for the derivation of field-based SSDs, since they lack sufficient data densities to estimate NOECs or EC_x (with x representing a low-effect level, e.g., 5 or 10) for a representative fraction of the field population. Moreover, the eventual aim of SSD usage is the protection of field communities, rather than of species presented in the SSD.

Therefore, four case studies with different chemicals were selected for which no field SSD could be derived, but which could be used to compare the laboratory-derived SSD with the absence or occurrence of ecological effects at explicit exposure intervals evaluated under semifield conditions.

9.4.1 LINURON IN AQUATIC MICROCOSMS AND MESOCOSMS

9.4.1.1 Compound

Linuron is a herbicide with a specific, photosynthesis-inhibiting TMoA. In aquatic ecosystems, the major proportion of the compound is bioavailable, since the sorption to (in)organic matrix elements is low. The half-life of the compound under aquatic field circumstances is 7 to 12 days (Crum et al., 1998).

9.4.1.2 Laboratory SSDs

An overview of the ecotoxicological profile of the test substance was obtained using the database AQUIRE (U.S. EPA, 1997). Additional data were obtained from compilations of Crommentuijn et al. (1997a), Snel et al. (1998), Mayer and Ellersieck (1986), and Van Wijngaarden and Brock (1999). Acute EC_{50} values were used for the construction of the SSD curve instead of chronic NOEC values, because only a very limited number of NOEC values was available. The compiled data were processed in the same way as described above for the chlorpyrifos example.

The laboratory SSDs are given in Figure 9.4. The curves show that the sensitivity of primary producers is high, whereas the sensitivity of invertebrates and fishes is much lower. The SSD curve for photosynthesis inhibition is very steep, which can be expected from the TMoA of linuron. The photosynthesis curve was composed using data on five macrophyte species and one algal species. There is apparently very little difference between the primary producers regarding the ability of linuron to reach the receptor sites. The SSD curve that is based on the growth inhibition of (other species of) primary producers showed approximately the same value for the 50th percentile, but a very different slope. The divergence of the slopes indicates that adverse effects on a component of physiological performance (like photosynthesis inhibition) need not be similar to realized ecological performance (like growth).

9.4.1.3 Case Study Data

Using a laboratory-based SSD curve based on acute EC_{50} data, two semifield experiments were analyzed to evaluate both the effects of an acute and chronic exposure regime.

Acute Exposure. The acute exposure regime of linuron was evaluated in outdoor mesocosms. Of the ten available mesocosms, two replicates received a nominal dosage of 0, 0.5, 5, 15, or 50 µg/l linuron. The compound was applied three times, taking into account a 4-week interval, and the mesocosms were slowly flushed between each application with clean water originating from an uncontaminated freshwater reservoir. The species composition of the phytoplankton, periphyton, and

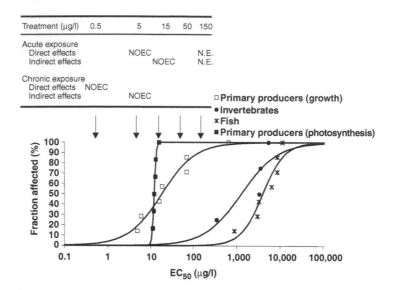

FIGURE 9.4 SSD curves for acute toxicity (individual points are LC_{50}/EC_{50}) for the herbicide linuron, divided into four groups of water organisms. Lines represent the results of the logistic regressions on these data. Arrows indicate the concentrations tested in the semifield tests. The table above the figure summarizes the NOECs reported for direct and indirect effects under an acute and a chronic exposure regime (see Table 9.1 for more details). N.E. = treatment level not evaluated.

invertebrate communities were monitored in time, together with chlorophyll-*a* content, various physicochemical parameters, and macrophyte biomass. The reader is referred to Crum et al. (1998), Kersting and van Wijngaarden (1999), and Van Geest et al. (1999) for more information.

Table 9.1 summarizes the NOECs that were recorded for the different endpoints, with a subdivision of the endpoints into direct (TMoA-related) and indirect effects. Generally, NOECs were considered valid when calculated for two consecutive sampling dates. The two highest treatment levels (15 and 50 µg/l) induced minor changes in structural endpoints and major changes in functional endpoints. The predominantly direct effects that were observed could easily be explained by the photosynthetic inhibiting TMoA of the chemical. No consistent effect of linuron was reported for the lower treatment levels (Van Geest et al., 1999).

Chronic Exposure. The semifield experiment for the evaluation of chronic exposure to linuron consisted of indoor microcosms (1 m³), which represented macrophyte-dominated drainage ditches. The systems were treated chronically for 4 weeks with 0, 0.5, 5, 15, 50 or 150 µg/l linuron, with two replicates per concentration. The species composition of the phytoplankton, periphyton, and invertebrate communities were monitored in time, together with chlorophyll-*a* content, various physicochemical parameters, and macrophyte biomass. The reader is referred to Van den Brink et al. (1997) and Cuppen et al. (1997) for a more detailed presentation and evaluation of the results.

TABLE 9.1
NOEC Values (µg/l) on Structural and Functional Endpoints Obtained from Semifield Experiments Performed with the Photosynthesis Inhibiting Herbicide Linuron

Direct Effects		Indirect Effects	
Acute exposure	**NOEC**	**Acute exposure**	**NOEC**
Structural endpoints	5[a]	Structural endpoints	15
Chlamydomonas (phytoplankton)	5[a]	*Anax imperator* (Odonata, invertebrates)	15
Coleochaete (periphyton)	15[a]	Tubificidae (invertebrates)	15
Myriophyllum spicatum (macrophyte)	15		
Functional endpoints	5[a]		
(mediated by direct effects)			
DO, pH, primary production	5[a]		
Chronic Exposure	**NOEC**	**Chronic Exposure**	**NOEC**
Structural endpoints	0.5	Structural endpoints	5
Chroomonas (phytoplankton)	0.5	Rotatoria	5
Cocconeis (periphyton)	0.5	*Chlamydomonas*	15
Elodea (macrophyte; bioassay)	0.5	Chlorophyll-a periphyton	15
Elodea (macrophyte; standing stock)	15	Cladocera	15
Cocconeis (phytoplankton)	50	Ostracoda	15
Phormidium (phytoplankton)	50	*Asellus aquaticus* (Isopod)	50
		Copepoda	50
		Chlorophyll-a phytoplankton	
		Chlorophyll-a neuston	
Functional endpoints	0.5	Functional endpoints	≥150
(mediated by direct effects)			
DO, pH	0.5	Decomposition	≥150
Alkalinity, conductivity	5		
Nitrate	15		

Note: The acute exposure was evaluated in outdoor mesocosms, the chronic exposure in indoor microcosms. The effects are subdivided into direct and indirect effects on the basis of the TMoA of the compounds, and into effects on structural and functional endpoints. Within each group, the effects are ordered from top to bottom according to increasing NOECs.

[a] Slight short-term effects observed only.
DO = dissolved oxygen.

Table 9.1 shows a summary of the NOECs that were reported for structural and functional endpoints. The lowest NOEC reported for direct effects is 0.5 µg/l. The direct effects of the herbicide on both the functional and the structural endpoints begin to manifest themselves between 0.5 and 5 µg/l. Indirect effects start to occur between concentrations of 5 and 15 µg/l for Rotatoria (Cuppen et al., 1997). More pronounced secondary effects on several endpoints are reported for the two highest treatment levels only, and thus they appear between a concentration of 15 and 50 µg/l.

9.4.1.4   Comparisons and Discussion

The comparison between the laboratory SSD and the effects observed in the semifield studies is given in Figure 9.4, in which the arrows indicate the treatment/effect levels. The NOECs of the most sensitive endpoint for direct and indirect effects are indicated above the figure for both the acute and the chronic study.

Among-Field Study Comparisons. Comparisons among the field studies suggest the repetition of a trend often found on the basis of laboratory toxicity data, namely, a relationship between chronic and acute effect levels. In the acute study, the NOEC for direct effects was 5 µg/l, and 15 µg/l for the indirect effects (see Table 9.1). In the chronic study, the direct effects start to occur at the concentration above 0.5 µg/l, indirect effects above 5 µg/l. It is striking that $NOEC_{Community}$ values for linuron, which are based on the most sensitive ecological endpoint, differ by a factor of 10 when comparing the studies that evaluate an acute and a chronic exposure. A factor 5 difference between the $NOEC_{Community}$ on comparison of a single and a chronic application has been previously observed in field studies for atrazine (Brock et al., 2000a). Although the number of cases, and with that the foundation of a fixed value, is very limited thus far, a fixed factor of 10 between acute and chronic $NOEC_{Community}$ values may be used in the construction of "chronic SSDs" from acute data (see Chapter 8 for more details). An extensive literature review on $NOEC_{Community}$ values, determined for various compounds in the field rather than in the laboratory, is needed to confirm this. Substantial confirmation of this factor by more studies would allow for predictions of chronic effects at the ecosystem level in the field from acute laboratory-based SSDs. This would be highly advantageous in the practical application of SSDs, since acute toxicity values are generally more available compared with chronic ones, while decision making often concerns chronic effects.

Acute Study and SSD. Direct acute effects of the short-term exposure to linuron start between 5 and 15 µg/l (see Table 9.1). These concentrations correspond with the 19th and 44th percentiles of the SSD based on acute EC_{50} values. These percentages seem rather high; the EC_{50} of at least 1 out of 5 primary producers would be exceeded, and yet still no significant effect becomes apparent! The first reason for this may be that the percentage of 19 is solely based on extrapolation; no EC_{50} lower than 5 µg/l was found in the literature. The lowest EC_{50} found was 5 µg/l for the green alga *Ankistrodesmus falcatus*. Second, some algal taxa are known to show reduced sensitivity to linuron after prolonged exposure. In our chronic study, the green alga *Chlamydomonas* sp. even increased in numbers after the treatment (see Table 9.1), but is known to be susceptible to linuron in short-term tests (Van den Brink et al., 1997). The duration of laboratory tests performed with algae, normally 3 days, may thus be too short to recognize reduced sensitivity and therefore overestimate the density effect of linuron. In addition, Snel et al. (1998) demonstrated that the inhibition of photosynthesis by linuron is a reversible process, and recovery may occur fast when the concentration of linuron drops below the critical threshold level. It is also remarkable that indirect effects of linuron were hardly observed in the acute, outdoor study. Some minor indirect effects could, however, be observed at 50 µg/l, corresponding with the 75th percentile of the SSD curve based on acute toxicity data.

Chronic Study and SSD. The occurrence of direct effects in the chronic study, starting between 0.5 and 5 µg/l, corresponds with the 2nd and 19th percentiles of the SSD curve based on acute EC_{50} values of primary producers. The occurrence of observed indirect effects, starting between 5 and 15 µg/l, correspond with the 19th and 44th percentiles of the SSD, respectively. When a high percentage of the primary producers is affected by the herbicide, one may indeed expect indirect effects due to the release of nutrients or changed food conditions. The decrease in macrophytes in the two highest treatment levels of the chronic study particularly resulted in a cascade of various indirect effects (see Table 9.1) (Cuppen et al., 1997). The macrophytes were apparently only adversely affected by a long-term exposure regime. This can be explained from the TMoA of the chemical, namely, that linuron inhibits photosynthesis and, with that, the synthesis of sugars. Since the inhibition of photosynthesis by linuron is a reversible process (Snel et al., 1998), only a chronic inhibition will lead to the death of primary producers such as macrophytes. A short-term inhibition will only force the macrophytes to use their sugar reserves. Since the half-life of linuron is relatively short (7 to 12 days) in the acute field study, the physiological performance of the macrophytes could be restored relatively fast, so no long-lasting effects were present. To compare the results from a chronic field study with an SSD curve based on acute toxicity data, one can use an extrapolation factor for the differences between acute and chronic laboratory toxicity values. Generally, there is a fixed factor of about 10 for the difference between acute EC_{50} and chronic NOEC values (Chapter 8). Using an extrapolation factor of 10, the chronic direct effects correspond with the 19th percentile of the SSD based on chronic NOECs extrapolated from the acute EC_{50} values.

The comparison between field effects and laboratory-derived SSD curves revealed several aspects of the SSD concept for linuron that were not encountered in the example with chlorpyrifos (example 1). This relates to different aspects, associated with the TMoA (e.g., growth inhibition vs. mortality, reversibility of the process, decrease of sensitivity over time). However, it can be concluded from the two field studies that effects due to an acute exposure start to emerge at concentrations that correspond with the lower part ($NOEC_{Community}$ = 19th percentile) of the SSD curve based on acute toxicity data, and that the effects due to chronic exposure relate to the lower percentage values ($NOEC_{Community}$ = 2nd percentile) of the SSD curve constructed from acute EC_{50} data. In both cases, concentrations related to the 10th percentile or lower, based on acute and extrapolated chronic toxicity data (derived from acute EC_{50} values by applying a safety factor of 10), seem to have no effect on field communities in cases of short-term and long-term exposure, respectively.

9.4.2 CARBENDAZIM IN AQUATIC MICROCOSMS

9.4.2.1 Compound Characteristics

Carbendazim is a benzimidazole fungicide that is supposed to inhibit the cell division of microorganisms (Delp, 1987). In the aquatic environment, however, microorganisms are not affected at realistic concentrations (Chandrashekar and Kaveriappa, 1994). Contrary to the example concerning linuron (Section 9.4.1), the TMoA of carbendazim to other aquatic organisms is largely unknown. Carbendazim has a relatively low lipophilicity (log K_{ow} = 1.5) and is relatively persistent, considering the half-life variability (DT_{50} 124 to 350 days; Tomlin, 1997). Therefore, the comparisons were made on the basis of chronic toxicity data.

9.4.2.2 Laboratory SSDs

An overview of the ecotoxicological profile of the test substance was obtained using the database AQUIRE (U.S. EPA, 1997), appended with data from Crommentuijn et al. (1997a) and Van Wijngaarden et al. (1998). Laboratory SSD curves were constructed for different groups of organisms, although the TMoA for nontarget organisms is scarcely known.

Acute Toxicity Data. Figure 9.5 shows the laboratory-based SSD curves for invertebrates, fish, and algae, based on only 13 EC_{50} values. There was no clear distinction between arthropods and nonarthropods within the invertebrate group, so these data were lumped together. Fish proved to be the most sensitive group, although this conclusion is based on three values only. The invertebrates appeared to be a factor of three less sensitive than fish, with the flatworm *Dugesia lugubris* the most sensitive representative of the group. Algae were the least sensitive group, but the algal SSD is also based on three values only.

Chronic Toxicity Data. In addition to the acute toxicity data, chronic NOEC values for invertebrates could also be found (Van Wijngaarden et al., 1998). Figure 9.6 displays the SSD curves for the invertebrates based on acute EC_{50} and on chronic NOEC values, respectively.

The SSD curve based on chronic NOEC values shows lower percentile values at a given concentration, but runs parallel to the acute SSD curve. The 10th percentile for the SSD curves based on acute EC_{50} and chronic NOEC values corresponds to 11 and 1.4 µg/l, and that of the 50th percentiles to 231 and 18 µg/l, respectively.

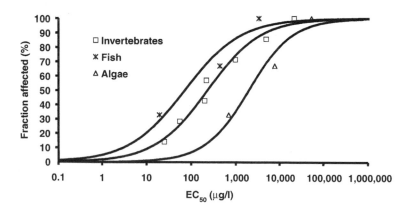

FIGURE 9.5 SSD curves for acute toxicity (individual points are LC_{50}/EC_{50}) for the fungicide carbendazim, divided into three groups of water organisms. Lines represent the results of the logistic regression on these data.

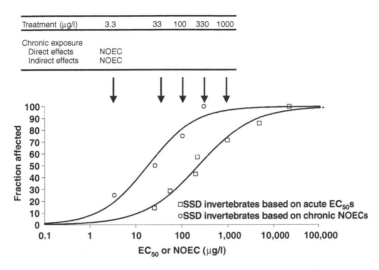

FIGURE 9.6 SSD curves for acute toxicity and chronic toxicity of the fungicide carbendazim with respect to aquatic invertebrates. The individual points represent the acute LC_{50}/EC_{50} and chronic NOEC values found in the literature. Lines represent the results of the logistic regression on these data. Arrows indicate the concentrations tested in the microcosm experiment of the case study. The table above the figure summarizes the NOECs reported for direct and indirect effects under a chronic exposure regime (see Table 9.2 for more details).

9.4.2.3 Field Study Data

A microcosm experiment was executed in accordance with an experimental design and test systems similar to the chronic linuron experiment (Section 9.4.1). The tests concerned carbendazim concentrations of 0, 3.3, 33, 100, 330, and 1000 µg/l. For a more detailed presentation and evaluation of the results, see Cuppen et al. (2000) and Van den Brink et al. (2000). Table 9.2 summarizes the effects found in the microcosm experiment in terms of NOECs.

No consistent NOECs lower than 3.3 µg/l were recorded. Direct effects became manifest following a treatment with 33 µg/l. Several "wormlike" taxa belonging to the groups of flatworms, leeches, and oligochaete worms showed altered abundance values, together with two crustacean taxa. At this treatment level, indirect effects in the form of increases of several snail taxa were also observed, indicating food-web changes due to increased food abundance for these species. Effects on functional parameters were observed only at high concentrations (>100 µg/l; Table 9.2).

9.4.2.4 Comparisons and Discussion

The arrows in Figure 9.6 represent the treatment levels that were evaluated in the microcosm experiment. The laboratory-based SSD curves predict major adverse effects on the invertebrate species in the field test following the treatments with high concentrations of carbendazim. The lowest treatment level (3.3 µg/l) resulted in no observable adverse effects in the microcosms and corresponds with the 19th percentile of the chronic NOEC SSD curve (based on invertebrates) and with the

TABLE 9.2
NOEC Values (µg/l) on Structural and Functional Endpoints Obtained from a Semifield Experiment Performed with the Fungicide Carbendazim

Direct Effects		Indirect Effects	
Chronic Exposure	**NOEC**	**Chronic Exposure**	**NOEC**
Structural endpoints	3.3	Structural endpoints	3.3
Acroperus harpae (Cladocera)	3.3	*Lymnaea* juvenile (Mollusca)	3.3
Alboglossiphonia heteroclita (Hirudinea)	3.3	*Physella acuta* (Mollusca)	3.3
Dero (Oligochaeta)	3.3	*Physa fontinalis* (Mollusca)	3.3
Dugesia lugubris (Turbellaria)	3.3	*Chlamydomonas* (phytoplankton)	100
Dugesia tigrina (Turbellaria)	3.3	*Cryptomonas* (phytoplankton)	100
Gammarus juvenile (Amphipoda)	3.3	*Cyclotella* (phytoplankton)	100
Alonella exigua (Cladocera)	33	*Monoraphidium* (phytoplankton)	100
Bithynia juvenile (Mollusca)	33	*Stephanodiscus* (phytoplankton)	100
Bithynia tentaculata (Mollusca)	33	*Elodea* (macrophyte; standing stock)	100
Gammarus pulex (Amphipoda)	33	*Chilomonas* (phytoplankton)	330
Simocephalus vetulus (Cladocera)	33	*Testidunella parva* (Rotatoria)	330
Stylaria lacustris (Oligochaeta)	33		
Bithynia leachi (Mollusca)	100		
Cyclopoida (Copepoda)	100		
Nauplius (Copepoda)	100		
Achnanthes (phytoplankton)	330		
Asellus aquaticus (Isopod)	330		
Asellus juvenile (Isopod)	330		
Keratella quadrata (Rotatoria)	330		
Lecane (Rotatoria)	330		
Proasellus coxalis (Isopod)	330		
Proasellus meridianus (Isopod)	330		
Trichocerca (Rotatoria)	330		
		Functional endpoints	100
		Long-term decomposition	100
		DO/pH metabolism	≥1000

Note: A chronic exposure to carbendazim was evaluated in indoor microcosms. The effects are subdivided into direct and indirect effects on the basis of the TMoA of the compounds, and into effects on structural and functional endpoints. Within each group, the effects are ordered from top to bottom according to increasing NOECs.

4th percentile of the acute EC_{50} SSD curve (based on invertebrates). It should be noted that these percentile values were based solely on the extrapolation of laboratory toxicity data, for which no single NOEC and EC_{50} value lower than 3.3 µg/l was found.

The 33 µg/l treatment level was the lowest concentration at which significant effects were demonstrated in the field tests, and this concentration is associated with the 19th percentile of the acute EC_{50} SSD curve and the 62nd percentile of the chronic NOEC SSD curve. Thus, whereas the direct effects of a chronic treatment of linuron emerged between the 2nd and 19th percentile of the acute EC_{50} SSD curve, those of carbendazim emerged between the 4th and 19th. These values are in good agreement with each other, which suggests the results of acute laboratory tests can be used to predict the direct effects that occur in the field as a result of chronic exposure. However, the values do not correspond for the indirect effects. Indirect effects of linuron occurred between the 19th and 44th percentile of the acute EC_{50} SSD, whereas carbendazim treatment led to indirect effects between the 4th and 19th percentile. This discrepancy can be explained by the fact that indirect effects are a result of ecological interaction, an aspect that is not modeled within the SSD concept.

9.4.3 CHLORPYRIFOS IN TERRESTRIAL FIELD STUDIES

9.4.3.1 Compound Characteristics

As mentioned when discussing example 1, chlorpyrifos is an organophosphorus compound that displays broad-spectrum insecticidal activity against a number of nontarget arthropods, as well as arthropod pest species. Soil surface half-lives are in the order of a few days to several weeks. The compound has an average soil and sediment sorption coefficient (K_{oc}) of $8.5 \ 10^3 \ 1 \cdot kg^{-1}$. Chlorpyrifos is characterized as relatively immobile in soil (Racke, 1993). The comparisons concern acute toxicity data.

9.4.3.2 Laboratory SSDs

The EC_{50} data for terrestrial invertebrates were obtained from Hamers et al. (1996a). The duration of the tests performed with arthropod species varied between 1 and 3 days, where all the tests performed with nonarthropod invertebrate taxa lasted 14 days. Data were obtained for six arthropod taxa, as well as for six nonarthropod taxa. Figure 9.7 shows the resulting SSDs, which indicate that arthropods are three orders of magnitude more sensitive to chlorpyrifos than other invertebrates. The four most sensitive arthropod taxa are all Collembolans. The 50th percentile of the arthropod SSD corresponds with 0.5 mg chlorpyrifos/kg soil, the 50th percentile of the nonarthropod SSD with 430 mg chlorpyrifos/kg soil.

9.4.3.3 Field Study Data

Dr. Geoff Frampton (University of Southampton, U.K.) kindly provided the field data used in this section. Three experiments, two of which have a similar experimental setup, were reanalyzed on the basis of the original data.

All of the experiments were set up as follows: one or four field plots were sprayed with chlorpyrifos and one or four others were left unsprayed and used as controls. The experiments were performed in four fields at "Mereworth" in southeastern England (51°16' N 0°23' E, sandy clay loam; Frampton, 1999) and in two fields in central England: "Near Kingston" (53°N 1°W, stony sand; Frampton, 2000) and "Field 5" (52°N 1°W, calcareous clay; Frampton, 2000).

Since the actual exposure in the field is a crucial aspect of the comparisons between laboratory SSDs and field data, the average exposure concentration in the

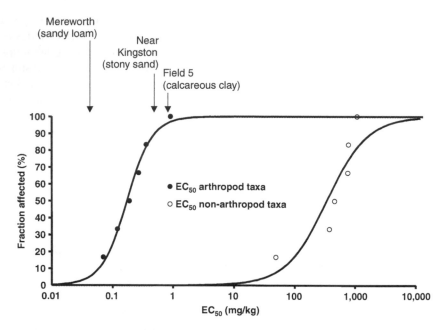

FIGURE 9.7 SSD curves based on LC_{50}/EC_{50} values for the insecticide chlorpyrifos of terrestrial arthropod and nonarthropod invertebrate taxa. The arrows indicate the estimated concentration in the soil in the field experiments (mg/kg); the accompanying name indicates the field.

fields was estimated as follows. The application rate, expressed in g/ha, was recalculated to the soil concentration (mg/kg) using the following formula:

$$C_{soil} = (dose * 1,000,000)/(10,000 * depth * B_d) \qquad (9.2)$$

where
    C_{soil} = soil concentration, in mg/kg
    dose = application rate, in kg/ha, corrected by fraction reaching the soil
    depth = depth of soil layer concerned, in m
    B_d = bulk density of soil, in kg/m³

A default value of 1400 kg/m³ was used for the bulk density; the default for the soil depth in which the compound is expected to be present is 0.05 m for soils that are not plowed. These calculations resulted in estimated soil concentrations of 0.041, 0.514, and 0.771 mg chlorpyrifos/kg soil for the treated plots of the Mereworth, Near Kingston, and the Field 5 fields, respectively.

Mereworth. The experiment at Mereworth was carried out in 1994 and comprised the application of 480 g of active ingredient (a.i.) chlorpyrifos/ha to winter wheat in summer (Frampton, 1999). This dose was applied to one plot each in four fields, while four unsprayed plots served as controls. It was estimated that 6% of the active

TABLE 9.3
Proportions (%) of the Total Number of Species Encountered
That Were Reduced by Chlorpyrifos to 0, 25, or 50%
of the Control Abundance for the Three Fields

| | Mereworth | Near Kingston | | Field 5 | |
	Collembola	All Taxa	Collembola	All Taxa	Collembola
			Effects Observed in the Field, %		
0%	9	40	50	34	50
25%	45	40	50	38	63
50%	73	60	50	47	88
		Percentage of Arthropod Taxa for Which EC_{50} Is Exceeded (as indicated by the laboratory SSD)			
SSD %	5	90	96		

Note: The lowest row indicates the percentiles of the SSD (Figure 9.7) corresponding with the calculated exposure concentrations in the field soils (see text).

ingredient reached the soil surface (Wiles and Frampton, 1996). Ground-dwelling Collembola were sampled using a suction sampler 10 days after the treatment.

A total of 11 different Collembola species were identified. Since it was not possible to test the significance of the effects of the treatment for the other experiments (see below), all the effects were expressed as percentage reductions in the collembolan abundance in the treated plots when compared to the controls (Table 9.3). The abundance of 1 of the 11 species was reduced to zero in the treated plots, 5 species were reduced to abundance values lower than 25% of their abundance in the controls, and 8 to values below 50%.

Near Kingston. The experiment at Near Kingston was part of the SCARAB research project funded by the U.K. Ministry of Agriculture, Fisheries and Food (Frampton, 2000), as were the Field 5 experiments. The aim of SCARAB was to investigate the responses of farmland arthropods to two pesticide regimes, current farm practice and a reduced-input approach (the latter implied no applications of insecticides); these regimes were applied to a range of fields during 1990 to 1996. The data of the Near Kingston and Field 5 fields, presented here, encompassed a time span in which the fields were only sprayed with chlorpyrifos.

The field was divided into two plots, one of which was sprayed with an application of 720 g a.i. chlorpyrifos/ha. It was estimated that ~50% of the active ingredient reached the soil surface (standing crop was spring barley: G.K. Frampton, personal communication). A total of ten arthropod species (two Collembola and eight non-Collembola) were identified in suction samples taken 24 days after the application. Four of the species had an abundance value of zero in the treated plot, and six had abundance values less than 50% of the control (see Table 9.3). Only one of the two Collembola species was affected: its abundance was reduced to zero in the treated plot (see Table 9.3).

Field 5. The experimental setup that was used for Field 5 resembled that of Near Kingston. It was estimated that 75% of the active ingredient reached the soil surface (standing crop was grass: G.K. Frampton, personal communication). In total, 32 different arthropod species (8 Collembola and 26 non-Collembola) were identified in suction samples taken 44 days after the application. Of these 32 species, 11 were reduced to zero in the treated plot, 12 had abundance values equaling 25% of the control or lower, and 15 had abundance values equal to or lower than half those of the control (see Table 9.3). Four of the Collembola species were reduced to zero, 5 were reduced to 25% of the abundance in the control plot, and 7 to half (see Table 9.3).

9.4.3.4 Comparisons and Discussion

The arrows in Figure 9.7 indicate the calculated exposure concentrations in the three experiments. The calculated exposure for the Mereworth fields corresponds with the lower part of the arthropod SSD curve, whereas that of the two other fields corresponds with the upper part. The SSD curve thus predicts that only a small proportion of the total species in the Mereworth fields would be affected by the chlorpyrifos application, whereas the majority of the species would be affected in the other two fields. This prediction is only partly confirmed by the results of the sampling of the arthropods in these fields (see Table 9.3).

Mereworth. The results for the Mereworth fields show that a few species were reduced to zero abundance (9%), and that the majority was reduced to 25 to 50% of the corresponding control abundance. The SSD, however, predicted that only 5% of the species are affected at $L(E)C_{50}$ level. A reason for the apparently stronger field effects can likely be found in the heterogeneous distribution of chlorpyrifos over the soil surface, so that some individuals were exposed while others were not. This is supported by the fact that only a very small percentage of the total dosage reached the ground (6%) due to the crop. The fraction of the species that is reduced to zero, however, corresponds rather well with the percentage of the species for which the EC_{50} is exceeded in standard laboratory tests (9 vs. 5). For a proper comparison, however, the fraction of the species that is reduced in abundance by 50% should be compared with the percentages of the SSD curve, as the latter is also based on EC_{50} values, and one should use measured field concentrations rather than estimated values.

Near Kingston and Field 5. The observed effects of chlorpyrifos in Near Kingston and Field 5 were roughly half those expected from the SSD curve. This discrepancy can, again, be explained by differences in actual exposure. In addition, because the treated fields were not sampled immediately after spraying (but rather 10 to 44 days later), some mobile, ground-dwelling arthropods may have recolonized the treated field plots before sampling.

   It is clear that the field experiments discussed here are only partly suited to confirm laboratory-based SSD curves. However, one should keep in mind that the experiments were performed for other reasons (see above). For a proper confirmation of terrestrial SSD curves, there is clearly a need for terrestrial ecotoxicity data in which exposure is properly quantified; total compound concentrations in soil are probably not sufficient for a proper SSD-based risk assessment.

9.4.4 COPPER IN A TERRESTRIAL FIELD

9.4.4.1 Compound

Like zinc (see Section 9.3.2), copper is a heavy metal with many biological functions and its biological availability is also determined by both substrate characteristics and the organisms' regulatory characteristics (Janssen et al., 1997a,b; Peijnenburg et al., 1999a,b). In view of the multiyear field study, chronic toxicity data were collected. As in the zinc example, all available laboratory toxicity data were used in a single sensitivity distribution because it is hypothesized that copper has no specific TMoA. This procedure is also used by the derivation of the EQC in the Netherlands (Crommentuijn et al., 1997b).

9.4.4.2 Laboratory SSDs

The data for the construction of the SSD curve for the effects of copper on terrestrial organisms were obtained from the compilation of Crommentuijn et al. (1997b). NOEC toxicity values were found for five macrophytes, one nematode, five oligochaetes, one collembolan, and one acarid mite. As was the case in Section 9.3.2, no geometric mean NOEC values were calculated from the multiple entries for some species, in view of the range of soil characteristics present in the tests. Terrestrial invertebrates are apparently slightly more sensitive to copper than are terrestrial plants (mean NOEC values 412 and 500 mg Cu/kg, respectively). For copper, the range of pH values in the tested plant series was more or less similar to the ranges used in the invertebrate tests; the pH values for all tests ranged from 5.0 to 5.6 for plants, and 5.1 to 7.3 for invertebrates. Because invertebrates and plants appeared to be approximately equally susceptible a "mixed" curve was constructed.

The resulting SSD curve is given in Figure 9.8. The 10th and 50th percentiles of the curve based on all NOECs available are 38 (95% CI: 32 to 44) and 210 (196 to 226) mg Cu/kg dry wt, respectively. The toxicity value that was found for the nematode *Caenorhabditis elegans* was on average 374 mg Cu/kg dry wt, although (as in Section 9.3.2) the NOEC ranges from 210 to 890 mg Cu/kg soil, depending on soil type and pH. The NOECs of this species fall well within the center of the SSD curve, but are mentioned separately because the case study pertains to copper effects on a nematode community.

9.4.4.3 Field Effects

The field experiment with copper is described in detail by Korthals et al. (1996), who kindly provided the original data for the further statistical analyses required for this confirmation study.

The case study data pertain to a multiyear study on the chronic effects of copper on a natural nematode species assemblage in an agricultural field plot. The experiment consisted of a factorial design, studying the effects of copper treatment (0, 250,

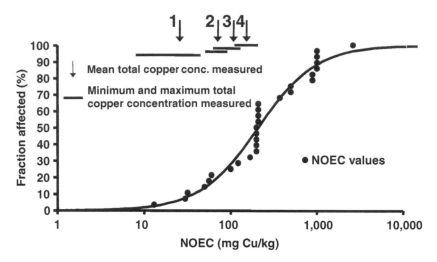

FIGURE 9.8 SSD curves for copper chronic toxicity (individual points are NOECs) of terrestrial organisms. The lines represent the results of the logistic regressions on these data. Arrows indicate the geometric mean copper concentrations of the copper treatment levels in the field study; the stripes indicate the minimum and maximum values measured at different pH levels in separate replicates (1 = nominal treatment equivalent to an original application of 0 kg Cu/ha 10 years before sampling; 2 = 250 kg/ha; 3 = 500 kg/ha; 4 = 750 kg/ha).

500, or 750 kg Cu/ha) and pH manipulation (4.0, 4.7, 5.4, and 6.1). Each combination of the two treatments was applied to eight agricultural plots (6 × 11 m), resulting in a total of 128 plots. The nematode community was sampled 10 years after the treatment to investigate the results of long-term exposure.

In contrast to the example given in Section 9.4.3, exposure measurements were made at the time of sampling. Table 9.4 presents the measured total and the 0.01 M CaCl$_2$-extractable Cu concentrations (in mg Cu/kg dry wt) in the samples, of which the latter is often assumed to be associated with the copper fraction available for uptake by organisms. Obviously, low soil pH is associated with lower extractability of copper.

Evident effects of both the copper treatment and the pH were found (Korthals et al., 1996), indicating adverse effects of copper on nematodes, and indicating that these effects are strongly influenced by the pH. Soil acidity probably influenced the response of the nematodes to copper, but probably also induced effects when no copper was added, as shown in the copper control treatments. For confirmation of the SSD concept, further analyses are needed to disentangle the effects of copper and pH and to try to calculate a NOEC$_{Community}$, taking into account the pH regime. Furthermore, this NOEC has to be expressed as a total copper concentration, because the SSD curve could only be derived using total copper concentrations. To these ends, multivariate techniques, principal component analysis (PCA, see Ter Braak, 1995; Kedwards et al., 1999), and Monte Carlo permutation tests were applied.

Quantification of pH and Copper Effects. PCA was used to display the combined effects of Cu and pH on the nematode community. PCA, in short, results in

TABLE 9.4
Geometric Mean Total (A) and 0.01 *M* CaCl$_2$-
Extractable (B) Copper Concentration
(mg Cu/kg dry wt) for the Upper 10 cm
of Soil for All Treatment Combinations

A.	Soil pH			
Total Cu	4.0	4.7	5.4	6.1
0	24	25	25	25
250	64	77	73	64
500	99	104	106	118
750	133	150	158	167

B.	Soil pH			
Extractable Cu	4.0	4.7	5.4	6.1
0	0.8	0.4	0.2	0.2
250	3	1.8	0.8	0.5
500	7.6	3.2	1.7	1
750	13.9	6.2	2.4	1.4

a diagram providing an optimal two-dimensional summary of the data. Both sample sites and nematode species are displayed. Sample sites positioned near each other are shown to have a similar community composition, whereas that of the sample sites positioned far apart is shown to be very different. Similarly, species placed near a sample site are shown to be relatively abundant at that site, whereas the species that are placed very far from a sample site show a low abundance (see Ter Braak, 1995, for formal interpretation rules). Figure 9.9 shows the PCA diagram, in which only the passive explanatory variables used in the PCA (arrows) are displayed, along with the treatment means (line connections in order of increasing Cu for each pH level). The first and second axis explained 23 and 9% of the total variance, respectively. The diagram shows that the differences in the nematode community due to the pH treatment are mainly displayed on the first axis and that pH is thus the dominant variable. The differences due to the Cu treatment are displayed on both the first (for the pH 4.0 and 4.7 treatments) and the second (pH 5.4 and 6.1 treatments) axes. This indicates, as expected, that the effects of copper strongly and systematically depend on the soil pH. The total number of nematodes found in the different treatments is strongly correlated with the soil pH. The total nematode density is negatively correlated with the copper level. Since the correlation is strongest for the extractable fraction, this suggests not only that copper is toxic to nematodes, but that the extractable fraction is likely related to actual exposure.

Disentangling Copper and pH Effects and Estimating the NOEC$_{Community}$. To separate the effects of copper from pH effects, a partial PCA was subsequently applied by introducing the pH regime as nominal covariables. Figure 9.10 shows the results. Evidently, the first axis mainly displays the differences between the

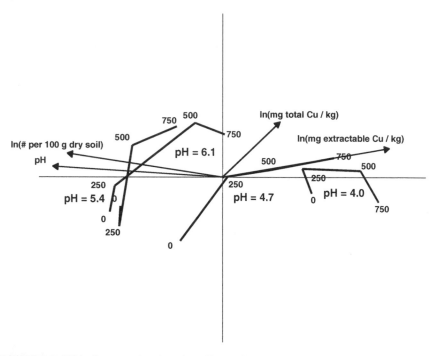

FIGURE 9.9 PCA diagram showing the effects of copper and pH regimes on a nematode community in agricultural soil. The first and second axes explain 23 and 9% of the total variance, respectively. Points indicate (averaged) treatment responses and are linked by lines within each copper treatment, whereas arrows indicate measured variables. Nominal copper (in kg/ha) is used as labels. The species are not shown for the sake of clarity.

copper treatments. The higher the concentration applied during treatment, the farther away the treatment is displaced from the control (left). Effects increase with increasing concentrations of copper. The species found on the left side of the diagram are indicated to be abundant in the controls and scarce in the highest treatment, indicating that they are the most sensitive species. No species are found on the extreme right side of the diagram, so no taxon could abundantly profit from the copper treatment.

To test the level of significance of the community responses to the Cu treatment, two kinds of permutation tests were performed (Verdonschot and Ter Braak, 1994). The first concerned testing the effects of copper against the control for each pH level, and the second concerned testing the effects of copper against the control using the pH as a covariable. The latter tests allow for more permutation possibilities and are the most effective. The first series of tests resulted in significant effects for all 500 and 750 copper treatments, and only one significant outcome for the 250 treatment (Table 9.5). This indicates that the 250 kg/ha treatment represents the turning point between no effects on a community level and significant effects. At the time of sampling, this corresponds with a copper concentration ranging from approximately 64 to 77 mg total Cu/kg dry wt. This level identifies the concentration at which effects are starting to emerge (e.g., slightly higher than the NOEC$_{Community}$) on the basis of total copper concentrations.

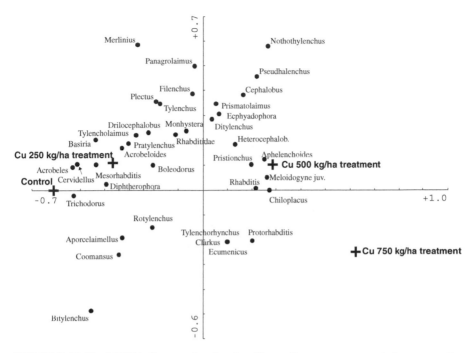

FIGURE 9.10 Partial PCA diagram showing the effects of copper on a nematode community in agricultural soil, separate from the effects of pH. The first and second axes explain 73 and 4% of the variance captured by the copper treatment, respectively.

TABLE 9.5
Results of Permutation Tests (*p*-values) on the Response of a Nematode Community to Copper per pH Level (four first columns) and Using the pH Treatments as a Block Structure (overall test in last column)

Cu at pH	4.0	4.7	5.4	6.1	Overall
250	>0.05	>0.05	>0.05	>0.05	0.032
500	0.005	≤0.001	≤0.001	0.028	≤0.001
750	≤0.001	≤0.001	≤0.001	≤0.001	≤0.001

9.4.4.4 Comparisons and Discussion

The comparison between the laboratory SSD and the field effect data is shown in Figure 9.8. The laboratory SSD curve is indicated, together with the treatment levels. Both the geometric means and the minimum–maximum intervals of the total copper concentrations that were measured are shown. The total copper concentrations in the field show ranges of this kind because of the effect of the pH on the leaching of copper from the topsoil; the lowest total copper concentrations are on the left side of each bar at the lowest pH. The extractable copper (and probably the actual exposure) is highest there (see Table 9.4). Note that the ranges, 10 years after application, show a minimum–maximum overlap as a consequence of leaching at a low pH.

The comparison of the laboratory SSD with the field data shows that the control field soils apparently contained a total copper concentration that relates to the 1.5th and 12th percentile range of the laboratory SSD. This is associated with the background concentration of copper that was originally present, and that likely became available following reduced pH treatments (right side of the control bar). In the control, at a low pH, the effects would be predicted from the laboratory SSD. Although a density reduction in the control was indeed found at a low pH, it is not clear whether this directly (and only) relates to the release of copper or other metals, as direct pH effects cannot be excluded as an alternative explanation.

The second treatment level indicates a concentration slightly higher than the community NOEC. Comparison between the community NOEC (somewhat lower than 64 to 77 mg total Cu/kg) and the laboratory SSD curve, irrespective of pH, suggests that community effects grossly start to occur concurrent with the steep rise of the SSD curve, in the same concentration range. The concentrations of 64 and 77 mg Cu/kg, at which transient effects were recorded, correspond with the 18th and 22nd percentile of the SSD curve based on all NOEC data. Apparently, despite the large uncertainties regarding the representativeness of the laboratory data for the field community with respect to actual exposure, the start of the community response is relatively well predicted by a low percentile of the laboratory SSD curve. We also considered (data not shown) the laboratory-based SSD and the semifield data for which the pH regime were similar (namely, near neutral) to limit exposure uncertainties. Also in this case, the comparisons reinforced the conclusion, although with less data to construct the laboratory SSD.

9.5 GENERAL DISCUSSION

9.5.1 CONFIRMATION

Under ideal conditions, the sensitivity distribution derived from laboratory toxicity data should exactly match the sensitivity distribution of populations in toxicant-exposed field ecosystems. This would follow from a situation in which the input data are completely field-relevant for each separate species, in which no indirect effects of the toxicant are present and in which there are no statistical flaws because of the processing of laboratory and field data. This would be the characteristic of a model that has reached the level of an almost mechanistic description of the outside living world. For the SSD concept, however, this level is beyond reach.

In attempts to establish a match, it should be noted that numerical models in an ERA do not fall into categories such as "proven" or "false" (Suter, 1993), as is the case in natural laws. This poses a problem regarding what can be scientifically done to investigate truth in SSD predictions. According to logical interference (Oreskes et al., 1994), a model like the SSD concept can neither be verified (i.e., its full truth has been demonstrated, meaning full reliability in any case) nor validated (i.e., it has a logically valid construction and, more trivially, its computer codes contain no known or detectable flaw). The former would require a theoretically impossible study design; the latter specifically points at uncertainties that remain in the input data and the data processing that cannot be fully solved as yet. What we tried to do in this chapter was to seek confirmation, which we consider the maximum feasible option.

In accordance with confirmation theory, the greater the number of confirming observations (modeling results match with empirical observations), the more probable it is that the model is not flawed. In serious cases, mismatching can cause the rejection of the whole model. It can also initiate model adaptation when logical omissions or inappropriate input data are identified. Finally, it can lead to the setting of limitations on model application if the model is useful for a certain application, but not for others. In connection with this, one should consider the consequences of the two ways in which the SSD-based results are used (the derivation of EQCs and ERA, respectively). The examples contain information with different consequences, as shown below.

Based on the foregoing, the general discussion of the findings of the examples will subsequently (1) address the uncertainties that one encounters when applying the SSD concept in the six examples, (2) address the scientific lessons learned from the apparent role of these uncertainties and possible adaptations of the general SSD concept, and, thereafter, (3) address the current implications of these lessons for the two frequently used applications of the SSD concept. In view of the differences encountered among the examples concerning aquatic and terrestrial data sets and case studies, the conclusions and recommendations may differ for the various compartments.

9.5.2 Uncertainties and Assumptions of the Concept

Table 9.6 provides a list of uncertainties relating to the SSD concept. This list is an extension of the list already suggested by Van Straalen and Denneman (1989). A subdivision is made of issues relating to the input data, the statistics, and the output data. We have noted that most of these issues are not explicitly addressed in many SSD applications, let alone proven to be valid. For example, it is clear that many issues concerning the input data should be addressed to improve the ecological relevance of the pertinent NOEC input value, as an accurate predictor of the performance of a species in a contaminated field situation. Thus, it should be noted that there is no such thing as "a field situation." This is particularly the case for soils, as each field site has unique characteristics in terms of either the biotic or abiotic conditions. Also, to improve the accuracy in ERA, the laboratory–field extrapolation should take the local site characteristics into account.

As a consequence of our limited knowledge, one tends to make assumptions instead of conducting a proper laboratory–field extrapolation. Among others, Versteeg et al. (1999) summarized the four most important assumptions in the SSD concept.

Concerning uncertainties regarding the input data, the authors mentioned (1) that it is assumed that the laboratory sensitivity of a species approximates its field

TABLE 9.6
List of Uncertainties That May Influence Matching between the Laboratory SSD Curve and the Field SSD Curve, or between the Laboratory SSD and the Magnitude and Type of Field Effects, All Regarding Structural Toxicity Endpoints[a]

SSD-Stage	Issue Types	Specific Issues of Concern	Expected Effects
Input data	Lab–field extrapolation	*Abiotic aspects*	*Expected difference*
		Compound availability	laboratory > field
		Heterogeneity of compound distribution	laboratory < field
		Presence of mixtures	laboratory < field
		Variable exposure conditions	laboratory < field
		Presence of other stress factors	laboratory < field
		Biotic aspects	
		Behavioral avoidance	laboratory < field
		Exposure duration	?, variable
		Relationship tested endpoint with population fitness	?, variable
		Presence of other stress factors	laboratory < field
		Adaptation	laboratory < field
		Costs of adaptation	?, variable
		Species interactions	laboratory < field
			Effect on output
Statistics	Model choice	Choice of statistical model	?
	Data aggregation	Use of taxon-specific toxicity data	Field relevance ↑
		Use of TMoA specific curves	Field relevance ↑
	Data availability	Higher numbers of input data	Field relevance ↑
		Higher numbers of taxa represented	Field relevance ↑
Output data	Field relevance	Integrated effect of the above issues for a compound or field site	Depends on net numerical effect of above expectations

Note: Expectations are given as frequently mentioned statements in the existing literature on the factor mentioned in laboratory-field extrapolation: "variable" = statements variable (e.g., due to different effects observed for different species); "?": expectation not explicitly stated; "↑": output probably has a higher field relevance.

[a] Density, etc., extended from Van Straalen and Denneman, 1989; Posthuma et al., 1998.

sensitivity, which means that the chosen laboratory test endpoint (a *component* of physiological or ecological performance, like "clutch size" or "growth") should represent "realized ecological performance" (meaning: maintaining a viable population as the net effect of *all* performance components). The truth of this assumption, for the terrestrial compartment, is thereby also affected by issues like biological availability of the compound, which may differ between laboratory and field (Sections 9.3.2, 9.4.2, and 9.4.4). It should also be acknowledged that indirect effects (ecological interactions) are not modeled by SSD and therefore SSD may underestimate or overestimate effects occurring in the field.

Concerning the statistical issues, the authors mentioned (2) that it is assumed that the distribution is well modeled by the selected statistical procedure, and (3) that the sample of the species on which the SSD is based is a random one. The latter can be questioned because the currently prescribed biotests have been selected mainly according to practical considerations or to the trophic position of a species rather than to taxonomic differences in sensitivity.

Finally, concerning the output (and its use in environmental decision making), they mentioned the assumption (4) that the protection of the prescribed percentile of species ensures an appropriate protection of field ecosystems.

These uncertainties and assumptions play a key role when seeking an answer to the questions on overall truth in the SSD concept, and they can be used as a guide toward improvements or limitations that should be considered when SSD data are used.

9.5.3 GENERAL SCIENTIFIC LESSONS FROM THE EXAMPLES

9.5.3.1 The Concept Itself

The level of truth in SSDs depends on aspects relating to the concept itself and those relating to the data that are used when applying the concept. As a matter of cause, the concept itself is scrutinized first, by investigating whether the examples contain data that would suggest complete inappropriateness of the method for its purposes. The first example, on the acute effects of chlorpyrifos in an aquatic system, is the most suitable for scrutinizing the concept itself, since the data uncertainties in this example are (by far) the smallest.

This example shows that, although SSD-modeled relationships are nonmechanistic, the field effects (the field SSD) confirm the prediction (the laboratory SSD). However, a prerequisite for this is that one applies an appropriate distribution of species over taxon-specific curves. The knowledge on the TMoA and the taxa potentially exposed served as a guideline in this distribution of data over separate SSDs.

The results of the example in Section 9.3.1 suggest that the SSD concept may be an "intrinsically" correct or accurate approach to predict field effects on a statistical basis. The most accurate predictions of field effects are likely to be generated for chemicals (applied separately) with a specific TMoA that have a fast effect on the target species groups. This means that the TMoA and exposed taxa should be taken into account in SSD modeling, and that the exposure period in the field should be so short that the ecological compensation mechanisms (recolonization, adaptation, etc.) are relatively unimportant. As yet, we are not aware of any other study in which the intrinsic (numerical) correctness of the SSD concept (as quantitatively addressed by full-curve SSD comparisons as executed in this study) has been applied. Most of the arguments stated in literature so far, either in favor of the SSD concept or in denunciation of it, have been supported by, in themselves, valid scientific arguments, but these regard only *parts* of the concept. As a consequence of the specific effects of uncertainties, like EQCs below the natural background concentration of metals (e.g., Hopkin, 1993), many of the arguments denouncing aspects of the SSD approach showed numerical outputs that were evidently wrong or scarcely believable. These arguments generally did not address the possibility that SSD output may, *overall*, be a useful approximation of truth in cases where the uncertainties are small.

9.5.3.2   Dependence on Uncertainties in Input Data, Statistics, or Field Characteristics

SSD-based predictions will likely become less accurate the more the compound and situation characteristics deviate from the ideal situation (contrasting to the example in Section 9.3.1). Decreased accuracy is to be expected when fewer toxicity data are available, the TMoA is less clear, and the more the actual exposure conditions in the field deviate from the laboratory settings. Of the aspects of reduced accuracy listed in Table 9.6, which are substantiated by the examples?

Input Data. Uncertainties in the input data may be manifold, although the central assumption in this respect is that "the laboratory sensitivity of a species approximates its field sensitivity." For the example in Section 9.3.1, Van Wijngaarden et al. (1996) showed that the similarity of SSDs for the arthropods can probably be attributed to the similarity of EC_{50} values of the same arthropod species in laboratory and field conditions; apparently, assumption (1) mentioned above (the laboratory sensitivity of a species approximates its field sensitivity) is confirmed in this example.

Environmental variables, such as temperature, water chemistry, and spatial and temporal exposure variations, did not change the overall response of the species in this example, at least not to a large extent. This is despite the fact that chlorpyrifos showed a sharp vertical concentration gradient in the water column during the first few days (Crum and Brock, 1994).

For various other examples, there are problems regarding the definition of "sensitivity" in the SSD context. In existing SSD literature, sensitivity is often not defined beyond the operational definition of a test endpoint, such as "$NOEC_{Reproduction}$" or LC_{50}. Reviewing the examples regarding the definition of *sensitivity* shows an array of possibilities and associated problems.

First, sensitivity data do not necessarily relate to NOEC values only, as is often thought in the historical context of the choices that are made for the EQC derivation. In view of former decisions in this interface between science and applications, particularly when establishing protective EQCs, chronic NOECs were historically preferred. However, the SSD concept can encompass other test endpoints (e.g., EC_x values) to construct SSDs, and both the availability of input data and the self-explanative and statistical power of the SSD outputs based on, e.g., EC_{50} values may be larger.

Second, the examples in Sections 9.3.2, 9.4.3, and 9.4.4 show that the sensitivity for soil organisms is often a composite characteristic, the value of which is determined by both "intrinsic" (species) sensitivity and environmental characteristics (that influence actual exposure). As an example, the chemical availability of zinc in field soils (as assessed by partition coefficients K_p in contaminated field soils) ranges across various orders of magnitude, mostly as an effect of the soil acidity (Janssen et al., 1997a). The discrepancy between laboratory SSDs and field SSDs observed in Section 9.3.2 (zinc in terrestrial systems) shows that, regarding the terrestrial use of the SSD concept, the key issue of bioavailability should be addressed quantitatively before considering denouncing the SSD concept from the discrepancy. The actual exposure must be defined in terms of the bioavailable fraction of the compound under both laboratory and field conditions. Concerning the example, it should also be noted that one should collect more toxicity data on the populations of interest before a proper evaluation of the predictive power of the SSD concept can be made.

In general, one should take the influence of environmental conditions on species sensitivities into account when constructing laboratory SSDs. The quantitative importance of this is higher for soils than for water. At present, methods to account for environmental influences on exposure grossly exist for organic contaminants (equilibrium partitioning hypothesis), and such methods are under development for metals (e.g., Janssen et al., 1997a,b). If the proposed correction methods seem inappropriate for ERA at a known contaminated site, it may be worth the effort to select only the toxicity data with similar soil characteristics, as mentioned in the terrestrial example with copper (Section 9.4.4), by using only soils with a pH similar to the studied field soil. However, for the derivation of EQCs, when generic protection is the aim of SSD application, all of the toxicity data may be of use, although the resulting SSD for soils should then be interpreted as the "distribution of sensitivities of species in various environments." This SSD can yield a generally protective EQC, as long as both the species and the soil conditions represent the range of ecosystems to be protected.

Third, the example concerning linuron in particular shows that there may be a difference between the laboratory-based toxic effects on a selected physiological performance characteristic (like photosynthesis activity) and the "realized ecological performance," that is, the performance on a population level. It is seldom explicitly shown that both have a one-to-one relationship. Often, the relationship is unclear, and numerically relevant discrepancies have been shown between frequently measured test endpoints and population-level effect endpoints (e.g., Van Straalen et al., 1989; Kammenga et al., 1997). If the protection aim is to assess the long-term impacts of exposure, then it is preferable that test endpoints are selected for each species that has a proven relationship with the population performance.

Fourth, the examples concerning chlorpyrifos and linuron demonstrate most clearly that it may be advantageous to construct SSDs while taking knowledge on the TMoA into account. Various examples show data reinforcing this idea, and the correspondence of the SSD curves to the data may be significantly improved when TMoA-related SSDs are derived, rather than "mixed" ones. All the examples suggest this, although in the case of limited data availability, or an unprecisely known TMoA, the advantage may be guided only by trial and error (for example, carbendazim), or it is limited by the availability of only small data sets. In the latter case, the level of improvement that can be expected from constructing various SSDs is exchanged for a reduced statistical accuracy per SSD.

Often, the availability of input toxicity data is limited, which in turns limits the possibility of constructing TMoA and taxon-specific SSDs. However, the existence of acute-to-chronic ratios in the laboratory may offer the solution to this data limitation (see, among others, Chapter 8). As shown in the examples, the laboratory-based approximation of the acute-to-chronic ratio also seems to hold in the scarce field studies for which acute and chronic data are available (Section 9.4.1).

Statistical Issues. The central assumption here is that "the sample of the species [for the laboratory SSD] is a random or representative one." Randomness can, however, not be tested on the basis of the examples. Randomness should, however, not be confused with the idea that the TMoA should serve as a guideline when constructing SSDs. It has been shown in the examples that the differences between laboratory-derived and field-derived SSD curves increase when the laboratory toxicity values of all taxa are taken into account simultaneously, irrespective of TMoA (see Figure 9.2). TMoA seems to be a key issue in constructing field-relevant SSDs. The example in Section 9.3.1, and to a lesser extent the other examples, strongly suggests that the above-mentioned assumption is likely to be confirmed only when the evaluation is restricted to randomly chosen members of the susceptible group, as were evidently the arthropods in the case of chlorpyrifos.

It should be noted that the laboratory SSD curve in the example given in Section 9.3.1 was based on EC_{50} values for 37 different taxa, and the field SSD curve on 14 taxa. This amount of data is usually not available. It remains to be investigated whether the assumption of "randomness" within TMoA-related groups is confirmed when only little data are available, and whether field SSDs and laboratory SSDs are also similar when limited data are available. In view of this assumption, it is expected that (at least for pesticides) the set of existing laboratory toxicity data is biased to sensitive organisms, because these are normally evaluated first. However, a representative set of field toxicity data in the form of EC_{50} values for different taxonomic groups is scarce, due to the large concentration range needed for their calculation. According to Hamers et al. (1996a), the confirmation of the SSD concept using a broad range of studies has been shown to be particularly problematic for soils.

Output Data. It is explicitly clear from the examples in Sections 9.4.1 to 9.4.4 that the output data of SSDs should be handled with care, in particular with respect to the distinction of direct and indirect effects. The examples suggest that direct field effects are often associated with the lower end of the laboratory SSD curve (when derived in a proper way, and with limited uncertainties), but that indirect effects are unpredictable using SSD. The latter logically follows from the fact that indirect effects are not explicitly modeled in the SSD concept. The prediction of indirect effects, given a prediction on direct effects, should follow from ecological models such as food-web models (e.g., Traas et al., 1998a). Such judgment may show that the actual effects will be greater than the predicted percentiles values, or that ecological compensation mechanisms will likely counteract the direct effects (see also Chapter 15).

Examples 3 and 4 show that, for the aquatic environment, the occurrence of direct effects corresponds with the lower part of the SSD curve (usually above the 5th to 10th percentile) when the exposure regime is the same in the laboratory and the field (acute field NOECs vs. acute laboratory EC_{50} values and chronic field NOECs vs. chronic laboratory NOECs), and when based on sensitive endpoints. The example with linuron additionally shows that there may be a relation between $NOEC_{Community}$ values based on a short-term and long-term exposure, i.e., there may be a percentage value that is protective for a specific field community in the majority of the exposure regimes.

It is evident, however, that the data that have been presented in the examples only concern a few cases, with a few compounds under few conditions. To support the idea that the low percentiles of SSDs relate to the initiation of observable community effects, a more thorough literature review is required. However, in view of the lack of indications otherwise in the aquatic examples studied, the comparisons suggest that (1) it is likely that a fixed (low) percentile value of the SSD curve (e.g., 5th) may be protective for field situations, and (2) that safe threshold levels based on acute exposure regimes (laboratory data) may be used to set protective values for chronic exposures. A similar conclusion on the protection level relating to aquatic EQCs derived by low-percentile SSD values was reached by Versteeg et al. (1999).

For the terrestrial compartment, the problem concerning the correct protective percentile level is more complicated. The terrestrial cases stress the need for additional well-designed field experiments and accompanying laboratory tests, to enable a confirmation study with limited effects of uncertainties. For terrestrial studies, the term *well-designed* pertains mainly to the correct incorporation of the bioavailable fraction of the compound and to the spatial heterogeneity of this fraction, and accounts for both field and laboratory tests. The last example on the effects of copper on a nematode community (Section 9.4.4) shows, however, that if a field experiment is conducted in a suitable way for a real laboratory vs. field SSD comparison, as in the aquatic example, community effects also start to emerge at the lower part of the SSD curve (16th percentile). Similar observations were made in Posthuma et al. (1998a); Posthuma et al. (2001). The observations indicate that, although the evidence is scarce, the effects observed in terrestrial laboratory tests may be predictive for a terrestrial field situation.

Although circumstantial evidence for the aquatic compartment is relatively strong, it is evident for the terrestrial compartment that the SSD concept can neither be firmly denounced nor proved true on the basis of scientific arguments. From the case studies it can be concluded that for soils the SSD concept should be handled with (more than now) explicit attention for exposure concentrations, exposed groups, and TMoA.

9.5.4 CONSEQUENCES FOR THE APPLICATION OF SSDs

The issues mentioned above to improve predictive accuracy lead to various conclusions on data quality (e.g., the ecological relevance of input data should be maximized wherever possible) and data processing (preferred TMoA-related curves). However, daily practice currently poses the problem of a gross data limitation that already exists, without considering these issues. What are the consequences of the lessons learned from the confirmation attempts for the practice of deriving EQCs or in ERA? Are they similar?

9.5.4.1 Consequences for EQC Derivation

It is current practice in the EQC derivation to require a certain taxonomic variability in input data before a single (and thus composite) SSD curve is produced, irrespective of the TMoA (specific or aspecific) of the compound. The requirements are explained in the chapters on EQC derivation in Section IIIA of this volume. These requirements have been formulated in the past, e.g., to avoid a fully dung-worm-determined set of soil quality criteria. Then standardized toxicity assay protocols for soil invertebrates were only available for the worm species *Eisenia andrei* or *E. fetida*. For the derivation of EQCs, the implications of using proven ecologically relevant input data for constructing TMoA-related SSDs would require a major change in one's way of thinking, yet would lead to results that have the firmest scientific foundation. However, because the scientific foundation is not the sole matter to take into account in EQC derivation, less fundamental changes can be envisaged in the interface between science and policy. The maintenance of the unimodal model as it is currently applied to "mixed" data might suffice, in view of the apparent robustness of lower-percentile estimates of hazardous concentrations such as HC_5, also when bimodality in the data is evident (Aldenberg and Jaworska, 2000). This, however, may highly depend on the number of toxicity data available (Newman et al., 2000). In addition, the data set may be subject to manipulation, e.g., by supplying more toxicity data of less sensitive species.

This chapter did not yield conclusive evidence — neither for North America, nor for Europe — to suggest changing the previously chosen 5th percentile as the percentile that likely makes the gross distinction between "non or low community level effects" and the evident likeliness of increased community effects. On the basis of the examples, this conclusion seems appropriate and useful in the daily practice of the derivation of EQCs, and a similar conclusion was also formulated by Versteeg et al. (1999). One should, however, base this percentile on toxicity measures reflecting the persistence of the compound and the duration of the stress.

Finally, it can be recognized that a set of EQCs cannot offer the same level of protection in all imaginable situations, if only one value per compound is given nationwide. For such generic usage, the value is often chosen at such a (low) level that it likely protects communities even under worst-case conditions. If the use of a single generically low value meets difficulties in practice, then systems with nongeneric aspects can be designed. This can be done, e.g., by (once again) looking at the TMoA-specific curves from which an HC value is derived, by taking into account the species groups that occur locally, or by modulating the EQC as a function of the soil characteristics (as is done in the Netherlands; see Chapter 12) or exposure duration (see the example in Section 9.4.1).

9.5.4.2 Consequences for ERA

In the case of ERA, the assessment problem is more clearly defined than in EQC derivation. There is a water body or a soil in which problems are suspected because of the presence of toxic compounds or the future release of toxic compounds. In this case, the available data should be used to construct SSDs that have the greatest resemblance to the local situation. These SSDs do not necessarily or even pertinently need to be similar to those used in the derivation of EQCs. This can be illustrated using Figure 9.4. By using the "mixed" TMoA SSD curve, one would envisage that, at a local ambient concentration of 100 µg/l linuron, the *direct* effects would pertain to (say) only 50% of all the species, while based on separate curves, the herbicide would directly affect approximately 100% of the algae and subsequently cause severe indirect effects due to food shortage for all of the dependent species. The latter effects are not uncovered by the SSD, since the TMoA-specific curves would (in this case) predict an effect ranging between 0 and 20% in invertebrates and fish.

In ERA, the obvious aim is to assess a situation rather than a compound, and this (again) focuses attention on the problem of dealing with the common situation of site contamination with mixtures. In contrast to the summation of PEC/PNEC (predicted environmental concentration/predicted no-effect concentration) ratios, the combination of percentiles (as dimensionless estimates of per-compound effects) certainly seems advantageous (see Chapter 15). However, a confirmation of the rules of calculus for coping with different kinds of mixtures seems an almost impossible task in view of the problems encountered in this study for single-compound cases only. It is therefore preferable in the near future that the presentation of percentiles affected per compound or per TMoA should be shown separately, even when a composite percentile affected is given after mixture calculus. The application of the latter step might unnecessarily enlarge or reduce uncertainties (when mixture-related assumptions are wrong), but a presentation of only one integrated percentile affected certainly hides the basic data needed for logical reasoning on indirect effects not covered in the SSD concept.

ACKNOWLEDGMENTS

The research reported in this chapter was financed by the Dutch Ministry of Agriculture, Nature Management and Fisheries, within the framework of program 359; by the Ministry of Agriculture, Fisheries and Food of the U.K. (MAFF project: Interspecific variation in sensitivity of aquatic organisms); by the Directorate of Soil of the Dutch Ministry of Housing, Spatial Planning and the Environment, within the framework of RIVM project M711701 (1998), "Validity of toxicity and risk limits for soils"; and by RIVM within the framework of RIVM project S/607501 (1999/2000), "Ecosystem risk assessment." Colleagues of the Department of Ecology and Ecology (Vrije Universiteit, Amsterdam) and the Netherlands Organisation for Applied Scientific Research (TNO-MEP) are gratefully acknowledged for their stimulating discussions. Dr. Gerard W. Korthals kindly provided the original data of the copper/pH experiment, and Dr. Geoff K. Frampton the data of the terrestrial chlorpyrifos example.

Section III

Applications of SSDs

This section presents various applications of species sensitivity distributions (SSDs) to illustrate the ways in which SSDs are currently used in practice. It has two subsections, A on derivation of environmental quality criteria and B on ecological risk assessment of contaminated ecosystems. The first subsection starts with a description of the true start of adopting SSD-based methods in an international regulatory context. Further, the subsection presents four examples of implementation of SSDs in the derivation of environmental quality criteria, two from North America, and two from Europe. The second subsection presents six examples of applications of SSDs in ecological risk assessment that illustrate the range of environmental problems that can be tackled by SSD-based methods, alone or combined with other methods. The chapters show how SSDs can function in a range of applications, from formal tiered risk assessment schemes to life cycle assessments of manufactured products. The chapters presented here were meant to present the range of applications of SSDs, without attempting complete coverage of all SSD applications.

A. Derivation of Environmental Quality Criteria

10 Effects Assessment of Fabric Softeners: The DHTDMAC Case

Cornelis J. van Leeuwen and Joanna S. Jaworska

CONTENTS

Abstract — DHTDMAC was a test case for the ecotoxicological risk assessment of chemicals. High political and economic stakes were involved. There is no doubt that the (inter)national discussions on DHTDMAC accelerated the mutual acceptance of the new extrapolation methodologies to assess environmental effects of chemicals based on Species Sensitivity Distributions. These discussions went through a three-step process of (1) confrontation, (2) communication, and (3) cooperation. From a general perspective, the cooperation evolved to European Union (EU)-approved risk assessment methodologies. In a more limited sense, the DHTDMAC case resulted in the development and marketing of a new generation of fabric softeners that are readily biodegradable.

10.1 INTRODUCTION

DHTDMAC, dihydrogenated-tallow dimethyl ammonium chloride (Figure 10.1), a quaternary ammonium surfactant, has been used as a fabric softener, to the exclusion of almost all other substances, in the household laundry rinsing process. Consequently, the chemical has been widely dispersed and may have contaminated the aquatic and terrestrial environment even after sewage treatment. The technical-grade product

1-56670-578-9/02/$0.00+$1.50
© 2002 by CRC Press LLC

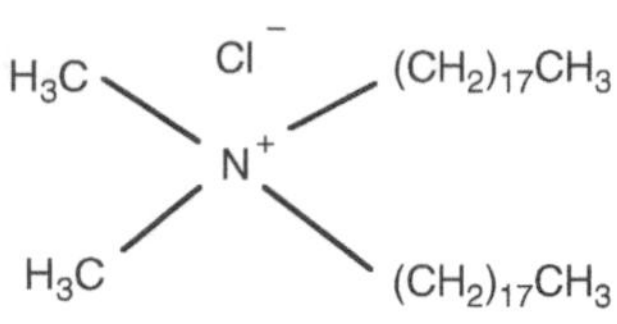

FIGURE 10.1 Chemical structure of DHTDMAC.

contains impurities such as mono- and trialkyl ammonium compounds with varying carbon chain lengths from C_{14} to C_{18}. The C_{18} variety is the most abundant. In the Netherlands about 2000 tonnes/year (as active ingredient) were used in the early 1990s. For the whole of Europe the amount used was approximately 50,000 tonnes/year.

In 1990, the use of fabric softeners became a political issue as a result of a discussion in the Dutch Parliament. This discussion was the result of disagreements between the detergent industry and representatives of the Dutch Ministry of the Environment (VROM) regarding the conclusions of a report prepared by the Dutch Consultative Expert Group Detergents–Environment (DCEGDE, 1988). An alternative risk assessment on DHTDMAC, including the comments of the detergent industry and a reaction by the representatives of VROM, was published in a Dutch journal (Van Leeuwen, 1989). This article catalyzed policy discussions and attracted public attention in the media. In the end, fabric softeners containing DHTDMAC were classified as dangerous for the environment. In the discussions and publications in the 1990s the acronym DTDMAC was most often used, which actually refers to DHTDMAC but with some unsaturated bonds in the alkyl chains.

As a result of risk management discussions between the Netherlands Association of Detergent Industries and VROM (VROM/NVZ, 1992; De Nijs and de Greef, 1992; Roghair et al., 1992; Van Leeuwen et al., 1992a) and to reduce the uncertainties in risk assessment for this type of compound, additional research on DHTDMAC was conducted at the National Institute of Public Health and the Environment (RIVM) in the Netherlands. The studies comprised (1) exposure modeling of DHT-DMAC in the Netherlands, (2) chemical analyses of the substance in effluents, sewage sludge, and surface waters, and (3) assessment of ecotoxicological effects.

The DHTDMAC case was the first case in which extrapolation methodologies based on Species Sensitivity Distributions (SSDs) were applied in risk assessment of industrial chemicals in the European Union (EU). But the DHTDMAC case was more. It was a classical clash between (1) science (ecotoxicological extrapolation methodology and SSDs), (2) environmental policy (the application of the precautionary approach; i.e., how to deal with uncertainties in risk assessment), and (3) the economy (the high market value of the fabric softeners for the chemical industry in the Netherlands and Europe). After this debate, a constructive cooperation followed between industry, VROM and RIVM. This chapter describes these risk evaluations of DHTDMAC and the cooperative actions. Note that the prediction of environmental concentrations is also subject to recent modeling development (e.g., Feijtel et al., 1997; Boeije et al., 2000), but the description of that subject in detail is beyond the scope of this chapter.

10.2   DHTDMAC BEHAVIOR IN WATER

DHTDMAC is a difficult substance to assess because of (1) its extremely low water solubility (<0.52 pg/l), (2) its high adsorptivity (with strong ionic and hydrophobic interactions), (3) its tendency to form complexes with anionic substances and minerals, and (4) the formation of precipitates. As is evident from the high variability in the available data sets, all these properties have implications for the estimation of physicochemical parameters, bioavailability, ecotoxicity, and monitoring. For example, reported sorption coefficients to suspended solids vary between 3,833 and 85,000 l/kg (Van Leeuwen, 1989; ECETOC, 1993a). The rate of decomposition of DHTDMAC greatly depends on the presence of sediment, microbial adaptation, and the type of dosing. Degradation is likely to be slow in surface water, where the concentrations are generally lower than those used in laboratory biodegradation tests. Studies with similar cationic surfactants have led the Dutch Consultative Expert Group Detergents–Environment (DCEGDE, 1988) to the conclusion that degradation will probably fail in surface water that has not been adapted; however, after adaptation the substance becomes inherently, completely biodegradable (ECETOC, 1993a). In 1990, no data were available on the anaerobic degradation in aquatic sediments.

The laboratory results on aquatic toxicity of DHTDMAC are highly dependent on test conditions, sample preparation, and the presence of impurities. Compared with other surfactants, the chemical appears to be relatively toxic to algae when tested in reconstituted water. In natural waters, effects may be observed at concentrations two to three orders of magnitude higher. In reconstituted water, the lowest no-observed effect concentration (NOEC) was observed with *Selenastrum capricornutum* (0.006 mg/l). In treated sewage effluents diluted in river water the NOEC for *Selenastrum* was 20.3 mg/l (Versteeg et al., 1992). Because of the extremely low solubility of DHTDMAC in the reconstituted water experiments, isopropanol was used as a carrier solvent. At this moment there is limited understanding of the physical form of DHTDMAC in this toxicity test. However, opinions have been expressed that this may have a strong impact on the results. In addition, MTTMAC (the derivative mono-tallow trimethyl ammonium chloride) is present in the reconstituted water studies with commercial-grade DHTDMAC and its contribution to toxicity should be taken into account because it is more toxic than DHTDMAC, but readily biodegradable.

10.3   EFFECTS ASSESSMENT OF DHTDMAC

What follows here is a summary of the work done by the Ministry of VROM and RIVM as published in 1992 (Van Leeuwen et al., 1992a). There were two major discussions at that time: (1) a discussion about the validity of the input data (the results of the toxicity tests) and (2) a discussion about the effects assessment (extrapolation) methods. This is why different sets of toxicity data were used (Table 10.1) and why different effect assessment methods were applied on these data (Table 10.2).

The results of the ecotoxicity studies from Roghair et al. (1992), the Dutch Consultative Expert Group Detergents–Environment (DCEGDE, 1988) and Lewis and Wee (1983) are summarized in Table 10.1. The NOECs are nominal concentrations

TABLE 10.1
NOEC Values (mg/l) Used to Calculate MPC and NC for DHTDMAC According to Various Risk Assessment Methods[a]

Species	Set A	Set B
Gasterosteus aculeatus	0.58*	—
Pimephales promelas[b]	0.23	0.053
Chironomus riparius	1.03*	—
Daphnia magna[b]	0.38	—
Lymnaea stagnalis	0.25*	—
Scenedesmus pannonicus	0.58*	—
Selenastrum capricornutum	0.71[c]	0.020[d]
Microcystis aeruginosa	0.21[c]	0.017[d]
Navicula seminulum	—	0.023[d]
Photobacterium phosphoreum	4.27*	—
Nitrifying bacteria	2.31*	—

Note: Set A are tests carried out in surface water, whereas the data presented in Set B are results of toxicity tests carried out in standard water without suspended matter. The data derived from Roghair et al. (1992) are nominal concentrations expressed as the active ingredient as indicated by an asterisk (*). The remaining data are taken from the Dutch Consultative Expert Group Detergents–Environment (DCEGDE, 1988) and Lewis and Wee (1983).

[a] Set A was used for the MPC and NC calculations using methods 1, 3, 4, and 5. Set B was used for the risk assessment according to method 2 (Van der Kooy et al., 1991).
[b] NOECs are based on measured concentrations of DHTDMAC in water. The test with *D. magna* was carried out with DSDMAC (distearyl dimethyl ammonium chloride).
[c] This is an algistatic concentration. The actual NOEC value is therefore lower.
[d] NOEC values for algae were obtained from the EC_{50} values divided by a factor of three.

that have been corrected for the DHTDMAC content of the technical-grade product that was tested. The results show that the algae *Microcystis aeruginosa, Selenastrum capricornutum,* and *Navicula seminulum* are the most sensitive and the bacteria the least sensitive. The differences in toxicity to the fish species *Gasterosteus aculeatus* and *Pimephales promelas,* the midge larva *Chironomus riparius,* the crustacean *Daphnia magna,* and the water snail *Lymnea stagnalis* are very small.

All the tests done with surface water (Table 10.1, Set A) produced higher NOEC values than the tests done with standard water without suspended material (set B). This can be easily explained by the adsorption of cationic surfactants to suspended matter which results in a reduced biological availability. The same has been observed

TABLE 10.2
MPCs and NCs for DHTDMAC Calculated
Using the Data in Table 10.1

Method	MPC	NC
1. Hansen (1989); U.S. EPA (1984b)	21	0.21
2. Van der Kooy et al. (1991)[a]	16	0.16
3. Van Straalen and Denneman (1989)	63	0.63
4. Van de Meent et al. (1990b)[b]	27–100	0.27–1.0
5. Van de Meent et al. (1990b)[b]	18–90	0.18–0.9

Note: NC is 1% of MPC (VROM, 1989b). Values are given in μg/l and represent "total" concentrations of DHTDMAC in surface water.

[a] A suspended matter content of 30 mg/l, a solids–water partition coefficient (K_{sw}) of 8.5×10^4 l/kg, and a correction factor of 0.8 for combined toxicity were used. As dissolved concentrations were not determined in the tests with algae, the lowest NOEC from the study was divided by 3 for the calculations.
[b] The interval represents the confidence interval of the calculated 95% protection level of the species. The upper limit is the median value. The lowest value represents the lower limit of the 95% confidence interval.

in a study by Lewis and Wee (1983), who demonstrated a variation in toxicity to algae of 200 to 2600 μg/l due to varying amounts of suspended matter in the water. Similar observations have been made by Pittinger et al. (1989). Therefore, by carrying out studies with surface water containing suspended matter (1 to 4 mg/l), the reduced biological availability and therefore reduced toxicity was taken into account. It is important to note that the OECD guidelines (OECD, 1984) for mimicking river water suggested much higher values of suspended solids (10 to 20 mg/l) as well as 2 to 5 mg/l of dissolved organic carbon.

The data presented in Table 10.1 were used to calculate the maximum permissible concentrations (MPC) and the negligible concentrations (NC, see Chapter 12 for explanation) for DHTDMAC according to five different effects assessment methods. The results of these calculations are shown in Table 10.2.

Method 1: The method entails to applying a safety factor of ten to the lowest NOEC. It is used in the United States to calculate concern levels (U.S. EPA, 1984b) and by the EU for the risk assessment of new and existing chemicals (CEC, 1996).

Method 2: The Dutch Ministry of Transport and Public Works used this method. It is applied to the lowest NOEC (expressed as dissolved concentration) obtained from experiments carried out with at least the following group of species: fish, crustacean, mollusks, and algae. If nominal concentrations rather than measured concentrations are given, the NOEC should be corrected for this. The combined toxicity of similar substances should also be taken into account. The "dissolved" concentrations in water

are then converted to "total" concentrations (dissolved + adsorbed), assuming a suspended matter concentration in surface water of 30 mg/l and an experimental or
estimated sediment–water partition coefficient (VROM, 1989b).

Method 3: This is the Van Straalen and Denneman (1989) method, reviewed by
the Health Council of the Netherlands (1989) and proposed in Premises for Risk
Management (VROM, 1989b). According to this method, the 95% protection level
for species is calculated under the assumption that the SSD can be described by a
log-logistic function.

Method 4: This is Van Straalen and Denneman's method as modified by Van
de Meent et al. (1990b). In this method the 95% protection level of the species is
calculated using Bayesian statistics. This method also provides a median value and
an estimate of the confidence limits of the 95% protection level.

Method 5: This method is described in detail by Van de Meent et al. (1990b).
It differs from method 4 only in the selection of data:

1. If more than one toxicity study is done with the same species and different
   toxicological criteria, the lowest NOEC is used.
2. If several toxicity studies are done with the same species and the same
   toxicological criterion, the geometric average of these values is used.
3. The lowest NOEC for each taxonomic group (fish, insects, crustaceans,
   mollusks, green algae, blue-green algae, bacteria, etc.) is used. More
   specifically, in the case of DHTDMAC, the tests with *Photobacterium
   phosphoreum* and *G. aculeatus* were excluded (Figure 10.2).

At that time is was concluded that the results of the various risk calculations for
cationic surfactants were remarkably close, and were equivalent to the variation in
the reproducibility of toxicological experiments. It was also not possible to make a

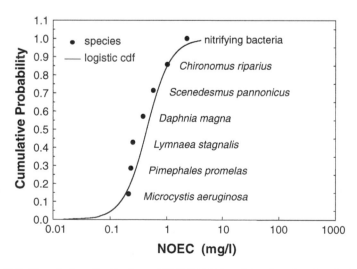

FIGURE 10.2 Cumulative distribution of DHTDMAC toxicity data fitted to logistic model
of log-transformed data.

definitive choice about the preferred extrapolation method. Further international discussions were needed on these methods. From a risk management point of view, the risk assessment problem was solved in a practical manner. To arrive at an MPC it was proposed to use the average of the results of the different extrapolation methodologies and the MPC was set at 50 µg/l (Van Leeuwen et al., 1992a). A year later, the Van Straalen and Denneman method was refined by Aldenberg and Slob (1993) who introduced confidence limits to the HC_5. The method of Aldenberg and Slob (1993) was officially adopted by the Dutch authorities and is still in use today.

10.4 RISK MANAGEMENT

On the basis of single-species laboratory toxicity data and various extrapolation methods, an MPC of 50 µg/l and an NC of 0.5 µg/l (Van Leeuwen et al., 1992a) were derived. In the same assessment, exposure calculations, assuming no degradation, indicated a median concentration of 3 µg/l and a 90th percentile of 45 µg/l. In 1990, concentrations of 6 to 25 µg/l were measured in the Rhine, Meuse, and Scheldt Rivers (Van Leeuwen et al., 1992a). Model predictions indicated that in approximately 30 to 40% of the surface waters considerably higher DHTDMAC concentrations were expected to occur (Van Leeuwen et al., 1992a).

At the same time, industry initiated its own risk assessment, including generation of additional data, and reached different conclusions due to differences in accounting for degradation, solubility, and, most importantly, bioavailability. Using a similar modeling approach as van Leeuwen et al. (1992a) but with in-stream removal, Versteeg et al. (1992) concluded that the median environmental concentration of DHTDMAC was 7 µg/l and the 90th percentile was 21 µg/l. Furthermore, Versteeg et al. (1992) used a novel approach to calculate a chronic "practical" NOEC that addressed the difference between bioavailability in laboratory studies and in the real environment. In these experiments continuous activated sludge units were fed with sewage dosed with DHTDMAC and the chronic toxicity tests were performed with the effluent. The lowest NOEC of 4.53 mg/l, found for *Ceriodaphnia dubia*, demonstrated a marked attenuation of toxicity in the presence of suspensed solids and in the absence of MTTMAC. VROM concluded that this approach transferred the problems from the water phase into suspendend solids and sediments phases and that this could not be the objective of sound environmental policy. On the basis of these results (and disagreements), which were discussed in the Dutch Parliament in spring 1990, the Netherlands Association of Detergent Industries agreed to replace DHTDMAC by chemicals of lower environmental concern within a period of 2 years. By the end of 1990 (Giolando et al., 1995), almost all DHTDMAC had already been replaced by a readily biodegradable substitute: DEEDMAC (diethyl ester dimethyl ammonium chloride).

10.5 DISCUSSIONS ABOUT THE SELECTION OF
SPECIES AND TESTING FOR ECOTOXICITY

The use of extrapolation techniques is based on the recognition that not all species are equally sensitive. Furthermore, it is assumed that by protecting the structure of ecosystems (i.e., the qualitative and quantitative distribution of species) their functional characteristics will also be safeguarded. Differences in sensitivity are the results of true interspecies variability (e.g., uptake-elimination kinetics, biotransformation, differences in the receptors, repair mechanisms), as well as variability in the experimental design (experimental errors and the composition of test media, e.g., pH, salinity, suspended matter, duration of the test, etc.). Van Straalen and Van Leeuwen (Chapter 3) discuss these aspects in more detail. In the case of DHTDMAC, discussions took place regarding all these aspects, i.e., the exclusion of the Microtox test and the exclusion of very susceptible species. It was clear to everybody that the exclusion of very susceptible and very tolerant species had a great impact on the value of the MPC. This extrapolation methodology demonstrated the great influence of aspects that have nothing to do with the statistical extrapolation technique, but everything to do with ecotoxicological test design and practical aspects of testing, e.g., the low solubility of DHTDMAC, the presence of suspended matter in the test media, and the density of algae (bioavailability of DHTDMAC), the presence of toxic impurities (MTTMAC), the minimal number of single-species toxicity tests necessary to predict effects at the ecosystem level, and the selection of these species (ecosystem sampling). The essence was a discussion about the limitations of single-species toxicity testing for predicting effects at the ecosystem level from a theoretical as well as a practical point of view.

10.6 DISCUSSIONS ABOUT THE EXTRAPOLATION METHODOLOGY

Adopting the percentage of "unprotected" species or the implementation of the 95% protection level as the MPC was probably one of the biggest mistakes in communicating extrapolation methodologies to the scientific and regulatory community. Many people interpreted this as if 5% of the species were sacrificed with each chemical that came on the market. This also resulted in discussion in the Dutch Parliament within the framework of the National Environmental Policy Plan (VROM, 1991). In retrospect, it would have been better to promote that the policy objective is to prevent ecosystems against the adverse effects of chemicals and that a "statistical cut-off value" of 5% is needed to obtain the MPC.

At the time of the DHTDMAC debate, the extrapolation methodologies were not yet validated in terms of MPCs derived from field studies. The development of validation activities was certainly stimulated by the DHTDMAC discussion (Emans et al., 1993; Versteeg et al., 1999). Lively discussions were generated on all other aspects, such as the minimal number of entry points (the sample sizes), their representativeness, the shape of the SSDs (e.g., the logistic, normal, and triangular distribution), the statistical verification of the assumed distribution (see Figure 10.2), the ecological relevance of this approach, and the fact that the whole idea was new. However, the main impact was not that this new methodology was scientifically discussed, but that it was applied and could have enormous economic consequences for the detergent industry. It was new and paradigm-breaking.

10.7 COMMUNICATION AND VALIDATION: THE DEVELOPMENT OF A COMMON RISK ASSESSMENT LANGUAGE

The extrapolation methodologies were discussed in three consecutive workshops on application of risk assessment to management of detergent chemicals organized by the Association Internationale de la Savonnerie et de la Detergence (AIS, 1989; 1992; 1995). In the third workshop, the Aldenberg and Slob (1993) model was accepted and applied for effects assessment of linear alkyl sulfonates (LAS), alcohol ethoxylates (AE), alcohol ethoxylated sulfates (AES), and soap to freshwater eco-systems (Van de Plassche et al., 1999a). It was concluded that the uncertainty in the risk quotient was largely due to a lack of chronic toxicity data.

The discussions on extrapolation, which became a real issue because of discussions in Dutch Parliament and because of the DHTDMAC case, were brought to the attention of the OECD Hazard Assessment Advisory Body. The OECD organized a workshop, led by the U.S. EPA in collaboration with VROM, in Arlington, Virginia in 1990. The workshop brought together representatives from industry (mainly the detergent industry), academia, and regulatory agencies. The main outcome of this workshop was the transatlantic agreement on extrapolation factors, the comparison of statistical extrapolation methodologies used in the United States, Denmark, and the Netherlands, and a thorough discussion on the role of field tests including the need to establish a comprehensive database of existing ecosystem studies with the aim of validating the statistical extrapolation methods. It became apparent that the statistical approaches used in Denmark (based on the lognormal distribution: Wagner and Løkke, 1991), the Netherlands (based on the log-logistic distribution: Aldenberg and Slob, 1993) and the United States (based on the log-triangular distribution; Erickson and Stephan, 1988) resulted in very comparable MPCs. The recommendation to compare field tests with extrapolated single-species studies was actively followed by several regulatory agencies, including the detergent industry. The results of this work were later presented at the SETAC workshop on freshwater field tests in Potsdam (Belanger, 1994; Van Leeuwen et al., 1994).

Belanger (1994) reviewed the literature of nine surfactants tested in microcosm, mesocosm, and field tests and compared these results with chronic single-species toxicity. The comparisons he made for LAS and DHTDMAC resulted in conservative estimates of the MPCs when these were based on extrapolated single-species tests. The differences, however, were within one order of magnitude. Van Leeuwen et al. (1994) presented work carried out at the RIVM (Emans et al., 1993; Okkerman et al., 1993). Only very few reliable field studies ($n = 6$) were available at that time. A comparison was made between the MPCs from field and extrapolated single-species studies for 23 data pairs (including the less reliable studies). The MPC based on field studies was generally higher than the MPC based on single-species tests, but the geometric mean of extrapolated single-species MPCs did not differ significantly from the geometric mean of the MPCs based on field studies. This was the case both for the Aldenberg and Slob method and the Wagner and Løkke method with 50% confidence for the extrapolated MPCs. Similar activities were carried out in

TABLE 10.3
Final MPC and NC Expressed as Dissolved Concentrations in μg/l for LAS, AE, AES, and Soap

Surfactant	MPC Based on Single-Species Data	Range of Field NOECs	Final MPC	NC
LAS	320	250–500	250	2.5
AE	110	42–380	110	1.1
AES	400	190–3700	400	5
Soap	27	—	27	0.27

Source: Van de Plassche, E. J. et al., *Environmental Toxicology and Chemistry,* 18, 2653, 1999. With permission.

the cooperative project between VROM and the Dutch Soap and Detergents Association (NVZ) on four major surfactants (LAS, AE, AES and soap). The work was recently published (Van de Plassche et al., 1999a). The comparison of the field studies and extrapolated single-species toxicity data are given in Table 10.3.

Recently, Versteeg et al. (1999) worked further on validation of the extrapolation approach. They summarized the chronic single-species and experimental ecosystem data on a variety of substances ($n = 11$) including heavy metals, pesticides, surfactants, and general organic and inorganic compounds. Single-species data were summarized as genus-specific geometric means using the NOEC or EC_{20} concentration. Genus mean values spanned a range of values with genera being affected at concentrations above and below those causing effects on model ecosystems. Geometric mean model ecosystem effect concentrations corresponded to concentrations expected to exceed the NOEC of 9 to 52% of genera.

This analysis, like the previous ones, suggested that laboratory generated single-species chronic studies can be used to establish concentrations protective of model ecosystems, and likely whole natural ecosystem effects. Further, the use of the "5% of genera affected" level is conservative relative to mean model ecosystem data, but is a fairly good predictor of the lower 95% confidence interval on the mean model ecosystem NOEC. From these validation studies the following conclusions are drawn:

1. Field studies can play an important part in elucidating the role of environmental factors that may modify exposure and susceptibility of species. Field studies, however, do have quite a number of disadvantages related to costs, standardization (mutual acceptance of data), and statistical design. Therefore, these studies should not be seen in isolation from each other, but should be incorporated in a tiered scheme of testing.
2. The refined extrapolation methods of Aldenberg and Slob and Wagner and Løkke seem to be a good basis for determining "safe" values, provided that at least four NOECs, and preferably many more, are available for different taxonomic groups.

3. Available data support the view of Crossland (1990) that "toxicity can be measured in the laboratory and the results of laboratory tests can be extrapolated to the field without great difficulty, provided that the exposure of the organism can be predicted."

10.8 CURRENT ACTIVITIES

Quaternary ammonium compounds continue to be scrutinized in Europe. Despite the significant decrease in use, down to 684 ton/year in 1998 for the whole of Europe, DODMAC (dioctadecyl dimethyl ammonium chloride), the main component in commercial DHTDMAC, is on the EU First Priority List of Existing Chemicals for risk assessment (RA) under the European Existing Chemicals Regulation (793/93). Using EUSES (based on the EU Risk Assessment Technical Guidance Document; CEC, 1996), deterministic RA has been conducted, indicating that the sediment compartment is critical.

ECETOC reassessed DHTDMAC using a probabilistic approach (Jaworska et al., 1999). The outcome of this analysis was that the aquatic and sediment compartments are not a cause for concern at current levels of use. Refinement of the sediment effect assessment would be required to increase the nominal safe usage threshold of this material. Again, MPC uncertainty was determined as the most influential parameter affecting the exposure/effect ratio. The lack of chronic toxicity data delayed reaching consensus between regulators and industry. Currently, additional chronic sediment toxicity data are generated and a final risk assessment report will be published by ECETOC.

11 Use of Species Sensitivity Distributions in the Derivation of Water Quality Criteria for Aquatic Life by the U.S. Environmental Protection Agency

Charles E. Stephan

CONTENTS

Abstract — The U.S. EPA has used three different procedures to calculate percentiles of species sensitivity distributions (SSDs) for use in the derivation of water quality criteria for the protection of aquatic life. In the first procedure, the average of the logarithmic variances for a variety of pollutants was used with the appropriate value from Student's t-distribution to calculate the desired percentile from the mean toxicity value for any pollutant of concern. The second procedure performed extrapolation or interpolation using fixed-width intervals and cumulative proportions. In the third procedure the log-triangular distribution was fit to the four mean acute values nearest the 5th percentile to extrapolate or interpolate to the 5th percentile. This procedure was

1-56670-578-9/02/$0.00+$1.50
© 2002 by CRC Press LLC

the basis for the development of "aquatic life tier 2 values" and was used in the development of the equilibrium-partitioning sediment guidelines for nonionic organic chemicals. During the work with SSDs a variety of recommendations evolved regarding data sets, the level of protection, and the calculation procedure.

11.1 BACKGROUND

Although the U.S. Environmental Protection Agency (U.S. EPA) has not used the term *species sensitivity distribution* (SSD) in its work on water quality criteria for aquatic life, this concept has been important since the agency decided that such criteria should be derived using written guidelines. Prior to the development of written guidelines, aquatic life criteria for the U.S. EPA, such as those in the "Red Book" (U.S. EPA, 1976), were derived using the "ad hoc approach." The ad hoc approach consisted of reviewing all data available concerning the toxicity of a pollutant to aquatic life and then using the data as deemed best by those selected to derive the criterion for that pollutant. The ad hoc approach allowed substantial inconsistencies among aquatic life criteria regarding how toxicity data were used and regarding the level of protection provided. This approach might also be called the "lowest number approach" or the "most sensitive species approach" because most of the criteria were derived to protect the most sensitive species that had been tested. This approach is usually criticized as resulting in criteria that are too low, but the resulting criteria can be too high if, for example, the most sensitive tested species is not as sensitive as one or more untested important species (Stephan, 1985).

11.2 INITIAL WORK

Late in 1977, David J. Hansen at the EPA laboratory in Gulf Breeze, Florida suggested to Donald I. Mount at the EPA laboratory in Duluth, Minnesota that the ad hoc approach for deriving aquatic life criteria for the U.S. EPA should be replaced by an approach based on written guidelines. In the new approach, guidelines describing the methodology to be used to derive aquatic life criteria would be written before criteria were derived so that, to the extent possible, all aquatic life criteria would be derived using the same methodology. The guidelines were intended to provide a systematic means of interpreting a variety of data in an objective, consistent, and scientifically valid manner and were to be modified only if sound scientific information for an individual pollutant indicated the need to do so (U.S. EPA, 1978a). Mount convinced the U.S. EPA to accept the idea of written guidelines and then formed an EPA aquatic life guideline committee consisting of Hansen; Gary A. Chapman at the EPA laboratory in Corvallis, Oregon; John (Jack) H. Gentile at the EPA laboratory in Narragansett, Rhode Island; and Mount, William A. Brungs, and Charles E. Stephan at Duluth.

This guideline committee began work in January 1978 and the first version of the aquatic life guidelines was published for comment in the *Federal Register* a few months later (U.S. EPA, 1978a,b). These guidelines provided that, after a policy decision was made concerning the percentage of species in an aquatic ecosystem that should be protected, "sensitivity factors" would be used to derive criteria that would protect the desired percentage. Because the policy decision concerning the level of protection had not yet been made, example sensitivity factors were derived to protect 95% of the species, using the average of the logarithmic variances for a variety of pollutants and the appropriate value from Student's t-distribution. For example, the sensitivity factor for acute toxicity to freshwater fishes was derived from logarithmic variances that described the dispersions of the acute sensitivities of freshwater fishes to each of several pollutants. The factor was divided into the geometric mean LC_{50} of freshwater fishes for each pollutant for which an aquatic life criterion was to be derived.

Similar factors were calculated for chronic toxicity to freshwater fishes and for acute and chronic toxicity to freshwater invertebrates; comparable factors were calculated for saltwater species when sufficient data were available. The calculation and use of sensitivity factors assumed that species sensitivities to a pollutant were lognormally distributed and that the logarithmic variance of a specific kind of data (e.g., acute toxicity to freshwater fishes) was the same for all pollutants. Despite the limitations that the logarithmic variances were averaged across pollutants and that the data sets for most pollutants contained test results for only a few species, this procedure for calculating sensitivity factors applied normal distribution theory to the data that were available concerning the sensitivities of species to a variety of pollutants. These same factors were used in the second version of the guidelines (U.S. EPA, 1979), where they were called "species sensitivity factors."

11.3 THE 1980 GUIDELINES

The third version of the guidelines was published as part of an announcement of the availability of 64 water quality criteria documents (U.S. EPA, 1980). This version contained two major changes related to the determination of the 5th percentile: minimum data requirements (MDRs) were imposed and a different calculation procedure was specified. The MDRs were imposed to ensure that, at a minimum, the data set contained a specified number and diversity of taxa, including a few specific taxa that were known to be sensitive to a variety of pollutants. Results of acute toxicity tests with a reasonable number and variety of aquatic animals were required "so that data available for tested species can be considered a useful indication of the sensitivities of the numerous untested species" (U.S. EPA, 1980). Tests with taxa that were known to be sensitive to one or more kinds of pollutants were required to increase the chances that the criteria derived from the smallest allowed data sets would be adequately protective. Although this requirement would bias the data sets for some pollutants, the degree of bias would decrease as the number of taxa in the data set increased.

Although freshwater and saltwater species were still considered separately, freshwater fishes and invertebrates were now considered together. Therefore, a single 5th percentile was calculated for acute toxicity to freshwater animals and it was used to protect 95% of the fishes and aquatic invertebrate species in aggregate. The final acute value (FAV) equaled the 5th percentile unless the FAV was lowered to protect an important species. The relationship between the 5th percentile and the FAV was explained as follows (U.S. EPA, 1980):

If acute values are available for fewer than twenty species, the Final Acute Value probably should be lower than the lowest value. On the other hand, if acute values are available for more than twenty species, the Final Acute Value probably should be higher than the lowest value, unless the most sensitive species is an important one.

The special consideration afforded important species was intended to protect a species that was considered commercially or recreationally important even if it were below the 5th percentile.

The procedure used to calculate the 5th percentile in the third version of the guidelines consisted of the following steps:

1. A species MAV (SMAV) was derived for each species for which one or more acceptable acute values were available for the pollutant of concern.
2. The log(SMAV) values were ranked and assigned to intervals with width = 0.11.
3. Each nonempty interval was assigned a cumulative proportion P and a log concentration C.
4. The 5th percentile was computed by linear interpolation or extrapolation using the P and C for the two intervals whose P values were closest to 0.05.

This procedure was later replaced because the calculated cumulative probabilities were positively biased, the procedure was quite sensitive to experimental variation, and the relationship of P to C was not linear in the available data sets. In addition, the interval width of 0.11 was not necessarily always appropriate and a small difference between two data sets could result in a large and/or anomalous difference between the estimates of the 5th percentile (Erickson and Stephan, 1988).

11.4 THE 1985 GUIDELINES

The fourth version of the guidelines was made available for public comment in 1984 (U.S. EPA, 1984c) and in 1985 the fifth (and current) version was published (U.S. EPA, 1985a,b). A slightly more detailed version of the MDRs, which now mentioned amphibians in addition to fishes and aquatic invertebrates, was used in both the fourth and fifth versions of the guidelines. In addition, it was decided that 95% of the taxa should be protected because 90 and 99% resulted in FAVs that seemed too high and too low, respectively, when compared with the data sets from which they were calculated. Of the numbers available between 90 and 99, 95 is near the middle and is an easily recognizable number (Stephan, 1985; U.S. EPA, 1985a). Klapow and Lewis (1979) had used a value of 90%, but they applied it to all available toxicity data for all species.

Both the fourth and fifth versions used a new procedure for calculating the 5th percentile; this new procedure was developed to be as statistically rigorous and appropriate as possible (Erickson and Stephan, 1988). A rationale was developed for assuming that an available set of MAVs is a random sample from a statistical population of MAVs. Therefore, the 5th percentile applies to a hypothetical population of MAVs, not to MAVs for taxa in any particular field situation, which is the basis for the following sentence in the 1985 guidelines (U.S. EPA, 1985a: p. 2):

Use of 0.05 to calculate a Final Acute Value does not imply that this percentage of adversely affected taxa should be used to decide in a field situation whether a criterion is too high or too low or just right.

Examination of available sets of MAVs indicated that the log-triangular distribution fit the data sets better than the tested alternatives and that this distribution should be fit to the four MAVs nearest the 5th percentile because these MAVs provide the most useful information regarding this percentile. Thus, these four MAVs received a weight of 1 whereas all other MAVs received a weight of 0. In addition, to compare procedures, FAVs were calculated for 74 actual data sets using the old procedure (U.S. EPA, 1980), the new procedure, and several modifications of the new procedure. The old procedure produced an FAV that was within a factor of 1.4 of the FAV produced by the new procedure for about 80% of the data sets; of the differences larger than a factor of 1.4, the new procedure produced the higher FAV in about 80% of the cases.

One of the alternative procedures that was tested was very similar to the "sensitivity factor" procedure used in the first and second versions of the guidelines; this and all other procedures that gave equal weight to all of the MAVs were rejected because they resulted in inappropriately low FAVs for positively skewed data sets and inappropriately high FAVs for negatively skewed data sets (Erickson and Stephan, 1988). Further, it was concluded that recommendations concerning calculation of the 5th percentile were the same whether the MAVs were for species or families. Thus, even though MAVs were for species in the third version of the guidelines and for families in the fourth version, MAVs were for genera in the fifth version.

The resulting recommended procedure used extrapolation or interpolation to estimate the 5th percentile of a statistical population of genus MAVs (GMAVs) from which the available GMAVs were assumed to have been randomly obtained. The available GMAVs were ranked from low to high and the cumulative probability for each was calculated as $P = R/(N + 1)$, where $R = $ rank and $N = $ number of GMAVs in the data set. The calculation used the log-triangular distribution and the four GMAVs whose P values were closest to 0.05. This procedure has been applied to data sets for 12 metals, 9 chlorinated pesticides, ammonia, atrazine, chloride, chlorine, chlorpyrifos, cyanide, diazinon, nonylphenol, parathion, pentachlorophenol, and tributyltin (Erickson and Stephan, 1988; U.S. EPA, 1999a,b).

The estimate of the 5th percentile is usually determined by interpolation when the data set contains more than 20 GMAVs but is often determined by extrapolation when fewer than about 20 GMAVs are in the data set. When determined by extrapolation, the estimate is lower than the lowest GMAV, which it should be when the data set is small. However, in some cases in which the four lowest GMAVs in a small data set are irregularly spaced, the estimate might be considerably lower than the lowest GMAV. Of course, increasing the number of GMAVs in the data set decreases concerns regarding extrapolation, in addition to decreasing concerns regarding bias.

11.5 RELATED DEVELOPMENTS

The use by the U.S. EPA of SSDs in the derivation of water quality criteria for aquatic life aided in the development of the concept of "aquatic life Tier 2 values"

(U.S. EPA, 1995a). A minimum of eight GMAVs was required in the 1985 guidelines so that the four GMAVs used in the calculation of the 5th percentile would all be below the 50th percentile to limit the amount of extrapolation. In some situations, however, it is desirable to be able to derive statistically sound aquatic life benchmarks when data are available for fewer than eight genera of aquatic organisms. The Tier 2 procedure specified that, if the aquatic life data set for a pollutant did not satisfy all eight of the MDRs for calculation of an FAV but did contain a GMAV for one of three specified genera in the family Daphnidae, a secondary acute value (SAV) could be calculated by dividing the lowest GMAV in the data set by a secondary acute factor (SAF), whose magnitude depended on the number of MDRs that were satisfied. Several sets of factors were statistically derived by sampling data sets used in the derivation of aquatic life criteria (Host et al., 1995), and one of these sets was selected for use as SAFs (U.S. EPA, 1995b).

The use by the U.S. EPA of SSDs in the derivation of aquatic life criteria also aided in the development of the equilibrium-partitioning sediment guidelines (ESGs) for nonionic organics (U.S. EPA, 1999c). Normalization was used to determine whether SSDs for individual pollutants differed between freshwater and saltwater taxa and between benthic genera and all of the genera used in the derivation of the FAV in aquatic life criteria. This analysis demonstrated that, for a nonionic organic pollutant, (1) a separate water quality criterion did not have to be derived for benthic organisms, and (2) data sets could be combined for derivation of a single water quality criterion that was applicable to both freshwater and saltwater aquatic life. When test results can be combined for different kinds of species, the data set is larger, which makes it easier to satisfy MDRs, reduces concern about bias, provides a better estimate of the 5th percentile, and allows the resulting benchmarks to have broader application.

11.6 RECOMMENDATIONS CONCERNING DATA SETS

During the work with SSDs the following recommendations evolved regarding the data sets to which SSDs are applied:

1. Each possibly relevant test result should be carefully reviewed to decide whether it should be included in the data set. Some aspects of the review should be organism-specific and some should be chemical-specific. An important caveat is that the review should not unnecessarily reject data for resistant taxa. Because low percentiles are of most interest, "greater than" values are acceptable for resistant species.
2. Selection of the MDRs should address the minimum required number of MAVs, the breadth of the taxa for which data should be available, and whether data should be available for specific taxa that are sensitive to many pollutants.
   a. Selecting the minimum required number of MAVs should take into account the percentile(s) to be calculated. If the minimum required number of MAVs is low, it will increase the probability that low

percentiles will be calculated by extrapolation, which results in bench-marks that have greater uncertainty than benchmarks obtained by inter-polation. However, increasing the minimum required number of MAVs will tend to increase the cost of satisfying the MDRs.

b. If amphibians, fishes, and aquatic invertebrates are to be protected by the same benchmark, the data set should be required to contain test results for all three kinds of animals. For each pollutant, it might be wise to determine whether there is an indication that one particular kind of aquatic animal (e.g., amphibians, benthic organisms) is more sensitive (and therefore less protected) than other kinds of animals.

c. Requiring that the data set include taxa that are known to be sensitive to some pollutants will bias the data set for some pollutants, but will increase the probability that percentiles calculated from small data sets are adequately protective. The amount of bias will decrease as the number of MAVs in the data set increases.

11.7 RECOMMENDATIONS CONCERNING THE LEVEL OF PROTECTION

Also during the work with SSDs the following recommendations evolved regarding the level of protection:

1. Selection of a very low percentile will mean that most benchmarks will be calculated by extrapolation, which will make the numerical value of the benchmark quite dependent on the calculation procedure used.

2. If a species is considered so important that it should be protected even if its SMAV is below the selected percentile, it is probably reasonable to require that the data for such a species be very reliable before a benchmark is lowered to protect that species. In addition to protecting commercially and recreationally important species, the U.S. EPA (1994a) suggested that, on a site-specific basis, it is appropriate also to protect such other "critical species" as species that are listed as threatened or endangered under Section 4 of the Endangered Species Act and species for which there is evidence that loss of the species at the site is likely to cause an unaccept-able impact on a commercially or recreationally important species, a threatened or endangered species, or the structure or function of the aquatic community. Because, for example, adult rainbow trout might be considered "critical" at a site, but rainbow trout eggs might not be con-sidered "critical" at the same site, it might be more appropriate to use the term "critical organism" rather than "critical species."

3. Selection of the percentile should take into account such implementation issues as whether one benchmark will be used to protect against both acute and chronic effects or whether one benchmark will be used to protect against acute effects and another benchmark will be used to protect against

chronic effects. In addition, decisions concerning the level of protection should take into account the way in which the benchmark will be used. For example, will the benchmark be used as a concentration that is not to be exceeded at any time or any place? If exceedences are allowed, will the magnitude, frequency, and/or duration of exceedences be taken into account?

4. Decisions concerning acceptable levels of protection are neither toxicological nor statistical decisions; such decisions should be made by risk managers, not risk assessors. Nevertheless, because a risk management decision is more likely to be appropriate if it is based on a good understanding of the relevant issues concerning risk assessment, toxicologists and statisticians should try to ensure that risk managers understand the relevant issues concerning use of SSDs. For example, statisticians and toxicologists should carefully explain to risk managers that, regardless of how it is selected, a percentile in a hypothetical population of MAVs is not likely to correspond to the same percentile in a population of MAVs for taxa in a specific field situation or across a range of field situations for the following reasons:

   a. The organisms used in a toxicity test might not have been the age or size of the species that is most sensitive to the pollutant of concern. Thus, the SMAV might not adequately protect the species.

   b. The SMAVs available for a genus might not be good representatives of the genus and so the GMAV might be biased low or high. Thus, the GMAV might overprotect or underprotect the genus.

   c. If the MDRs require taxa that are known to be sensitive to some kinds of pollutants, data sets for some pollutants are likely to be biased toward sensitive species, but the degree of bias is likely to decrease as the number of MAVs in the data set increases.

   Unless species are selected from a field population using an appropriate procedure (e.g., using random or stratified random sampling), use of the resulting benchmark(s) to protect field populations requires a leap of faith that the distribution of the sensitivities of tested species is representative of the distribution of the sensitivities of field species.

5. If it is possible that the selected level of protection might vary from one risk manager to another or from one body of water to another, statisticians and toxicologists can provide flexibility in two ways:

   a. Provide concentrations that correspond to a variety of percentiles that might be selected.

   b. Provide an equation that is believed to fit acceptably over the range of percentiles that might be selected.

6. Statisticians and toxicologists should also make it clear that use of SSDs in the derivation of aquatic life benchmarks rests on the assumption that selecting a percentile is an appropriate way of specifying a level of protection. This is a fundamental assumption regardless of whether the hypothetical population of MAVs does or does not correspond well with MAVs for the group of species in any small or large geographic area.

11.8 RECOMMENDATIONS CONCERNING THE CALCULATION PROCEDURE

In addition, the following recommendations evolved regarding the procedure used to calculate the percentile:

1. The acceptability of a calculation procedure should depend on its statistical properties, not on whether it gives low or high benchmarks on the average.
2. Determining whether the MAVs in the data set should be for species, genera, or families should consider that the higher the taxon, the smaller the number of MAVs that can be derived from an existing set of data. In contrast, the lower the taxon, the more likely that there will be more than one MAV for taxa that are taxonomically similar and therefore are likely to have similar sensitivities.
3. If the MAVs in the data set are for species, at least two important issues should be addressed in the derivation of each SMAV.
   a. Will data quality affect the derivation of the SMAV? For example, will some acceptable data be given more weight than other acceptable data in the derivation of the SMAV?
   b. Will the derivation of the SMAV consider that, on a pollutant-specific basis, different life stages of a species might have different sensitivities?
   If the MAVs in the data set are for higher taxa, these same issues can affect the derivation of MAVs, but an additional issue is, for example: Should a GMAV be derived directly from a combined consideration of all the acute values for the genus or should the GMAV be derived from SMAVs that were derived separately for each species?
4. Because the benchmarks of interest to most risk managers are in the range of the sensitive taxa, it is important that the calculation procedure be appropriate in this range (Erickson and Stephan, 1988). To ensure that the calculation procedure is appropriate in the range of sensitive taxa, the procedure should not allow MAVs for resistant taxa to impact the calculation of low percentiles.
5. Although it would be possible to fit different models to different data sets, such a curve-fitting approach ignores the effect of random variation on data sets. If one model is to be fit to all data sets, a model should be selected to give a good average fit over a range of data sets (Erickson and Stephan, 1988).
6. Even if there are many MAVs in the data set, low percentiles cannot be estimated well if there are large gaps between the MAVs in the range of a percentile of interest.
7. The variation in benchmarks that can result from use of different calculation procedures should be examined by comparing results calculated using two or more reasonably acceptable procedures. Confidence limits calculated using any one procedure do not account for differences between calculation procedures.
8. Because the calculation procedure can only partially overcome the limitations of a small data set, the number of MAVs in the data set should be increased if the uncertainty is too great.

ACKNOWLEDGMENTS AND DISCLAIMER

I thank Gary Chapman, Russ Erickson, Dave Hansen, Don Mount, and several reviewers for many helpful comments. This document has been reviewed in accordance with U.S. Environmental Protection Agency policy and approved for publication. Mention of trade names or commercial products does not constitute endorsement or recommendation for use.

12 Environmental Risk Limits in the Netherlands

Dick T. H. M. Sijm, Annemarie P. van Wezel, and Trudie Crommentuijn

CONTENTS

1-56670-578-9/02/$0.00+$1.50
© 2002 by CRC Press LLC

Abstract — In the Netherlands, environmental risk limits (ERLs) are used as policy tools for the protection of ecosystems. Species Sensitivity Distributions (SSDs) play an important role in deriving ERLs, which are subsequently used by the Dutch government to set environmental quality standards (EQSs) for various policy purposes. This chapter aims to make transparent how the ERLs are derived and for which purposes they are used. The information may thus be useful for interested parties in other countries for developing their own ERLs, by adoption of one or more of the methodologies, or by providing insight into the procedure. The chapter provides an overview of the methodologies that are used for deriving the ERLs. SSDs are preferred over other methods, such as using safety factors. In addition, it will show which type of information is needed as input for SSDs and for deriving ERLs. Reference is made as to where to find the numerical values for both ERLs and EQSs.

12.1 INTRODUCTION

12.1.1 FOCUS, AIM, AND OUTLINE

The focus of this chapter is on deriving environmental risk limits (ERLs) for the protection of ecosystems in the Netherlands and the use of species sensitivity distributions (SSDs) in this procedure. ERLs are used in the Dutch environmental policy for different purposes. This chapter aims to make transparent how the ERLs are derived and for which purposes they are used. The information may thus be useful for interested parties in other countries in developing their own ERLs, by adoption of one or more of the methodologies, or by providing insight into the procedure.

The major aim of this chapter is to provide an overview of the methodologies that are used for deriving the ERLs. It will show that SSDs are preferred over other methods, such as using safety factors. In addition, it will show which type of information is needed as input for SSDs and for deriving ERLs. Reference is made as to where to find the numerical values for both ERLs and environmental quality standards (EQSs).

In the introduction, the policy background (Section 12.1.2), the relationship between ERLs and EQSs in the Netherlands (Section 12.1.3), and the use of EQSs within the Dutch environmental policy (Section 12.1.4) are described. The methodologies on deriving ERLs are described in Section 12.2. In Section 12.3 some examples are provided and reference is made to the current set of ERLs and EQSs in the Netherlands; the section finishes with some concluding remarks.

12.1.2 POLICY BACKGROUND

As described in the Third National Environmental Policy Plan (VROM, 1998), environmental policy in the Netherlands has taken two tracks: the source-oriented track and the effects-oriented track.

In 1990, the Dutch government formulated the premises that underlie the source-oriented track in the First National Environmental Policy Plan (VROM, 1989a). The following statements apply:

- Unnecessary environmental pollution should be avoided;
- The stand-still principle, i.e., there should be no further damage to the environment adopted; and
- Abatement at the source is preferred over treatment at a later stage.

The aim of the *source-oriented* track is to reduce the emission of (potentially) hazardous substances from point and nonpoint sources. The means to reach that goal is by applying the best available techniques or best environmental practice. The governmental tool used to enforce emission reduction policies is licensing. Local authorities need to inspect the records of any emission.

Source-oriented environmental standards for substances are thus, in essence, emission standards. The source-oriented track, however, cannot adequately deal with the possible adverse effects that substances may have on organisms in their environment. For example, the available techniques may not result in sufficient emission reductions. Another example is sources that are nonpoint, from transboundary deposition, or are difficult to identify. There may even be a burden of historical significance; for example, polychlorinated biphenyls (PCBs) are no longer used in the Netherlands, but still pose a potential problem.

Hence, there is a need for an *effects-oriented* track. The premise of the effects-oriented track is that exposure to substances should not result in "adverse" effects on humans and ecosystems (VROM, 1994). EQSs indicate the level where "adverse" effects may be expected. The concentration of a substance in an environmental compartment thus needs to be related to the EQSs.

Systematic national, regional, and local monitoring programs serve to determine the concentration of substances in the various environmental compartments. If the concentrations of the substances in any environmental compartment exceed the EQSs, (additional) measures need to be taken to identify (other) sources, and to further reduce and control these emissions. Hence, the effects-oriented track may be used to evaluate the source-oriented track.

In addition to evaluating the source-oriented track, the EQSs are used as a policy goal, i.e., levels in the environment that currently exceed the EQSs should be reduced to a level below the EQSs in a given period of time. These actions must be taken to meet the general basis of the overall environmental policy, which is sustainable development (VROM, 1989a). Sustainable development means that the quality of the environment is guaranteed for the next generation and beyond: exposure to substances should not result in "adverse" effects on humans and ecosystems.

12.1.3 ERLs and EQSs in the Netherlands

We have already mentioned ERLs and EQSs. The former are scientifically underpinned and are used as advisory values to set EQSs by the government. The government may take into consideration the advice of consulting parties, such as the Dutch National Health Council or the Dutch Soil Protection Technical Committee, when setting the EQSs. In addition, when setting the intervention value, additional socioeconomic factors are taken into account. Table 12.1 shows the relationship between the different ERLs and EQSs.

The various ERLs are

- The negligible concentration (NC) for water, soil, and sediment;*
- The maximum permissible concentration (MPC) for water, soil, and sediment;
- The ecotoxicological serious risk concentration for soil, sediment, and groundwater (SRC_{ECO}).

Each of the ERLs and the corresponding EQS represents a different level of protection, with increasing numerical values in the order Target Value < MPC** < Intervention Value. The EQSs demand different actions when one of them is exceeded, which is explained elsewhere (VROM, 1994).

The target value and MPC are based on the NC and the MPC, respectively. The target value is based solely on ecotoxicological data. For soil, there is no EQS at the level of the MPC.*** The intervention values for soil and groundwater are based on the lowest value of two underlying ERLs: one based on ecotoxicological data, the other based on human toxicological data and a human exposure model (Figure 12.1) (Swartjes, 1999). In the present chapter, only the derivation of the ecotoxicologically related ERLs are discussed.

EQSs have different regulatory contexts. The intervention value and target value for soil and groundwater have a legal context (VROM, 2000), and indicate that soil cleanup must be considered when the intervention value is exceeded, or indicate the value where negligible effects are to be expected in case of the target value. The

---

* Except for a few volatile substances, ERLs for air are not derived (yet), because of a lack of data, while the available data are difficult to interpret (Rademaker and van Gestel, 1993).
** A complicating factor is that the term MPC is used both as an ERL and as an EQS. For historical reasons, however, the same abbreviation is used.
*** Because the policy-related MPC is a level that needs to be targeted in the short term, the MPC is considered to be less relevant for soil being a static compartment where a dilution of the substance or a reduction in the concentration of the substance is not to be expected in the short term.

TABLE 12.1
ERLs and the Related EQS That Are Set by the Dutch Government for the Protection of Ecosystems

Description	ERL	EQS	Implications of Exceeding EQS and Actions Required
The NC represents a value causing negligible effects to ecosystems. The NC is derived from the MPC by dividing it by 100. This factor is applied to take into account the possible combined effects of the many substances encountered in the environment.	NC (for water, soil, and sediment)	Target Value (for water, soil, sediment and groundwater)	Ecosystems may not be fully protected Actions: • Regular monitoring • If needed, site-specific risk assessment and (further) reduction of emissions
A concentration of a substance in water, soil, or sediment that should protect all species in ecosystems from adverse effects of that substance. Pragmatically, a cutoff value is set at the 5th percentile if an SSD of NOECs is used in the refined effects assessment. This is the HC_5^{NOEC}.	MPC (for water, soil, and sediment)	MPC (for water and sediment)	Ecosystems are not fully protected Actions: • Regular monitoring • If needed, site-specific risk assessment and (further) reduction of emissions
A concentration of a substance in the soil or groundwater at which soil or groundwater functions will be negatively affected or are threatened to be negatively affected. It is assumed that adverse effects on both ecotoxicological functioning and the structure of a soil ecosystem occur when 50% of the species and/or 50% of the microbial and enzymatic processes are possibly affected.	SRC_{ECO} (for soil and groundwater)	Intervention Value (for soil, sediment, and groundwater)	Unacceptable risk to humans or the environment Actions: • Subsequent actual risk assessment • If needed, followed by cleanup of the site • Cleanup must reduce concentrations to level of target value

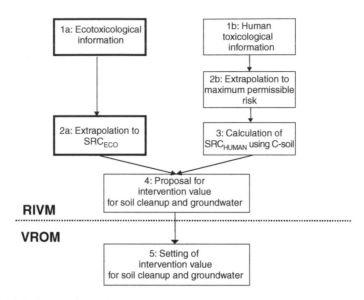

FIGURE 12.1 Schematic outline of the procedure to derive the intervention value for soil, including sediment, and groundwater. The methodology used in the bold boxes is outlined in this chapter. See the appendix to this chapter for the maximum permissible risk. Above the dashed line, the ERLs are derived by RIVM. Below the dashed line the intervention value is set by VROM. C-soil is a software program to determine the SRC_{HUMAN}.

EQSs, MPC, and target value for water and sediment have a nonlegally binding status, but the regional authorities use them effectively for granting permits.

In the following sections the starting points for each of the ERLs will be briefly described. Each of the ERLs relates to a single substance, unless otherwise stated.

12.1.3.1 Ecotoxicological Serious Risk Concentration

The SRC_{ECO} represents a level in the soil, sediment, or groundwater at which soil, sediment, or groundwater functions will be negatively affected or threatened to be negatively affected. It is assumed that adverse effects on both ecotoxicological functioning and the structure of a soil ecosystem occur when 50% of the species and/or 50% of the microbial and enzymatic processes are possibly affected (Denneman and Van Gestel, 1990). The intervention value for soil, sediment, or groundwater can be based on serious risks for the soil, sediment, or groundwater ecosystem but can also be determined by other adverse effects such as on human health (VROM, 1990).

The SRC_{ECO} for soil, sediment, or groundwater is derived in the project "Intervention Values for Soil Clean-up and Groundwater." A schematic outline of the procedure to derive the intervention value for soil, sediment, and groundwater is presented in Figure 12.1.

12.1.3.2 Maximum Permissible Concentration

The MPC is set at a level that should protect all species in ecosystems from adverse effects of a substance (VROM, 1990). Pragmatically, a cutoff value is set at the

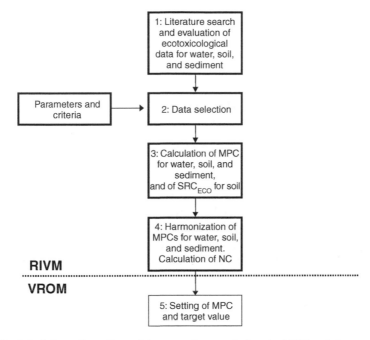

FIGURE 12.2 Schematic outline of the procedure to derive the MPC and the target value. The methodology used in the bold boxes is outlined in this chapter. Above the dashed line the ERLs, i.e., MPC, NC, and SRC_{ECO}, are derived by RIVM. The SRC_{ECO} is calculated not only for soil, but also for sediment and groundwater. Below the dashed line the MPC and target value are set by VROM.

5th percentile if an SSD of NOECs (no-observed-effect concentrations) is used in the refined effects assessment (Section 12.2.4.1). This is the hazardous concentration for 5% of the species (HC_5^{NOEC}) (Van de Meent et al., 1990a; Aldenberg and Slob, 1993).

A schematic outline of the procedure to derive the MPCs is presented in Figure 12.2. MPCs are determined for the individual compartments: water, soil, and sediment. To account for intercompartmental exchange processes that might occur if disequilibrium existed, harmonization of ERLs by equilibrium partitioning is included (DiToro et al., 1991). MPCs for water, sediment, and soil are derived in the project "Setting Integrated Environmental Quality Standards."

12.1.3.3 Negligible Concentration

The NC represents a value causing negligible effects to ecosystems. The NC is derived from the MPC by dividing it by 100. This factor is applied to take into account the possible combined effects of the many substances encountered in the environment (Könemann, 1981; Deneer et al., 1988a; VROM, 1989b). NCs for water, sediment, and soil are derived in the project "Setting Integrated Environmental Quality Standards" on the basis of harmonized MPCs (see Figure 12.2).

12.1.4 EQSs in the Dutch Environmental Policy

When levels of substances in the environment exceed any of the individually derived EQSs, distinct programs follow, which will be briefly explained in the following sections.

12.1.4.1 Intervention Value and Target Value

The intervention value is used for the risk assessment of historically polluted sites and for curative purposes, i.e., as part of the process in deciding when a polluted site needs cleanup. When the intervention value for soil, sediment, or groundwater is exceeded, this implies that there is "serious risk for soil, sediment or groundwater contamination." This causes a potential unacceptable risk to humans or to the environment (VROM, 1990). In principle, there is thus a need for cleanup. However, a subsequent actual risk assessment is required. This must take into account specific local conditions, actual exposure routes, the function and surface area of the threatened soil, sediment, or groundwater, and the magnitude of the contamination. The actual risk assessment determines the urgency to clean up the site. In this chapter, only the derivation of the ecotoxicologically related ERLs is discussed.

Target values indicate the soil, sediment, or groundwater quality at which the risks of adverse effects are considered to be negligible. To prevent unnecessary pollution, target values are embedded into specific regulations.

12.1.4.2 MPC and Target Value

For water, the MPC should not be exceeded. The target values indicate the final level to be reached in the Netherlands. The MPC and target values do not differentiate between ephemeral streams and mainstem rivers; they are generic values. However, in some cases, for example, for some metals and for organotin substances, there is a differentiation for marine water and fresh water. Data from national and regional monitoring, and other measurement programs, are compared with these EQSs. No specific information is given by the authorities on how many cases the monitored concentrations may exceed the MPC. This is evaluated on a case-by-case basis. When the MPC of a substance is exceeded, the compound is regarded as a "substance of concern," and, as such, recommendation for regular monitoring in relevant water bodies or effluents is put in place. Additional monitoring, or site-specific risk assessments, may result in recommendations to the local or regional authorities to reduce point source emissions further. MPCs are also used as a base to set permit emissions to water.

A long-term strategy to reach the target value is the responsibility of regional authorities and, for example, should be laid down in regional water management plans. National or supranational (e.g., European Union) policy objectives may provide further boundary conditions for the regional strategies. Once in every 4 to 8 years the effectiveness of the overall strategy, national and regional, is evaluated in national policy documents of the Ministries of Transport and Public Works, and of Housing, Spatial Planning and Environment (e.g., VROM, 1998).

For sediment, the EQSs include the intervention value, the MPC, and the target value. The MPC and target value are used to evaluate the quality of the sediment compartment and are used in the same way as described for the water compartment. In addition, the intervention value, MPC, and target value are embedded in a system to evaluate the environmental quality of the dredging material from harbors to differentiate between different classes of material. In that case, these values are no longer referred to as EQSs, but as product quality standards.

12.2 DERIVING ENVIRONMENTAL RISK LIMITS

A schematic outline of the methodology to derive ERLs is presented in Figure 12.2, which includes four steps. Steps 1 to 4 (Figure 12.2) are followed separately for each chemical or group of chemicals if the MPC and NC for water, sediment, and soil are derived. Steps 1 to 3 (Figure 12.2) are followed for each substance or group of substances if the SRC_{ECO} for soil is derived. In this section each of the four steps is described.

12.2.1 LITERATURE SEARCH AND EVALUATION (STEP 1)

12.2.1.1 All Environmental Compartments

Sources used for the collection of single-species ecotoxicity data and for data on soil–water and sediment–water partition coefficients, are in-house and external documentation centers and libraries, as well as bibliographic databases (e.g., Biosis, Toxline, and Chemical Abstracts). In Section 12.2.3 several criteria and parameters used in the evaluation are briefly described.

12.2.1.2 Water

For water, chronic and acute toxicity data are sought in the different bibliographic databases, and subsequently tabulated. Distinct tables for freshwater and marine species are produced. Table 12.2 provides the information on relevant experimental conditions and results that are collected for aquatic species. Information on the species is required to relate possible species-specific toxicity. Information on the purity of the substance is required to indicate the effect concentration to the active ingredient. Information on the test conditions is required to evaluate the effect concentration as reported. For example, a nominal concentration is usually an overestimate of the actually measured concentration. The expression of endpoint is important since it is used to further "normalize" all endpoints into one similar endpoint (see Section 12.2.2).

When no or only few toxicity data are available, and the substance exerts its toxicity via a nonspecific toxic mode of action (Verhaar et al., 1992), QSARs (quantitative structure–activity relationships) are used to estimate aqueous toxicity. QSARs for 12 aquatic species of different taxonomic groups are available (Van Leeuwen et al., 1992b), from which NOECs can be derived and subsequently used in deriving ERLs (Van de Plassche and Bockting, 1993).

TABLE 12.2

Information on Relevant Experimental Conditions and Results That Are Collected for Acute or Chronic Toxicity Studies on Aquatic Species

Organism	• <u>Species</u>, Taxon, Strain, Scientific Name, <u>Age, Weight, Size, and Life Stage</u>
Substance	• <u>Purity</u> (or active ingredient, e.g., analytical grade or in percentage)
Test water	• Natural water, tap water, reconstituted water, artificial medium
Test conditions	• Is substance <u>analyzed</u> or is concentration based on <u>nominal</u> concentration?
	• Flow-through, static, or semistatic experiment, etc.
	• <u>Duration</u> of the test (hours, days, or months)
	• <u>Type of endpoint</u> (growth, reproduction, mobility, mortality)
Results	• <u>Expression of endpoint</u> (LC$_x$, EC$_x$, NOEC, etc.)
	• <u>Reference</u> of the study

Note: The <u>underlined</u> parameters are essential. It must be noted that separate procedures exist for marine water and for fresh water. However, when the data indicate that there are no statistical differences between the two, data from marine water and from fresh water are combined to derive a single environmental risk limit for water.

12.2.1.3 Soil

For the terrestrial environment, effects data on microbiological processes and enzymatic activities are sought, in addition to toxicity data on all terrestrial species. The data on microbial and enzymatic processes are commonly expressed as a NOEC or as an EC$_x$ value ($x = 0$ to 100%). Because many different soils are used for the many terrestrial toxicity tests, normalization of the terrestrial test results takes place (Denneman and Van Gestel, 1990). All data on the sensitivity of species are recalculated for a standard soil: for example, a soil that contains 10% organic matter (H) and 25% of clay (L). The following equation is used for normalization of studies with metals (see also Table 12.3):

$$EC_{x(\text{ssoil})} = EC_{x(\text{exp})} \frac{R_{(\text{ssoil})}}{R_{(\text{exp})}} \qquad (12.1)$$

where

$EC_{x(\text{ssoil})}$ = effect concentration; normalized NOEC or LC$_{50}$ for standard soil (or sediment), in mg/kg

$EC_{x(\text{exp})}$ = effect concentration; NOEC or LC$_{50}$ for soil (or sediment) as used in the experiment, in mg/kg

$R_{(\text{ssoil})}$ = reference value for standard soil or sediment ($H = 10\%$, $L = 25$), in mg/kg

$R_{(\text{exp})}$ = reference value for soil or sediment used in the experiment ($H = y\%$, $L = z\%$), in mg/kg

TABLE 12.3
Empirical Reference Lines for Calculating the Background Concentration for Different Dutch Soils and Sediments

Metal or Metalloid	Reference Line for Soil or Sediment	Cb
Antimony (Sb)	3.0	3.0
Arsenic (As)	$15 + 0.4 (L + H)$	29
Barium (Ba)	$30 + 5L$	155
Beryllium (Be)	$0.3 + 0.033L$	1.1
Cadmium (Cd)	$0.4 + 0.007(L + 3H)$	0.8
Chromium (Cr)	$50 + 2L$	100
Cobalt (Co)	$2 + 0.28L$	9.0
Copper (Cu)	$15 + 0.6(L + H)$	36
Lead (Pb)	$50 + L + H$	85
Mercury (Hg)	$0.2 + 0.0017(2L + H)$	0.3
Molybdenum (Mo)	0.5	0.5
Nickel (Ni)	$10 + L$	35
Selenium (Se)	0.7	0.7
Thallium (Tl)	—	1.0
Tin (Sn)	$4 + 0.6L$	19
Vanadium (V)	$12 + 1.2L$	42
Zinc (Zn)	$50 + 1.5(2L + H)$	140

Note: H = percentage of organic matter in soil or sediment (based on dry weight), L = percentage of clay content in soil or sediment (based on dry weight); Cb = the background concentration for standard soil or sediment (in mg/kg dry weight), where $H = 10\%$, and $L = 25\%$.

Source: De Bruijn and Denneman, 1992.

The reference values for metals in soil are based on so-called reference lines (Table 12.3). These reference lines were derived by correlating measured ambient background concentrations from various, relatively unpolluted sites in the Netherlands to the percentage clay and the organic matter content of these soils (Edelman, 1984; De Bruijn and Denneman, 1992).

For organic substances, the literature results are normalized on the basis of the organic matter content:

$$\text{EX}_{x(\text{ssoil})} = \text{EX}_{x(\text{exp})} \frac{H_{(\text{ssoil})}}{H_{(\text{exp})}} \tag{12.2}$$

where

$\text{EC}_{x(\text{ssoil})}$ = effect concentration: normalized NOEC or LC_{50} for standard soil (or sediment), in mg/kg

$\text{EC}_{x(\text{exp})}$ = effect concentration: NOEC or LC_{50} for soil (or sediment) as used in the experiment, in mg/kg

TABLE 12.4
Information on Relevant Experimental Conditions and Results That Are Collected for Acute and Chronic Toxicity Studies on Terrestrial Species

Organism	• Species or process, taxon, strain, age, weight, length, or life stage
Substance	• Purity (e.g., analytical grade or in percentage)
	• For metals and other naturally occurring substances: added concentration corrected for background?
	• Substance added in solution?
Soil	• Type of soil according to American soil-type classification, and sample depth
	• Soil characteristics (organic matter content, clay content, pH, CEC)
Test conditions	• Is substance analyzed?
	• Temperature
	• Soil-to-water ratio
	• Duration of the test (hours, days, or months)
	• Type of endpoint (growth, shoot growth, reproduction, number of young, cocoon production, sperm production, etc.)
Results	• Expression of endpoint (EC$_x$, NOEC, etc.)
	• Recalculation of endpoint in standard soil
	• Reference of the study

Note: The underlined parameters are essential.

$H_{(ssoil)}$ = organic matter content of standard soil (or sediment) ($H = 10\%$), in mg/kg

$H_{(exp)}$ = organic matter content of soil or sediment used in the experiment ($H = y\%$), in mg/kg

It must be noted that for soils with a low organic matter content, i.e., $H < 2\%$, H is set at 2%. Similarly, for soils with a high organic matter content, i.e., $H > 30\%$, H is set at 30%. However, for polycyclic aromatic hydrocarbons (PAHs), for soil with an organic matter content of less than 10% or more than 30%, the percentage of organic matter is set at 10 and 30%, respectively.

When no data on terrestrial species are available the equilibrium partitioning method (EqP method) is applied to derive ERLs for soil (Section 12.2.4.5). In the latter case, soil–water partition coefficients are required.

The results of terrestrial toxicity tests are divided into species and processes. Table 12.4 provides the information on relevant experimental conditions and results that are collected for terrestrial species.

12.2.1.4 Sorption Coefficients

Sorption coefficients (K_p) are derived from batch experiment studies (Bockting et al., 1992; 1993). Only studies in which the humus or organic matter content or organic carbon content is reported are accepted. Organic carbon content is derived from the

TABLE 12.5
Information on Relevant Experimental Conditions and Results That Are
Collected to Use Equilibrium Partition Coefficients, i.e., Both for
Soil–Water and Sediment–Water Partition Coefficients

Substance	• <u>Purity</u> (e.g., analytical grade or in percentage)
	• Organic carbon normalized sorption coefficient (K_{oc})
	• <u>Added concentration corrected for background</u>?
	• Substance added in solution?
	• Check for mass balance
Soil	• Type of soil according to American soil-type classification, and sample depth
	• <u>Soil characteristics</u> (organic matter content, pH, CEC)
Test conditions	• Is substance <u>analyzed</u>?
	• <u>Soil-to-water ratio</u>
	• <u>Equilibration</u> time, before adding the test substance
	• <u>Duration</u> of the test (hours, days, or months)
Results	• <u>Log K_{oc}</u> (for organic substances)
	• <u>Log K_p</u> (for metals and metalloids)
	• <u>Reference</u> of the study

Note: The <u>underlined</u> parameters are essential.

organic matter content by dividing it by 1.7. Table 12.5 provides the information on relevant experimental conditions and results that are collected for sorption coefficients, i.e., both for sediment–water and soil–water sorption coefficients.

12.2.1.5 Sediment

For sediment, in principle, effects data on microbiological processes, enzymatic activities, and benthic species are combined. The data are recalculated for standard sediment, i.e., a sediment that contains 10% organic matter (H) and 25% of clay (L). The same equations for soil are used for the normalization of studies with metals (Equation 12.1) and organic substances (Equation 12.2).

However, in almost all cases no sediment toxicity data are available. Therefore, for sediment, the EqP method is almost always applied to derive ERLs for sediment (Section 12.2.4.5). To apply the EqP method, the sediment–water partition coefficients are required.

When no or only few experimental data are available, the organic carbon normalized partition coefficient, K_{oc}, for organic substances can be estimated using the regression equations provided by Sabljic et al. (1995) and DiToro et al. (1991). Both references give empirical formulas from which a log K_{oc} can be derived from the log K_{ow}. The log K_{ow} is derived from the MEDCHEM (1992) database, where the so-called star values are preferred. If this star value is not available, the ClogP method is used to estimate the log K_{ow}. For metals, the partition coefficients are normalized on standard soil or sediment, i.e., containing 10% organic matter and 25% clay.

12.2.2 DATA SELECTION (STEP 2)

For toxicity data in addition to partition coefficients, a selection is made for the further use in the extrapolation step.

12.2.2.1 Toxicity Data

Toxicity data are selected to obtain one single reliable toxicity value for each compound and species. Exposure of the species will depend on the environmental compartment in which they reside and on the testing guidelines. One value per species is required as input in the subsequent extrapolation method (Section 12.2.4). Therefore, acute and chronic toxicity data are weighed over the species as follows (Slooff, 1992):

- If for one species several toxicity data, based on the same toxicological endpoint, are available, these values are averaged by calculating the geometric mean.
- If for one species several toxicity data, based on different toxicological endpoints, are available, the lowest value of all is selected. The lowest value is determined on the basis of the geometric mean, if more than one value for the same parameter is available.
- In some cases, data for effects on different life stages are available. If from these data one distinct life stage is demonstrated to be the most sensitive, this result will be used in the extrapolation.

Microcosm or mesocosm studies were evaluated in some cases (e.g., for pesticides; Crommentuijn et al., 1997a), but were not yet taken into account — first, because of a lack of guidance to interpret these data for the sake of deriving generic ERLs; second, because, in the case of pesticides, generically derived values were within one order of magnitude with comparable values from mesocosm studies. The use of these studies will be further discussed in future activities. In the Netherlands no hardness-related ERLs are derived. However, a distinction is made between freshwater and marine water, i.e., the procedure to derive ERLs for water follows two routes, one for fresh water, and one for marine water (it must be noted that separate procedures exist for marine water and for fresh water). However, when the data indicate that there are no statistical differences between the two, data from marine water and from fresh water are combined to derive a single environmental risk limit for water.

The following procedure is used to convert available toxicity data into NOECs:

- The highest reported concentration, not statistically different from the control at $p < 0.05$, is regarded as the NOEC.
- The highest concentration showing less than 10% effect is considered to be the NOEC if no statistical evaluation is possible.
- If only a lowest-observed effect concentration (LOEC) is reported, the LOEC is converted into an NOEC as follows:

- LOEC > 10 to 20% effects: NOEC = LOEC/2.
- LOEC ≥ 20% effects and distinct concentration–effect relationship is available: NOEC = EC_{10}.
- LOEC ≥ 20% effects and no distinct concentration–effect relationship is available: LOEC is 10–50% effects: NOEC = LOEC/3; LOEC ≥ 50% effects: NOEC = LOEC/10.
- The "toxische Grenzkonzentration"(TGK) or "toxic threshold" (Bringmann and Kühn, 1977) is regarded as a NOEC.
- If a "maximum acceptable toxicant concentration" (MATC) is presented as a range of two values, the lowest is selected as NOEC; if an MATC is presented as one value, the NOEC = MATC/2.

For soil, toxicity data on terrestrial species as well as on microbial and enzymatic processes may be available. The latter toxicity data describe the performance of a process by an entire microbial community. The process is thus likely to be performed by more than one species. Under toxic stress, the functioning of the process may be taken over by less sensitive species. It is concluded that effects on species and effects on processes are quite different, and the results of ecotoxicological tests with microbial processes cannot be averaged with single-species tests, because of the fundamental differences between them (Van Beelen and Doelman, 1996). Therefore, the data for species and processes are not combined and are selected separately.

For microbial and enzymatic processes more than one value per process is included in the extrapolation method. NOECs for the same process but using a different soil as substrate are regarded as NOECs based on different populations of bacteria or microbes. Therefore, these NOECs are treated separately. Only if values are derived from a test using the same soil is one value selected.

The selection of data will result in a set of toxicity data, which is then used for extrapolation.

12.2.2.2 Partition Coefficients

For organic substances, the mean log K_{oc} from all available experimental partition coefficients is calculated. This value is converted into the K_p value for a standard soil or sediment by multiplying it by 0.0588 (= organic carbon content of the standard soil):

$$K_{p_{(ssoil)}} = K_{oc} * f_{oc} \qquad (12.3)$$

where
$Kp_{(ssoil)}$ = partition coefficient for standard soil, in l/kg
K_{oc} = organic carbon-normalized partition coefficient, in l/kg
f_{oc} = fraction organic carbon of standard soil (= 0.0588)

For metals, empirical distribution coefficients are sought, e.g., for suspended matter–water, sediment–water, and soil–water.

12.2.3 CRITERIA AND PARAMETERS

12.2.3.1 Ecotoxicological Endpoints

With respect to the ecotoxicity studies from the literature, only relevant ecotoxico-logical endpoints are included, i.e., those that affect the species at the population level. In general, these endpoints are survival, growth, and reproduction. For terrestrial species microbial mediated processes and enzyme activities are also taken into account.

The endpoints are commonly expressed as an acute LC_{50} or EC_{50} for short-term tests with duration of 4 days or less, or as a chronic NOEC for long-term tests with a duration of more than 4 days. For microorganisms and algae, NOECs may be derived from experiments lasting less than 4 days. The decision whether the test is acute or chronic depends on the species that is tested. For example, a 16-h test with protozoans is considered an acute test, whereas a fish test lasting 28 days is considered a chronic test.

Occasionally, other ecotoxicological endpoints are accepted. This is the case when the endpoint is considered ecologically relevant, e.g., immobility in tests with daphnids. To date, no methodology is available to evaluate studies in which carcinogenicity and mutagenicity are taken as endpoints. In those cases it is still not clear whether species are affected at the population level.

12.2.3.2 Test Conditions

In general, studies must be conducted according to accepted international guidelines, such as the OECD guidelines (OECD, 1984). If study designs deviate from those guidelines, they may still be accepted as relevant studies. The following quality criteria are taken into account in evaluating the studies:

- The purity of a test substance should be at least 80%.
- Studies using animals from polluted sites are rejected.
- In aquatic studies, concentrations that exceed ten times the aqueous solubility are rejected.
- A maximum of 1 ml/l of solvent used for application of the test substance in aquatic studies is accepted; the OECD guidelines accept a maximum of 0.1 ml/l. Exceeding the OECD value must be mentioned in the footnote of the summarizing tables.
- Solvent use in terrestrial studies may not exceed a value of 100 mg/kg if the solvent was not allowed to evaporate from the soil before the test animals were introduced. If the animals were introduced after evaporation of the solvent, the initial solvent use may reach values of 1000 mg/kg.
- The recovery of the substance in aquatic studies needs to be 80% or more.

12.2.3.3 Secondary Poisoning

Some contaminants accumulate through the food chain and thus exert toxic effects on higher organisms, such as birds and mammals (see also Section 12.2.4.4). An indication for the bioaccumulative potential of a substance is obtained from its physicochemical properties, such as the log K_{ow}, the aqueous solubility, and the bioconcentration factor (BCF). If the substance is potentially bioaccumulative, toxicological data on the sensitivity of birds and mammals and BCFs for worm, fish, and mussel need to be sought. The substances for which this step is required are organic substances with a log $K_{ow} > 3$ and a molecular weight < 600. For metals this is considered on a case-by-case basis.

12.2.3.4 Sorption Coefficients

For sorption coefficients (K_p), only the results from batch experiments are considered reliable (Bockting et al., 1992; 1993). Also studies are considered reliable if performed according to the OECD guidelines (OECD, 1984). Only studies in which the humus or organic matter content or the organic carbon content is reported are accepted. In addition, well-performed field data may also be accepted.

12.2.4 CALCULATING ENVIRONMENTAL RISK LIMITS (STEP 3)

The extrapolation methods that are used for effect assessment, and thus for deriving the ERLs are the "refined effect assessment" (Section 12.2.4.1), and the "preliminary effect assessment" (Section 12.2.4.2). The former, i.e., SSD, is preferred over the latter and applied if chronic toxicity data for more than four different taxonomic groups are available. The latter is applied if chronic data for fewer than four species of different taxonomic groups, fewer than four data on different processes, or only acute data are available. In the case substances are transformed relatively fast, studies on these substances are evaluated on a case-by-case basis.

For naturally occurring substances, such as metals, the "added risk approach" is applied (Section 12.2.4.3). For both organic substances and metals that potentially accumulate through the food chain, the ERLs for direct exposure, based on single-species toxicity data and an ERL for secondary poisoning are derived. The secondary poisoning approach is described in Section 12.2.4.4. If, for soil or sediment, no toxicity data are available, ERLs are derived on the basis of ERLs on aquatic toxicity data and applying the equilibrium partitioning method (Section 12.2.4.5). Probabilistic modeling (Section 12.2.4.6) has recently been used for substances that accumulate through the food chain, such as pcBs.

12.2.4.1 Refined Effect Assessment

The refined effect assessment or statistical extrapolation method is based on the assumption that the sensitivities of species in an ecosystem can be described by a statistical frequency distribution (SSD). This SSD describes the relationship between the concentration of the substance in an environmental compartment and the fraction of species for which the NOEC will be exceeded. This method is applied provided that at least four NOEC values of species of different taxonomic groups are available. For a detailed overview of the theory and the statistical adjustments since its introduction, the reader is directed to the original literature (Kooijman, 1987; Van Straalen and Denneman, 1989; Wagner and Løkke, 1991; Aldenberg and Slob, 1993; Aldenberg and Jaworska, 2000).

The method of Aldenberg and Slob (1993) is used for deriving SRC_{ECO} and MPC values if NOECs for four or more different taxonomic groups are available. For aquatic species, freshwater and marine data are combined if there are no differences in sensitivity between the different aquatic species. This is tested with an unpaired t-test. Prior to this, differences in variance are tested by an F-test. If there are significant differences observed, the unpaired t-test is performed with a Welch correction for differences in variance. When there is a statistically significant difference, distributions for freshwater and marine species are estimated separately.

The statistical extrapolation method assumes that the NOECs that are used for estimating the distribution fit the log-logistic distribution. An advantage of the log-logistic distribution is that it allows the analytical evaluation of the cumulative distribution by integration:

$$PAF(x) = 1 / \left(1 + \exp\left((a - x)/\beta \right) \right) \qquad (12.4)$$

where
  $PAF(x)$ = potentially affected fraction of all possible species at concentration x
  α = mean value of the log-logistic distribution
  β = scale parameter of the log-logistic distribution
  x = $\log(c)$ = logarithm of the concentration

The log-logistic distribution can be characterized by α and β. The α indicates the mean value of the distribution that determines the location of the distribution on the concentration axis $\log(NOEC)$ or $\log(c)$. The β is the scale parameter of the distribution that determines the width or shape of the distribution and is equal to approximately one half times σ ($\beta = \sigma \sqrt{3}/\pi$).

$PAF(x)$ has a value between 0 and 1 and is the fraction of species that have $\log(NOEC)$ values smaller than x. The concentration that corresponds with a 50% protection level or PAF = 0.5 is the HC_{50} and is called the SRC_{ECO}. The HC_{50} can be derived from the same sensitivity distribution as used for deriving the MPC. For the HC_{50} estimation, the frequency distribution is not required, since it is only necessary to derive α, which is the same as calculating the geometric mean of the underlying data. The concentration that corresponds with a 95% protection level or PAF = 0.05 is the HC_5 and is called the MPC. Rewriting Equation 12.4 into Equation 12.5 can derive the HC_5 and HC_{50}:

$$x = \alpha - \beta \ln\left((1 - PAF)/PAF \right) \qquad (12.5)$$

where
  x = HC_5 or HC_{50} for PAF = 0.05 or 0.5, respectively

The current method to calculate the HC_5 and HC_{50} makes use of the lognormal species sensitivity distribution (Aldenberg and Jaworska, 2000) instead of a log-logistic distribution. The differences between these two distributions are small. The advantage of analytical tractability of the logistic distribution is outweighed by the statistical advantage of using the lognormal distribution. It is now possible to routinely calculate the 90% confidence interval of the HC_5 and HC_{50} values and the confidence interval of the PAF at a given environmental concentration (PEC).

The HC_5 and HC_{50} are calculated as (Aldenberg and Jaworska, 2000):

$$\log HC_p = x - k \cdot s \qquad (12.6)$$

where
   x = mean of log-transformed data
   k = extrapolation constant depending on protection level and sample size (Aldenberg and Jaworska, 2000)
   s = standard deviation of log-transformed data

12.2.4.2 Preliminary Effect Assessment

The preliminary effect assessment method is a method in which assessment factors are applied to the toxicity data available. The magnitude of this factor depends on the number and type of toxicity data. The factors and conditions used for deriving MPCs are shown in Tables 12.6 through 12.8 for aquatic MPCs, terrestrial MPCs, and secondary poisoning MPCs, respectively. For deriving MPCs the method is often referred to as the modified EPA method (Van de Meent et al., 1990a).

It was decided to use the assessment factors from the Technical Guidance Document (TGD) of the European Union (CEC, 1996) from now on, because of the harmonization of the project "Setting Integrated Environmental Quality Standards" with the framework of admission of plant protection products and biocides (Kalf

---

TABLE 12.6
Modified EPA Method to Derive the Maximum Permissible Concentration for Aquatic Ecosystems in Case Fewer Than Four Studies on Ecotoxicological Effects on Four Taxonomic Groups Are Available

Available Information	Assessment Factor
Lowest acute $L(E)C_{50}$ available or QSAR estimate for acute toxicity	1000
Lowest acute $L(E)C_{50}$ available or QSAR estimate for acute toxicity for minimal algae/crustacean/fish	100
Lowest NOEC available or QSAR estimate for chronic toxicity[a]	10
Lowest NOEC available or QSAR estimate for chronic toxicity for minimal algae/crustacean/fish	10

[a] This value is compared to the extrapolated value based on acute $L(E)C_{50}$ toxicity values. The lowest is selected.

TABLE 12.7
Modified EPA Method to Derive the MPC for Terrestrial Ecosystems in Case Fewer Than Four Studies on Ecotoxicological Effects on Four Taxonomic Groups Are Available

Available Information	Assessment Factor
Lowest acute $L(E)C_{50}$ available or QSAR estimate for acute toxicity	1000
Lowest acute $L(E)C_{50}$ available or QSAR estimate for acute toxicity of microbe-mediated processes, earthworms, or arthropods and plants	100
Lowest NOEC available or QSAR estimate for chronic toxicity[a]	10
Lowest NOEC available or QSAR estimate for chronic toxicity for three representatives of microbe-mediated processes or of earthworms, arthropods, and plants	10

[a] This value is compared to the extrapolated value based on acute $L(E)C_{50}$ toxicity values. The lowest is selected.

TABLE 12.8
Modified EPA Method for Birds and Mammals to Derive the MPC for Secondary Poisoning in Case Fewer Than Four Studies on Ecotoxicological Effects on Four Taxonomic Groups Are Available

Available Information	Assessment Factor
Fewer than three acute LC_{50} values and no chronic NOECs available	1000
At least three acute LC_{50} values and no chronic NOECs available	100
Fewer than three chronic NOECs available[a]	10
Three chronic NOECs available	10

[a] This value is compared to the extrapolated value based on acute $L(E)C_{50}$ toxicity values. The lowest is selected.

et al., 1999). The schemes with assessment factors used are shown in Table 12.9 for the aquatic compartment and in Table 12.10 for the terrestrial compartment.

Some modifications have been applied to the original schemes for the purpose of the project "Setting Integrated Environmental Quality Standards":

- First, the classification in taxonomic groups is used instead of the original classification in trophic levels, because this classification is used throughout the whole derivation of MPCs.
- Second, for terrestrial data a comparison with equilibrium partitioning is made in all cases of preliminary risk assessment.
- A third minor modification is that as input for one species the geometric mean of several toxicity data based on the same toxicological endpoint is taken instead of the arithmetic mean.

TABLE 12.9
Assessment Factors Provided in the CEC (1996) to Derive a Predicted No-Effect Concentration (similar to MPC) for Water

Available Data	Additional Criteria	MPC Based On	Assessment Factor
$L(E)C_{50}$ values for base set, i.e., algae, *Daphnia*, and fish		$L(E)C_{50}aqua_{min}$	1000
Base set + 1 NOEC (not algae)	NOEC from same taxonomic group as $L(E)C_{50}aqua_{min}$ (fish or *Daphnia*)?		
	Yes	$NOECaqua_{min}$	100
	No. $L(E)C_{50}aqua_{min}/1000 < NOECaqua_{min}/100$	$L(E)C_{50}aqua_{min}$	1000
	No. $L(E)C_{50}aqua_{min}/1000 \geq NOECaqua_{min}/100$	$NOECaqua_{min}$	100
Base set + 2 NOECs	NOEC from same taxonomic group as $L(E)C_{50}aqua_{min}$?		
	Yes	$NOECaqua_{min}$	50
	No	$NOECaqua_{min}$	100
Base set + 3 NOECs	NOECs for algae, *Daphnia*, and fish?		
	Yes	$NOECaqua_{min}$	10
	No. NOEC from same taxonomic group as $L(E)C_{50}aqua_{min}$	$NOECaqua_{min}$	10
	No. NOEC not from same taxonomic group as $L(E)C_{50}aqua_{min}$	$NOECaqua_{min}$	50
Field data or model ecosystems	Expert judgment	As appropriate	Case-by-case

For the aquatic compartment it is required that the base set be complete, i.e., acute toxicity studies for algae, *Daphnia*, and fish. However, for more hydrophobic compounds (log $K_{ow} > 3$) short-term toxicity data may not be representative, since the time span of an acute test may be too short to reach a toxic internal level. In those cases, the completeness of the base set is not demanded and an assessment factor of 100 may be applied to a chronic test, which should not be an alga test if this is the only chronic test available.

The modified EPA method should be used if the base set is incomplete and the tables of the TGD cannot be applied. According to the modified EPA method an assessment factor of only ten should be applied to the lowest NOEC, whereas the highest assessment factor in the TGD method to apply to a chronic NOEC, if the base set is complete, is 100. To eliminate this inconsistency when the base set is incomplete, a factor of 100 or 1000 will be applied to the lowest NOEC or $L(E)C_{50}$, respectively, to derive the MPC.

If data are available for terrestrial species as well as processes, the data are considered separately and MPC are derived for both.

The factors and conditions used for deriving SRC_{ECO} values are shown in Table 12.11. In principle, an acute-to-chronic ratio (ACR) of 10 is always applied

TABLE 12.10

Assessment Factors Provided in the CEC (1996) to Derive a Predicted No-Effect Concentration (similar to MPC) for Soil

Available Data	Additional Criteria	MPC Based On	Assessment Factor
≥ 1 L(E)C$_{50}$		L(E)C$_{50}$terr.$_{min}$	1000
1 NOEC, no L(E)C$_{50}$ values		NOECterr.$_{min}$	100
1 NOEC, $\geq$ 1 L(E)C$_{50}$ values	L(E)C$_{50}$terr.$_{min}$/1000 < NOECterr.$_{min}$/100	L(E)C$_{50}$terr.$_{min}$	1000
	L(E)C$_{50}$terr.$_{min}$/1000 $\geq$ NOECterr.$_{min}$/100	NOECterr.$_{min}$	100
2 NOECs	NOEC from same taxonomic group as L(E)C$_{50}$terr.$_{min}$?		
	Yes	NOECterr.$_{min}$	50
	No	NOECterr.$_{min}$	100
3 NOECs	NOEC from same taxonomic group as L(E)C$_{50}$terr.$_{min}$?		
	Yes	NOECterr.$_{min}$	10
	No	NOECterr.$_{min}$	50
Field data or model ecosystems	Expert judgment	As appropriate	Case-by-case

TABLE 12.11

Assessment Factors Used to Derive the SRC$_{ECO}$ for Soil and Groundwater

Available Data	Additional Criteria	SRC$_{ECO}$ Based On	Assessment Factor
Only L(E)C$_{50}$ values and no NOECs		Geometric mean of L(E)C$_{50}$ values	10
$\geq$ 1 NOEC available[a]	Geometric mean of L(E)C$_{50}$ values/1000 < geometric mean of NOECs/10	Geometric mean of L(E)C$_{50}$ values	1
	Geometric mean of L(E)C$_{50}$ values/1000 $\geq$ geometric mean of NOECs/10	Geometric mean of NOECs	10

[a] This value is subsequently compared with the extrapolated value based on acute LC$_{50}$ toxicity values. The lowest is selected.

to the acute toxicity data to compare acute L(E)C$_{50}$ values with chronic NOECs. The data for the terrestrial compartment are always compared with those derived from the SRC$_{ECO}$ for the aquatic compartment by equilibrium partitioning.

12.2.4.3 The Added Risk Approach

The added risk approach, which is modified from Struijs et al. (1997), is used to take natural background concentrations into account when calculating ERLs for naturally occurring substances. The approach starts with calculating a maximum permissible addition (MPA) on the basis of available data from laboratory toxicity tests. This MPA is considered to be the maximum concentration to be added to the background concentration (Cb). Hence, the MPC is the sum of the Cb and the MPA:

$$MPC = Cb + MPA \qquad (12.7)$$

The MPA is calculated using a similar approach as the MPC for substances having no natural background concentration (Sections 12.2.4.1 and 12.2.4.2). With regard to the bioavailable fraction of the metals in laboratory tests, we assume that the metals that are added to the test medium are fully bioavailable, i.e., the bioavailable fraction of the added metal in the laboratory tests is 100%. The background concentration is thus always part of the ERL, and therefore the ERL cannot approach zero. The implicit assumption is that the background concentration has resulted in the biodiversity of ecosystems or serves to fulfill the need for micronutrients of species in the environment.

The NC is defined as the background concentration (Cb) plus the negligible addition (NA): NC = Cb + NA, where NA = MPA/100.

For practical reasons, the background concentration is added to the MPA or NA, since it thus can be compared with monitoring data. The background concentration and the MPA are independently derived values.

The theoretical description of the added risk approach as described by Struijs et al. (1997) includes a further refinement by allowing the bioavailable fraction of the background concentrations to vary between 0 and 100%. However, to which extent the metals are bioavailable is not relevant, since any potential adverse or positive effect of metals originating from the background is considered desirable, because of its contribution to biodiversity. Besides, at this moment there is insufficient information available to derive the bioavailability of the background concentrations for metals. When bioavailability is theoretically varied, the resulting MPCs do not differ greatly from the MPC when bioavailability is set at 0% (Crommentuijn et al., 2000a).

The added risk approach is not used in deriving SRC_{ECO} values, since it is decided that when the SRC_{ECO} determines the intervention value, it is always used as a trigger value. Site-specific information on the background concentrations is required for determining the urgency.

12.2.4.4 Secondary Poisoning

Most environmental risk limits are derived from concentrations in water, air, sediment, or soil. In the environment some species, in particular those higher in the food chain, are additionally or mainly exposed to substances via their food. Through the food, these species may accumulate potentially toxic substances to high concentrations. This process is called secondary poisoning. Several models have been developed for estimating the risk for secondary poisoning. These models describe simple food chains and relate the effect concentrations for birds and mammals to effect concentrations in water or soil. A key element in these models is that the concentration in the food of the higher organisms is divided by the BCF or biota-to-sediment (or soil) accumulation factor (BSAF) to obtain a concentration in the relevant compartment, e.g., water or soil.

The following two food chains are taken as representative examples. The first is a simple aquatic food chain: water → fish or mussel → fish-eating bird or mammal. The second is a simple terrestrial food chain: soil → worm → worm-eating bird or mammal (Romijn et al., 1993; 1994; Luttik et al., 1993). The MPC derived for secondary poisoning is defined as

$$MPC_{water/soil} = \frac{NEC_{bird/mammal} \cdot f}{BCF_{fish/worm/mussel}} \qquad (12.8)$$

where

$MPC_{water/soil}$ = maximum permissible concentration for water or soil

$NEC_{bird/mammal}$ = no-effect concentration for birds or mammals; this value is obtained from extrapolation of effect data in a way comparable to that for deriving the MPC for direct exposure

$BCF_{fish/worm/mussel}$ = bioconcentration factor for fish, worm, or mussel

f = factor that takes into account the caloric content of the food, which is 0.32 for fish as food, 0.20 for mussel as food, and 0.23 for worm as food

The factor of 0.23 in the formula is used to correct for differences in caloric content of the food used in the laboratory for birds and mammals and the caloric content of worms (Romijn et al., 1993; 1994; Luttik et al., 1993). The simple food chain as proposed by Romijn et al. (1993) is used for deriving ERLs (Van de Plassche, 1994).

The MPC, which includes secondary poisoning, is compared with that for direct exposure. Usually, the lowest MPC of the different MPCs calculated is taken as the respective MPC for water or soil. However, the uncertainties in derivation of both the MPCs and the BCFs or BSAFs are taken into account.

Secondary poisoning is generally not taken into account in deriving SRC_{ECO} values, since it was decided that for the purpose of the intervention values, i.e., pollution at local sites, the geographic areas of interest are too small with regard to protection of ecosystems.

SSDs were also used in taking into account secondary poisoning if a sufficient amount of data were available. The use of probabilistic modeling offers a further tool to do so (Section 12.2.4.6).

12.2.4.5 Equilibrium Partitioning Method

The EqP method was originally proposed by Pavlou and Weston (1984) to develop sediment quality criteria for organic substances. Shea (1988) and DiToro et al. (1991) have described the concept in detail. Three assumptions are made when applying this method. First, it is assumed that bioavailability, bioaccumulation, and toxicity are closely related to the pore water concentrations. Second, it is assumed that sensitivities of aquatic organisms are comparable with sensitivities of organisms living in the sediment. Third, it is assumed that equilibrium exists between the chemical sorbed to the particulate sediment organic carbon and the pore water, and that these concentrations are related by a partition coefficient (K_{oc}).

For metals, a similar approach is adopted, but with two modifications. First, empirical distribution coefficients are used for describing the relationship between the metal sorbed to the particulate sediment or soil, and the pore water. Second, the metal concentrations are not normalized to organic carbon, but are normalized to a standard soil or sediment. One of the current discussions on sediment water quality criteria is that the bioavailability of metals in sediment is highly affected by the acid-volatile sulfide (AVS) content in relation to the simultaneously extracted metals (SEM). If the ratio SEM:AVS is smaller than 1, the metals would not be bioavailable and would not cause any deleterious effects (Swartz et al., 1985; DiToro et al., 1990; 1992; Allen et al., 1993; Ankley et al., 1996). There, thus, seems to be reasons to indicate that the SEM:AVS concept may be used for evaluating *site-specific* toxicity of metals. However, there are a number of comments on the SEM:AVS concept, which limits its use for a *generic* approach. Metals are bioavailable in situations when the SEM:AVS < 1 (Ankley, 1996). The quality of both the SEM data and the AVS data are currently under speculation. The experimentally determined SEM values may underestimate the actual concentration of metals (Cooper and Morse, 1998), whereas the AVS values from pooled sediment samples may overestimate the actual AVS concentration in the top, aerobic sediment layer (Van den Berg, G.A. et al., 1998). In addition, relative to the SEM:AVS concentrations, sediment guidelines based upon dry weight–normalized concentrations were equally or slightly more accurate in predicting both nontoxic and toxic results in laboratory tests (Long et al., 1998). Finally, the proposed SEM:AVS concept requires further research to better implement its significance (Ankley et al., 1996; Ankley, 1996; Mayer et al., 1996):

- For benthic organisms that have a habitat at or slightly above the sediment surface where aerobic conditions prevail, and the AVS content will be very low;
- To protect aquatic systems from metal release associated with sediment suspension;
- For the transport of metals into the food web either from sediment ingestion or the ingestion of contaminated benthos;
- For organisms that are capable of actively extracting substances from sediments, such as polychaetes or bivalves, even under anaerobic conditions of the sediment (Griscom et al., 2000), that may produce ligands for (essential) metals, to accelerate uptake.

These latter findings currently limit the value of the SEM:AVS ratio for generic risk assessment.

The arguments and assumptions to adopt the EqP method are that many terrestrial and benthic organisms will be exposed mainly via the water or pore water. For example, soft-bodied organisms such as earthworms and enchytraeids will be mainly exposed via the pore water. However, other species may take up substances from their food or directly from the soil or sediment, in which case there may be reasons not to adopt the EqP method. The reason for using empirical partition coefficients for metals is that the concentration of a metal in the pore water depends on, e.g., pH, in addition to the organic matter content (Janssen et al., 1997a).

The environmental risk limit for terrestrial and sediment species using the EqP method is derived using the following equation:

$$\text{ERL}\left(\text{sed/soil}_{\text{EP}}\right) = \text{ERL}\left(\text{water}\right) * K_{p_{(\text{ssoil/ssed})}} \tag{12.9}$$

where
  $\text{ERL(sed/soil}_{\text{EP}})$ = environmental risk limit for terrestrial species using the EqP
                     method, in mg/kg
  ERL(water) = environmental risk limit for aquatic species, in mg/l
  $K_{p_{(\text{ssoil/ssed})}}$ = partition coefficient for standard soil or standard sediment, in
                     l/kg

12.2.4.6 Probabilistic Modeling

Probabilistic modeling is a new approach used for deriving ERLs for substances that accumulate through the food chain, i.e., probabilistic food web modeling (Traas et al., 1996; Jongbloed et al., 1996; Van Wezel et al., 2000a). The approach as described by Van Wezel et al. (2000a) includes two major parts. First, all toxicity data for aquatic organisms, mammals, and birds are recalculated into an equivalent toxic concentration in (the organic carbon of) sediments or soils. For that reason biomagnification factors are included to recalculate the concentration from predator to food. BSAFs are included to recalculate the concentration from organism to the organic carbon of sediment or soil. In this way, all types of studies are readily compared on the same concentration axis and are integrated into one MPC. Second, probabilistic techniques are used to incorporate the spread in the data. Thus, probability distributions of sediment or soil concentrations associated with adverse effects are obtained. The contribution of the underlying parameters to the variance in the resulting probability distribution is also determined. MPCs are based on the 5th percentile of the combined distributions. The basis assumption is that the substances in the food chain are in equilibrium with the organic carbon–normalized concentration in the bed sediment.

Probability distributions can be fitted for all the toxicity data together, or for only the data on birds and mammals. This distinction is made to evaluate potential bimodal distributions, e.g., for example, when aquatic species are far less sensitive than birds and mammals. The goodness-of-fit of the probability distributions is tested by the Kolgomorov–Smirnov test. If the s-value is lower than 0.1, the distribution is accepted as a valid basis upon which to build the MPC. If not, the most sensitive probability distribution is chosen as a basis to derive the MPC.

The selected 5th percentiles are transformed into a concentration in mg/kg organic carbon content, and subsequently set as MPC. This procedure is analogous to the procedure as described by Aldenberg and Slob (1993), with the difference that probability distribution is used as the input instead of single values.

For a summary of the whole procedure as described in the preceding text, see Figure 12.3.

12.2.5 HARMONIZATION (STEP 4)

Harmonization of ERLs is undertaken because substances in the environment distribute over the different environmental compartments, after initially emission into one or more of these compartments. For example, PAHs will be mainly emitted to air and water, via combustion sources, but will ultimately accumulate in soils, sediments, vegetation, and biota. Because of these intercompartmental exchange processes, harmonization of the individual ERLs is needed.

Harmonization of SRC_{ECO}s for soil with other compartments is currently not done. Harmonization, however, would take place in the following two cases:

1. The intervention value for groundwater is derived from that of soil, i.e., by dividing by $(10 \times K_p)$, or
2. If insufficient data for the SRC_{ECO} for soil are available, it is derived from ecotoxicity data for water, using the EqP method.

Independently derived ERLs can be harmonized not only for different compartments, but, if human toxicological risk limits are available, these can also be harmonized with ERLs.

When independently derived MPCs for water, soil, and sediment are available, MPCs for water, soil, and sediment are harmonized. This is achieved by calculating the ERL for sediment or soil from the ERL for water, by applying the EqP method (Section 12.2.4.5.). In principle, the lowest value of the independently derived MPC and the MPC resulting from Equation 12.8 is then taken as the harmonized MPC. However, the uncertainties in both the MPCs and the partition coefficient are taken into account.

The NC is calculated as MPC/100 for substances without a natural background, and is calculated as (Cb + MPA/100) for substances with a natural background.

12.3 EXAMPLES AND CURRENT ERLs AND EQSs

This chapter describes the procedure and methodologies that are used in the Netherlands to derive ERLs and their implementation in Dutch environmental policy. In this section a few examples are provided on how to calculate ERLs (Section 12.3.1), i.e., following steps 1 to 4 from Section 12.2. In Section 12.3.2 reference is made to the currently existing ERLs and EQSs. Section 12.3.3 presents some concluding remarks.

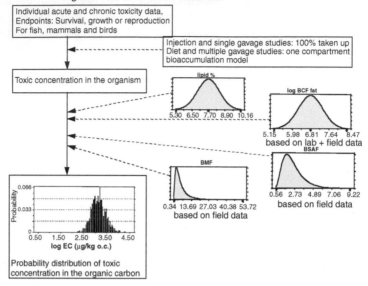

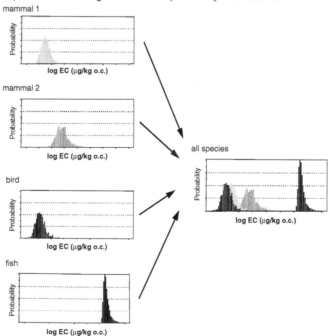

FIGURE 12.3 Overview on the steps taken in the derivation of MPCs using probabilistic modeling with PCBs as an example: Steps 1 (top) and 2 (bottom).

12.3.1 EXAMPLES

For each of the situations as described in Section 12.2.4, examples are provided.

12.3.1.1 Refined Effect Assessment

For deriving the MPC for water for atrazine (Crommentuijn et al., 1997a), a total of 25 studies on chronic toxicity were found to meet the quality criteria. Some of the studies were combined because the same species was used. These studies include data on Cyanophyta ($n = 2$), Algae/Chlorophyta ($n = 6$), Crustacea ($n = 7$, number of species: 4), Insecta ($n = 2$), Pisces ($n = 7$, number of species: 4), Amphibia ($n = 1$). NOEC values ranged from 3 to 2500 µg/l. By using statistical extrapolation with SSDs, the MPC = 2.9 µg/l, which is equal to the MPC, as is the EQS (IWINS, 1997).

12.3.1.2 Preliminary Effect Assessment

For deriving the MPC for water for mevinfos (Crommentuijn et al., 1997a), no chronic ecotoxicity data were available. Acute ecotoxicity data were available for crustaceans ($n = 6$), insects ($n = 1$), and fish ($n = 1$). The LC_{50} and EC_{50} values ranged from 0.16 to 130 µg/l, with a lowest EC_{50} of 0.16 µg/l for *Daphnia pulex*. The resulting MPC for water is calculated by dividing 0.16 by a factor 100 (see Table 12.6), i.e., 0.0016 µg/l. The EQS is set at 2 ng/l (IWINS, 1997). In this case the value was rounded off for simplicity (IWINS, 1997).

12.3.1.3 Added Risk Approach

For deriving the MPC for cadmium in soil (Crommentuijn et al., 1997b), information on the background concentration Cb, and on the HC_5 for soil species and soil processes are required. There were 35 chronic ecotoxicity data on terrestrial species available, which resulted in 13 different entries for refined effect assessment. There were 95 toxicity data on terrestrial microbial processes and enzyme activities available, which resulted in 72 different entries for refined effect assessment. The HC_5 for species and processes were thus derived using statistical extrapolation, resulting in 0.76 and 11 mg/kg, respectively. The lowest value is selected as MPA. The background concentration (Cb) is found to be 0.8 mg/kg. The MPC for soil is thus 1.6 mg/kg (i.e., 0.8 + 0.76). In this case, the value was rounded off for simplicity (IWINS, 1997).

12.3.1.4 Secondary Poisoning

For deriving the MPC for aldrin in water (Van de Plassche, 1994), the HC_5 for aquatic species was derived as well as the HC_5 for secondary poisoning. The HC_5 for aquatic species was derived using statistical extrapolation and was 0.029 µg/l. The HC_5 that was derived taking into account secondary poisoning, and primarily dominated by toxicity data on birds and mammals, was 0.018 µg/l. The MPC for water was selected as 0.018 µg/l. A significantly different value was set by the Dutch government, i.e., 0.9 ng/l (IWINS, 1997). This example clearly shows that the Dutch government has a right to set an EQS different from its underlying ERL. The reason for doing so was that the Dutch government selected a different method for including secondary poisoning.

12.3.1.5   Equilibrium Partitioning Method

For almost all substances, MPCs in sediment have been derived following the EqP method. For example, the MPC for carbaryl in sediment is derived using a sediment–water partition coefficient of 10.7 l/kg and the MPC for carbaryl in water of 0.23 µg/l, and thus results in 2.5 µg/kg (Crommentuijn et al., 1997a). The MPC, as EQS, for sediment is set at 3 µg/kg (IWINS, 1997). In this case the value was rounded off for simplicity (IWINS, 1997).

12.3.1.6   Probabilistic Modeling

Only for PCBs (Van Wezel et al., 2000a) is probabilistic modeling used to derive MPCs. After lipid normalization, toxicity data for birds, mammals, and aquatic organisms were recalculated into equivalent "toxic" concentrations in organic carbon (oc) of soil and sediment. Accumulation through the food chain is thus taken into account. The MPC for PCB-153 is thus derived as 151 µg/kg oc. Currently, no EQS has been set for the PCBs; the advice of the Dutch National Health Council and the Dutch Soil Protection Technical Committee remains outstanding.

12.3.1.7   Harmonization

For harmonization, MPC for soil (or sediment) that are derived following the EqP method is compared with the MPC for soil (or sediment) based on the terrestrial (or benthic) species or processes. For example, the MPC for propoxur for soil based on terrestrial species is 4200 µg/kg. Following the EqP method, the MPC is 0.013 µg/kg (MPC for water of 0.01 µg/l × soil–water partition coefficient of 1.32 l/kg). It is decided to set the MPC for soil based on that derived following the EqP method, despite the large differences in values. The main reason is that the MPC for water is primarily based on ecotoxicity data for insects, and the data set, which was used to derive the MPC for soil, did not include any insects. Since propoxur is an insecticide, it was found important to base the MPC on data on insects.

12.3.2   Current ERLs and EQSs

In the last decade, the National Institute of Public Health and the Environment has derived ERLs for approximately 200 substances, which belong to several chemical classes, such as metals, PAHs, organic substances, pesticides, etc. The Dutch government has used most of them to set EQSs. For some of the recently derived ERLs, EQSs have not been set. One reason is that they have yet to be placed on an agenda. Another reason is that the government has asked for additional, independent advice from the Dutch National Health Council and the Dutch Soil Protection Technical Committee on the methodology that is used to derive ERLs, i.e., for the PCBs. The

TABLE 12.12
References for Numerical Values of the ERLs and EQSs in the Netherlands

ERLs

For MPCs and NCs for:	Crommentuijn et al. (2000b)
• Chlorophenols	
• Several volatile compounds	
• Substances with a potential for secondary poisoning	
• PAHs[a]	
• Pesticides	
For MPCs and NCs for (trace) metals and a metalloid	Crommentuijn et al. (2000c)
For MPCs and NCs for anilines	Reuther et al. (1998)
For MPCs and NCs for PCBs	Van Wezel et al. (2000a)
For MPCs and NCs for a series of trace metals and an organosilicon substance	Van de Plassche et al. (1999b)
For MPCs and NCs for phthalates	Van Wezel et al. (2000b)
For SRC$_{ECO}$ for:	Swartjes (1999)
• Metals	
• Inorganic contaminants	
• Aromatic contaminants	
• PAHs	
• Chlorinated hydrocarbons	
• Pesticides	
• Other pollutants	

EQSs

For MPCs and target values for different substances with ERLs before 1997	Crommentuijn et al. (2000b,c)
For intervention values and target values for different substances with SRC$_{ECO}$ before 2000	VROM (2000)

[a] A more extensive publication on deriving the MPC and NC of PAHs can be found in Kalf et al. (1997).

ERLs can be found in the RIVM reports or in the literature; the EQSs can be found in IWINS (1997) and VROM (2000) or elsewhere (Table 12.12).

12.3.3 CONCLUDING REMARKS

Although uncertainty exists in deriving ERLs, state-of-the-art methods have been used when possible. Some of these uncertainties are simply based on too little information on the ecotoxicity of a given substance. Other uncertainties lie in the fact that there is no conclusive approach available for any of the methods. In addition, one must realize that the ERLs are derived for generic purposes, and, therefore, for site-specific risk assessment, site-specific conditions need to be taken into account. For example, bioavailability of substances in the field may be different from that in

TABLE 12.13

Evaluation of the Underlying Basis for the MPCs That Were Derived before 1997 by RIVM and Served to Set EQSs by the Dutch Government

Method	Water	Soil	Sediment	% of Total
Organic Substances and Pesticides ($n = 150$)				
Refined effect assessment	67	0	0	15
Preliminary effect assessment	85	49	0	30
EqP method	0	100	148	55
Metals ($n = 18$)				
Refined effect assessment	12	5	0	31
Preliminary effect assessment	6	4	0	19
EqP method	0	9	18	50

Sources: Crommentuijn et al., 2000b,c.

laboratory tests and may thus result in misinterpretation of the risk of concentrations found in the field, etc.

An evaluation of the MPCs that were derived before 1998 shows that for organic substances and pesticides 15% of the MPCs was derived using the refined effect assessment method, 30% was derived using the preliminary effect assessment, and 55% was derived following the EqP method (Table 12.13). For metals, 31% of the MPCs was derived using the refined effect assessment method, 19% was derived using the preliminary effect assessment, and 50% was derived using the (modified) EqP method (Table 12.13). We would like to reemphasize that the uncertainty of the MPCs generally increases in the following order: refined effect assessment < preliminary effect assessment < the EqP method. This order is thus based purely on the number of available ecotoxicity studies. Chapman et al. (1998) grossly corroborate this relative order in uncertainties. Many of the ERLs may thus be better based if more ecotoxicological data were available.

One of the aspects that promises better insight in deriving ERLs and their use in environmental policy is the use of probabilistic modeling, which is currently receiving serious attention. A further outlook on future developments is provided in Chapter 22.

Given all the uncertainties and the scientific developments, the Dutch government reevaluates the EQSs every 5 years, if needed, or at shorter time intervals if there is sufficient scientific reason to do so (IWINS, 1997).

Despite the great interest in environmental chemistry and ecotoxicology over the last decades there is still a huge lack of data. The reader is challenged to discuss the presented procedure and methodologies and may use the procedure and methodology as a basis for deriving regional or national environmental quality standards.

APPENDIX: HUMAN TOXICOLOGICAL RISK LIMITS AND INTEGRATION WITH ERLs

In addition to ERLs, human toxicological risk limits are derived in the Netherlands: the human toxicological maximum permissible risk (MPR). In general, in the human toxicological evaluation of chemical substances, a distinction is made between two standard approaches (Janssen and Speijers, 1997). The nonthreshold approach is chosen for compounds that, on the basis of available evidence, are regarded as genotoxic carcinogens. In this case the MPR is defined as the exposure level with an excess lifetime cancer risk of 10^{-4} (VROM, 1989b). For other compounds, a threshold approach is used. In that case the MPR is defined as an acceptable daily intake (ADI) or tolerable daily intake (TDI) (VROM, 1989b). The human toxicological risk limits are integrated with the ERLs in two different ways:

1. In the project "Intervention Values for Soil Clean-up and Groundwater," the intervention value for soil cleanup and groundwater is based on either the human toxicological serious risk concentration (SRC_{HUMAN}) or the ecotoxicological serious risk concentration (SRC_{ECO}) (see Figure 12.1). The SRC_{HUMAN} and the SRC_{ECO} are expressed as a concentration in the soil and are distinctly derived values (Swartjes, 1999). The SRC_{HUMAN} is based on the MPR. The MPR is converted into a concentration in the soil using the model C-soil (Van den Berg and Roels, 1991; Van den Berg, 1994). In principle, the lower of the two, the SRC_{HUMAN} or the SRC_{ECO}, is chosen as proposed for the intervention value.
2. In the project "Setting Integrated Environmental Quality Standards," all MPCs and NCs are harmonized for the different environmental compartments (water, sediment, and soil). For volatile substances, MPCs and NCs have also been harmonized with human toxicological risk limits (Van de Plassche and Bockting, 1993). For these volatile substances, it is assumed that inhalation through air is the predominant route of exposure for humans. Harmonization of the human toxicological risk limits for air with those for soil and water is done using the multimedia fate model Simple-Box (Van de Meent, 1993). On the basis of the MPC for soil and water, the resulting equilibrium concentration in air is calculated and compared with the human toxicological risk limit for air. In principle, the MPC for water and soil has to be adjusted if the corresponding equilibrium concentration in air exceeds the human toxicological risk limit (Van de Plassche and Bockting, 1993; Mennes et al., 1998). However, the uncertainties in the MPCs in the human toxicological risk limits and in the input parameters of the model are taken into account. Harmonization of MPCs for water, sediment, and soil with human toxicological risk limits based on multiroute exposure patterns is the subject of continuing study (Mennes et al., 1995; 1998).

13 A Rank-Based Approach to Deriving Canadian Soil and Sediment Quality Guidelines

Connie L. Gaudet, Doug Bright, Kathie Adare, and Kelly Potter

CONTENTS

Abstract — This chapter discusses approaches used in Canada to incorporate information on species contaminant sensitivity distributions into the derivation of national environmental quality guidelines for the protection of soil- and sediment-associated biota. The current sediment and soil quality protocols for development of guidelines illustrate what is termed here as a "rank-based" approach to national guideline derivation. Although the sediment and soil quality databases differ significantly, they both are based on an interpretation of the distribution of species, and in the case of sediment, community-level sensitivity to contaminants using specified percentiles of no-effects and/or effects databases as an estimate of the concentration of a substance in soil or sediment at which few or no appreciable ecological effects are expected to occur. The rationale and support for this rank-based approach are discussed.

13.1 INTRODUCTION

Since the mid-1980s, Environment Canada has been developing environmental guidelines to protect the quality of Canada's fresh and marine waters (CCREM,

1987). More recently, Canada has released soil and sediment quality guidelines (CCME, 1999). In this chapter we describe efforts to derive soil and sediment quality guidelines based on Species Sensitivity Distributions (SSDs). Development of national guidelines has attempted to meet the needs of environmental managers across Canada and reflects a carefully considered balance between data availability and quality, environmental complexity, and the need to provide generic benchmarks for quality of the soil and sediment environments that will be generally protective of the biological components of sediment and soil systems.

13.2 CANADIAN SOIL QUALITY GUIDELINES

13.2.1 OVERVIEW OF THE CANADIAN APPROACH

The CCME (Canadian Council of Ministers of the Environment) "soil quality criteria" were implemented initially in 1991 and have been applied extensively in the investigation and remediation of contaminated sites since then. In 1996, after several years of discussion across Canadian jurisdictions, the CCME released a national protocol for the development of ecologically based and human health–based soil quality guidelines. Guidelines developed using this protocol were intended to replace the 1991 criteria with scientifically defensible values (CCME, 1996). Canadian Soil Quality Guidelines are currently based on protection of four major uses of land in Canada: agricultural, residential/parkland, commercial, and industrial. Table 13.1 shows ecological receptors and exposure pathways considered important for each land use. This chapter addresses only the interpretation of ecological data, and further information on human health protection can be found in CCME (1996).

Soil quality guidelines for protection of ecological receptors rely on the published results of toxicity tests based on sensitive endpoints for key plant and animal receptors associated with soil. Endpoints considered acceptable for guideline derivation are normally those considered critical to the maintenance of soil-associated plants and animals, such as mortality, reproduction, and growth (Table 13.2). To

---

TABLE 13.1
Receptors and Exposure Pathways Considered in the Derivation of Environmental Soil Quality Guidelines

Land Use	Soil Quality Guidelines for Soil Contact	Soil Quality Guidelines for Soil and Food Ingestion
Agriculture	Soil nutrient cycling processes	Grazing animals (primary consumers)
	Soil invertebrates	Nongrazing wildlife (secondary and
	Crops/plants	tertiary consumers for bioaccumulative
	Livestock/wildlife	substances)
Residential/parkland, commercial, and industrial	Soil nutrient cycling processes	Not applicable
	Soil invertebrates	
	Plants	
	Wildlife	

TABLE 13.2
Example of Some Selected Plant and Invertebrate Toxicological Studies for Cadmium

Species	Effect	End Point	Concentration (mg/kg)	Form of Cd (exposure period)	Soil pH	Test Substrate[a]	Extraction Method
Pine	Shoot dry weight	EC_{50}	20	$CdCl_2$ (12 weeks)	6.0	Sandy loam 1.5% O.M.	Nominal
	Root dry weight	EC_{15}	20				
Lettuce	Seedling emergence	NOEC	44	$CdCl_2$ (120 hours)	4–4.2	Artificial soil	$HNO_3/H_2O_2/$ HCl digest
	Seedling emergence (29%)	EC_{25}	94				
	Seedling emergence	LOEC	102				
		EC_{50}	143				
Beech	Shoot elongation	EC_{60}	20.2	$CdNO_3$ (21 months)	4.8–3.7	Mix of sand, peat, and forest soil	NH_4Ac extraction
Corn	Yield	EC_{20}	5.4	$CdNO_3$ (2 months)	5.3	Acid sandy clay 1.5% O.M.	Aqua regia extraction
Blazing-star	Shoot growth	EC_{80}	30	$CdCl_2$ (6 weeks)	4.8	Sandy mesic typic 1.93% O.M.	HNO_3 extraction
Thimbleweed	Shoot growth	EC_{30}	30				
Little bluestem	Shoot growth	EC_{25}	27.57	$CdCl_2$ (12 weeks)	4.8	Sandy mesic typic 1.93% O.M.	Nominal
	Total weight	EC_{25}	21.75				
	Root growth	EC_{25}	18.62				
Earthworm	Growth	EC_{50}	33	$CdCl_2$ (12 weeks)	6.7–6.8	Artificial soil 20% clay, 10% peat, 69% sand	Nominal
	Survival	LC_{50}	253				
	Sexual development	NOEC	<10				

Note: The table is illustrative only and is not intended to reflect actual data used in deriving the Canadian Soil Quality Guideline for Cadmium (in CCME, 1999).

[a] O.M. = organic matter.

derive guidelines that would be generally protective across the broad range of sensitivities of soil-associated biota (plants and animals), toxicological data for relevant species and endpoints are ranked by increasing chemical concentration (CCME, 1996). For agricultural and residential/parkland land uses, the effects and no-effects data distributions are combined and the 25th percentile of the data distribution is chosen as an estimate of the concentration of a chemical in soil at which few significant effects are observed in the soil biota.

The 1996 protocol specifies minimum criteria that must be respected in terms of the distribution of effects and no-effects data for this approach to be used. For commercial and industrial land uses, the level of ecological protection is relaxed. For these land uses, the 25th percentile of the effects data distribution is selected as an estimate of the concentration at which some significant effects are expected to occur in soil-associated biota, but not at the level of median lethality in the population. An uncertainty factor may be applied to the point estimates to derive the final soil guideline.

The protocol is intended to evolve as knowledge in the area of soil ecotoxicity evolves, and will continue to be applied to the development of soil guidelines in light of professional judgment. For example, recent work on developing a national standard for petroleum hydrocarbons in soil has led to the establishment of an ecotoxicological database for petroleum hydrocarbon constituents that has enabled a more complete examination of SSDs and the ways in which these distributions can be interpreted to develop national soil guidelines (unpublished data, 2000). The quality and quantity of data for plants and invertebrates have, for example, enabled a reexamination of the utility of combining effects and no-effects distributions in generating soil guidelines. Future updates to the Canadian Soil Quality Protocol, likely to be completed in 2001/2002, will reflect the current work with petroleum hydrocarbons.

13.2.2 DISCUSSION OF RANK-BASED APPROACH

Experience in Canada has suggested that the ability to define an environmentally protective threshold based on direct soil contact is severely limited by the availability of adequate soil toxicity data of reasonable quality, depending on the contaminant of interest (Table 13.3). The rank-based approach described above is deemed to be appropriate for the meta-analysis of often-disparate data types. The intent of the guideline is to base contaminated site remediation decisions on the larger SSD, as opposed to, for example, the lowest LOEC (lowest-observed-effects concentration) value from the literature.

Peer-reviewed studies on the toxicity of a contaminant of concern to different biota (animals and plants) generally use vastly different experimental methodologies (including soil type, method of contaminant introduction, exposure period, and toxicological response) and provide different measurement endpoints, variously cited as no-observed-effects concentration (NOEC) or LOEC or as disparate mortality or effects levels (LC_x or EC_x, respectively). For agricultural and residential/parkland land uses, the recommended soil guideline value is intended to represent the soil concentration for a given contaminant that falls at the upper end of the range where

TABLE 13.3
Partial Summary of Existing Canadian Soil Quality Guidelines for Specific Substances and the Availability of Ecotoxicity Data for Derivation of a Value for the Protection of Soil Invertebrates and Plants

Substance	Soil Quality Guideline for Soil Invertebrates and Plants[a] (mg/kg)	Year Guideline Released	Availability of Adequate Published Ecotoxicity Data
Arsenic (inorganic)	17	1997	Sufficient ecotoxicity data exist per CCME screening protocols
Barium	NC[b]	1999	Data were insufficient to calculate an SQG[c]
Cadmium	10	1997	Sufficient ecotoxicity data exist
Chromium (total)	64	1997	Sufficient ecotoxicity data exist
Chromium (hexavalent)	NC	1999	Data were insufficient to calculate an SQG
Copper	63	1999	Sufficient ecotoxicity data exist
Cyanide (free)	0.9	1997	Sufficient ecotoxicity data exist
Lead	300	1999	Sufficient ecotoxicity data exist
Mercury	12	1999	Sufficient ecotoxicity data exist
Nickel	50	1999	Sufficient ecotoxicity data exist
Thallium	1.4	1999	Sufficient ecotoxicity data exist
Vanadium	130	1997	Sufficient ecotoxicity data exist
Zinc	200	1997	Sufficient ecotoxicity data exist
Benzene	NC	1997	Data were insufficient to calculate an SQG
Ethylbenzene	NC	1997	Data were insufficient to calculate an SQG
Toluene	NC	1997	Data were insufficient to calculate an SQG
Xylene	NC	1997	Data were insufficient to calculate an SQG
Pentachlorophenol	11	1997	Sufficient ecotoxicity data exist
Phenol	20	1997	Sufficient ecotoxicity data exist
Benzo(a)pyrene	NC	1997	Data were insufficient to calculate an SQG
Naphthalene	NC	1997	Data were insufficient to calculate an SQG
DDT (total)	12	1999	Sufficient ecotoxicity data exist
Tetrachloroethylene	NC	1997	Data were insufficient to calculate an SQG
Trichloroethylene	NC	1997	Data were insufficient to calculate an SQG
PCBs	33	1999	Sufficient ecotoxicity data exist
Ethylene glycol	NC	1999	Data were insufficient to calculate an SQG

[a] Based on agricultural, residential, or commercial land uses.
[b] NC = not calculated.
[c] SQG = Soil Quality Guideline.

an absence of effects has been observed in experimental studies (no-effects data distribution) and at the lower end of the range where effects have been observed (effects data distribution). The rank-based approach is implicitly based on the notion

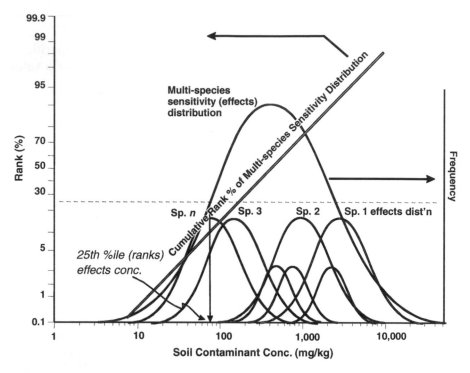

FIGURE 13.1 Conceptual model for cumulative distributions on intra- and intertaxon variability in sensitivity to contaminants in soil.

that, where sufficient toxicity data exist, it is possible to identify the breakpoint in soil concentration beyond which risks are substantially increased regardless of the extent of the disparity of the underlying data types. Conceptually, this is shown in Figure 13.1.

As shown in Figure 13.1, it might be expected that for each species the concentration of soil contamination at which a standardized toxicological response is observed would vary among individuals. The sensitivity of individuals within a single population may vary based on differences in age, nutritive status, disease, or some other attribute. As a starting point, it is assumed that each individual SSD follows a lognormal distribution (shown in the figure as a normal distribution on a contaminant concentration scale expressed logarithmically). In addition, it is often assumed that each different species (or other taxonomic grouping) exhibits a unique contaminant sensitivity distribution. The aggregate of the intrataxon and intertaxon contaminant sensitivity (the sum of the areas under all individual lognormal sensitivity distributions) itself forms a larger overall lognormally distributed species sensitivity distribution, and theoretically encompasses the range of concentrations of a contaminant in soil at which effects are encountered. In addition to this overall statistical distribution derived from meta-analysis of the effects data, there also exists an SSD

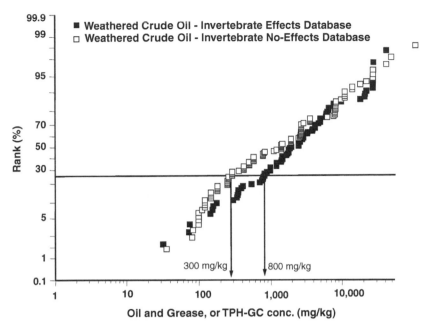

FIGURE 13.2 Ranks data for toxicity of weathered crude oil to soil invertebrates (with comparison of effects and no-effects data distribution).

based on the overall no-effects database, again comprising a species distribution based on intra- and intertaxon variation.

If there are merits to theoretical assumptions regarding the lognormal distribution of species sensitivity, then rank-based approaches as defined in CCME protocols initially appear to be intuitively less satisfactory than a formalization of the use of probability distribution estimates. The following, however, shows that in practice the two approaches are highly similar.

Figure 13.2 illustrates the distribution of data extracted from the published literature (primarily from Saterbak et al., 1999; Wong et al., 1999) on the toxicity of weathered crude oil in various surface soils to soil invertebrates, including earthworms. A major portion of the data was derived from studies on field-collected soils from crude oil–contaminated sites in the United States, subjected to varying degrees of weathering and/or bioremediation. No attempt was made to remove "redundant" toxicity endpoints; the overall SSD includes a range in the magnitude of toxicological responses to the same species based on differences in exposure time, experimental regime, soil type, crude oil characteristics, and possibly even the presence of cocontaminants. The figure also includes toxicity data from different species.

The individual soil invertebrate data points were divided into the effects and no-effects data sets, and each of these subsequently ranked from lowest to highest based on the corresponding soil concentration. The rank percentile (on the y-axis) was subsequently plotted against the corresponding reported soil "exposure" concentration of weathered crude oil. This graphical representation of the ranked data is functionally equivalent to the CCME (1996) protocol for deriving soil quality guidelines described above, although it looks separately at effects and no-effects distributions. The 25th percentile of the ranked data for the no-effects and effects database (horizontal line) was 300 and 800 mg/kg, respectively.

The advantage of plotting the data as shown in Figure 13.2 is that it allows better scrutiny of the underlying data distribution. Data points plotted as their rank percent in the database tend to follow a straight line when plotted along the abscissa with a probability-type scale. The fact that the data approximate a straight-line distribution when the soil concentrations are plotted along the ordinate using the logarithmic scale suggests that the sensitivity of the invertebrate species tested adheres to a lognormal distribution, as might be predicted.

This example is provided not to summarize any conclusions regarding the toxicity of weathered crude oil to soil invertebrates, but rather to demonstrate that the data set used would have yielded a similar SSD regardless of whether a rank-based, probability distribution, or bootstrapping technique was used. For this particular data set, it was interesting to note that the no-effects and effects apparent SSD exhibited an almost complete overlap, and a total range encompassing more than three orders of magnitude of the soil concentration of weathered crude oil.

If the reconstructed SSD adequately represented the underlying theoretical SSD in an accurate, unbiased manner, then it would be possible to examine the predictive relationship between soil concentration and overall magnitude of organismic or suborganismic response to a contaminant. Unfortunately, this is not the case, since the data available to construct SSDs rarely, if ever, represent a random or demonstrably representative subsample of all individuals or species potentially at risk in the geographic region for which generic environmentally quality guidelines are intended to apply. The problem with bias in the available data, therefore, is likely to exert a larger influence on environmental protection management targets subsequently established than the derivation method used.

13.3   CANADIAN SEDIMENT QUALITY GUIDELINES

Chemical contaminants in sediments have been associated with a wide range of impacts on the plants and animals that live within and upon bed sediments. Both acute and chronic toxicity of sediment-associated chemicals to algae, invertebrates, fish, and other organisms have been measured in laboratory toxicity tests (Thomas et al., 1986; Kosalwat and Knight, 1987; Dawson et al., 1988; Long and Morgan, 1990; Burton, 1991, 1992; Burton et al., 1992; Lamberson et al., 1992). Field surveys have identified the subtler effects of environmental contaminants, such as the development of tumors and other abnormalities in bottom-feeding fish (Malins et al., 1984; 1985; Couch and Harshbarger, 1985; Goyette et al., 1988). Sediment-associated chemicals also have the potential to accumulate in the tissues of aquatic organisms (Foster et al., 1987; Knezovich et al., 1987). Elevated tissue concentrations in benthic or other aquatic organisms can result in the bioaccumulation of chemicals in higher levels of the aquatic food web (Government of Canada, 1991a,b). Bioaccumulation in the context of this document is defined as the process by which substances are accumulated by aquatic organisms from all routes of exposure.

13.3.1 Overview of the Canadian Approach

The protocol for deriving sediment quality guidelines evolved over several years based on the need to address the broad range of effects as described above, and the practical limitations posed by data availability. Available approaches used to develop environmental quality guidelines were reviewed, including sediment background, spiked-sediment toxicity test, water quality guidelines, interstitial water toxicity, equilibrium partitioning, tissue residue, benthic community structure assessment, screening-level concentration, sediment quality triad, apparent effect threshold, International Joint Commission sediment assessment strategy, and National Status and Trends Program (MacDonald et al., 1992).

Based on discussions across federal, provincial, and territorial governments, a modified National Status and Trends Program approach was adopted by Canada. This approach involves the compilation of data on species and community sensitivity to a range of chemicals. As for the soil guidelines described earlier, data are taken from published studies in the literature. Unlike soil, however, the characterization of effects is based not only on the results of spiked sediment toxicity tests conducted under controlled conditions, but also on field studies (co-occurrence data consisting of matching sediment chemistry and biological effect data) (Long and Morgan, 1990; Long, 1992; MacDonald, 1993; Long et al., 1995), and equilibrium partitioning theory. Toxicological studies address both species- and community-level effects. For example, the database may include entries based on species richness or total abundance in benthic communities, as well as histopathological disorders in demersal fish, and more traditional endpoints associated with spiked sediment toxicity tests (EC_{50} or LC_{50} concentrations). The sediment effects database, originally developed by Long and Morgan (1990), has been updated and expanded to include data from additional sites (including available Canadian data), various biological endpoints (particularly chronic effects), and information on more chemicals (MacDonald, 1993).

To develop national sediment quality guidelines for discrete substances, acceptable toxicological data are compiled on a chemical-by-chemical basis and are sorted according to ascending chemical concentrations. With some exceptions noted above, the data tables are similar to the ranked SSDs described earlier for soil quality guidelines. Each entry consists of the measured chemical concentration, location, analysis type (or approach), test duration, endpoint measured, species, and life stage tested, whether associated biological effects or no biological effects were observed, and the study reference (Table 13.4). Entries for which no effects were observed comprise the no-effect data set and are indicated as NE (no effect, i.e., nontoxic, reference, or control), NG (no gradient), SG (small gradient), or NC (no concordance). Data on characteristics of the sediment and overlying water column are also summarized, where available. To derive a guideline, the effect data set for the chemical under consideration must contain at least 20 entries; and the no-effect data set for the chemical under consideration must contain at least 20 entries.

For each chemical, a threshold effect level (TEL) is calculated as the square root of the product (i.e., the geometric mean) of the lower 15th percentile concentration of the effect data set (the E_{15}) and the 50th percentile concentration of the no-effect

TABLE 13.4

Summary of a Portion of the Available Data on the Effects of Sediment-Associated Cadmium (mg/kg) in Marine and Estuarine Ecosystems

Cadmium (conc. ± SD)	Effect/ No Effect	Area	Analysis Type	Test Type	Endpoint Measured	Species	Life Stage	TOC (%)	Ref.
1	NE	San Francisco Bay, CA	COA	10-d	Least toxic (13.6 ± 7.76% mortality)	Amphipod	Adult	1.4 ± 0.79	Chapman et al. 1987
1	NE	San Francisco Bay, CA	COA	10-d	Least toxic (4.63 ± 2.91% avoidance)	Amphipod	Adult	1.44 ± 0.74	Chapman et al. 1987
1	NE	San Francisco Bay, CA	COA	48-h	Least toxic (18 ± 8.01% abnormal)	Mussel	Larva	1.2 ± 0.38	Chapman et al. 1987
1	NE	San Francisco Bay, CA	COA	48-h	Least toxic (17.3% mortality)	Mussel	Larva	1.25	Chapman et al. 1987
1	NE	San Francisco Bay, CA	COA	4-wk	Least toxic (116 ± 4.3 young produced)	Tigriopus californicus (copepod)	Adult	1.23 ± 0.09	Chapman et al. 1987
1	NE	Burrard Inlet, BC	SQO	—	Sediment quality objective	Aquatic biota	—	—	Swain and Nijman 1991
1	NG	San Francisco Bay, CA	COA	10-d	Moderately toxic (28.3 ± 7.51% mortality)	Amphipod	Adult	2.01 ± 0.98	Chapman et al. 1987
1	NG	San Francisco Bay, CA	COA	10-d	Most toxic (95% mortality)	Amphipod	Adult	4.03	Chapman et al. 1987
1	NG	San Francisco Bay, CA	COA	10-d	Highly toxic (37% avoidance)	Amphipod	Adult	4.03	Chapman et al. 1987
1	NG	San Francisco Bay, CA	COA	48-h	Moderately toxic (25.1 ± 6.61% abnormal)	Mussel	Larva	1.26 ± 0.17	Chapman et al. 1987
1	NG	San Francisco Bay, CA	COA	48-h	Highly toxic (66.8% abnormal)	Mussel	Larva	3.59	Chapman et al. 1987
1	NG	San Francisco Bay, CA	COA	48-h	Moderately toxic (57.1% ± 13.6% mortality)	Mussel	Larva	1.14 ± 0.33	Chapman et al. 1987
1	NG	San Francisco Bay, CA	COA	4-wk	Moderately toxic (94.9 ± 10.1 young produced)	Tigriopus californicus (copepod)	Adult	2.87 ± 1.07	Chapman et al. 1987

		Location	Method	Duration	Biological response	Species	Life stage	Value	Reference
1	SG	Curtis Creek, Baltimore, MD	COA	10-d	Significantly toxic (55% mortality)	Hyalella azteca (amphipod)	Juvenile	—	McGee et al. 1993
1.01 ± 1.09	a	Laboratory	SSTT	10-d	LC_{50}	Rhepoxynius abronius (amphipod)	—	—	Ott 1986
1.04 ± 1.21	NE	Long Island Sound, NY, CT	COA	10-d	Not significantly toxic (23 ± 4.24% mortality)	Ampelisca abdita (amphipod)	Subadult	2.46 ± 1.22	Bricker et al. 1993
1.08 ± 1.2	NE	Puget Sound, WA	COA	2-d	Not significantly toxic (6.67 ± 8.07% abnormal development)	Dendraster excentricus (echinoderm)	Embryo	1.51 ± 0.33	Pastorok and Becker 1990
1.1 ± 2.0	NC	Southern California	COA	—	Low abundance (57.6 ± 13.6 N/0.1 m²)	Benthic species	—	—	Word and Mearns 1979
1.1	SG	Puget Sound, WA	COA	2-d	Significantly toxic (3.8% abnormal chromosome)	Dendraster excentricus (echinoderm)	Embryo	1.5	Pastorok and Becker 1990
1.11 ± 0.355	SG	Long Island Sound, NY, CT	COA	—	—	Microtox (Photobacterium phosphoreum)	—	2.51 ± 0.45	Bricker et al. 1993
1.12 ± 0.777	NE	Long Island Sound, NY, CT	COA	48-h	Significantly toxic (EC_{50}: 0.014 = 0.006 mg dw/mL)	Mulinia lateralis (bivalve)	Larva	2.12 ± 1.04	Bricker et al. 1993
1.13 ± 0.867	SG	Long Island Sound, NY, CT	COA	48-h	Significantly toxic (96 ± 1.66% normal development)	Mulinia lateralis (bivalve)	Larva	2.52 ± 0.997	Bricker et al. 1993
1.14 ± 0.155	SG	Long Island Sound, NY, CT	COA	48-h	Significantly toxic (68.5 ± 11.4% mortality)	Mulinia lateralis (bivalve)	Larva	2.33 ± 0.364	Bricker et al. 1993
1.2 ± 1.0	a	Fraser River Estuary, BC	COA	—	Sediments devoid of feral clams	Macoma balthica (bivalve)	—	1.95	McGreer 1982
1.2	a	San Francisco Bay, CA	AET	10-d	San Francisco Bay AET	Rhepoxynius abronius (amphipod)	Adult	—	Long and Morgan 1990
1.2 ± 0.36	NE	Pensacola Harbor and Bay, FL	COA	10-d	Not significantly toxic (9 ± 1.73% mortality)	Nereis virens (polychaetes)	Adult	—	EG & G Bionomics 1980

[a] Adverse biological effect and concordance between observed biological response and measured chemical concentration.

AET = apparent effect threshold; COA = co-occurrence analysis; NC = no concordance; NE = no effect; NG = no gradient; SG = small gradient; SSTT = spiked-sediment toxicity test.

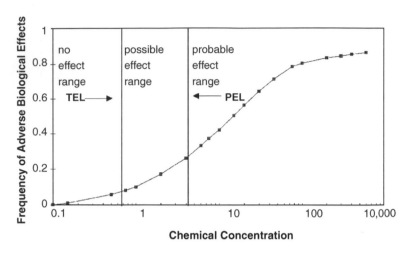

FIGURE 13.3 Conceptual example of effect ranges for a sediment-associated chemical.

data set (the NE_{50}). This rank-based approach uses all acceptable species sensitivity data to determine a range of sediment chemical concentrations that is dominated by no-effect data entries (i.e., adverse biological effects are never or almost never observed below the TEL). If the uncertainty associated with the TEL is high, a safety factor may be applied to the TEL. Otherwise, the TEL is considered representative of the concentration below which adverse effects to sediment-associated biota are not anticipated.

A second sediment quality assessment value, the probable effect level (PEL), is also recommended to represent the lower limit of the range of chemical concentrations that are usually or always associated with adverse biological effects (Figure 13.3). The PEL is calculated as the square root of the product (i.e., the geometric mean) of the 50th percentile concentration of the effect data set and the 85th percentile concentration of the no-effect data set. This value, in conjunction with the TEL, is used to identify ranges in chemical concentrations associated with adverse biological effects and the incidence of adverse effects within each of these concentration ranges. This evaluation provides a means of estimating the likelihood of observing similar adverse effects at sites with sediment chemical concentrations that fall within the defined concentration ranges. The establishment of ranges in chemical concentrations also allows the estimation of the likelihood of adverse biological effects calculated on the basis of the frequency distributions of the toxicity data for each chemical. Within each of the concentration ranges, the incidence of adverse biological effects is quantified by dividing the number of effect entries by the total number of entries, and expressing this ratio as a percentage. The guidelines for cadmium in marine sediments reported by MacDonald (1994; TEL = 0.676 mg·kg^{-1}; PEL = 4.21 mg·kg^{-1}) illustrate this calculation. In this example, only 5.6% of the cadmium concentrations within the no-effect range (0 to 0.68 mg·kg^{-1}) were associated with adverse biological effects (MacDonald, 1994). This suggests that there is a low probability of observing adverse effects when cadmium concentrations fall within this range. In the possible- and probable-effect range for cadmium, the incidence of adverse biological effects was 20.1 and 70.8%, respectively. MacDonald (1994) calculated a TEL and PEL for fluoranthene of 0.11 and 1.49 mg·kg^{-1}, respectively. The incidence of adverse biological effects was 9.5, 20.2, and 79.7% in the no-effect, possible-effect, and probable-effect ranges. The positive correlation observed between the frequency of effects and chemical concentrations for both cadmium and fluoranthene inspires confidence in the guideline values established for these chemicals. These examples demonstrate how analysis of the distribution of observed biological effects within each of the concentration ranges provides a means of estimating the relative reliability (i.e., degree of certainty) of the guidelines derived. In comparison, for mercury 7.8, 23.6, and 36.7% adverse effects were observed in the no-effect, possible-effect, and probable-effect ranges, respectively (the TEL = 0.13 mg·kg^{-1} and the PEL = 0.696 mg·kg^{-1}) (MacDonald, 1994). In this way, the use of both the TEL and the PEL helps distinguish sites and chemicals of little toxicological concern, of potential toxicological concern, or significantly hazardous to exposed organisms.

Two slightly different methods have been reported to derive guidelines using the information evaluated and compiled for each chemical in the guideline derivation tables. Long and Morgan (1990) derived guidelines on the basis of the effect data set only. Guidelines were calculated as the lower 10th percentile of the effects range and the 50th percentile (median) of the effects range. The lower threshold value represents a concentration above which adverse effects on sensitive life stages and/or species are expected to begin, and the higher value represents a threshold concentration above which adverse effects on most species are frequently or always observed. The calculation of percentiles of the data tends to minimize the influence of single (potentially outlier) data points on the development of guidelines (Klapow and Lewis, 1979). The method used by MacDonald (1993) to derive guidelines considered both the effect data set and the no-effect data set, and has been adopted in the Canadian approach.

Despite the slight differences in calculating the two guideline values using these methods, the agreement between the Long and Morgan approach and the Canadian approach is very good (on average, they vary within a factor of two) (Long et al., 1994).

13.3.2 FIELD VALIDATION OF THE SEDIMENT QUALITY GUIDELINE APPROACH

At present, we are aware of only one viable method for making accurate estimations of the level of environmental protection afforded in ecosystem(s) of interest from SSDs derived from the major portion of toxicity data that is currently available. This is through the careful design and conductance of field-based follow-up studies, where population- and ecosystem-level responses are associated after the fact with predicted risk levels based on SSDs. The following example is based on field validation of environmental protection goals based on Canadian Marine Sediment Quality Guidelines, rather than soil quality guidelines, because we are not aware of any appropriate examples from a terrestrial environment in Canada.

A study was conducted in 1993 of contaminant levels in surface sediments of Esquimalt Harbour, British Columbia, Canada, and the associated effects on sediment

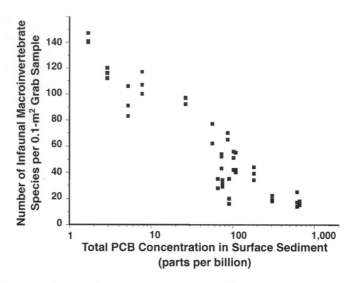

FIGURE 13.4 Relationship between concentration of PCBs (as total Aroclor concentrations) in sediment from Esquimalt Harbour, BC, and reduction in benthic infaunal biodiversity, 1993.

infauna were evaluated (Bright et al., unpublished). There was an excellent linear relationship between benthic biodiversity in the harbor, measured either as species richness or multivariate scores, for a variety of metals and organic contaminants, which were themselves highly intercorrelated in their spatial distribution. Figure 13.4 shows the clear inverse linear relationship between the total polychlorinated biphenyl (PCB) concentration in sediment and the benthic infaunal species richness. Table 13.5 summarizes the strength of the predictive linear relationship between species diversity and various contaminants and, based on the resulting regression equations, the level of contamination that was associated with a 10, 50, and 90% reduction in species richness within the harbor. Finally, the observed relationship between sediment contamination and biodiversity was examined in light of the existing CCME sediment quality guidelines, derived from other marine invertebrate toxicity data.

The example shows the value of field validation studies for establishing ecologically relevant levels of environmental protection. The acquisition of similar data for terrestrial sites is essential for the setting of ecosystem-based environmental protection guidance, and for periodically revising generic, environmentally protective soil contaminant threshold levels derived from SSDs.

13.4 DISCUSSION

The Canadian approach to sediment quality guidelines reflects one way in which a rank-based approach to interpretation of SSDs can be used to derive national benchmarks of environmental quality. This approach has been extensively reviewed by experts from across North America, has been described orally in numerous technical and scientific forums, and has been described in various publications (Long and

Morgan, 1990; Long, 1992; MacDonald, 1993; Long and MacDonald, 1992; Long et al., 1994). Guidelines developed using this approach (Long and Morgan, 1990; Long et al., 1994) have been used by the National Oceanic and Atmospheric Administration to identify priority regions (ones having the highest probability for observing adverse biological effects) within which surveys have subsequently been implemented to further investigate sediment quality conditions. The guidelines have also been used in assessing hazardous waste sites, dredged material, and monitoring data in the United States (Long and MacDonald, 1992). The Florida Department of Environmental Protection has derived sediment quality assessment guidelines on the basis of this approach for use in identifying regional priorities (MacDonald, 1993). The International Council for Exploration of the Sea (Study Group on the Biological Significance of Contaminants in Marine Sediments) has also elected to adopt this approach in the development of guidelines for participating nations (Long and MacDonald, 1992).

The rank-based approach described above is used currently in the development of interim sediment quality guidelines in Canada. Ideally, sediment guidelines should be developed from detailed dose–response data that describe the acute and chronic toxicity of individual chemicals in sediments to sensitive life stages of sensitive species of aquatic organisms. A detailed understanding of the factors that influence toxicity would also support site-specific sediment quality assessments by providing a basis for evaluating the applicability of guidelines under site-specific conditions (e.g., total organic carbon, or TOC, grain size, acid-volatile sulfide) (DeWitt et al., 1988; DiToro et al., 1990; 1991; Landrum and Robbins, 1990; Swartz et al., 1990; Carlson et al., 1991; Loring and Rantala, 1992; Ankley et al., 1991; 1993).

Currently, it is the opinion of Environment Canada that this ideal approach is not supported by adequate scientific information to facilitate the development of national guidelines. To date, only a limited number of controlled laboratory studies have been conducted to assess the effects of sediment-associated chemicals on estuarine, marine, and freshwater organisms (Long and Morgan, 1990; Burton, 1991; MacDonald, 1993). Additional research into sediment-spiking methodologies is also required before the results of spiked-sediment bioassays can be used to generate dose–response data appropriate for guideline development. However, other types of data that are routinely collected throughout North America contribute to our understanding of the toxic effects of these chemicals. Specifically, a wide variety of sediment toxicity tests have been conducted to assess the biological significance of concentrations of chemicals in sediments from specific geographic locations. These toxicity tests include those performed on benthic organisms (e.g., bivalve mollusks, shrimp, amphipods, polychaetes, nematodes, chironomids) and on pelagic organisms (e.g., sea urchin larvae, oyster larvae, luminescent bacteria). Further, numerous field studies have been conducted that assess the diversity and abundance of benthic infaunal and epifaunal species. Comprehensive data on the concentrations of chemicals in these sediments have also been collected for many of these field studies (Long and Morgan, 1990; MacDonald, 1993). Specific characteristics of the sediments and the overlying water column also aid in the interpretation of the corresponding toxicity data for the studies for which this information has been collected.

TABLE 13.5

Estimated Contaminant Concentrations Associated with Levels of Environmental Protection (maintenance of biodiversity) of Benthic Macroinvertebrates in Esquimalt Harbour, British Columbia, and Comparison with Canadian Sediment Quality Guidelines Based on the Geometric Mean of the 25th Percentile of the Effects Database and 75th Percentile of the No-Effects Database

Substance	Relationship between Conc. of Substance in Sediment and Benthic Species Richness in or near Esquimalt Harbour {Spp. Richness = x(log₁₀[substance]) + y}				Concentration[a] Leading to a Percent Reduction in the Average Spp. Richness of the Reference Sites (by % reduction and severity of impact)			CCME Draft Interim Sediment Quality Criteria		NOAA Effects-Ranges		Washington State Sediment Management Standards	
	x	y	R^2	Probability	10% ↓	50% ↓	90% ↓	TEL[b]	PEL[c]	ER-L[d]	ER-M[e]	SQSs[f]	IZ-Max[g]
LPAH	−58	222	0.70	<0.001	110 ppb	690 ppb	4300 ppb	—	—	—	—	5200	5200
HPAH	−59	257	0.82	<0.001	400 ppb	2400 ppb	14,500 ppb	—	—	—	—	12,000	12,000
TPAH	−59	264	0.79	<0.001	530 ppb	3200 ppb	19,000 ppb	—	—	4000	35,000	—	—
ΣPCBs	−47	141	0.85	<0.001	6.3 ppb	60 ppb	570 ppb	21.5	189	50	400	—	—
As	−78	134	0.76	<0.001	2.5 ppm	9.6 ppm	40 ppm	7.24	41.6	33	85	57	93
Cr	−390	813	0.35	<0.001	66 ppm	87 ppm	110 ppm	52.3	160	80	145	270	270
Cu	−77	193	0.68	<0.001	15 ppm	57 ppm	230 ppm	18.7	108	70	390	390	390
Fe	−502	2330	0.28	<0.001	27,000 ppm	33,000 ppm	41,000 ppm	—	—	—	—	—	—
Pb	−53	142	0.65	<0.001	5.3 ppm	39 ppm	290 ppm	30.2	112	35	110	450	530
Sb	−55	69	0.65	<0.001	0.24 ppm	1.6 ppm	11 ppm	—	—	2	25	—	—
Zn	−160	402	0.69	<0.001	73 ppm	140 ppm	280 ppm	124	271	120	270	410	960
Ni	n.s.	n.s.	0.01	0.52	n.a	n.a.	n.a.	15.9	42.8	30	50	—	—
Co	n.s.	n.s.	0.00	0.96	n.a.	n.a.	n.a.	—	—	—	—	—	—

Note: ns = not significant. Regression coefficients, therefore, are not provided; na = not applicable; — = no value provided; all sediment quality values are given in parts per million dry weight for inorganic elements, and parts per billion dry weight for organics.

[a] Concentration of substance in sediment in ppm dry weight for inorganic elements, or ppb dry weight for PAHs and PCBs.

[b] TEL = threshold effects level.

[c] PEL = probable effects level.

[d] ER-L = effects range — low.

[e] ER-M = effects range — median.

[f] SQSs = Washington State Marine Sediment Quality Standards, Chemical Criteria.

[g] IZ-Max = Puget Sound Marine Sediment Impact Zones, maximum chemical criteria.

These field studies, which report matching sediment chemistry and biological effect data (i.e., data are collected from the same locations at the same time), provide information relevant to the guideline derivation process.

The sediment quality guideline protocol is practical in that it can be implemented in the short term using existing data. It is scientifically defensible because it is supported by a weight of evidence of the available toxicological data on sediment-associated chemicals. For sediment quality guidelines to be effective in Canada, they must be formulated from an understanding of biological effects (preferably cause-and-effect relationships, including data on sensitive endpoints like growth, reproduction, and genotoxicity), and they should account for the factors (e.g., TOC) that influence the bioavailability of sediment-associated chemicals. Since information on the toxicity of field-collected sediments is used in deriving guidelines, the various factors that affect the bioavailability of chemicals are implicitly considered, as well as the effects of mixtures of chemicals. The guideline derivation procedure is also applicable to all classes of chemical and mixtures of chemicals that are likely to occur in Canadian sediments.

13.5 CONCLUSIONS

The Canadian approach to interpretation of SSDs in development of soil and sediment quality guidelines has been designed to meet the immediate need of environmental managers across Canada to provide nationally consistent benchmarks for evaluating the ecological significance of sediment and soil contamination. Both the soil and sediment approaches assume that there is a point in the distribution of ranked species sensitivity data that represents the environmental concentration below which the incidence of effects is considered acceptable or insignificant (e.g., a threshold effects concentration or level). For both soil and sediment, current approaches are based to some extent on a consideration of the quality and quantity of available ecotoxicological data, and Canadian approaches to estimating these national values are expected to evolve as more data become available.

Of the types of uncertainty associated with establishing accurate SSDs as listed in Table 13.6, all may be substantially reduced as new knowledge emerges. In particular, there is a need to undertake toxicity testing with a much broader range of species, including species that are native to the ecosystems of concern, as opposed to standardized test organisms where the relative sensitivity of standardized and native species has not been established.

Few approaches based on SSDs adequately account for the major sources of intraspecific variation. In many cases, the original intent of the researchers who generated the relevant toxicity data was to examine specific influences on a species' response (for example, soil type). The study intent and major conclusions are often ignored, however, when screening for toxicity endpoints or reduced to a lower, averaged, or other singular estimate. Finally, the effect of multiple toxicant interactions or toxicant–stressor interactions on the response of an organism is an active area of research. Ongoing and new research it is hoped will increase our collective confidence that we have constructed accurate and meaningful SSDs.

TABLE 13.6
Uncertainty Analysis for the Use of Literature-Derived Toxicity Data to Reconstruct Generically Applicable SSDs

Major Issue	Associated Confidence Level When Constructing SSDs	Likely Effects on SSD Bias and Accuracy
Interspecific variations in sensitivity	Moderate to low	Narrow suite of species examined produces unrealistically narrow SSDs
		Data for the few taxa available is biased high or low relative to the larger group of potential ecological receptors; policy targets are based on a sample SSD that departs from the true SSD
Intraspecific variations	Low	Failure to account for known or at present unrecognized factors that increase or decrease toxicological response of a species will render environmental protection targets either insufficient or overly protective
		Sensitive life-stages or members of the population may experience higher than expected risks, leading to population-level effects
Variation associated with the environment	Low	Failure to account for known or at present unrecognized factors that increase or decrease toxicological response of a species will render environmental protection targets either insufficient or overly protective
Variation based on stressor interactions	Low	Focus on one stressor/contaminant at a time may lead to unappreciated and elevated risks from integrated exposures

14 Ecotoxicological Soil Quality Criteria in Denmark

Janeck J. Scott-Fordsmand and John Jensen

CONTENTS

Abstract — Danish environmental authorities have established ecological quality criteria (EQC) with the aim of protecting organisms in the terrestrial and aquatic environment from adverse effects of pollution. The terrestrial ecotoxicological EQCs for soils were based on Species Sensitivity Distribution (SSD) extrapolations, on a safety assessment factor–based approach, and on a direct expert evaluation. This chapter reviews the background, the derivation of 34 ecotoxicological soil quality criteria (ESQC), and the administrative implementation of these criteria. The algorithm used for setting ESQC criteria was based on a direct evaluation (in all cases) in combination with two extrapolation methods (not for all) of data obtained from the open literature. The direct evaluation consisted of an expert evaluation of toxicity, bioaccumulation fate, and background levels. The two extrapolation methods were a statistical sensitivity distribution–based method and an application factor method. Calculation examples of the extrapolation methods have been given and the present chemical specific ESQC have been listed. The ESQC derivation and implementation have also been regarded in historical and environmental policy context. The Danish ESQC are considered as guidance values above which possible ecotoxicological effects may occur in soils.

14.1 HISTORICAL BACKGROUND

Controlling and combating water pollution dates back to the early 1970s, whereas soil pollution has primarily received attention within the last decade. In 1992, on request of the Danish Environmental Agency, the National Environmental Research Institute (NERI) and the Water Quality Institute (VKI) evaluated the principles and the state of the art for deriving ecotoxicological soil quality criteria (ESQC). Based

1-56670-578-9/02/$0.00+$1.50
© 2002 by CRC Press LLC

on this work, they formed a number of recommendations (for further details see Pedersen and Pedersen, 1993). It was regarded possible to establish at least two different types of criteria: one set of criteria protecting all target groups, i.e., organisms living in the soil except vertebrates, against any adverse effects caused by chemicals and another set considering the current or future use of the land and only protecting a selected group of the organisms. In the succeeding work, as described in this chapter, only the criteria protecting all groups of organisms were considered. To derive these criteria two methods were proposed. The first included the use of species sensitivity distributions (SSDs) and the other included the use of such assessment factors as known in risk assessment of chemicals within the European Union (EU) (CSTE/EEC, 1994). The development of compound-specific ecotoxicological quality criteria for both the aquatic and the terrestrial environment were initiated based on the above recommendations. For the aquatic environment, data were extrapolated to quality criteria using only the assessment factor method according to the OECD guidelines (Pedersen, 1994), because the assessors for this compartment determined there was insufficient evidence for using the SSD-based method. This displays the various opinions by different experts on the SSD-based approach. As SSDs were not used for the water quality criteria these will not be further discussed in this context. For the terrestrial environment the ecotoxicological-derived criteria were based on both the assessment factor and SSD-based extrapolations combined with a direct expert evaluation of all the data (Scott-Fordsmand et al., 1996). ESQC were produced for 34 chemical compounds in the period of 1994 to 1996 (Scott-Fordsmand and Bruus-Pedersen, 1995; Jensen and Folker-Hansen, 1995; Jensen et al., 1997).

14.2 DERIVATION OF ECOTOXICOLOGICAL SOIL QUALITY CRITERIA

The goal of the ecotoxicological-derived SQC was to protect the function and structure of the ecosystems against effects caused by pollutants. Ideally, adequate data for the effects of a compound should be obtained alone and in combination with other compounds.

Whether the function of the ecosystem was protected when all species were protected was difficult to assess and was (and still is) debated in the literature. Van de Meent et al. (1990a,b) argue that the function of the ecosystem is guaranteed when the species are protected. This has been questioned by Forbes and Forbes (1993) and Smith and Cairns (1993), who argue that structure and function of an ecosystem are often uncoupled, and thus toxicity data for species cannot be used to predict safe levels for ecosystems. At the time of ESQC derivation, however, little knowledge was available to prove or disprove such statements. As only limited knowledge was available about the effects of pollutants on the function of the ecosystem, it was decided that the Danish ESQC should aim at protecting the structure of the ecosystem and in this way one hopes they also protect the function.

As it was not possible to find data on tests of all organisms and functions in an ecosystem against all possible compounds and combinations of these, extrapolation

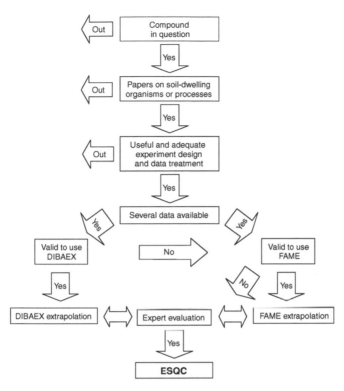

FIGURE 14.1 Presentation of the algorithm used for derivation of the Danish ESQC.

from simple laboratory tests to the ecosystems was attempted. This involved a direct expert evaluation of the test data and two extrapolation methods as implemented in an ESQC algorithm (Figure 14.1).

The algorithm relied on open literature–derived data (reports, reviews, and papers in scientific journals), included selection of data (NOEC, LOEC, and EC_x values), and treatment of selected data by extrapolation methods combined with an expert evaluation (see Figure 14.1). Only papers describing effects of singular added compounds were used and no mixture studies or studies involving carriers such as sewage sludge etc. were included. Nor were results based on experiments with aquaculture, pure sand cultures, or feeding studies included. The extrapolation models used were a simple FActorial application MEthod (hereafter termed FAME) and a DIstribution-BAsed EXtrapolation (an SSD approach hereafter termed DIBAEX).

In all cases an expert evaluation of the obtained ESQC data was carried out, as neither of the extrapolation methods were thought to be well validated. The direct expert evaluation also aimed at a no-effect level for the ecosystem; therefore, the lowest no-effect and effect values reported for individual species were evaluated, taking background concentrations, fate, and bioavailability of the compound in question into consideration.

For data sets of fewer than five values the FAME approach was used, dividing the lowest-effect or no-effect level by a factor between 10 and 1000, depending on

TABLE 14.1
Application Factors for the Determination of Soil Quality Criteria, as adopted from CSTE/EEC

Information Available	Application Factor
The lower end of the acute $L(E)C_{50}$ range, when the data available are few, or the range of organisms narrow, bearing in mind that outlier values may be due to error or experimental conditions that deviate too much from real-world conditions	1000
The lower end of the range of acute $L(E)C_{50}$ values, when there is an extensive database covering a (phylogenetically) wide range or the lower end of the chronic $L(E)C_{50}$ or NOEC values when few data are available	100
The lower end of (apparent) chronic NOEC data determined by a sufficient and representative number of tests	10

Source: Pedersen, F. and Pedersen, M. B., Miljøprojekt 247; Miljøstyrelsen, Copenhagen, Denmark, 1993. With permission.

the quality of the test data set (Table 14.1). The factors were adopted from the method suggested for water quality objectives by the Scientific Advisory Committee on Toxicity and Ecotoxicity of Chemicals (CSTE) within the DG-Environment in the EU (CSTE/EEC, 1994). These safety factors were based on water toxicity data, where the differences between no-effect and effect levels were observed to vary between 2 and 10,000, depending on the compound used (U.S. EPA, 1984b). The aim of the factors involved was to include interactive effects of chemicals, intra-/interspecific differences, differences between chronic and acute effects, and differences in effect concentrations between field and laboratory studies.

In cases when five or more data were present the DIBAEX using Wagner and Løkke (1991) was also applied to derive a protection concentration, *Kp*, for a certain fraction of the soil organisms (95%, with a 95% predefined certainty). The DIBAEX-based extrapolation used here was based on three assumptions:

1. Interspecies variation in sensitivity could be described by a lognormal function (this was tested by a Kolomogorov–Smirnov test).
2. The toxicity data used represent the sensitivity range of the species in the ecosystem. It has been stressed that different taxonomic groups should be included in the data set to construct a representative set of test organisms. On the other hand, large taxonomic distances should be avoided because they are assumed to have different distribution curves (Løkke, 1994). Nevertheless, data were often pooled as in some cases an insufficient number of data were available for making SSDs for specific classes of organisms.
3. The protection of a certain fraction of the species secures the function and structure of the ecosystem. The SSD approach does not consider effects on higher trophic levels, but this was evaluated for each compound in the direct expert evaluation. At the time (1994), the Danish authorities did not find that there was a scientific basis for including models for

assessing secondary poisoning. In cases of persistent and biomagnifying compounds this information was included in the final expert evaluation and would generally lead to a conservative derivation of criteria.

The results of the two extrapolation methods were compared (Figure 14.2). The values calculated by the two methods were only to be considered as estimates of a predicted no-effect concentration (PNEC), although this was formally not the case for the DIBAEX approach. The DIBAEX approach only protects 95% of the species in theory. In all cases, outcome was compared with a direct expert evaluation, in cases with large disagreement between the extrapolation methods, or if the extrapolated PNECs were below assumed background or essential concentrations, the emphasis was placed on the direct evaluation of the data. For organic compounds the ESQC was typically recommended at the nearest order of magnitude but for metals the ESQC was more precise.

The above-mentioned methods required representative and ecologically relevant data. In the case of benzo(a)pyrene, virtually no data were available and an attempt was made to apply aquatic results by converting these to soil-relevant conditions. Such conversions were done using the octanol–water partitioning coefficient (P_{ow}) and the organic carbon content of a typically Danish soil. This was done under the assumption that the bioavailable fractions were identical to the soil–water soluble fraction and that terrestrial and aquatic organisms were equally sensitive. These conversions were done as a last resort and were considered less reliable.

14.3 SPECIFIC ECOTOXICOLOGICAL SOIL QUALITY CRITERIA

Currently, ESQCs have been established for a wide range of compounds — both organic and inorganic (Table 14.2). The organic compounds were assessed using the FAME and DIBAEX approach (when data were sufficient) combined with a direct expert evaluation. For metals, both of the extrapolation methods generally computed PNEC (FAME)/Kp (DIBAEX) values that were below background concentrations for Danish soils or at concentrations considered essential for some organisms (see Figure 14.2). Hence, metals in the terrestrial environment were assessed on the basis of a direct expert evaluation of the available data, considering fate, background levels, and bioavailability. The two extrapolation methods were disregarded for metals. The ESQC was not designed for any specific soil type as this was considered impossible at the time.

At the time of deriving the ESQCs in Denmark the extrapolation methods were much discussed (as they still are) (Forbes and Forbes, 1993; Smith and Cairns, 1993; Hopkin, 1993; Van Straalen, 1993a). Many limitations and problems with the extrapolation methods could be listed; for example:

The obvious problem that DIBAEX did not (and did not even attempt to) protect all the species;

That not all species were protected induced the potent problem of a key species being eliminated at concentrations below the criteria;

COPPER

FAME:

For copper the lowest of NOEC values found was 10 mg/kg. However, since only a narrow range of invertebrates is represented an application factor of 100 is used. Thus using the factorial method the no-effect level is predicted at:

0.1 mg Cu/kg

This is below what is considered the essential requirement level (5–10 mg/kg) for many organisms, and therefore this extrapolation is rejected.

DIBAEX:

For copper 43 NOEC values were found for microorganisms, invertebrates, and plants. For species of the same genus closely related species such as earthworms, only the lowest NOEC is used, giving a data set of 23 NOEC values. The data are lognormal-distributed and the average of the logarithm$_{10}$ (χ_m) for the 23 values is 1.836, and the standard deviation (S_m) is 0.617. From Table 1 in Wagner and Løkke (1991) k is found to be 2.329, when m=23, p=0.05, and δ=0.05. The protection concentration (K_p), which protects 95% of the species in 95% of the cases, is calculated using the following formula:

$K_p = \exp(\chi_m)/\exp(S_m \cdot k)$

$K_p = \exp(1.836)/\exp(0.617 \cdot 2.329) = 2.5$ mg/kg

The use of the method presented by Wagner and Løkke (1991) predicts a protection level of 2.5 mg/kg. This level is lower than the lower end of what is assumed to be the Danish background level (6–20 mg/kg), and the minimum requirement level for many soil-living organisms is considered to be approximately 5–10 mg/kg; hence this method is rejected. In the case of copper it is necessary to make a direct data evaluation taking the bioavailability into consideration.

ESQC for Copper:

Based on an expert evaluation of all the collected toxicity data, background concentrations, fate, and bioavailability, the SQC for copper was set at 30 mg/kg.

PENTACHLOROPHENOL

FAME:

NOEC values were found for all groups of organisms. However, only few NOEC for plants and invertebrates were available. The lowest NOEC was 2 mg/kg. Applying the highest assessment factor this resulted in a PNEC of 0.002 mg/kg.

DIBAEX:

A total of 15 data was used in the SSD calculations. This was 7 NOEC values and 8 extrapolated values. The 8 effect data (EC$_{50}$) were expoleted to NOEC values by a division of 3. The average of the logarithm$_{10}$ (χ_m) to the 15 NOEC values is 0.81 and the standard deviation (S_m) is 0.79. From Table 1 in Wagner and Løkke (1991), the value of k is found to be 2.566, when m=15, p=0.05, and δ=0.05. The protection concentration (K_p) is calculated as:

0.06 mg/kg

using the formula presented by Wagner and Løkke (1991).

ESQC for pentachlorophenol:

Based on the two estimated PNECs and considering the fact that PCP is anthropogenic, chlorinated, hazard classified as very toxic, and relatively slowly degradable, the SOC was set at 0.001 mg/kg.

FIGURE 14.2 The use of FAME and DIBAEX, using NOEC values or extrapolated NOEC values for soil-dwelling organisms; for further details, see references Scott-Fordsmand and Bruus Pedersen (1999), Scott-Fordsmand et al. (1996), and Jensen and Folker-Hansen (1995).

TABLE 14.2
ESQC in Denmark

Inorganic Compounds[a]	ESQC (mg/kg dry soil)	Organic Compounds[b]	ESQC (mg/kg dry soil)
Arsenic	10.0	Anionic surfactants (general)	5.0
Cadmium	0.3	Benzo(*a*)pyrene	0.01
Chromium (III)	50.0	Chlorobenzene	
		Mono-	0.1
		Di-	0.1
		Tri-	0.001
		Tetra-	0.001
		Penta-	0.001
		Hexa-	0.001
Chromium (VI)	2.0	LAS	5.0
Copper	30	Nonylphenol	0.01
Lead	50.0	Pentachlorophenol	0.005
		Sum of other chlorophenols	0.01
Mercury	0.1	Phthalate	
		Di-methyl-	0.1
		Di-ethyl-	0.1
		Di-butyl-	0.1
		Di-iso-octyl-	1.0
		Di(2-ethylhexyl)-	1.0
		Di-*n*-octyl-	1.0
Molybdenum	2.0	PCBs	0.01
Nickel	10.0	Polynuclear aromatic	1.0
Selenium	1.0	hydrocarbons (PAH)	
Silver	1.0		
Thallium	0.5		
Tin	20.0		
Zinc	100.0		

[a] For detailed ESQC consideration for each compound, see references Scott-Fordsmand and Bruus Pedersen (1999), Scott-Fordsmand et al. (1996), Jensen and Folker-Hansen (1995), and Jensen et al. (1997).
[b] Cyanide, dioxins, and white spirit: insufficient data.

Sources: Scott-Fordsmand and Bruus Pedersen, 1995; Jensen and Folker-Hansen, 1995; Jensen et al., 1997.

The fact that the choice of factors in the FAME approach, attempting to cover all kinds of interactions, was based on aquatic data;

The statistical problem that both methods relied on NOEC and LOEC values, which may not be good estimates of "real" no-effect and lowest-effect levels (Laskowski, 1995).

More problems with the methods could be mentioned (Scott-Fordsmand et al., 1996). In spite of the limitations and the problems with the algorithm used, it was considered the most suitable approach at the time and thus was chosen for the establishment of ESQC in Denmark. In the future, the advances made for these extrapolation methods should be included in the algorithm for derivation of ESQC.

14.4 IMPLEMENTATION IN ENVIRONMENTAL POLICY

The Danish ESQC are considered as guidance values above which possible ecotoxicological effects may occur in soils. As seen from the above, they are of a general nature and not separated into different soil types or use of the land. Hence, they were not specifically developed with risk assessment of contaminated sites in mind but as general criteria of natural habitat land.

The typical risk assessors of contaminated sites in Denmark are private consultants with little or no ecological expertise. The assessors very often face a "So what?" question in situations where the ecotoxicological criteria are exceeded. Until recently, no overall framework on how and when to use the ecotoxicological criteria in connection to contaminated sites have been available for the risk assessor. The new Danish Soil Act still does not contain any mandatory enforcement concerning protection of soil ecosystems. With regard to contaminated sites protection of human health and groundwater are prioritized (Danish Ministry of Energy and Environment, 1999). However, in the Technical Guidance Documents on contaminated sites related to the Soil Act (Danish Environmental Protection Agency, 1998a,b), the ESQC are referred to as a support tool for the various regulations dealing with soil contamination. These guidance documents include for the first time the outline of an overall framework for how to proceed when the ESQC are exceeded. In cases where soil concentrations are above the criteria, a site-specific risk assessment may be proposed. It is, however, only an overall nonmandatory framework, listing some of the available tools in a tiered approach. In addition to ESQC, the tools include bioassays such as earthworm or springtail toxicity tests and field monitoring using biodiversity or indicator species as measures. The interpretation of the results from bioassays or field monitoring is still open to the assessor.

In addition to their application for contaminated sites, ESQC have so far been used as critical limits in the calculations of critical loads of lead and cadmium in Denmark (Bak and Jensen, 1999). Furthermore, in line with their application within the critical load concept, the criteria have been used for evaluating risk of contaminants dispersed with sewage sludge. The cutoff value for linear alkylbenzene sulfonates (LAS) in sludge and compost present in the Danish sludge ordinance (Danish Ministry of Energy and Environment, 1996) was derived on the basis of the ESQC for LAS. In Denmark, there are at present no attempts to further develop/derive ESQC.

ACKNOWLEDGMENTS

This chapter was undertaken as a part of the Centre for Biological Processes in Contaminated Soil and Sediment (BIOPRO) of the Danish Strategic Environmental Research Programme.

B. Ecological Risk Assessment

15 Probabilistic Risk Assessment Using Species Sensitivity Distributions

Keith R. Solomon and Peter Takacs

CONTENTS

Abstract — Ecological risk assessment (ERA) is the estimation of the risk of adverse effects to communities of species in habitats exposed to toxic substances. ERA includes considerations for the responses of highly varied receptor organisms as well as spatial and temporal differences in the exposures that these organisms experience. ERAs can be conducted for many reasons, ranging from the need for simple ranking systems to the need for more complex probabilistic methods. ERA is normally conducted in a series of tiers, each level with greater realism of exposure and toxicity measures and greater certainty in the values used. Early tiers in the process may make use of simple hazard quotients (HQs), whereas higher tiers make use of probabilistic procedures in which the likelihood of exposure and effect are considered. Probabilistic ERA (PERA) is based in the ecological principles of redundancy and resiliency, makes use of more of the information available to the risk assessor, and avoids the use of unrealistic worst-case scenarios. This chapter discusses approaches to risk assessment that have led to the use of species sensitivity distributions (SSDs) for establishing environmental criteria, describes their use in ecological risk assessment, and illustrates these with examples drawn from recent PERAs for pesticides and other substances affecting the ecosystem through toxicological mechanisms. The chapter focuses on more detailed issues related to the basic principles and methods. This chapter also discusses some of the basic considerations for the use of PERA in the new methods for pesticide risk assessments, such as those proposed by the Ecological Committee on FIFRA Risk Assessment Methodologies (ECOFRAM).

15.1 INTRODUCTION

Over the last five decades, the use of risk assessment to protect human health has expanded, but it has been only during the last two decades that ecological risk assessment (ERA), especially in aquatic environments, has come to the forefront of evaluating environmental pollution issues. ERA includes assessment of the potential impacts of a wide range of potential stressors, including those related to exposures to substances. In the context of the toxicology of substances, the aim of ERA is the estimation of the risk of adverse effects (such as lethality, reduced survival, and altered reproductive potential) to communities of species in habitats exposed to the substances and the probability of adverse effects posed by the introduction of new substances. It can also be used for the prioritization of pollutants for regulatory purposes, as well as in the development of environmental quality guidelines.

ERAs can be conducted for many reasons ranging from the need for simple ranking systems to the need for more complex probabilistic methods (Figure 15.1). This chapter discusses approaches to risk assessment that have led to the use of

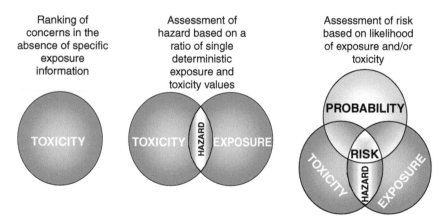

FIGURE 15.1 Illustration of the types of approaches to ERA.

species sensitivity distributions (SSDs) for establishing environmental criteria, describes their use in ecological risk assessment, and illustrates these with examples drawn from recent probabilistic ERAs (PERAs) for pesticides and other substances affecting the ecosystem through toxicological mechanisms. The mechanics of PERA have been described (Cardwell et al., 1993; Baker et al., 1994; Jongbloed et al., 1996; Klaine et al., 1996a; Parkhurst et al., 1996; Solomon, 1996; Traas et al. 1996; Giesy et al., 1999; Campbell et al., 2000), and the intent of this chapter is to focus on more detailed issues related to the basic principles and methods. This chapter also discusses some of the basic considerations for the use of PERA in proposed new methods for pesticide risk assessments (ECOFRAM, 1999a,b).

15.1.1 TIERS IN RISK ASSESSMENT

For logistical reasons, it is frequently necessary to divide complex tasks into smaller components that can be more easily managed or divided among workers. Nowhere is this more true than in ERA, where relationships between components of the ecosystem can be very complex. The use of tiers or steps in the process of risk assessment is one method used to reduce complexity and narrow the focus of risk assessments to the key issues and has been recommended frequently for use in ERA (U.S. EPA, 1992; 1998c; Suter, 1993; Baker et al., 1994; ECOFRAM, 1999a,b). The use of tiered approaches in risk assessment has several advantages for the risk assessor and those proposing the activity that causes the risk. The initial use of conservative criteria allows substances that truly do not present a risk to be eliminated from the risk assessment process, thus allowing the focus of expertise to be shifted to more problematic substances or situations. As one progresses through the tiers, the estimates of exposure and effects become more realistic as uncertainty is reduced through the acquisition of more or better quality data. Tiers are normally designed such that the lower tiers in the risk assessment are more conservative (less likely to pass a hazardous chemical), while the higher tiers are more realistic with assumptions more closely approaching reality. Because the lower tiers are designed to be protective, failing to meet the criteria for these tiers is merely an indication that an assessment based on more realistic data is needed before a regulatory or risk management decision can be reached. The Ecological Committee on FIFRA Risk Assessment Methodologies (ECOFRAM), the joint U.S. EPA, industry, and academia working group established to suggest new method for risk assessment of pesticides, has recommended that a flexible system of tiers be used in pesticide risk assessment (ECOFRAM, 1999a,b). The tiers consist of varied and progressively more sophisticated sets of tools than can be selected to answer specific questions as these are revealed in the ERA process. Some of these tools are deterministic; others are probabilistic.

15.1.2 CLASSIFICATION AND SCORING SYSTEMS

The simplest forms of risk assessment are the widely used scoring or classification systems. Examples of these include the U.S. Department of Defense (USDOD) prioritization model for waste sites, the U.S. EPA Office of Toxic Substances (OTS) chemical scoring system, and the U.S. EPA hazard ranking system (U.S. EPA-HRS) (Suter, 1993). The International Joint Commission has used a scoring system (IJC, 1993) to identify candidate substances for virtual elimination, and Environment Canada has developed a set of criteria for selecting substances that are potentially hazardous to the environment for their Priority Substances List (PSL) (Environment Canada, 1994). The basic principle of a scoring system is to assign a rank or priority to a substance. This is usually accomplished by assigning a score to several of the properties of the substances being assessed, manipulating these scores in some way, and then using the scores to rank (and select) some of these substances for further action. Some scoring systems use a single criterion for a property (greater than or less than), whereas others may use multiple criteria, which are assigned numerical scores. Correctly used, scoring systems have been employed to rank substances in order of priority for further assessment. This is usually carried out in the initial stages of risk assessment. Further assessment is normally required because the scoring systems commonly make use of worst-case data; they cannot always handle missing values, weighting, or scaling in clear or appropriate ways; and they take no consideration of exposures other than through estimates of total production, use, or release of the substance into the environment. The rank numbers produced from combinations of scores have no meaning in the real world; their only use is to allow prioritization of substances for a more detailed assessment. Unfortunately, scoring systems are often incorrectly used in place of a full risk assessment.

15.1.3 The Quotient Method of Hazard Assessment

The most widely used method in ERA is the hazard quotient (HQ) method, by which the environmental concentration of a stressor, either measured or estimated, is compared to an effect concentration such as the median lethal concentration (LC_{50}) or no-observed-adverse-effect concentration (NOAEC) (Urban and Cook, 1986; Calabrese and Baldwin, 1993). These are simple ratios of single exposure and effects values and may be used to express hazard or relative safety. For example:

$$\text{Hazard} \approx \frac{\text{Exposure concentration}}{\text{Effect concentration}} \quad \text{or Margin of Safety} \approx \frac{\text{Effect concentration}}{\text{Exposure concentration}}$$

The calculation of HQs has normally been conducted by utilizing the effect concentration of the most sensitive organism or group of organisms and comparing this to the greatest exposure concentration measured or estimated in the environmental matrix. Calculation of the HQ may be made more conservative by the use of an uncertainty (application) factor (CWQG, 1999), such as division of the effect concentration by a number, for example, 20. This is done to allow for unquantified uncertainty in the effect and exposure estimates or measurements. In this case, if the HQ is greater than 1, a hazard exists. Under the ecological risk assessment guidelines for pesticide risk assessment currently used by the U.S. EPA (Urban and Cook, 1986), the HQ [incorrectly referred to as the risk quotient (RQ)] is compared to a level of concern (LOC). Different LOCs are used for different classes of organisms, depending on the nature of the effect measure or whether endangered species are likely to be affected (Urban and Cook, 1986). For example, for aquatic organisms, LOCs range from 0.05 for acute risks to endangered species to 1 for chronic risks based on the no-observed-effects concentration (NOEC) as the measure of effect. In many hazard assessments, some form of uncertainty factor is incorporated into the calculation of the HQ, either explicitly as part of the calculation itself or in the criteria for acceptance of the HQ. Because they frequently make use of worst-case data, they are designed to be protective of almost all possible situations that may occur. However, reduction of the probability of a type II error (false negative) through the use of very conservative application factors and assumptions can lead to the implementation of expensive measures of risk mitigation for stressors that pose little or no threat to humans or the environment (Lee and Jones-Lee, 1995; Moore and Elliott, 1996). A common error in the interpretation of HQs is the assumption that the HQ itself is proportional to the "risk." As the concept of risk should always incorporate an element of probability, the HQ is biased because it assumes that the conditions of the HQ exist on every occasion and in every location. In addition, the HQ is based on a point estimate of effect (LC_{50} or NOAEC) and does not consider the relationship between the concentration and the effect. Depending on this relationship, usually expressed as the slope of the concentration or dose–response line, the same HQ may have different interpretations. For example, different HQs result from the use of different measures of effect such as the LC_{50} and the LC_5, the concentration at which 5% of the test organisms would respond (Figure 15.2). Thus, the HQ approach is only useful for early tiers, preliminary risk assessments, or as a screening tool.

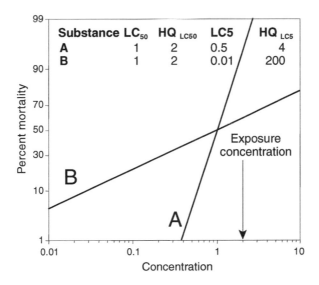

FIGURE 15.2 Illustration of the relationship between slope of two hypothetical concentration–response lines and HQs at different intensities of effect.

15.1.4 PROBABILISTIC APPROACHES

Probabilistic approaches to risk assessment have been suggested for upper tiers in the ERA process (Baker et al., 1994; ECOFRAM, 1999a,b). These methods use distributions of species sensitivity combined with distributions of exposure concentrations to better describe the likelihood of exceedences of effect thresholds and thus the risk of adverse effects. Using distributions of species sensitivities is not a new idea. Distributional approaches have been used in the regulation of food additives for the protection of human health for several years (Munro, 1990). From the environmental point of view, distributions of LC_{50} values were used to distinguish between resistant and susceptible populations of animal ectoparasites (Solomon et al., 1979), but the notion of using these distributions for setting environmental quality guidelines originated from early work in the Netherlands (Van Straalen, 1982, personal communication; Kooijman, 1987) and the United States (U.S. EPA, 1985a). Other authors who have expanded upon the probabilistic risk estimation process include Aldenberg and Slob (1993), Okkerman et al. (1991; 1993), Van Straalen and Denneman (1989), and Wagner and Løkke (1991). The suggestion to compare distributions of species sensitivity directly to distributions of exposure concentrations (Cardwell et al., 1993) was recommended for pesticide risk assessment by the Aquatic Risk Assessment Dialogue Group (Baker et al., 1994), demonstrated for metals and other substances (Parkhurst et al., 1996), and incorporated in a computer program (The Cadmus Group, Inc., 1996a). In this form, PERA has been used in a number of risk assessments for pesticides and other substances (Klaine et al., 1996a; Solomon et al., 1996; 2001; Cardwell et al., 1999; Giesy et al., 1999; Hall et al., 1999; Campbell et al., 2000; Giddings et al., 2000; 2001; Hendley et al., 2001; Maund et al., 2001; Travis and Hendley, 2001), and has been recommended for regulatory risk assessment of pesticides (ECOFRAM, 1999a,b). The general concepts as they apply to ecological and human health risk assessment have been reviewed and extensively discussed (Forbes and Forbes, 1993; 1994; Suter, 1993; Balk et al., 1995; Anderson and Yuhas, 1996; Burmaster, 1996; Power and McCarty, 1996; Richardson, 1996; Solomon, 1996; Bier, 1999; Roberts, 1999), and this discussion is extended in other chapters in this book.

There have been several major driving forces behind the shift toward PERA:

1. A consensus that worst-case scenarios can often overestimate exposure;
2. Probabilistic methods give more usable information to environmental managers and regulators;
3. Demands by registrants and users of pesticides to adopt more realistic standards; and
4. To "scientifically prove" the justification of these standards in order to reduce the cost of implementation and use of risk-mitigating measures (Suter, 1993; Paustenbach, 1995; Richardson, 1996).

The major advantage of PERA is that it uses all relevant single species toxicity data and, when combined with exposure distributions, allows quantitative estimations of risks. In addition, the data may be revisited, the decision criteria become more robust with additional data, and the method is transparent, producing the same results with the same data sets. The method does have some disadvantages. More data are usually needed, although these are mostly low-cost studies. For new or poorly studied substances where there are few environmental concentration data, models that have not been widely verified or validated have to be used to estimate exposures. It is not easily applied to highly bioaccumulative substances where exposure is via the food chain as well as the environmental matrix.

15.2 EXPRESSION OF RISKS IN PERA

Assessment endpoints and measures of effect can be defined at all levels of organization in ecosystems, from the individual to the community. However, these are not necessarily of equal importance (Suter, 1993). In contrast to human health protection, individual organisms in the ecosystem are generally regarded as transitory and, because they are usually part of a food chain, are *individually* unessential for maintaining ecosystem function (Suter, 1993). A self-maintaining or reproducing population is persistent on a human timescale and can be easily conceptualized by humans as being in need of protection. Thus, most assessment measures in ERA are defined at the population, rather than at the level of the individual organism. Only in the case of the protection of rare, endangered, or long-lived species are organisms in the environment afforded similar protection to that enjoyed by individual humans. Generally, ERA is aimed at protection of populations, or the functions of communities and ecosystems. This acknowledges the fact that populations are less sensitive than their most sensitive member and, similarly, that functions of communities and ecosystems are less sensitive than those of their most sensitive components. Effects on a population are not necessarily of concern (to the ecosystem) as long as the functions of the population can be taken over by other populations. In this context, function is the interaction of the population with other populations or with the abiotic environment. Functions in ecosystems are normally related to energy and nutrient flow: production and consumption of biomass, controlling the abundance of other species, providing food to predators, processing organic detritus, and mineralizing organic compounds (Suter, 1993).

Functional redundancy is essential to the continuance of ecosystems in the face of natural stressors, such as the effects of winter in temperate climates. Redundancy is the result of selection imposed by fluctuating and unpredictable environmental conditions. Most ecosystems exhibit functional redundancy, where multiple species are able to perform each critical function (Walker, 1992; 1995; Baskin, 1994). Functional redundancy is particularly relevant to the justification for the use of PERA. It is the basis for being able to tolerate effects in some sensitive populations because these are unlikely to impair the functions of the ecosystem as a whole. This was the principle upon which the U.S. water quality guidelines were developed (U.S. EPA, 1985a). As is illustrated in Figure 15.3, there is a general relationship between exposure concentration and the impact of any substance; however, there are deviations from this general rule. For example, functions may be maintained where few species are affected, but as the number of species affected increases, indirect effects amplify the effects of the substance to greater than predicted levels. Redundancy of

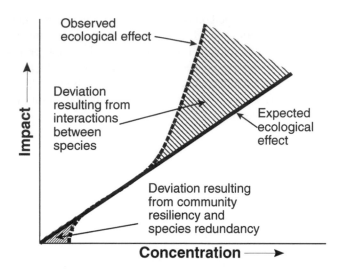

FIGURE 15.3 Illustration of ecosystem resiliency in response to stressors and effects caused by interactions between organisms. (Redrawn from a paper napkin original by David Hansen.)

function has been observed in a number of experimentally manipulated systems ranging from terrestrial (Tilman, 1996; Tilman et al., 1996) to aquatic (Stephenson et al., 1986; Giddings et al., 1996; Solomon et al., 1996). The exact nature of the relationship between function and structure in ecosystems has been debated (Wardle et al., 2000) and undoubtedly varies with the type of system and functional classes it contains; however, thresholds in structural changes below which functions are unaffected are consistently observed. These observations support the concept that, in ERA, some effects at the level of the organism and population can be accepted, provided that these effects are restricted on the spatial and temporal scale. In other words, they do not affect all communities all of the time, and keystone organisms are not adversely affected. In jurisdictions where risks and benefits are considered together in decision making, it has become increasingly recognized that assessment measures should be at the functional level of populations and the community and that some effects on populations and species diversity may also be tolerated if there is a counterbalancing benefit. In other jurisdictions, particularly where consideration of risk–benefit is not mandated, assessments continue to be directed toward the protection of populations.

It is important to recognize that, although all PERA methods are essentially similar, they may be used for different purposes. In some uses, PERA methods are used to set environmental guidelines and criteria. In others, PERA is used to assess risks in situations where exposures are known, and their significance is being assessed. In the first use, the objective is to determine a single specific "protective" criterion, whereas, in the second, the task is to determine which of a series of measured or estimated concentrations is causing unacceptable risks.

15.2.1 CRITERIA SETTING

For the setting of criteria, an *a priori* decision must be made regarding what level of protection is acceptable. Normally, this level of protection is expressed as a percentage of species for which protection should be sought (Forbes and Forbes, 1993; Balk et al., 1995). Suggestions for these vary from 100% of the species 95% of the time (Kooijman, 1987) to 95% of species 95% of the time, that is, a hazardous concentration (HC_5) that will exceed no more than 5% of the species NOECs (Van Straalen and Van Rijn, 1998). The U.S. EPA method of setting water quality criteria (U.S. EPA, 1985a) uses a 95% protection level but applies this to acute and chronic effects, as final acute and final chronic (FAV and FCV, respectively). Similar techniques have been recommended for the determination of water quality criteria in the North American Great Lakes Initiative (U.S. EPA, 1995c). The report of the Aquatic Risk Assessment Dialogue Group (Baker et al., 1994) suggested that protection of 90% of species be used in assessing risks from pesticides; however, this was in a context different from that of criteria setting (see further discussion below). Based on the theoretical concept of the tails of the distribution extending to infinity, setting the criteria of protection at 100% of the species would be impractical and is counter to the biological evidence for thresholds of toxicity, below which no organisms would be affected. In setting *a priori* criteria, there is no generalized guideline regarding which species should be included in the SSD. Attempts to select "representative species" by including sensitive, moderately sensitive, and tolerant species could result in a distribution that is more uniform than the assumed distribution and thus lead to an overestimate in the variance in sensitivity (Forbes and Forbes, 1994). This would be more likely to occur when data for only a few species are available (as is often the case) with the result that extreme values are overrepresented. This reduces the slope of the cumulative frequency distribution plot and could lower the criterion value (e.g., 5th centile) to an unrealistic level (Forbes and Forbes, 1994; Balk et al., 1995). Theoretically, species should be selected randomly for inclusion in sensitivity distributions. This is consistent with the original underlying probabilistic assumption; however, it is not possible from a practical standpoint as some organisms cannot be tested under laboratory conditions. These constraints present interesting challenges that are discussed in more detail in other chapters.

15.2.2 RISK ASSESSMENT

Another use of PERA is in assessing risks from situations that already exist. For example, where a substance is being released, or is about to be released, into the environment and a risk assessment is needed to make a risk–benefit regulatory decision. In this case, no predefined percentage of species to protect is needed as this will vary from one situation to another, depending on other lines of evidence, such as the types of organism most likely to be affected or the toxicological properties of the substance. Unlike the process of criteria setting, it may be very appropriate to exclude certain types of organisms from the SSDs or, based on biological knowledge, to tolerate more frequent exceedences of species sensitivity values for some taxa than for others. Because the potential adverse effects of measured or estimated exposures are being assessed in this method, the combination or segregation of exposure data sets adds significant utility to the risk assessment process. It allows more realistic toxicity and exposure information to be applied to ranking of exposure scenarios for the purposes of mitigation or regulatory decision making. Because risk assessment considers both the likelihood of exposure concentration occurring and the likelihood of species being sensitive, risk can be expressed as a joint probability. For example, $n\%$ of species will be affected $x\%$ of the time or in $y\%$ of the locations, depending on the type of exposure data collected. These probabilities can be expressed as the probability of exceeding a fixed criterion on the SSD, such as, for example, the 10th centile of the distribution of all species or a distribution of inherently more sensitive species (Solomon et al., 1996). Another method of presenting these joint probabilities is in the form of an exceedence curve, called a joint probability curve (JPC) in ECOFRAM (1999a). This format was suggested in the Aquatic Ecological Risk Assessment (AERA) program (The Cadmus Group, Inc., 1996a), recommended for displaying risks by the Aquatic Working Group of ECOFRAM (ECOFRAM, 1999a), and has been used to display risks for pesticides such as chlorpyrifos (Giesy et al., 1999) and diazinon (Giddings et al., 2000). The derivation of the exceedence profile (JPC, Figure 15.4) is relatively simple and offers a useful tool for communication of risks as it allows what-if questions to be addressed and gives the risk assessor and risk manager a method for assessing the effects of

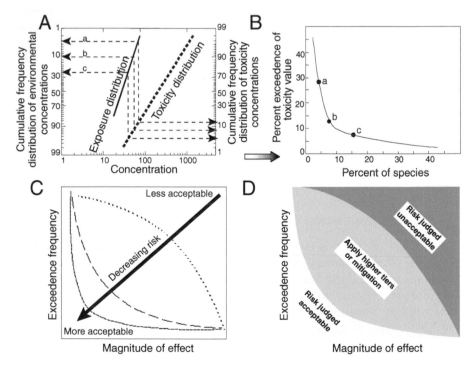

FIGURE 15.4 Presentation of exceedence probabilities (A) as a continuum of likelihoods in an exceedence profile (B) and the use of these curves in decision making (C and D). (Adapted from ECOFRAM, 1999a.)

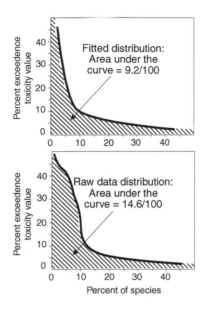

FIGURE 15.5 Illustration of the use of area under the exceedence profile as an indicator of estimated total risk for the purposes of ranking risk.

changes in assumptions, such as the choice of a different centile from the SSD (ECOFRAM, 1999a,b; Giesy et al., 1999).

In a strictly numerical sense, the area under the exceedence profile summarizes the ecological risk and can be used for objective ranking of risk scenarios where data are available from multiple sites or different exposures periods. Thus, for example, the area under the curve can be calculated and normalized to a total possible area of 100 (Figure 15.5). These estimated total risks can then be used for ranking purposes. This can be easily done with a spreadsheet from equations fitted to the species and concentration distributions or directly from raw data if they do not fit a specific distribution.

15.3 METHODS FOR CONDUCTING PERA

PERA will undoubtedly continue to evolve as risk assessors obtain more and more experience with these and related techniques. The objective of this section is to describe several practical issues in PERA and how these have been approached in published and proposed PERA methods. Some of these points were discussed at length in the workgroups of ECOFRAM, and the reader is referred to the draft final reports of ECOFRAM for more details (ECOFRAM, 1999a,b).

15.3.1 THE FRAMEWORK

As all other risk assessment processes, PERA is a subset within a larger framework of risk assessment. The framework and guidelines proposed by the U.S. EPA (U.S. EPA, 1992; 1998c) have been widely used for risk assessment and their similarity to those developed by other agencies (NRC, 1993; CEPA, 1997) makes them widely applicable. This section focuses on issues related to probabilistic analysis, but it is important to recognize the need for the preliminary steps recommended in the framework. The characteristics of the stressor substance that govern its environmental fate are important in determining the representativeness of any environmental samples that may be available and the need to consider long- or short-term exposures in toxicity characterization. For example, all pesticides have a clearly defined use pattern. Based on the distribution of target pests and their host crops, sales, or use of the product, a knowledge of the physical and chemical properties that determine off-target movement, and the features of the abiotic environment that interact with the pesticide, it is possible to determine the most likely locations where toxicologically relevant exposures will occur (Solomon, 1996; Giesy et al., 1999; Hall et al., 1999; Giddings et al., 2000; Hendley et al., 2001). Similarly, climate or use patterns may affect seasonal occurrence of the pesticide in the environment and will determine the need to consider seasonal biological cycles in terms of possible increased sensitivity of some life stages or predator–prey relationships to synchronous exposures (Giddings et al., 2000).

15.3.2 THE EXPOSURE PROFILE

Measuring exposure in environmental matrices is one of the critical components of risk assessment but can be subject to errors through improper sampling techniques and incorrect analysis. Obtaining an unbiased and representative sample from an environmental matrix may be very difficult and costly, yet it is probably the most important part of any exposure characterization. Sampling must consider both the temporal and spatial heterogeneity of the stressor. For example, the concentration of a pesticide may vary with water depth or distance from the shore immediately after a spray-drift contamination of water. Similarly, the concentration of a stressor in flowing water may decrease over distance from the source of contamination because of breakdown in the water, adsorption to sediments, or dilution from uncontaminated water entering downstream of the source. Concentrations in soil may vary with soil type, with the chemical and physical properties of the pesticide, and with climatic factors such as rainfall, percolation, or leaching through the soil.

The objective of sampling the environmental matrix is to obtain a characterization of exposure that will be useful in the risk assessment process. Even with a good sampling design to address spatial heterogeneity, temporal variations in concentrations may be very important in assessing risks in relation to duration of exposure and choosing the appropriate exposure time for the toxicity data (ECOFRAM, 1999a). For example, in a headwater stream system, the duration of peak exposure concentration may be short because of the hydraulics of the system and may be easily missed with a single daily grab sample. A continuous sampling system where daily integrated sampling was carried out would incorporate these exposures; however, peaks of duration less than 1 day could be obscured. Sampling intervals should be designed to take into account the known hydraulics and breakdown kinetics of the stressor in question. Thus, in small headwater streams, more frequent sampling with intervals of less than 1 day may be more appropriate. For slow-flowing rivers, or a rapidly degrading substance in a pond or reservoir, daily sampling may be adequate. For slowly degraded substances in stagnant pools, ponds, or reservoirs, even less frequent sampling may be sufficient.

Many historical data sets of measured environmental concentrations were not collected for the express purposes of probabilistic risk assessment. In many cases, limited resources have resulted in the collection of samples during those periods when greater concentrations would be expected. Thus, for a pesticide, periods of rainfall or the use season may be sampled more frequently than other times. Obviously, a simple distributional analysis of these data will be biased toward times when concentrations may be higher. Incorporating samples taken on short intervals with those taken on long intervals in the distribution results in more weight being assigned to the short-interval samples. Time-weighted mean concentrations (TWMCs) have been used as a method for unifying data sets with unequal sample intervals (Solomon et al., 1996). The time basis can be matched to that used for toxicity assessment (96 or 48 h; Solomon, 1996; Giesy et al., 1999), or simple daily TWMCs can be calculated. This method does cause some problems if samples are taken over periods longer than the TWMC interval. In this case, an assumption of concentration may need to be made during the period of no observation, and this may introduce an artifact in the data set. The artifact will likely only influence the lower concentrations but will have to be considered in selecting the distribution model, if one is used.

The concentration–time series of data that results from environmental sampling can be analyzed by means of postprocessor tools such as the Risk Assessment Tool to Evaluate Duration and Recovery (RADAR) developed as part of the efforts of ECOFRAM (1999a). This tool provides information on pulse height, width, and interpulse interval, which is particularly useful for assessing likely effects on classes of organisms with known recovery times and time–exposure responses.

In many risk assessments, the actual stressor concentrations in the environment cannot be measured, and risk assessors must use models to predict these concentrations. Models may be used in Monte Carlo simulations where measured or estimated distributions of input values are used to generate distributions of output values (Klaine et al., 1996a; ECOFRAM, 1999ab; Ritter et al., 2000). Output from Monte Carlo simulations is useful for distributional and probabilistic analyses and many possible time intervals can be utilized, depending on the model. The ability to select a model that will "detect" peak exposures of short duration that may be missed in environmental sampling is attractive; however, if the model is in error, the error will be propagated through the entire data set. Use of Monte Carlo analysis also requires additional information on the distributions of input values forcing the use of default or assumed values when insufficient data are available.

In terms of risk assessment for pesticides, several multicompartment models for estimating pesticide concentrations in environmental matrices are available. The most simplified of these is the Generic Expected Environmental Concentration model (GENEEC). GENEEC (version 1.3: Parker, 1999) mimics a Pesticide Root Zone Model/Exposure Analysis Modeling System (PRZM/EXAMS: Burns, 1997; Mullins et al., 1993) simulation of a generic 10-ha row crop field draining into a 1-ha, 2-m-deep farm pond. It incorporates spray drift to estimate the concentration in water at various times after a contamination event and has a choice of several crop types.

GENEEC is designed as a Tier 1 model. It is conservative and only gives one output for each use scenario. A more complex combination of EXAMS and PRZM has been used with a preprocessor called the Multiple Scenario Risk Assessment Tool (MUSCRAT; ECOFRAM, 1999a). MUSCRAT is an application program that links chemical, crop, soil, and climate databases and facilitates the creation of PRZM-3 and EXAMS-II input files, batch-processes multiple model simulations, and performs statistical analyses on predicted exposure concentrations for pesticides (ECOFRAM, 1999a). It gives multiple values as output, and these output values can be analyzed as distributions rather than as single deterministic values. MUSCRAT is designed to provide modeled data for use in higher tiers of the probabilistic risk assessment process as recommended by ECOFRAM (1999a).

The modeling process for estimating exposures in aquatic media was proposed by ECOFRAM to take place in a series of tiered steps of increasing complexity to more closely approximate actual field concentrations with each tier (ECOFRAM, 1999a). These tiers are linked to similar steps in the characterization of effects. The lower tiers produce more conservative numbers; however, it is recognized that the models may be more conservative than the actual values (Figure 15.6).

The application of distributional analysis to concentrations of substances in the environment must be done with consideration for the fact that these data are usually censored by the limits of analytical detection (Figure 15.7). In practice, all exposure concentration data below the limit of detection (LOD) or limit of quantitation (LOQ) are assigned a dummy value of zero. These data are used in the calculation of the total number of samples (n) but are not used to estimate centiles directly. The assumption here is that the values below the LOD lie on the same distribution as the values above the LOD. With recent advances in analytical chemistry, values

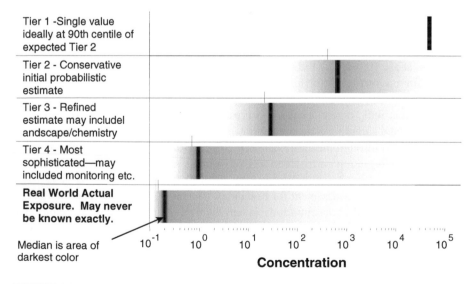

FIGURE 15.6 Illustration of the modeling scheme proposed for aquatic matrices. (Adapted from ECOFRAM, 1999a.)

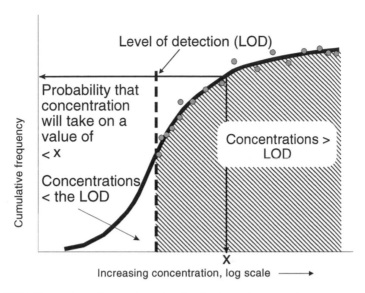

FIGURE 15.7 Illustration of censoring of a distribution of measured concentration data by the level of detection of the analytical method.

below the LOD are usually of little toxicological significance; however, the regression equation for the distribution may be used to estimate the concentration of data points below the LOD for the purposes of developing an exceedence profile plot. The substitution of a value of half the LOD for all the data points below the LOD, a practice used for estimating mean concentration of a data set, results in a biased data set that will be difficult to fit to any model.

Frequency of exposure is a very important consideration in PERA. The return frequency of an event (how often the event happens) is important in the choice of methods for probabilistic risk assessment and is related to the ecological cost of recovery from the event (Solomon, 1996). In assessing exposure, the return frequency protected against should be consistent with the resiliency of vulnerable populations. Resiliency is determined by life cycle characteristics, the reproductive capacity of the potentially affected organisms, and the ability of their populations (or their function in the ecosystem) to recover from the event. This should be addressed early in the problem formulation stage of the assessment as it affects the exposure and the effects analysis. The Aquatic Risk Assessment and Mitigation Dialogue Group (Baker et al., 1994) recommended conservative approaches to PERA, such as the use of low return frequencies (i.e., one or fewer occurrences in 30 years). This safeguards all organisms in situations where limited information is available on the toxic mode of action or the sensitivity of species. Where better information is available, more appropriate return frequencies may be used. For example, more frequent exposure to a stressor may be tolerated where a stressor affects organisms with short life cycles and high rates of reproduction. In temperate regions, many ecosystems undergo a period of dormancy and the system is, in a sense, reset seasonally by the winter. Thus, for some organisms, mechanisms for propagation beyond the winter reset already exist, and dormant stages are produced from which populations will develop in the next season. Similar mechanisms exist in environments where ephemeral water bodies are subjected to a seasonal drying out. Therefore, as many organisms in these regions undergo seasonal resets, a stressor return that occurs less frequently than once per season is likely to be tolerable from the viewpoint of the long-term productivity of the population and the sustainability of function in the ecosystem, especially if the effects are spatially restricted and repopulation can occur from refugia.

However, protection of longer-lived species without seasonal resets, such as some fish, birds, or mammals, may require the consideration of return frequencies of several years. From a practical point of view, the annual maximum concentration of the stressor is a useful starting point for the purposes of PERA, as it is a conservative return frequency for most organisms. The likelihood of annual maxima exceeding a toxicity threshold may then be assessed using a fixed centile from the SSD or an exceedence profile. As has been pointed out (ECOFRAM, 1999a), shorter return frequencies can be tolerated if organisms with different reproductive strategies are the most likely to be affected. For example, unicellular algae are among the most sensitive aquatic organisms to atrazine. These algae have rapid reproduction times and can quickly recover from effects, even if these were to occur with return frequencies of 30 days (Solomon et al., 1996). Similarly, many zooplankton species can recover from reductions in numbers of several orders of magnitude within a month or two (Kaushik et al., 1985; Stephenson et al., 1986). From a practical point of view, monthly return frequencies may be more important in assessing risks in these types of organisms.

Along with the consideration of return frequency, the time between events may also be important. This issue was specifically addressed in the development of the RADAR post-processing tool (ECOFRAM, 1999a). Again, it is well known that organisms may recover from exposure to a toxic substance if this substance does not cause mortality but merely acts as a stasis agent. For example, atrazine and the other triazine herbicides are reversible inhibitors of photosynthesis, and when an exposed plant is moved to uncontaminated environment, it will continue to grow and develop (Jensen et al., 1977; Klaine et al., 1996b), provided that the exposure time has not been so long that its internal energy reserves have been depleted to the point of lethality. Standardized toxicity assays for algae are based on growth as an endpoint, and concentrations of a substance that inhibit growth are often assumed to cause death. Concentrations of stressors that result in death have been shown to be different from those that result in the reduction of growth (Faber et al., 1997). Practically, for stressors such as these, PERA should be focused on the likelihood of concentrations exceeding a certain threshold value for a certain time and then followed by a time interval judged to be inadequate to allow for recovery to occur.

In practice, the likelihood of occurrence of exposures that meet certain criteria can be assessed. These criteria may, for example, consist of an effect concentration threshold, a duration, and an interval between exposures. As illustrated for a measured data set for atrazine in Lost Creek, Ohio (Solomon, 1996), no exposures exceeded the threshold of 5.6 µg/l for a period of more than 6 days, and many of these were followed by a considerable period of exposure less than the threshold

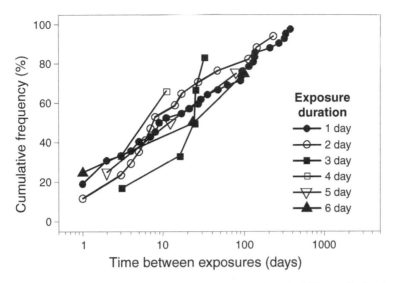

FIGURE 15.8 Likelihood of atrazine concentrations exceeding the 10th centile for the plant SSD for several combinations of duration and interexposure interval in Lost Creek, Ohio. (Based on a reanalysis of raw data from Solomon et al., 1996.)

(Figure 15.8). Comparison of these exposures to laboratory studies with algae that demonstrated full recovery after up to 14 days of exposure to atrazine (Klaine et al., 1996b) suggested insignificant risk to algal populations in this particular situation; however, the example illustrates how the method could be used.

15.3.3 THE TOXICITY PROFILE

Standardized test methods for measuring toxicity are routinely used and required by a number of regulatory agencies. The basic principle behind the use of standardized laboratory toxicity tests is not that the particular organisms in the tests are those that require protection in the environment but rather that these organisms act as surrogates for all those other organisms in the ecosystem that could be exposed but that cannot be tested in the laboratory. Test organisms are usually selected for ease of use and because historical testing has shown that the species is somewhat sensitive and would provide a reasonable worst-case measure of effect. To make the effect measure even more conservative, the tests are normally conducted under conditions where the exposures are maintained at a constant concentration, usually by continuous addition to a continuous-flow treatment system. Maintenance of constant exposure concentration is suitable for effluent testing; however, it may not be appropriate for short-lived substances used relatively infrequently in the environment.

The Framework for Ecological Risk Assessment (U.S. EPA, 1992; 1998c) defines a number of suitability criteria for effect measures (measurement endpoints). One of the most important of these is a clear, mechanistic, and preferably quantitative linkage between the measure of effect and the endpoint that is being assessed. Because the assessment measures for characterizing the risk of pesticides in the ecosystem are normally at or above the population level, measures of effect that are relevant to the sustainability of populations are more appropriate than those that may merely indicate exposure or adaptation to a stressor. Measures of effect that reflect survival (mortality), growth, development, or reproduction should thus be chosen over biomarker responses (as defined in Huggett et al., 1992) unless there is a clear linkage between the biomarker and a population- or community-level effect. Useful effect measures include acute and chronic LC and EC values (where the effect observed is clearly related to population sustainability).

The LC_{50} or EC_{50} is a commonly reported measure of effect. Knowledge of the concentration–response relationship can allow the entire range of concentration–responses to be estimated, but these data are commonly absent from many published papers and reports. It may be argued that the LC/EC_{50} is inappropriate for use in risk assessment as it implies that half the population of organisms could be affected. However, in some organisms, the LC/EC_{50} may be a useful or even a conservative effect measure from the point of view of the population. Studies of effects of insecticides on zooplankton and terrestrial arthropods have shown that LC_{50} values were a conservative predictor of effects at the population level. Populations of *Daphnia galeata mendota*, exposed to concentrations of the pyrethroid insecticide fenvalerate that caused some mortality in individuals, were able to sustain a rate of increase similar to those of unexposed controls (Day and Kaushik, 1987). Similar results have been reported for the insecticide dieldrin in *D. pulex* (Daniels and Allen, 1981). Populations of the aphid *Acyrthosiphon pisum*, exposed to the 72-h LC_{60} concentration of the insecticide imidacloprid, were able to maintain rates of population increase similar to untreated controls (Walthall and Stark, 1997). This lack of population-level responses, even at exposure concentrations above the LC_{50}, was attributed to compensatory mechanisms where the unaffected individuals are able to maintain heightened rates of reproduction due to decreased competition for limiting resources (Walthall and Stark, 1997).

To ensure that the results of a distributional analysis are as representative as possible of the range of sensitivities that exist in an environment being assessed, the data set should be as large as possible. As large toxicity data sets developed under specific good laboratory practice (GLP) and quality assurance (QA) guidelines are not normally required for regulatory purposes, data from other sources can be included in the distributional analyses. Inclusion of these data requires that some selection criteria be used to ensure the best quality of data. ECOFRAM has recommended that the Great Lakes Initiative Bioassay Quality Criteria (U.S. EPA, 1995c) be used. Protocols that do not conform to these criteria could be judged suitable based on supplemental knowledge of octanol–water partition coefficient (K_{OW}) and stability in the matrix (ECOFRAM, 1999a). For example, a test with a highly water-soluble and persistent substance under static conditions with nominal concentrations may be judged acceptable, whereas a similar test with a strongly adsorbed and nonpersistent substance would be judged less acceptable.

Species toxicity data reported as less than a value should not be included in the SSD (ECOFRAM, 1999a). If the only response concentration for a species is reported as a greater-than value, this datum should not be used in the distribution as it would result in a skewing of the distribution. However, it should still be used in the denominator (n) for the calculation of rank. Organisms that only respond to high concentrations would likely be judged to be at low risk and may be omitted from the risk assessment process when grouping is used to examine the toxicity data further. However, because of differences between laboratories and in experimental designs, it may be that effects in a species are reported at concentrations greater than the maximum used in other laboratories. ECOFRAM identifies these situations as requiring expert judgment in the formal sense (ECOFRAM, 1999a).

Where multiple toxicity data are reported for a single species and one or more of these are reported as greater-than values, these values should be ignored and the remainder treated as suggested below. Where responses are reported at concentrations above the solubility limit for the substance, they should be treated as greater-than values as above. Formulants added to the end-use pesticide products may increase water dispersability. Although this does not increase solubility, emulsified active ingredients present at concentrations greater than their water solubility may be as biologically available as they are below their maximum water solubility. In these cases, the data could be used. In the case of the insecticide permethrin, more than one third of the toxicity values reported in the literature (Solomon et al., 2001) were above the reported water solubility limit of 6000 ng/l. These concentrations probably do not reflect actual exposure concentrations for dissolved permethrin. It is appropriate to exclude these values from the regression, but to include them in the total number of data points (n) for the purpose of calculating ranks. When the permethrin LC_{50} values in excess of the solubility limit are excluded from the lognormal regression, the fit of the line to the data improved (Figure 15.9), and the slope and intercept of the line changed resulting in less conservative lower centiles.

Frequently, multiple reports of toxicity tests carried out on the same organism are found in the literature or even submitted to regulatory agencies. To avoid bias, an organism can only be represented once in a distribution and the following rules for dealing with these situations have been suggested (ECOFRAM, 1999a). If multiple data points for a single species include a life stage (i.e., larval or juvenile) that is known to be inherently more sensitive to the stressor, these data should be chosen over responses for less sensitive stages. If, after all the above criteria are applied, two or more data points are still available for analysis, the geometric mean of the remaining data points should be used to represent the sensitivity of that species. Although the geometric mean has been criticized for some uses (Parkhurst, 1998), toxicity data are essentially lognormally distributed and the geometric mean is more appropriate. The procedure has traditionally been used to estimate central tendency for toxicity data (U.S. EPA, 1985a) and gives a more conservative estimate than would the arithmetic mean.

Many substances such as pesticides have some degree of specificity in their mechanism of action. For example, herbicides may be selectively toxic to some groups of plants (weeds vs. corn) as well as being less toxic to animals and other organisms that do not possess the receptor system (e.g., photosynthesis) for the pesticide (Figure 15.10). Similarly, an insecticide that acts on the nervous system of insects is unlikely to be highly toxic to plants. However, organisms with well-developed nervous systems (insects, other arthropods, and vertebrates) are likely to be more sensitive. But, as is illustrated in the case of permethrin, even these organisms may show differences in sensitivity because of divergence in their physiology

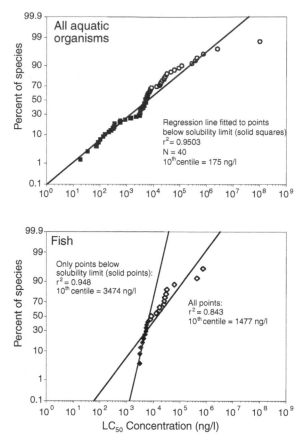

FIGURE 15.9 Distribution of LC/EC$_{50}$ values for the insecticide permethrin for all aquatic organisms and for fish with points above the solubility limit (6000 ng/l) excluded from regression. (Redrawn from ECOFRAM, 1999a.)

or biochemistry (Figure 15.11). Specificity of action may not always be the case. For example, some biocides, such as the chlorophenols, are similarly toxic to a wide range of organisms (Liber et al., 1994), and the grouping of all organisms together for distributional analysis may be appropriate. Thus, from a basic understanding of the mechanism of action of a pesticide, it may be possible to identify and group sensitive organisms. Given a range of exposure concentrations in the environment, these organisms are more likely to be adversely affected. This is helpful from the point of view of risk assessment as it allows the assessor to focus on the groups at higher risk and to devote less time and resources to groups that are exposed to very low or negligible risks. In addition, with a knowledge of the ecology of the potentially impacted groups of organisms or systems, it is possible to assess the likelihood that indirect effects will occur as a result of an effect on keystone groups of predator or prey/food organisms.

Although the mechanism of action of the pesticide is an important criterion for grouping of organisms, habitat may also be important. For example, there may also

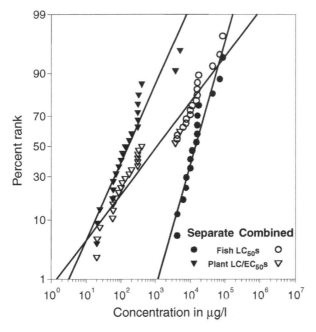

FIGURE 15.10 Illustration of the distribution of toxicity values for atrazine in fish and plants analyzed separately and combined into a single data set. (Data from Solomon et al., 1996.)

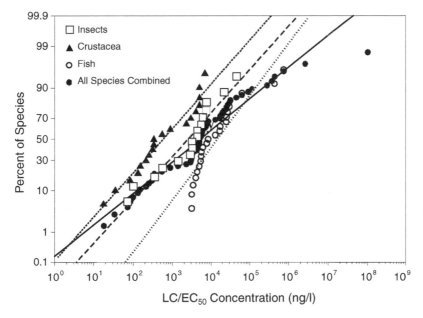

FIGURE 15.11 Distribution of permethrin LC_{50} values for fish, crustacea, and insects. (Redrawn from ECOFRAM, 1999a.)

be good mechanistic reasons to separate data for freshwater (FW) and saltwater (SW) organisms, where it is known that one group has an inherently different sensitivity because of interactions between salinity and the stressor of concern (Hall and Anderson, 1995). In other cases, this may not be necessary.

It is also possible to group organisms together on the basis of their reproductive strategy and life cycle. Thus, organisms that are able to recover rapidly from an adverse effect at the population level (reduction in numbers caused by mortality) may be considered differently from another group of organisms that may require a longer period of recovery. For example, most microalgae have short reproductive cycles and would be expected to recover from a decrease in population more rapidly than a population of fish subjected to a similar reduction. Thus, the frequency of occurrence and the intensity of the effect that could be tolerated would be different. This is also important when deciding how the exposure data should be analyzed in terms of frequency of occurrence.

Analysis of SSDs for a number of pesticides has shown that, when organisms of inherently differing sensitivity are grouped together, the resulting distribution did not display a good fit to the lognormal model, and larger data sets were required to describe the distribution adequately (Newman et al., 2000). However, when data were segregated into groups, such as fish, phytoplankton, etc., data sets fitted the lognormal model more closely, and fewer data points were required to describe the distribution adequately (Newman, personal communication). On this basis of the points discussed above, a number of potential criteria for grouping of effects data are suggested in Figure 15.12 (ECOFRAM, 1999a).

As discussed previously for exposure concentrations, it may be necessary to adjust toxicity data to match it more closely to the most likely exposure periods. If bioassay data for the appropriate time interval are available, they can be used, or the time–concentration function can be used to adjust toxicity data from longer time intervals to shorter time intervals (Giesy et al., 1999). Knowledge of the mechanism of action or appropriate toxicity tests that include observations after exposure has ceased should be considered to ensure that latency of toxic action does not occur (ECOFRAM, 1999a).

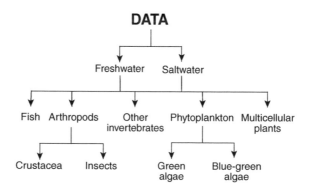

FIGURE 15.12 Possible groupings for toxicity/species sensitivity data sets. (Redrawn from ECOFRAM, 1999a.)

15.3.4 CALCULATING RANKS FOR THE DISTRIBUTION

Distributional analysis of toxicity and exposure data is based on the assumption that the data represent the universe of samples. Obviously, it is not possible to test all the species in the universe and, for this reason, an approximation is usually made. The same is true of exposure data since it is not practical or feasible to sample all possible locations or times. As it is unusual for sufficient toxicity and exposure data to be available to allow a cumulative frequency distribution of data to be plotted directly, an approximation to a frequency distribution is normally used (Parkhurst et al., 1996). This approximation assumes that the number of species tested (n) is 1 fewer than the number in the "universe." To obtain symmetrical graphical distributions (normal distributions), y-plotting positions are calculated as percentages using the formula $P = 100 \times i/(n + 1)$, where i is the rank number of the data point and n is the total number of data points in the set (Parkhurst et al., 1996). This gives an empirical cumulative probability based on the Weibull equation. Similar empirical probabilities can also be calculated using other formulas such as the Blom equation $P = (i - 0.375)/(n - 0.25) \times 100$ or the Hazen equation $P = (i - 0.5)/n \times 100$ (Cunnane, 1978). These two equations may be useful for small data sets as they compensate for the size of the data set (Cunnane, 1978).

15.3.5 CHOOSING THE DISTRIBUTION MODEL

PERA makes use of the general observation that many risk variables are characterized by lognormal distributions (Paustenbach, 1995; Burmaster and Hull, 1997; Murphy, 1998). These variables include nonequilibrium contaminant mixing in air, water, soil, and tissue as well as bioaccumulation, food consumption in fish, life span of organisms, and susceptibility to toxic substances. The lognormal distribution is governed by the central limit theorem, which predicts central tendency of repeated observations. However, most distributions appear lognormal until inspected more closely (Murphy, 1998). Burmaster and Hull (1997) point out that natural logs (ln), not common log ($\log_{10}$), should be used for plotting a lognormal distribution; however, in practice there is no difference between the two. To fit a lognormal distribution to data, a graphics program such as SigmaPlot™ (SPSS, 2000) can be used for good graphic output, and a built-in regression subroutine or spreadsheets can be used to conduct linear regressions on transformed data (sample spreadsheets for Excel™ and QuatroPro™ are available from ksolomon@tox.uoguelph.ca or jgiesy@aol.com and include the necessary routines to produce the exceedence profile). Although it has been suggested that the lognormal model be used for distributions of toxicity data (Burmaster and Hull, 1997), other models may be more appropriate for particular data sets. Other authors have used log-logistic distributions (Kooijman, 1987; Van Straalen, 1990), while ECOFRAM has suggested that maximum likelihood approaches may also be used where specific distributional models are unsuitable (ECOFRAM, 1999a). The assumption of a reasonable fit to a model makes calculations of exceedence probabilities relatively easy, but is not necessary for the concept of probabilistic risk assessment to be used. Centiles of distributions may be estimated from large data sets by simple ranking and interpolation, using the percentile function in a spreadsheet, or by using a descriptive model, such as a polynomial, to describe the relationship (Giesy et al., 1999). These methods are more suited to large data sets where extrapolation beyond the observations is not needed.

Confidence intervals may be useful for assessing uncertainty in exposure distributions and for point estimates on these distributions. As with other uses of point estimates, the use of the distributional approach without confidence intervals assumes a sample large enough that sampling error contributes relatively little to the uncertainty in the overall risk assessment, relative to other sources of uncertainty. Similar approaches can be applied to distributions of species sensitivity (The Cadmus Group, Inc., 1996a); however, a contrary argument is that this assumes that all species are equal in the sense of their role and function in the ecosystem and that they can be treated in a purely numerical fashion. Some species (keystone species) may be more important in the ecosystem than others, and to use confidence intervals these species would have to be weighted appropriately. The inherent difficulty in assigning weights suggests that the issue of confidence intervals in SSDs be applied on a case-by-case basis or dealt with descriptively through the use of other lines of evidence (Hall and Giddings, 2000).

15.4 OTHER LINES OF EVIDENCE

PERA is a purely numerical technique and it would be scientifically inappropriate to base a risk assessment solely on the results of a probabilistic assessment. This has been well illustrated in the use of other lines of evidence in addition to PERA in published risk assessments (Klaine et al., 1996a; Solomon et al., 1996; 2001; Cardwell et al., 1999; Giesy et al., 1999; Hall et al., 1999; Campbell et al., 2000; Giddings et al., 2000; 2001; Hendley et al., 2001; Maund et al., 2001; Travis and Hendley, 2001) and the recommendation to use PERA in the ecological risk assessment process (U.S. EPA, 1992; 1998c; NRC, 1993; Suter, 1993; ECOFRAM, 1999a,b).

In an ERA, there are several lines of evidence that should be considered, including observations from mesocosm and field tests as well as ecological knowledge on the role and function of sensitive species in the ecosystem. These are discussed in more detail below.

In the context of pesticides and their use in integrated pest management practices, there is a wealth of experience and understanding of the effects of selectively toxic substances on populations and communities. In particular, the life-table approaches used in pest management are useful in identifying stages in the life cycle of pests that are critical to sustainability of the population. Approaches similar to these have not been widely used in ERAs; however, both sciences could benefit from the increased transfer of information.

Laboratory bioassays carried out with small numbers of a single species, such as acute and chronic laboratory tests in organisms with a short life cycle, cannot take into account effects that involve interactions between populations of different species in communities or those that affect ecosystem function, such as changes in productivity or nutrient flow. A number of procedures have been proposed for ecosystem- and community-level tests, and there are numerous examples of their utility (Hill et al., 1994). Most of this work has been carried out in aquatic systems, but some terrestrial studies have also been carried out in the purely toxicological sense. The aquatic systems range from simple laboratory systems to complex flowing stream systems, usually referred to as mesocosms or microcosms. Microcosm and mesocosm studies provide effect measures that are closer to the assessment measures, for several reasons (Solomon et al., 1996).

Measurements of productivity in microcosms incorporate the aggregate responses of many species in each trophic level. Because organisms will likely vary widely in their sensitivity to the stressor, the overall response of the community may be quite different from the responses of individual species as measured in laboratory toxicity tests. Microcosm studies allow for the observation of population and community recovery from the effects of the pesticide and indirect effects of pesticides on other trophic levels. Indirect effects may result from changes in food supply, habitat, or habitat quality. Microcosm studies can be designed to approximate realistic stressor–exposure regimes more closely than standard laboratory single-species toxicity tests. Most studies, especially those conducted in outdoor systems, incorporate partitioning, degradation, and dissipation, important factors in determining exposure. These factors are rarely accounted for in laboratory toxicity studies, but may greatly influence the magnitude of ecological response. Where PERAs have included studies in the field or in microcosms, it has been consistently noted that effects in the field are rarely observed at concentrations equivalent to lower centiles of toxicity distributions (Solomon et al., 1996; Giesy et al., 1999; Hall et al., 1999; Versteeg et al., 1999; Giddings et al., 2000; 2001). In fact, ecologically significant effects are sometimes only observed at concentrations exceeding 25th centiles of laboratory-based acute toxicity values (Hall and Giddings, 2000; Giddings et al., 2001).

Although older microcosm and mesocosm studies were not set up as hypothesis-testing experiments, many of the data are very useful for interpreting effects of stressors at the community and ecosystem level. Most mesocosm studies are run for a limited period of time (usually one season) and three types of ecologically relevant observations at the population level have been made. These are (1) no effect, (2) effects with recovery in the period of observation, and (3) effects with no recovery observed in the period of observation. The first two are used as equivalents to ecosystem-level NOECs and NOAECs (Solomon et al., 1996; Giesy et al., 1999), ecologically acceptable concentrations (EAC, Campbell et al., 1999), or $NOEC_{community}$ (Van den Brink et al., 1996; 2000; Giddings et al., 2000; 2001). When several concentrations of the stressor have been tested, it is possible to construct a matrix that allows easy interpretation of responses (Giddings et al., 2000; 2001).

Another useful line of evidence in PERA is an assessment of the ecological role of those organisms that are most likely to be affected in a particular scenario. One of the central dogmas of PERA is that, by protecting species composition, community function is also protected, and a "low-level risk" such that a small percent of species are affected (e.g., 5 to 10%) will not "significantly" harm the community (Forbes and Forbes, 1993). In addition, it implies that, at these low risk levels, interactions among community and ecosystem members can be ignored and that ecological redundancy and resiliency will ensure that ecosystem functions continue.

For this to be true, the species impacted should not be keystone organisms (Calabrese and Baldwin, 1993; Forbes and Forbes, 1994). Added to this complexity is that these organisms should not be "societally important" (Suter, 1993), commercially valuable, or endangered. Traditionally, ecologists have identified keystone species by their effects on the species richness and composition of the community in which they live. Keystone species may also be those that have major consequences on ecosystem functions, but may or may not have significant influence over species composition. Keystone species may occur at any trophic level but are often top predators, such as the interesting killer whale–seal–sea otter triad observed in the Northwest Pacific (Estes et al., 1998). In addition, primary producers may also be important as fixers of energy and nutrients and as habitat modifiers (Lehtinen et al., 1988) and also serve as keystone organisms. The results of microcosm studies may allow interactive effects to be identified and may even reveal keystone species or groups; however, microcosms may not contain all the species in the ecosystems where the risk assessment is focused, and it may be necessary to extrapolate from the microcosm to these environments on the basis of similarity of taxa or ecological analogy. Clearly, the most sensitive organisms identified in the SSD, and the taxa to which they belong, must be assessed against the ecological knowledge of their function in the ecosystem, their reproductive strategies, their ability to recover and repopulate an ecosystem, and their ability to tolerate the types of exposures identified in the PERA. When this type of analysis was conducted for chlorpyrifos, diazinon, or the pyrethroids in aquatic ecosystems, the effects on sensitive taxa could be placed in an ecosystem context (Giesy et al., 1999; Giddings et al., 2000; 2001).

Difficulties may be encountered where keystone species already have been removed or substantially affected by anthropogenic stressors other than the one being assessed. Here, the argument that ecological functional redundancy can buffer any adverse effects of stressors becomes less meaningful. Ecosystems with impoverished biota are expected to show a much greater structure–function coupling when perturbed by subsequent stressors (Forbes and Forbes, 1994). It is generally agreed that environments that are at risk from current or proposed chemical pollutants are not in their natural state (Lloyd, 1992; Cairns, 1998; Leslie and Timmins, 1998) and that confounding stressors may be present as well. This must also be considered under other lines of evidence.

The presence of endangered species in the habitat exposed to the risk generates increased scrutiny by regulators and the public. The presence of an endangered organism in the lower centiles of an SSD would raise additional concerns; however, it should be recognized that the reason for endangerment and sensitivity to the stressor may not be correlated. The endangered species may not be sensitive to the stressor and this source of risk would therefore be an issue of low importance.

15.5 UNEXPLORED USES FOR PERA

Among some of the shortcomings of the probabilistic methods is that they do not consider secondary exposure to the stressor that may occur through food chain bioaccumulation or sediment intake. The problem with many chemical stressors is that the exposures used in laboratory testing situations are different from those occurring in the ecosystem. For example, measurements of concentration in sediments may not be available, even though this is the major route of exposure. Because of kinetic issues, laboratory tests may be too short term to allow a strongly bioaccumulative substance to attain maximum concentration in the receptor organism. Thus, bioassay results could underestimate the toxicity of the substance in relation to its effects in the environment. PERA can be applied to strongly sorbed substances if the relationship between concentrations is known and the relevant exposures can be estimated from models (Giesy et al., 1999); however, this introduces additional uncertainties into the assessment and makes the process more complex. Alternatively, if a certain route of exposure is identified as important, specific tests utilizing this route can be conducted, such as has been recommended for sediment-mediated toxicity of pesticides (ECOFRAM, 1999a).

PERAs can be conducted for pollutants that bioaccumulate by accounting for secondary exposure through food webs; such a methodology has been proposed by the Organization for Economic Cooperation and Development (OECD) (Balk et al., 1995). Use of bioaccumulation and food web models introduces additional complexity, thus making the risk assessment and, more important, its communication to risk assessors more difficult. Alternatively, effects concentrations measured in exposed test organisms can be used as a basis for comparison to concentrations of the same bioaccumulative substances in samples taken from organisms in the potentially affected environment. This does, however, require good data on the relationship between body burden and effects and a good database of concentrations measured in organisms collected from the environment.

Mixtures present a particular challenge in ERA. As has been pointed out in a number of PERAs already conducted, it is recognized that pesticides and other stressors seldom occur in isolation and that organisms are frequently exposed to mixtures of substances, often with different mechanisms of action (Solomon et al., 1996; Giesy et al., 1999; Lee and Jones-Lee, 1999). Where effluents have a constant composition, whole-effluent testing can be used as a physical model for assessment of a complex mixture; however, other substances, such as pesticides, are rarely used in consistent mixtures or applied in such a way that they will enter the environment in predictable combinations of concentrations. In addition, as pesticides are regulated as single substances, risk assessments have traditionally been conducted on one active ingredient only.

These same constraints apply to probabilistic risk assessments, which often have a narrowed focus to minimize the resources needed for such a complex activity. However, environmental monitoring has shown that pesticides do co-occur in the same location and time, and this has raised questions regarding risk assessments of mixtures. Where substances are known to act additively, it is possible to use the toxic equivalent (TE) or toxic unit (TU) approach to add concentrations and assess risks from the mixtures. This has been applied to several classes of compounds, such as the dioxins (Ahlborg et al., 1994; Parrott et al., 1995), chlorinated phenols (Kovacs et al., 1993), and polyaromatic hydrocarbons (Schwarz et al., 1995). This approach was used to assess the combined risk from atrazine and its metabolites (Solomon, 1999) using probabilistic approaches and could be used to assess risks from other pesticides with a common toxic mode of action, such as the organophosphorus or triazine pesticides. Traditionally, these equivalents have been based on responses measured in the same organism, for example, the laboratory rat. This is appropriate if the risks are to be assessed in the same organism or extrapolated to another (humans) with appropriate uncertainty factors. However, potencies measured in one animal may not be the same in another, and wide interspecific extrapolations, such as from rats to fish, may not be possible (Parrott et al., 1995). This situation becomes more complex when dealing with ERAs. If the TEs are based on responses measured in a single species, the relationship between the potency of the components in the mixture may be different from those in other organisms in the species distribution. Thus, as was the case with dioxins in fish and rats, extrapolation of the risk, whether assessed via HQ or PERA, could be incorrect. Alternatively, TEs could be based on point estimates of potency derived from SSDs. Again, use of these TEs would necessitate similar distributions (same slope when expressed as cumulative frequencies curves) and that the order of the species in the distributions was the same. If this were not the case, risk assessments may be incorrect. If the substances interact through response addition, similar approaches would be possible, although differences in exposure times may introduce additional complexity to the assessment.

If based on measured concentrations from the environmental matrix, the PERA of mixtures does have the advantage of being able to provide an assessment of the likelihood of co-occurrence of the substances, this is a considerable improvement over assuming that all the worst-case concentrations will be present in the same location and at the same time. Even though it may be difficult to quantify the total risk in a meaningful way, it would be possible to assess the effect of removal of one of the components of the mixture as a differential risk if measured concentrations were used. This could be useful in risk management decision making.

The above discussions on PERA of mixtures are based on the use of measured concentrations. Although good progress has been made on the use of models to estimate concentrations of some substances in the environment (see modeling discussion on pesticides above), we are not aware of models that have been successfully applied to predict concentrations of the components of mixtures.

15.6 CONCLUSIONS

PERA has been applied to a number of substances as part of higher-tiered, more realistic risk assessments. Although PERA provides tools to conduct risk assessments more thoroughly and to handle large data sets, other lines of evidence are also important in reaching the final conclusions. PERAs require a broad range of expertise and resources and are most easily carried out where good data sets for toxicity and exposure values are available. PERA is a significant improvement on the traditional HQ approach, but it will likely continue to evolve as the entire science moves forward and as the interpretation of the methods is tested. As has been pointed out, one of the major hurdles that PERA faces is its acceptance by the public and regulators (Solomon, 1996; Roberts, 1999). Risk managers will likely continue to demand, or at least interpret, probabilistic risk estimates as point estimates of high certainty (Chapman, 1995; Moore and Elliott, 1996; Richardson, 1996). Decision makers want to know whether it is safe or not and prefer being told what will happen, not what might happen (Morgan, 1998). Similarly, the public demands absolute safety but has less understanding of the science. In addition, judging from their addiction to lotteries and other games of chance, people have no concept of probability and also greatly misperceive risks to themselves and fellow humans (Slovic, 1987), never mind those to the environment.

From the scientific point of view, there are several interesting and challenging aspects to PERA, particularly with regard to modeling of exposure concentrations and the application of the technique to risk assessment of mixtures. However, the most challenging aspect of all is the educational effort that will be required to ensure that the method is understood and adopted, not only by the scientific community, but also by regulators and the public. To its advantage, PERA is data driven and thus can be perceived as more transparent; however, its application still relies on expert knowledge and interpretation in ecotoxicology and ecology. In fact, the application of PERA will require greater knowledge of ecology and ecosystem function than is currently required of simpler approaches to assessing risk such as the use of hazard quotients.

ACKNOWLEDGMENTS

The authors thank all their colleagues who acted as sounding boards for many of the ideas and concepts discussed in this chapter. As is evidenced by the length of the author lists in many of the PERA-based risk assessments, this is not a one-person task and we thank our co-workers for their contributions. Parts of this chapter are based on an M.Sc. thesis by Peter Takacs, defended in 1999 at the University of Guelph.

16 The Potentially Affected Fraction as a Measure of Ecological Risk

Theo P. Traas, Dik van de Meent, Leo Posthuma,
Timo Hamers, Belinda J. Kater,
Dick de Zwart, and Tom Aldenberg

CONTENTS

1-56670-578-9/02/$0.00+$1.50
© 2002 by CRC Press LLC

Abstract — Species sensitivity distributions (SSDs) can be used in ecological risk assessment to quantify ecological risk associated with concentrations of pollutants. This chapter proposes the potentially affected fraction (PAF) of species, related to the "forward" use of SSDs, as a quantitative measure of ecological risk to biotic communities. By using dimensionless PAF values instead of risk quotients, the ecological risk of a mixture of compounds can be calculated with aggregation protocols based on fundamental theory on mixture toxicity and the toxic mode of action (TMoA) of compounds in the mixture. The multisubstance PAF (msPAF) is proposed as a parameter to quantify the overall ecological risk of mixtures of toxicants for ecological communities. Three examples are shown to illustrate the use of PAF and msPAF in ecological risk assessment, both in a diagnostic and a prognostic context. PAF and msPAF calculations are shown for different types of risk assessment: comparisons between (1) sites, (2) compounds, and (3) risk due to pollutants in combination with other stress factors. Several numerically important factors that influence exposure, the calculation, and the interpretation of SSDs and PAF are taken into account: mixture toxicity, choice of input data, similarity of exposure quantification for SSDs and field data, and the choice between using one SSD to capture all relevant toxicity data and splitting the data over various SSDs according to taxonomic groups and modes of action.

16.1 INTRODUCTION

An important practical goal of ecotoxicology is to predict quantitatively how ecosystems respond to exposure to toxic substances. When ambient concentrations exceed environmental quality criteria, the ecological consequences are of interest for risk management (see, e.g., Chapter 12).

Ecotoxicologists use various methods to predict ecosystem responses to toxic stress. In recent risk assessments, ecological field observations on different taxonomic groups and on different spatial scales were compiled and compared to basic ecotoxicological data derived from laboratory testing (Suter et al., 1999; Solomon et al. 2001). Usually, site-specific evaluations are only executed when ecological risk is suspected on the basis of lower-tier results of an assessment procedure. Laboratory toxicity data can be used for such preliminary risk assessments. For many compounds, especially newly developed chemicals or biocides, it is very common to assess risks based on only laboratory-derived ecotoxicological data for a relatively small number of species.

Ecological risk assessment problems are highly variable, and thus a variety of methods are used, related to the specific assessment goals. Comparisons can be made between compounds to determine the relative ecological risk of different substances, products, or processes (as in life cycle assessment). Another application is to define total ecological risks of contaminants in different areas in order to protect sensitive species, or to prioritize area cleanup targets based on a ranking of total ecological risk. Another target is to compare risk between species groups to determine the most sensitive group, or to protect certain species. Any combination of these endpoints may occur in site-specific risk assessment, or in setting priorities for remediation between polluted sites.

The most common comparison in hazard assessment is based on the ratio of an environmental concentration to a target value such as a no-effect concentration (OECD, 1989). Although a certain ratio of environmental concentration to a target value gives a rough guide concerning risk, this ratio gives little information on environmental impacts other than that a ratio <1 implies no or negligible damage. If two substances, however, have the same ratio >1, their environmental impacts may be quite different (see Chapter 15), and the sum of these ratios has no toxicological meaning.

As an alternative, this chapter proposes to use species sensitivity distributions (SSDs) based on laboratory toxicity data to derive a measure of effect that can be used in ecological risk assessment. We believe the method is an improvement over current quotient methods, since it encompasses the often-observed nonlinearity of species sensitivity and, especially, it allows for aggregating risks over compounds in a mixture.

Conceptually, any set of toxicity test endpoints (NOEC, EC_{50}, etc.) can be used to generate SSDs. In this chapter, however, the no-observed effect concentration (NOEC) for relevant endpoints such as mortality, growth, and reproduction has been chosen to generate SSDs, since this test endpoint plays a central role in deriving environmental quality criteria for hazardous substances in the Netherlands (see Chapter 12).

The basic feature of the proposed method is that a probability distribution function is fitted to the available log-transformed NOEC data, assuming that the data set is a representative sample of the true sensitivity distribution. In the Netherlands, the 5th percentile of the NOEC distribution, the HC_5, has been selected as a threshold that is assumed to safeguard ecosystem structure and function within chosen limits of confidence (Kooijman, 1987; Van Straalen and Denneman, 1989; Aldenberg and Slob, 1993; Chapters 4, 5, and 12). Using this SSD, one can estimate not only the HC_5, but also the fraction of species exposed above their NOECs (or another relevant study endpoint) for a given environmental concentration (Figure 16.1, left and right panel, respectively). Thus, the derivation of environmental

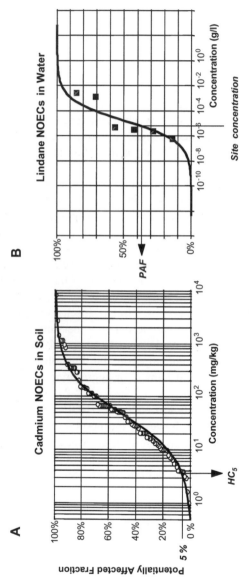

FIGURE 16.1 The use of SSDs for deriving environmental quality criteria (such as the 5th percentile of the sensitivity distribution, termed the HC_5) (A) and for risk assessment (B). The SSD is described by a logistic distribution of log toxicity data according to Aldenberg and Slob (1993).

quality criteria is linked to the method used to study effects of criteria exceedence, thus providing methodological transparency. Both forms of use of SSDs were introduced together in the original paper (Van Straalen and Denneman, 1989).

The fraction of species affected (y) that is calculated from the SSD given an environmental concentration (x) has become known as the potentially affected fraction (PAF) of species (Klepper and Van de Meent, 1997). We propose to use the PAF as a measure of ecological risk. Van Straalen (Chapter 4) defined ecological risk predicted with SSDs (the undesired event), as "a species chosen randomly out of a large assemblage is exposed to an environmental concentration greater than its no-effect level." For the PAF approach, as demonstrated by Aldenberg et al. (Chapter 5), this statement is equivalent to stating that *a certain fraction of species is expected to be (potentially) affected above its no-effect level at a given environmental concentration.*

This chapter presents the PAF as an index for ecological risk based on NOECs, which can take values between 0 and 1 (i.e., 0 and 100% of species affected). The PAF allows comparisons between substances, species groups, sites, and regions. The use of SSDs to estimate the PAF is illustrated for single toxicants and mixtures of chemicals. These examples are followed by a discussion on confirmation of PAFs with field data and the limitations of PAF-based assessments.

16.2 POTENTIALLY AFFECTED FRACTION CALCULATIONS

16.2.1 OVERVIEW

Although conceptually simple, the use of PAF for risk assessment must take into account several quantitatively important issues as suggested previously (U.S. EPA, 1985a; Van Straalen and Denneman, 1989; Smith and Cairns, 1993; Chapters 4, 5, and 21). How to deal with these issues is illustrated in selected case studies in Section 16.3.

16.2.1.1 Fitting the SSD Model

Different probability distributions have been proposed to construct SSDs, based on a sample of log toxicity data. The most common ones are the triangular (U.S. EPA, 1985a), the normal (Wagner and Løkke, 1991; Aldenberg and Jaworska, 2000), and the logistic distribution (Kooijman, 1987; Van Straalen and Denneman, 1989). Goodness-of-fit techniques can be employed to choose the best fitting distribution (Versteeg et al., 1999) or to decide on distribution-free resampling methods, such as bootstrapping (Newman et al., 2000; Chapter 7). The case studies presented here have used the logistic distribution for ease of calculation (Aldenberg, 1993) and mathematical clarity in msPAF calculations. For small data sets ($n < 10$), goodness-of-fit techniques generally do not distinguish between the normal and the logistic distribution (Aldenberg and Jaworska, 2000). The minimum size of the data set that can be employed for SSD analyses is disputed (see Chapters 4 through 8). As a pragmatic choice in the Netherlands, the minimum number of input data needed for the purpose of deriving environmental quality criteria is set at four (Aldenberg and Slob, 1993).

16.2.1.2   Similarity of Exposure Regime in Laboratory Tests and Field

The construction of an SSD from toxicity test data only makes sense conceptually when the exposure regime of the test is comparable to the exposure regime in the field problem that is to be assessed. This similarity relates to factors such as temperature, pH for dissociating compounds and metals, water hardness, salinity, and exposure time. For example, when long-term average concentrations of compounds in the field are of concern, chronic NOECs are typically used to construct SSDs. Solomon and Takacs discuss this matter in more detail in Chapter 15.

16.2.1.3   Bioavailability

The total amount of a substance is often not toxicologically meaningful since most substances are only partially available for uptake in biota. Particular care should be taken to define the available fraction in case of systematic differences in bioavailability between laboratory test systems used to generate data for the SSD and the field. Within an ecosystem, groups or organisms may be exposed differently (e.g., benthic organisms vs. pelagic organisms) and this should be accounted for by constructing SSDs that are relevant for the exposure route (e.g., DiToro et al., 1991).

16.2.1.4   Background Concentrations

Many organic chemicals are antropogenic, but for metals there has always been a certain amount present in the environment, at different levels depending on the geological origin of the sediments or soils. Naturally occurring concentrations usually are subject to metabolic regulation and homeostasis mechanisms, and should not be considered in risk comparisons. From a biodiversity viewpoint, high natural metal concentrations sometimes lead to specialized communities. This is a natural part of the variation in environmental factors that shapes communities and enhances biodiversity. In such cases, it is necessary to relate the PAF calculations to the effect of added metals, attributable to human activities.

16.2.1.5   Subgrouping of Data in Different SSDs

After compiling all relevant toxicological information, it can be useful to separate the data over more than a single SSD, e.g., for different taxonomic groups. With this procedure, one can identify sensitive groups or, alternatively, identify which ecological processes may be affected most in relation to typical features of the identified groups, such as ecological function, potential for population recovery, recolonization, etc. (Chapter 12; Crommentuijn et al., 1997; Beek and Knoben, 1997; Suter et al., 1999). A useful way to distinguish groups is to use knowledge on the toxic mode of action of compounds. For example, crustaceans are expected to be much more sensitive to organophosphate insecticides or pyrethroids than fish (see Solomon et al., 2001).

16.2.1.6   Mixture Toxicity

Mixtures are the rule rather than exception in field conditions. In the case of risk assessment for mixtures, the PAF values calculated for separate compounds can be aggregated to an overall value, for which the term *multisubstance PAF* (msPAF) is proposed. The aggregation protocol is based on assignments of toxic modes of action to the compounds in the mixture.

16.2.2 Mixture Toxicity and Multisubstance PAF

One of the most important potential improvements of the use of SSDs as opposed to safety factors or risk quotients is that an aggregated measure of ecological risk can be calculated for mixtures of chemicals, as first discussed in the context of SSDs by Hamers et al. (1996a). The protocols by which compound-specific PAFs are aggregated for mixtures of compounds are summarized below. The protocols are derived from common toxicological theories on joint effects of compounds (Plackett and Hewlett, 1952; Könemann, 1981; Greco et al., 1995, among others). They are applied *after* corrections for bioavailability differences among test media and the field (see above). This is motivated from the idea that the common theory has been derived in the framework of molecule–receptor interactions, which is best addressed when applying them at the level of bioavailable concentrations.

An aggregation protocol has been designed to aggregate single-substance PAF values to a single overall term of ecological risk of a mixture. The protocol is based on application of two toxicological concepts: concentration addition (termed Simple Similar Action by Plackett and Hewlett, 1952) and response addition (termed Independent Joint Action by Plackett and Hewlett, 1952). The concepts are summarized and the associated rules of calculus are introduced below.

16.2.2.1 Concentration Addition

The toxicological concept of concentration addition has been defined for compounds that have the same toxic mode of action and that show no toxicological interactions (Plackett and Hewlett, 1952). This concept is usually applied to characterize the response observed in single-species toxicity tests in which the response to several compounds with the same mode of action is considered. By definition, the joint effect of such compound mixtures can be calculated using (relative) concentration addition rules of calculus (Plackett and Hewlett, 1952; Deneer et al., 1988a; Altenburger et al., 2000; Chapter 15).

Neutral hydrophobic organic compounds are believed to act by a similar, nonspecific toxic mode of action called narcosis, which every hydrophobic compound can exert, but which can be masked by another, more specific toxic effect (Könemann, 1981). In other words, almost every hydrophobic chemical exerts at least a narcotic, nonspecific toxicity that contributes to the joint toxicity of mixtures, and this is often referred to as baseline toxicity (Verhaar et al., 1992; 1995). At the species level, toxicity equivalence factors (TEFs) have been used to express the toxicity of one compound as a fraction of another compound with the same toxic mode of action. Transfer of the TEF principle to SSDs by scaling compounds in a similar way results in so-called hazard units (Box 16.1). For compounds with the same toxic mode of action, a single SSD can be derived using (relative) concentration addition quantified by hazard units, representing the separate compounds and any mixture of these compounds. It is assumed that the toxic mode of action in this case applies to the SSD comprising all relevant species.

For compounds with a non-narcotic toxic mode of action, the situation is more complicated. Some species experience the same type of effect due to specific interactions at targets sites that only occur in a fraction of the species, and other species only experience narcosis, due to lack of specific target sites for the compound. This holds, for example, for organophosphate (OP) biocides. Each OP biocide acts by acetylcholinesterase inhibition, and thus all compounds in this group can be seen as fractions of each other provided that the exposed organisms have a receptor for this compound. For those organisms, concentration addition rules of calculus should be applied for the specifically working OP biocides. However, species that lack the target receptor are not sensitive for OP exposure, and will only experience narcotic baseline toxicity. This means that concentration addition for specific toxic modes of action is only appropriate within a single SSD if the SSD consists of species having the specific receptor. For the organisms without the receptor, it may be more useful to consider narcotic concentration addition. Moreover, compounds can be more toxic than expected from baseline toxicity, although the exposed species do not have the receptor for a compound. These complications are further elaborated in Chapter 22.

As a method to aggregate several PAF values from SSDs for compounds with a similar toxic mode of action to a single msPAF term, it is proposed to apply concentration addition rules. The cumulative density function (CDF) of (log) toxicity data is regarded as similar to the log-dose–response curve for single species. This is illustrated in Box 16.1 for NOECs, but the principle can be applied to other test endpoints as well.

BOX 16.1: Concentration Addition

PAF calculations for concentration addition are performed according to the following steps:

Laboratory Toxicity Data

- The NOECs are obtained from an appropriate data set for each compound. Whether NOECs are the appropriate endpoints to use in msPAF calculations remains to be established.
- The data are expressed in the units considered relevant for exposure (e.g., mg/l for aquatic organisms, or mg/l soil pore water for soil organisms likely exposed through the pore water).
- These data are scaled into dimensionless hazard units (HUs), with 1 HU defined as the concentration where the NOEC is exceeded for 50% of all species tested, i.e., the median of the toxicity data of the whole data set (Equation 16.1).

$$\overline{\text{NOEC}}_i^j = \frac{\text{NOEC}_i^j}{\bar{x}_i} \qquad (16.1)$$

- for $i = 1$ to n compounds and for $j = 1$ to m species, where

$\overline{NOEC}_i^j$ = the scaled NOECs in HUs ($\mu g \cdot l^{-1}/\mu g \cdot l^{-1}$)

$\bar{x}_i$ = the median NOEC for substance i

- This is analogous to toxic unit (TU) scaling for a single species where 1 TU is the concentration causing 50% effect, e.g., for mortality (Sprague, 1970).
- The SSDs for each compound are obtained by fitting a logistic model (or normal, etc., see Section 16.2.1) to the log toxicity data in HUs for each compound (see Chapters 4 and 5) or, alternatively, it can be found by shifting the CDF of log NOECs by subtracting $\log(\bar{x})$.
- Concentration addition in the context of SSDs implies that SSDs have similar variance for compounds with the same toxic mode of action. The hypothesis (e.g., for two compounds) $H_o{:}\sigma_A^2 = \sigma_B^2$ (lognormal distribution) is not rejected (i.e., SSDs for different compounds do not have significantly different variance). If they differ significantly, no rules of calculus have as yet been developed to take divergence in variance (equivalent to slope divergence in the classical concentration addition literature; see Bliss, 1939; Plackett and Hewlett, 1952; De March, 1987) into account. If they do not differ, this does not prove concentration additivity; it can also be a result of small sample sizes.

Exposure Data

- For each compound present in an environmental sample, the exposure concentration is recalculated to HU.

PAF Calculations for Concentration Addition (CA)

- The msPAF$_{CA}$ is read from the CDF by HU addition (non-log-transformed) for a single toxic mode of action (TMoA):

$$HU_{TMoA} = \sum_i HU \qquad (16.2)$$

- If the log toxicity data are fitted to a logistic distribution, the PAF of each compound i is given by

$$PAF_i(x) = \frac{1}{1 + e^{-(x-\alpha)/\beta}} \qquad (16.3)$$

with α the mean of log toxicity data and β equal to $(\sigma \cdot \sqrt{3})/\pi$ with σ the standard deviation and x the log of the exposure concentration.

- After recalculation of all compound concentrations to HUs, α = zero by definition and x is given by the sum of the HU$_{TMoA}$ of all compounds with the same toxic mode of action. The β is averaged over the compounds or, alternatively, can be derived from a database analysis for the toxic mode of action (Chapter 8).

- After substitution in Equation 16.3, the msPAF$_{CA}$ for a group of substances with the same toxic mode of action is calculated as

$$PAF_{TMoA} = \frac{1}{1 + e^{-\log(\Sigma HU_{TMoA})/\beta_{TMoA}}} \qquad (16.4)$$

16.2.2.2   Response Addition

For compounds that have different modes of action, and that do not interact at the target site of toxic action, mixture toxicity can be described by the term *response addition* (RA), initially referred to as Independent (Joint) Action (Bliss, 1939; Plackett and Hewlett, 1952). This concept is usually applied to characterize the response observed in single-species toxicity tests for chemicals with dissimilar modes of action. Response addition can be assessed using rules of calculus according to Box 16.2, in which a correlation of sensitivities to the different compounds is absent.

The outcome of exposure to compounds following the concept of response addition is dependent on the value of the covariation of sensitivities for the individual compounds in a mixture. This covariation is quantified by a correlation r, which is actually the rank correlation between the sensitivities of the tested organisms for separately applied compounds (Könemann, 1981). In the framework of SSDs, this correlation pertains to a similar rank correlation at the level of species sensitivities. For those effects where interactions do occur between the chemicals, mathematical theory is well developed, but toxicological confirmation is lacking or confusing (Greco et al., 1995). The question that remains is, what the value of r is when predicting the msPAF$_{RA}$ as an aggregation over single-substance PAFs and if we really need it. In the present msPAF$_{RA}$ calculations, we therefore pragmatically assume that all combinations of compounds for which effect addition is applied, a correlation coefficient of $r = 0$ applies. This assumption is supported quantitatively by the frequent observation that the joint effect of chemicals for which response addition certainly applies is in practice *under*estimated by response addition rules of calculus (at $r = 0$) and that complex mixtures often show a tendency to show a response similar to concentration addition (Warne and Hawker, 1995; Pedersen and Petersen, 1996; Deneer, 2000; Backhaus et al., 2000; Faust et al., 2000).

BOX 16.2: Response Addition

Laboratory Toxicity Data

- Laboratory toxicity data are selected and used to generate SSDs for each compound in the mixture as described for concentration addition in Box 16.1, in nonscaled units relevant to the route of exposure.

Exposure Data

- For each compound present in an environmental sample, the concentration is determined in the same units.

PAF Calculations for Response Addition

- The combination effect for compounds with different modes of action is calculated analogous to the probability of two nonexcluding processes (Hewlett and Plackett, 1979), where a species is affected by compound A and/or B:

$$P(A \cup B) = P(A) + P(B) - P(A \cap B) \qquad (16.5)$$

- Regarding the last term of Equation 16.5, there are various theoretical "extreme" possibilities, yielding three specific cases of Equation 16.5. These are related to the covariation of sensitivities, denoted by r. For the present use in SSDs, it is assumed that sensitivities are uncorrelated in response addition.
- When $r = 0$, then $p(A \cap B) = P(A) \cdot P(B)$, which leads to

$$\text{PAF}_{RA} = \text{PAF}_A + \text{PAF}_B - \left(\text{PAF}_A \cdot \text{PAF}_B\right) \qquad (16.6)$$

- To simplify the calculation of Equation 16.6 for complex mixtures and assuming $r = 0$, the complement of the PAF, the nonaffected fraction, is introduced. If $Q_A = 1 - \text{PAF}_A$, and $Q_B = 1 - \text{PAF}_B$, the nonaffected fraction of the mixture of A and B is a part Q_A of Q_B, i.e., the nonaffected fraction is given by

$$Q_{AB} = Q_A \cdot Q_B \qquad (16.7)$$

- The msPAF for two substances according to response addition is given by

$$\text{PAF}_{RA} = 1 - Q_{AB} = 1 - \left(Q_A \cdot Q_B\right) = 1 - \left(1 - \text{PAF}_A\right)\left(1 - \text{PAF}_B\right) \qquad (16.8)$$

- For example, if two compounds yield PAF values of 0.1 and 0.15, then $Q_{AB} = (1 - 0.1)(1 - 0.15) = 0.765$, and the msPAF$_{RA}$ for a mixture according to the complementary value is $1 - 0.765 = 0.235$. This is the same as when calculated following Equation 16.6, namely: msPAF$_{RA} = 0.1 + 0.15 - (0.1 * 0.15) = 0.235$.
- *Complex mixture calculus for response addition.* Equation 16.5 can be extended for more compounds assuming no correlation between substances. When a third compound is considered, this equation can become quite complex (Hamers et al., 1996a). By making use of the complement

of the PAF this is avoided; thus, the preferential calculus for msPAF_{RA} is given in Equation 16.8. For more than two chemicals, this leads to

$$\text{PAF}_{RA} = 1 - \prod_i \left(1 - \text{PAF}_i\right) \qquad (16.9)$$

for $i = 1$ to n substances, with PAF_{RA} representing the msPAF of various compounds calculated by response addition assuming $r = 0$.

___

16.2.2.3 Aggregation to msPAF

With the algorithms described in Boxes 16.1 and 16.2 we propose to aggregate individual PAFs to msPAF, based on the toxic mode of action of the substances in the mixture. The substances comprising the mixture are grouped in three different types: (a) one or more groups of compounds in which there is within-group concentration addition, (b) groups of compounds distinguished by between-group response addition, and (c) remaining compounds with a toxic mode of action that is unique for the given mixture. Calculation of overall msPAF of all the compounds from the three different types a to c then proceeds in the following order:

1. SSDs are calculated for each compound, yielding as many (single-substance) SSDs as there are compounds on the list.
2. Concentration addition (according to Box 16.1) is applied to each group of compounds with the same toxic mode of action, to aggregate single PAFs for compounds of type a, showing within-group concentration addition. This yields msPAF values based on concentration addition (msPAF_{CA}) for as many modes of action as are distinguished in the mixture.
3. Response addition according to Box 16.2 is applied to the msPAF_{CA} values obtained in step 2 with the remaining single PAF values (aggregating over types a to c).

The msPAF protocol proposed by Hamers et al. (1996a) applied concentration addition only to compounds with a narcotic toxic mode of action, whereas all other modes of action were treated following response addition. In current practice, concentration addition is also applied within groups of compounds that have non-narcotic toxic modes of action such as photosynthesis inhibition, acetylcholinesterase inhibition, etc. (see also Chapter 22).

16.3 CASE STUDIES USING msPAF AS A MEASURE OF ECOLOGICAL RISK

Selected case studies executed at the Dutch National Institute of Public Health and the Environment (RIVM) and the Dutch National Institute for Coastal and Marine Management (RIKZ) are used to illustrate the use of PAF and msPAF in risk assessment in Dutch landscape elements or ecosystems. The PAF and msPAF results were applied to comparative risk analyses, and addressed comparisons in space, in time, and between compounds. Each example starts with a short problem description, followed by methodological highlights and the results obtained.

16.3.1 SPATIAL AND TEMPORAL TRENDS IN METAL PAF VALUES IN AN ESTUARY

16.3.1.1 Problem Definition

A study on the Western Scheldt Estuary in the southwestern Netherlands was conducted to calculate and identify current ecological risks of cadmium, copper, zinc, and chromium, and risks related to different emission reduction or sediment removal strategies. The original data are summarized by Kater and Lefèvre (1996). Metals are present in the waters of this estuary at concentrations regularly exceeding Dutch water quality criteria. Spatial and temporal differentiation of risks, as well as scenario-based remediation efficacy studies, was needed.

16.3.1.2 Methodological Issues

General Features. Various choices were made regarding the issues mentioned in Section 16.2.1. These are (1) bioavailability is calculated with a water quality model; (2) background concentrations were not taken into account; (3) data for all taxonomic groups were combined in one SSD, but pertained to four groups: crustaceans, mollusks, annelids, and fish. Apart from PAFs for individual metals, an msPAF was calculated.

Laboratory Toxicity Data. The toxicological input data used to construct SSDs are based on laboratory toxicity experiments with saltwater organisms for cadmium ($n = 14$ species), copper ($n = 7$), and zinc ($n = 6$), with a test endpoint the NOEC for reproduction (Kater, 1995). This chronic test endpoint was chosen in view of the exposure durations relevant to the problem and its higher sensitivity compared to mortality.

Field Data and Bioavailability. Environmental concentrations in the Western Scheldt Estuary for the selected metals were calculated with a one-dimensional dynamic water quality model based on metal emissions and fate and transport in the Western Scheldt Estuary (Van Eck et al., 1995). The water quality model of the Western Scheldt is a one-dimensional, tidal averaged box model with a water column and a sediment layer (Van Eck et al., 1995). The Western Scheldt from Rupelmonde (B) to Vlissingen (Nl) was modeled as 14 compartments that were considered homogeneous. These 14 compartments were considered sufficient to describe the longitudinal gradient of the estuary. The risk analyses were performed for each compartment using the dissolved metal concentration in each compartment. Input for the water quality model were emissions at the model boundaries and along the estuary from (un)treated industrial and domestic wastewater discharges from canals and atmospheric deposition. Because properties like salinity, organic carbon concentrations, redox conditions, and phytoplankton activity determine the dissolved metal concentrations, water quality parameters were calculated first. The trace metal concentration was then calculated for three fractions: dissolved metal species, particle-bound metal (in equilibrium with each other), and particulate metal sulfides. The particle-bound metal fraction is calculated from the pH, alkalinity, and salinity dependent distribution coefficient K_d for the oxygenated lower estuary and the measured dissolved concentration. The oxidation of the suspended metal sulfides is described as a function of temperature and oxygen.

The transport of fluvial and marine particulates, important for calculating trace metal contents of suspended and bottom sediments, is based in the model on a one-dimensional sediment balance of the estuary and resuspension/sedimentation of the upper 0.5 m of sediment twice a week. The simulation period was 10 years, after which the metal concentrations reached equilibrium in all management scenarios. The concentrations in the 10th year were used to calculate the PAF for the four metals using the SSDs, except for the temporal analyses.

16.3.1.3   Results

Metal PAFs. The first results were based on the maximal concentrations during a year in each compartment when no emission reduction scenario is considered. The maximum concentrations are chosen for risk analysis because of the uncertainties in both water quality model results and in estimation of parameters (using NOECs for reproduction) used for risk analyses. By estimating the risk using the maximum concentration a worst-case situation is obtained.

For cadmium and chromium, the calculated PAF values were below 5% in all compartments. This is considered an acceptable risk in the policy framework, since it indicates concentrations below the Dutch maximum permissible concentration regarding ecological effects (Chapter 12). With regard to semifield data, these values have been found to indicate a situation where no or negligible ecological effects are expected (Okkerman et al., 1993; Versteeg et al., 1999).

For copper and zinc, the spatial analysis indicated that, proceeding upstream to downstream, PAF estimates drop (Figure 16.2). This is mostly due to the dilution effect of the estuary, where relatively clean seawater streams in twice every 24 h. Zinc is mainly discharged by industry near the beginning of the Scheldt Estuary (compartment 1) but also at the end of the estuary near the cities Vlissingen and Terneuzen (compartment 14). The main copper discharges are also from Belgian industry and wastewater and industry in Terneuzen.

The temporal trends analyses indicated that zinc and copper have a different emission profile and associated PAF trends over time (Figure 16.3). For copper, risks were highest in the summer, whereas for zinc highest risks are observed during winter. This has been explained by variation in the stability of copper and zinc sulfides during the year, which influences the water concentrations (Zwolsman et al., 1997), and thereby the risks.

Risk Management Scenarios. The contemporary PAFs were compared to three management scenarios: dredging, "current risk reduction strategy," and "best-available technology" (see Kater and Lefèvre, 1995, for further explanation). To decide which scenario most efficiently reduces the ecological risk of both copper and zinc,

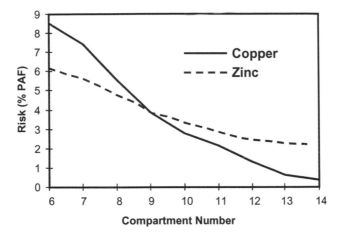

FIGURE 16.2 Spatial trends in PAF in the Western Scheldt Estuary due to copper or zinc. Results are shown for the various compartments for which concentrations were calculated, going from the beginning of the estuary (left) to the end of the estuary (right).

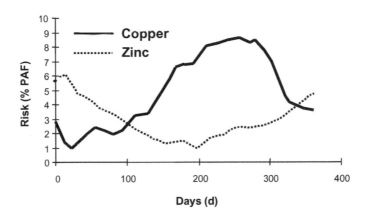

FIGURE 16.3 Temporal trends in highest estimated PAF values for copper or zinc in the Western Scheldt Estuary.

the msPAF was calculated for the copper–zinc mixtures for each management scenario using Equation 16.9 (Figure 16.4).

As an example, the msPAF for the no-action scenario is $1 - (1 - 0.08733)(1 - 0.006125)$, yielding an msPAF of 14.3%. The different panels show that the msPAF values calculated according to the response addition model are slightly below numerical addition of the separate PAF values of copper and zinc. It was found that the best reduction in PAF would be achieved in the best-available technology scenario for both copper and zinc (Figure 16.4). Dredging resulted in a small reduction of PAF values. Both the current risk reduction strategy and the best-available technology scenario lead to PAF levels below the environmental risk limit value of 5% for zinc and copper and for the calculated msPAF levels as well.

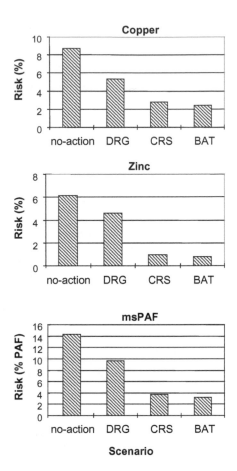

FIGURE 16.4 Maximum risk expressed as the maximum PAF or msPAF over 14 compartments in the Western Scheldt Estuary for different management scenarios: no-action, dredging (DRG), current risk reduction strategy (CRS), and best-available technology (BAT).

Use of Results. The Western Scheldt Estuary calculations demonstrated the potential of PAF analysis for the identification of compounds that cause the highest local risks, of compartments where the highest risk is expected, and of periods of the year where highest risks are expected. The results were presented as supportive information to determine risk reduction strategies, where measures can be considered on a cost–benefit basis. For various reasons, the present risk assessment based on PAF has not yet been used in environmental policy for the Western Scheldt Estuary.

16.3.2 SPATIAL TRENDS IN COMPLEX MIXTURE PAF VALUES IN SURFACE WATER ECOSYSTEMS

16.3.2.1 Problem Definition

Surface waters in the large river systems and lakes in the Netherlands are routinely monitored for approximately 140 compounds. Surface waters in rural areas in the

Netherlands, the large rivers, and Lake IJsselmeer contain a mixture of contaminants with concentrations of some constituents exceeding their water quality criteria. In the framework of National Environmental Outlooks (RIVM, 1997a; 1998), toxicant concentrations that exceed quality criteria have been a trigger for investigating the ecological damage of such events, as was the case for five pesticides and four metals. The monitoring data and the SSDs for these compounds have been used to calculate msPAF as a measure of ecological risk in these waters, with the option of identifying the most hazardous compounds.

16.3.2.2 Methododological Issues

General Features. Choices regarding the methodological issues mentioned in Section 16.2.1 are as follows:

1. The bioavailable fraction used for PAF calculations is the dissolved fraction. For major surface waters, dissolved metal concentrations were calculated using the means of the reported suspended matter concentrations at the monitoring sites and the water–particle partition coefficients given by Bakker and van de Meent (1997, appendix II).
2. Background concentrations in water were not taken into account.
3. Data for all taxonomic groups were lumped.
4. msPAFs were calculated according to *both* procedures of Section 16.2. The msPAF calculations have been made both for the theoretically preferred approach, and the alternative, to illustrate the magnitude by which different assumptions influence the numerical value of the end result (the msPAF).

Bioavailability of Organic Compounds. The bioavailability of organic substances in water is known to be influenced mainly by sorption of pollutants to suspended matter and dissolved organic carbon. For the current SSD-based evaluations, the dissolved concentration in water is taken as an estimate of the biologically available concentration in both laboratory and field conditions. Monitoring data were reported as total (dissolved + particle-bound) concentrations, and dissolved concentrations were approximated according to Equation 16.10:

$$C_{\text{dissolved}} = C_{\text{total}}/\left(1 + K_p \cdot \text{TSS}\right) \qquad (16.10)$$

where TSS is the concentration of total suspended solids in water (in kg/l), and K_p is the suspended solids–water partition coefficient (l/kg).

Laboratory Toxicity Data. The toxicological input data for five pesticides and four metals are derived from various studies for deriving environmental limits and database searches from AQUIRE (see Chapter 8) and Crommentuijn et al. (1997a,b). SSDs were constructed by fitting a logistic distribution to log NOEC data for growth, reproduction, or survival. This yields sufficient statistics to use for PAF calculations (Klepper and van de Meent, 1997, annex I; Chapter 8).

Exposure Data. Environmental concentrations of relevant compounds in major rivers, lakes, and canals were obtained from the monitoring authorities. For the year 1994, PAF and msPAF values were calculated for the metals and biocides based on the years' median concentrations at the sampling stations. Hence, the PAF values are estimates of the median ecological risk (as far as substances were included in calculations) in these water bodies.

Calculations. SSDs were calculated for all single metals and biocides based on a logistic distribution of log-toxicity data (NOECs). With these SSDs, median PAFs are calculated for all selected substances in regions of the river Rhine basin (Table 16.1).

For metals, we propose to aggregate compound-specific PAFs on the basis of response additivity, given the different metabolic functions and toxic effects of the essential and xenobiotic metals in organisms. However, for illustration of the alternative option that is defendable from the observations cited earlier, the msPAF values have also been calculated under the assumption of concentration additivity. (The Appendix to this chapter shows the calculation details.)

The selection of biocides monitored in the Rhine basin consists of two herbicides of different chemical classes (triazines and ureas), two organophosphate insecticides, and lindane, a central nervous system seizure agent. For illustration, msPAF for all pesticides was calculated by the theoretically preferred msPAF algorithms (Section 16.2.2.3) for within-group concentration addition for the herbicides and the organophosphate insectides, and between-group response addition to combine the PAFs of these two groups and the remaining substance lindane. These data were compared to calculus based on response addition only, assuming unique toxic modes of action for all compounds (see the Appendix to this chapter for calculation details).

16.3.2.3 Results

The SSD analysis show that the median msPAF of metals in 1994 is higher than for biocides, with the msPAF calculated from the preferred approach (response addition, Box 16.2) ranging between 7 and 27%. These results differ only slightly from the calculations based on concentration addition, with a maximum difference of only 4% in msPAF. Within the group of heavy metals, copper contributes most to the msPAF (see Table 16.1). Despite current metal emission reduction scenarios, prognosis of future metal concentrations in surface waters still leads to a slight increase of concentrations and PAFs for the year 2015 (RIVM, 1997a, results not shown).

PAF values for separate biocides are below 6%, with the highest risk associated with dichlorvos. Again, the difference between different msPAF algorithms is small. The difference between within-group concentration addition followed by between-group response addition (preferred) and response addition only (less preferred) is at most 2%. The overall msPAF is calculated by combining the preferred msPAF metals with the preferred msPAF biocides by response addition (see Table 16.1). The results show that the overall ecological risk is generally lower in the lakes IJsselmeer, Markermeer, and Randmeren than in the other locations of the Rhine basin.

TABLE 16.1
Ecological Risk of Heavy Metals and Biocides in Regions of the River Rhine Basin in the Netherlands (1994), Expressed as PAF (%) for Each Compound and Location

	Bovenrijn + Waal	Nederrijn	IJssel	Twenthekanalen	IJsselmeer	Markermeer	Ketelmeer	Randmeren E	Randmeren W	A'dam-Rijnkanaal
Metals										
Cadmium	1	1	1	1	0.2	0.1	1	0.2	0.2	0.4
Copper	12	22	15	24	10	10	20	11	6	16
Lead	0.1	0.1	0.2	0.1	0.0	0.0	0.1	0.0	0.0	0.1
Zinc	4	5	5	3	1	1	5	1	1	2
msPAF Metals (RA)	17	27	21	27	12	11	24	13	7	18
msPAF Metals (CA)	21	29	24	30.0	15	14	27	16	11	22
Biocides										
Atrazine	0.1	0.1	0.1	0.1	0.0	0.0	0.1	0.1	0.0	0.0
Dichlorvos	3	3	3	2	1	1	2	1	2	2
Diuron	0.0	0.2	0.5	0.4	0.1	0.3	0.3	1	0.2	0.2
Lindane	ND	ND	ND	ND	1	ND	ND	ND	ND	1
Mevinfos	0	0	0	0	0	0	0	0	0	0
MsPAF Biocides (CA+RA)	4	5	6	5	4	4	5	7	4	5
MsPAF Biocides (RA)	3	3	4	2	3	2	3	3	2	4
MsPAF All compounds	20	30	26	31	15	14	28	19	11	22

Note: msPAFs have been calculated for metals and biocides separately, according to the preferred and the alternative approach. For the metals, response addition is the preferred approach; for the biocides, concentration addition within compound groups with the same toxic mode of action, followed by response addition over these groups and over the rest of the compounds is preferred. The msPAF for all compounds is calculated by response addition of the preferred msPAF for metals and biocides (see text for further explanation). ND = not determined.

Use of Results. Environmental management interest was, first, on overall effects of environmental mixtures, and how they differ in space and time given current risk reduction management efforts for rivers. Second, identification of the most potent compounds was desired, to define future targets of risk management (RIVM, 1997a; 1998). As shown by Table 16.1, metal concentrations are higher in several regions in the Rhine basin in the Netherlands, such as the lake Ketelmeer and the location Twenthekanalen. Current emission reduction schemes hardly influence the values of PAF based on extrapolation to the year 2015 using water quality models (results not shown). This suggests that emission reduction schemes for metals need to be intensified to counteract the (long-range aerial) emissions that are the cause of the predicted concentration increase. Remedial action could also be focused more on improving water quality by taking action for compounds that are more easily subject to risk management, e.g., biocides.

16.3.3 SPATIAL DIFFERENTIATION OF MsPAF OF METALS IN SOILS FOR TERRESTRIAL PLANTS IN A MULTIPLE-STRESS CONTEXT

16.3.3.1 Problem Definition

Dutch topsoils contain metal concentrations ranging from background concentrations to concentrations exceeding the Dutch environmental quality criteria for metals in soils. This may pose problems for biota that live in or on the soil.

In the framework of National Environmental Outlooks (RIVM, 1997a; 2000) environmental risks were presented to the Ministry of the Environment based on msPAF calculations. Regarding metal concentrations in soil, msPAF values for soils were calculated, and the assessment was extended to an investigation of the empirical relationship between the current distribution of plant species over the Netherlands, metal contamination, and other soil factors that evidently also influence plant distribution. The msPAF of metals in Dutch soils and other soil factors, such as pH, were quantified and analyzed on a national scale. In view of evaluation of possible changes over time and of effects of environmental management activities, the analyses encompassed 1950, 1995, and 2030 ("hindcasting" and prognostic use of SSD-based methodology).

16.3.3.2 Methodological Issues

General Features. A risk assessment for the combined risk of cadmium, zinc, and copper was performed for terrestrial plants in the Netherlands. The area was divided into grid cells (0.25 × 0.25 km), to allow geographic information system (GIS)-based mapping. All relevant data were expressed for each grid cell.

Laboratory Data. Metal NOECs for terrestrial plants were taken from the inventories of Crommentuijn et al. (1997b; 2000c) and Efroymson et al. (1997). The metal concentrations added to the test soils were considered to be the bioavailable metal fraction, the fraction also derived for field data. This approach was preferred over recalculation to soil pore water (Klepper and van de Meent, 1997; Traas et al., 1998b) because many existing toxicity data could not be recalculated to soil pore water concentrations due to insufficient detail in the toxicity reports.

Field Data: Metals and Bioavailability. Available data of total metal concentrations in soils were used as primary input in the form of digital maps, based on interpolation of measurements (Tiktak et al., 1998; Otte et al., 1999). All concentrations were recalculated into exchangeable metal concentrations potentially available for plant uptake based on Tiktak et al. (1998). The exchangeable fraction was calculated using the following two equations, based on data of Janssen et al. (1997a) for Dutch metal-contaminated field soils:

$$X_{Inert} = 10^p \, (\%OC)^r \, (\%Clay)^s \, \left(X_{Total} \right)^n \tag{16.11}$$

with p, r, s, n regression coefficients, %OC the percentage organic carbon, %Clay the fine clay fraction (lutum), and X_{Total} the total metal concentration as determined by total extraction methods. The exchangeable metal is calculated from

$$X_{Exchangeable} = X_{Total} - X_{Inert} \tag{16.12}$$

Exchangeable metal concentrations for each grid cell were calculated for 1950, 1995, and 2030 (Tiktak et al., 1998; Otte et al., 1999). Combination of the laboratory toxicity data and the field concentration analyses allows for calculating msPAF values. The msPAF was calculated according to response addition (Equation 16.9).

Biological Data: Plants. Ecological fieldwork has yielded a database (>160,000 records) on plant species dispersion in the Netherlands. An empirical modeling approach enables the description of the suitability of soil types for the investigated plant species as a function of soil characteristics. This model is known as MOVE (De Heer et al., 2000), an acronym for "MOdel for the VEgetation." The environmental issues currently incorporated for this model are eutrophication, desiccation, acidification, and toxicity. These are modeled in a multivariate statistical regression method, yielding statistically significant regression formulae for 251 species. The model input mainly consists of groundwater levels, soil pH, nitrogen availability, risk of toxic compound mixtures (expressed in terms of msPAF of metals), and basic information on the vegetation structure. The direct output is the theoretical suitability of each grid cell for each plant species occurrence as a function of habitat factors. Suitability is defined as density above a minimum value (De Heer et al., 2000). The output is usually given for ecoregions with similar geological structure and ecology, presented as the fraction of grid cells in the specified region that can maintain a viable population for each species, which can be aggregated to the fraction of total species in an ecoregion.

16.3.3.3 Results

msPAF Values. The results of the msPAF analysis of metals in 1995 for plants is presented as a map of msPAF values for each grid cell in the Netherlands (Figure 16.5). Over all grid cells, msPAF values were in the range of 0 to about 50%. This indicates that locally, substantial NOEC exceedence is predicted due to

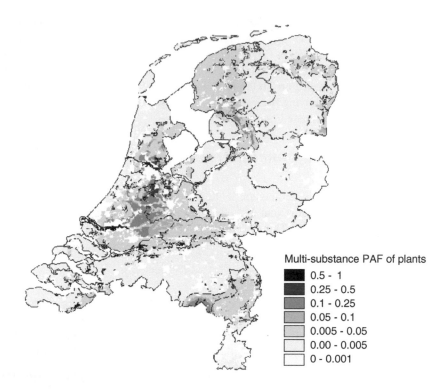

FIGURE 16.5 Geographical representation of the msPAF of the metals cadmium, copper, lead, and zinc in the Netherlands for the year 1995, based on SSDs composed of toxicity data of plants.

the combined risks of the metals. At most other sites risks are predicted to be negligible. High PAF values are mainly found in industrial areas with larger emissions of metals, such as in the south of the Netherlands near a former smelter, and in the west-central part of the Netherlands. It appears that soil type is an important determinant of metal PAF since certain clay and peat soils show higher PAF values than sandy soil.

Over time, physicochemical conditions in the soil change, and model results suggest that total metal concentrations slightly increase as a result of aerial deposition. This results in slight increases in predicted msPAF of metals for the year 2030 (results not shown). When looking at the grid cells for natural areas only, the msPAFs were (and remain) in the low range, from 0 to ~10%.

Distribution of Plant Species. The msPAF of metals is considered a relative measure of risk, useful to compare areas or trends. When msPAF is used as an input variable in MOVE to (statistically) attribute the variation in plant species distribution to soil factors, an indication is obtained of the statistical association between the msPAF and the distributions of species in the field.

Results of model-based exercises involving the metal msPAF analyses are shown in Figure 16.6 (De Heer et al., 2000). To construct this figure, the year 1950 was taken as a reference situation regarding species richness, which was defined as 100%.

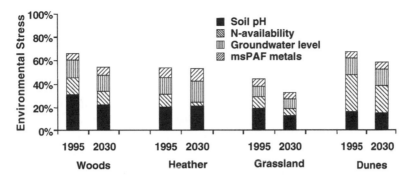

FIGURE 16.6 Multiple environmental stress on plant species due to environmental factors in 1995 and 2030 compared to 1950 (100%). In the assessment, the ecological risk of toxic substances was defined as the msPAF values for Cd, Cu, Zn, and Pb based on plant toxicity data.

The y-axis expresses the reduction in the proportion of grids "with occurrence" aggregated over all plant species relevant for an area type since 1950. A high value, as appearing, for example, for "woods," indicates that, averaged over the separate values for the plant species typical for woods, the successful maintenance of populations of such species is reduced by approximately 65%. Of all grids relevant for this vegetation type, 65% has on average become a less suitable habitat for forest plants. Large changes in habitat suitability have occurred since 1950 for all four vegetation types, as shown by Figure 16.6. The statistical analysis of the plant data shows that msPAF, like the other stress factors studied, explains a significant proportion of the variation in the plant occurrence data set.

Use of the Results. On the basis of a multiple regression analysis of habitat factors on species abundance, the relative importance of the different environmental stressors to the deterioration of ecosystems can be identified, provided that the statistical associations reflect mechanistic phenomena. This means that, when the last condition is met, the most effective measures for limiting species diversity reductions can be derived. The analysis can be applied to various geographical scales to address regional or national environmental issues (RIVM, 2000). Metals can be of importance in plant distribution, and this might trigger risk management decisions. However, metal concentration changes at the national scale are largely beyond direct policy control. Probably more important, the msPAF analyses used in the framework of large-scale monitoring data on both site characteristics and species compositions yields an interpretable result in the sense that msPAF is statistically associated with plant distribution in the field, albeit with unknown magnitude of effects on population occurrence and vitality. Such data are important to confirm the SSD concept (see Chapter 9). Plant community composition and population vitality also depends on other factors that have not been modeled, such as agricultural use of a soil or habitat loss. The exact field consequences of increased msPAF of metals for plants should be confirmed by further study. Further study is focused on biocides by applying similar calculus protocols.

16.4 DISCUSSION

This chapter proposes the use of SSDs in ecological risk assessment, by calculating the PAF and msPAFs on the basis of measured or predicted ambient concentrations of toxic compounds in the environment. The proposed approach is theoretically attractive compared with the alternative of risk quotients and is feasible. However, it is crucial to obtain values of PAF and msPAF that are relevant to field conditions. This discussion first addresses the issues that may qualify or disqualify PAF and msPAF as (at least) adequate *relative* measures of risk in contaminated environmental compartments, then discusses the current uncertainties in PAF and msPAF calculations, the use of PAF and msPAF in comparative contexts, and finally the relationships with community responses in the field.

16.4.1 SIMILARITY OF ENDPOINTS IN LABORATORY TESTS AND FIELD

In all the example PAF calculations shown, care was taken to collect laboratory toxicity data with toxicity endpoints that are maximally similar regarding the assessment problem under consideration. In all examples, the assessment endpoints concerned chronic effects of long-term exposures. For other assessments, however, different endpoints might be needed, such as acute lethality to quantify the risks of spills, incidental effluent releases, and peak concentrations of pesticides (see, e.g., Chapter 15). To be consistent with the toxicity data, exposure data may require preprocessing prior to use. Exposure concentrations are usually averaged (such as a moving average; see Solomon, 1996) but the maximum or median concentration during a certain period may also be used. The manipulation and use of exposure data are applied to improve the predictive ability of the SSD for field conditions, but this hypothesis must be tested more thoroughly (Chapter 9). The choice of input data for SSD analysis should be guided by the similarity to assessment endpoints and similarity with exposure data, but at this moment this issue strongly depends on expert judgment and the availability of data.

16.4.2 CORRECTING FOR BIOAVAILABILITY DIFFERENCES

In various examples, corrections were made to translate total measured concentrations into the exposure concentrations that are assumed to be bioavailable. The aim of the procedure is to construct an SSD and a data set of field data in the same relevant exposure units and exposure duration. Different methods have been used, including spatially explicit modeling of soil accumulation characteristics, as in the example on terrestrial plants. The choice whether and how to correct is context dependent, driven (or limited) by data availability, and dependent on expert judgment. Correction requires additional concepts beyond SSDs and additional measured parameters, both for the field and the laboratory data. Obviously, like the SSD concept itself, these concepts and their underlying theory require validation.

16.4.3 BACKGROUND CONCENTRATIONS OF NATURAL COMPOUNDS

Various examples show PAF calculations for heavy metals, and this has placed the issue of natural background concentrations into focus. One way of handling background concentrations has been given in the plant study, but its validity must be confirmed, conceptually and by comparison with field data. Alternatively, one could construct SSDs using toxicity data collected for species with similar background exposure. Species that originate from low-level background concentrations are likely to be toxicologically different from species typical from metalliferous regions or substrates. Currently, limitations in the number of existing toxicity data for soils in which background concentrations have been reported and limited knowledge on species minimum requirements for essential metals still pose problems in SSD interpretation for background concentrations.

16.4.4 AGGREGATION TO MSPAF

One of the most useful features of PAF concept is that it can be aggregated to reflect the risk posed by multiple substances, motivated by established toxicological principles. It should be noted that the principles of combination toxicology, derived from single species, are directly transposed to the level of biological communities, with the hidden assumption that joint effects transfer unchanged from species to community level. Whether this assumption holds true remains an open question. However, empirical data support the current pragmatic approach. Several authors have suggested that concentration addition grossly yields the best approximation of observed mixture effects in aquatic systems, despite the presence of evidently non-concentration addition combinations of compounds (e.g., Deneer, 2000). The example calculations shown in Table 16.1 for PAFs suggest that different aggregation protocols (based on all species lumped in SSDs) do not lead to large differences in predicted msPAF. Hypothetical msPAF calculations have shown that the results of msPAF calculations following either concentration or response addition (with $r = 0$) do not differ much when the slopes of the SSDs are in the range of 0.4 to 0.6 for the log-logistic model. In hindsight, the toxicological rationale could be improved by considering the msPAF for specific taxonomic groups. It can be easily envisioned that a lumped SSD or an SSD for a specific group of species with a target site for a toxic mode of action (e.g., algae and macrophytes for a herbicide) have quite different variances, thus influencing msPAF calculations. Taxonomic differentiation of SSDs would imply that SSDs are fitted to toxicity data for groups considering the toxic mode of action in relation to the presence of target sites of toxic action (see Chapter 22).

Another issue that deserves attention is the lack of statistical power in most NOEC tests (Van der Hoeven, 1998) or at low concentrations in general. Depending on the size of the type II error made in estimating the NOEC, the lack of statistical power in this concentration range may imply that an expected concentration- or response-additive effect may or may not be observed in true communities. Additional confirmation should determine whether the applied toxicological concepts are valid for the use with NOECs. In view of this issue, SSD and msPAF calculations might yield more profitable results for risk assessment when working in the concentration range where statistical power is higher, and where toxicological principles have been underpinned by data. Subsequently, the analyses should then be "translated" to the concentration level of interest. Although the principles of mixture toxicity for communities are as yet not well established by theoretical or practical evidence on community responses to joint exposure, the proposed msPAF aggregation may be a starting point for further conceptual development to address these phenomena.

16.4.5 Uncertainties in PAF Calculations

There is substantial uncertainty in the steps associated with PAF calculations (see Klepper et al., 1998). Although there is no universal procedure to develop a specific SSD-based ecological risk assessment, it is essential to take uncertainties explicitly into account. One of the options to present uncertainties is the use of confidence intervals. Although the confidence intervals of a PAF can be calculated quite easily (see Chapter 5), all presented examples use additional concepts and parameters, for example, related to exposure and bioavailability issues. This adds additional uncertainty to the model (Traas et al., 1996). Uncertainty analysis of SSD-based models should be further developed and a clear ecological interpretation of the confidence interval should be the eventual target.

16.4.6 PAF Applications

PAF applications in risk assessment and management have some advantages over the use of risk quotient methods. First, PAF calculations are based on the same principle that has been used to derive environmental quality criteria. This is a consistency argument. Second, it is powerful since it allows for comparisons of (aggregated) ecological risk over time and space, between taxa, and with other stressors (in very specific conditions). Furthermore, since input data relate to ambient concentrations, the PAF approach can be used not only diagnostically by using measured or interpolated environmental concentrations, but also prognostically by using predicted environmental concentrations generated by scenario analyses. These features allow for comparison of the efficacy of different risk management scenarios for a specific location or for emission reduction measures on a local, regional, and national scale. These comparisons do not necessarily require that the PAF be an absolute quantification of ecological effects. Finally, a useful feature of PAF is that there is no need to define a cutoff percentile as a threshold for ecological effects (Van Straalen and Denneman, 1989). In addition, they can deal with distributed exposure concentrations (Chapters 4, 5, and 15). Those features provide the potential for future developments.

16.4.7 Relationships between msPAF Values
and Community Responses

The value of PAF comparisons is determined by the interpretation and validity of the PAF as an indicator of ecological risk. PAFs are based on NOECs and the NOEC is an imprecise estimate of the concentrations at which effects occur (e.g., Kooijman and Bedaux, 1996). Since many tests have high variance due to low numbers of observation, lack of statistical significance can mask ecologically significant effects. It is therefore probable that ecological effects occur at a certain PAF (e.g., 20%), but the ecological implications of such a value are not clear. If NOECs are exceeded in the field, we may expect some toxic effects (such as increased respiration, decreased growth or fertility, possibly increased mortality), but this need not have any impacts on a population level. Even increased mortality could have little impact on a population that experiences high predation pressure. Confirmation studies could shed light on the implications associated with increased PAF levels.

Observing effects in the field may be difficult because of high variability and a lack of a reference situation. The expected effects may be masked if other, less sensitive species outcompete the affected species (Klepper et al., 1999) or effects may be larger than expected because of interactions between species not predicted by single-species toxicity tests (Chapter 15). Ecological interactions are an important point to consider when interpreting SSD-based output (see Chapter 22). The relation between calculated PAF values and ecological interactions has not yet been clearly established, although some clues can be found in data from mesocosm studies. For example, species shifts due to toxic effect of copper and zinc have been observed in nematode communities in contaminated experimental field plots, and could be related to calculated PAF values (Klepper et al., 1999). A range of response types has been described by Smit et al. (in press), clearly demonstrating the occurrence of ecological interactions.

In practice, different natural stressors may influence the effects of toxicants either directly by affecting the species' viability or through interactions with the compound. This is a factor that is not taken into accout when calculating msPAF. The example on plants touches upon this subject. A clear example of the role of additional stress in determining true field responses is provided by Stuijfzand et al., 1999). In *in situ* bioassays in the Meuse and the Rhine, a higher mortality of larvae was observed in spring and summer of 1996 for larvae of various caddisfly species (*Hydropsyche angustipennis*, *H. siltalai*, and *H. exocellata*) in the Meuse. This was attributed to higher toxicant concentrations in the Meuse due to the extremely low water discharge, but also to death by low oxygen conditions and high temperature.

Despite uncertainties and shortcomings, the current approaches that are used to quantify PAF values for toxicant-stressed ecosystems yield values that can be interpreted as relative measures of risk, useful in various risk management contexts. It remains to be established whether PAF values are adequate indicators of community responses, as tentatively shown in Chapter 8 of this book, especially in the context of varying levels and duration of exposure and possible adaptation of communities to other environmental stress.

16.5 CONCLUSIONS

The idea of PAF seems to offer a conceptual advantage in comparison with risk quotients where the ratio itself cannot be aggregated on the basis of a defendable, ecotoxicological concept. The outcome of PAF calculations should be confirmed and validated to improve ecological interpretation of the output, and associated methods such as correction for bioavailability and mixture toxicity should be further underpinned by original work in related disciplines.

ACKNOWLEDGMENTS

The authors thank Olivier Klepper for his contributions to the development of the PAF concept and its applications, Marijke Vonk for contributions to the plant study, and two reviewers whose comments were very helpful.

APPENDIX

16.A.1   MSPAF METALS BY RESPONSE ADDITION (PREFERRED)

The $msPAF_{RA}$ is calculated from the individual PAF values of Table 16.1 by using Equation 16.9. The PAFs for each metal can be calculated from the exposure concentrations and the SSD parameters α and β in Table 16.A.1, according to Equation 16.3. α is the mean of the log10 of toxicity data and β is related to the standard deviation σ of log toxicity data as: $\sigma \cdot \sqrt{3}/\pi$.

16.A.2   MSPAF METALS BY CONCENTRATION ADDITION

The msPAF of metals can also be calculated according to concentration addition ($msPAF_{CA}$). As explained in Box 16.1, the procedure for concentration addition requires scaling NOECs (Equation 16.1, Box 16.1) and recalculation of the exposure concentration into HUs. Since we already calculated α and β for the metals without

---

TABLE 16.A.1
Parameters (α and β) of the Logistic Distribution Fitted to the $\log_{10}$ of Toxicity Data (g/l) for Selected Compounds

	α	β
Cadmium	−4.80	0.57
Copper	−4.89	0.46
Lead	−3.84	0.38
Zinc	−4.06	0.38
Atrazine	−4.03	0.71
Dichlorvos	−4.80	0.71
Diuron	−5.06	0.35
Lindane	−4.77	0.84
Mevinfos	−4.15	0.64
Metals		0.53
Herbicides		0.56
OP compounds		0.71

Data, courtesy of Dick de Zwart; see Chapter 8.

the scaling to HUs a simple transformation of the exposure concentrations for metals C_{ENV} (Table 16.A.2) is sufficient:

$$HU = \frac{C_{ENV}}{10^{\alpha}} \qquad (16.A.1)$$

for $\log_{10}$ transformed NOECs. To calculate the msPAF according to concentration addition of metals, the sum of HUs for all metals is substituted in Equation 16.4 (Box 16.1) and the use of the average β for all metals combined (see Table 16.A.1).

16.A.3 MSPAF BIOCIDES BY CONCENTRATION ADDITION WITHIN TMoA FOLLOWED BY RESPONSE ADDITION (PREFERRED)

According to the proposed aggregation of msPAF (Section 16.2.2.3), msPAF within a mode of action is calculated for (a) herbicides and (b) organophosphate compounds by concentration addition as explained above. The two values for these groups and the PAF for lindane (with a third, different mode of action) are combined by response addition (Equation 16.9).

16.A.4 MSPAF BIOCIDES BY CONCENTRATION ADDITION

The msPAF of biocides calculated according to concentration addition follows the same rules as for concentration addition of metals using Equations 16A.1 and 16.4.

16.A.5 MSPAF ALL COMPOUNDS

The msPAF for all compounds is calculated by combining the two preferred msPAF values (msPAF$_{RA}$ for metals and msPAF$_{CA+RA}$ for biocides) by response addition using Equation 16.9.

TABLE 16.A.2
Median Concentrations (g/l) of Selected Compounds at Environmental Monitoring Stations in the Rhine Basin of the Netherlands

Compound	Bovenrijn + Waal	Nederrijn	IJssel	Twenthekanalen	IJsselmeer	Markermeer	Ketelmeer	Randmeren E	Randmeren W	Adam-Rijnkanaal
Cadmium	2.29E-08	3.16E-08	2.61E-08	2.35E-08	6.17E-09	2.90E-09	3.98E-08	6.01E-09	4.35E-09	1.13E-08
Copper	1.59E-06	3.37E-06	2.08E-06	3.79E-06	1.29E-06	1.21E-06	2.88E-06	1.45E-06	6.35E-07	2.24E-06
Lead	3.27E-07	2.34E-07	5.51E-07	2.17E-07	1.11E-07	1.28E-07	2.81E-07	1.31E-07	1.39E-07	2.77E-07
Zinc	6.06E-06	6.76E-06	7.1E-06	4.77E-06	1.6E-06	1.68E-06	6.48E-06	1.52E-06	2.02E-06	2.83E-06
Atrazine	9.99E-08	7.99E-08	8E-08	6.99E-08	6E-08	4.99E-08	7E-08	9.99E-08	6E-08	4E-08
Dichlorvos	5E-08	4E-08	6E-08	2E-08	1E-08	9.99E-09	3E-08	1E-08	2E-08	2.5E-08
Diuron	9.97E-09	5.99E-08	1.1E-07	9.99E-08	3.99E-08	6.99E-08	6.99E-08	2.49E-07	5.99E-08	5.5E-08
Lindane	ND	ND	ND	ND	2.95E-09	ND	ND	ND	ND	3.98E-09
Mevinfos[a]	1E-08	1E-08	1E-08	1E-08	1E-08	1E-08	1E-08	1E-08	1E-08	1E-08

[a] Detection limit.

Data kindly supplied by the Dutch Institute for Inland Water Management and Wastewater Treatment (RIZA) in the framework of the National Environmental Outlooks.

17 Methodology for Aquatic Ecological Risk Assessment

William J. Warren-Hicks, Benjamin R. Parkhurst, and Jonathan B. Butcher

CONTENTS

1-56670-578-9/02/$0.00+$1.50
© 2002 by CRC Press LLC

Abstract — The methods for assessing ecological risk to aquatic organisms exposed to toxic chemicals or pesticides are described in *Methodology for Aquatic Ecological Risk Assessment* (Parkhurst et al., 1996). The methodology consists of three tiers: (1) a screening-level risk assessment; (2) quantification of risks identified in the first tier as potentially significant using existing data; and (3) quantification of risks using new site-specific data. This chapter focuses on the development of species' effects distributions in a Tier 2 risk assessment. A mathematical basis for the underlying probability models and species sensitivity distributions (SSDs) was derived and is described in detail. The development and use of expected environmental concentrations (EECs), development and use of species-derived risk distributions, and risk characterization are discussed. Data from two case studies are used to demonstrate the use of SSDs in evaluating aquatic ecological risks. In the first case study, involving point-source discharges to a stream from a wastewater treatment plant, the risk evaluation software of the methods was used to calculate and then apply risk-based effluent limits for ammonia, silver, and mercury. Virtually all of the risk of the effluent was caused by ammonia. In 1996, the effluent would have passed the risk-based effluent limits for ammonia for all but 2 months. During these 2 months (January and September) only 6 to 7% of the aquatic species in the Jordan River was estimated to be adversely affected by chronic ammonia toxicity. Thus, the risks of these effluent limit exceedences to aquatic life in the Jordan River were relatively small. In the second case study, in which no point sources were present, the chemicals examined were first identified through a screening-level risk assessment. Diazinon was determined to pose the greatest risk, so this chemical was examined in a probabilistic risk assessment using the risk evaluation software of the methods. Results showed that up to 11% of the freshwater taxa and 13% of all arthropods could have been affected by chronic diazinon toxicity. These results demonstrate that the tiered methods described in Parkhurst et al. (1996), along with existing data, can be used to produce quantitative estimates of risks to aquatic organisms and communities.

17.1 INTRODUCTION

The *Methodology for Aquatic Ecological Risk Assessment* (Parkhurst et al., 1996) is the product of a 2-year project conducted by The Cadmus Group, Inc. and sponsored by the Water Environment Research Foundation (WERF). The methodology is intended for use by regulatory authorities or members of the regulated community who need to estimate the effects of toxic chemicals and pesticides on aquatic communities from

- new point or nonpoint sources of chemicals,
- improved wastewater treatment,
- an increase or decrease in discharge from an existing wastewater treatment facility,
- exposure to pesticides,
- a more stringent or less stringent numerical water quality standard, or
- hazardous waste site cleanup or remediation.

The methodology consists of three tiers: Tier 1, screening-level risk assessment (SLRA); Tier 2, quantification of risks identified as potentially significant in Tier 1, using existing data; and Tier 3, risk quantification using new site-specific data. An overview of each tier is provided in the following sections.

This chapter focuses on the development of species effects distributions in a Tier 2 risk assessment and provides only a very brief overview of the Tier 1 and Tier 3 approaches. Further details on all three tiers can be found in Parkhurst et al. (1996). After the methodology and underlying statistical approaches are described, the results of applying the procedures in a case study are presented.

A Windows-based software system is available for implementing the statistical approaches presented in this chapter. To obtain the software and user's manual, contact WERF at its Washington, D.C. office.

17.1.1 TIER 1: SCREENING-LEVEL RISK ASSESSMENT

The Tier 1 SLRA is designed to be the most cost-effective, environmentally protective, and scientifically conservative step in the risk assessment process. SLRAs identify potentially significant risks by distinguishing those toxic chemicals posing the greatest potential risk and eliminating those posing negligible risk. Tier 1 estimates the potential effects of toxic chemicals, singly or in combination, in effluents, wastewaters, receiving waters, sediments, and aquatic biota. This tier may also be used to rank the potential ecological risks to other resources, for example, to identify ecological receptors and sites (areas or habitats) at greater or lesser risk.

The Tier 1 SLRA for aquatic resources — the aquatic ecological risk assessment (AERA) — is based on the following four assumptions:

1. The exposure scenario is worst case (i.e., the most sensitive species are present when the chemical concentrations are maximum).
2. All of the chemical concentration is bioavailable.
3. A diverse, multitrophic-level, aquatic community is present, and the biological integrity of the community is sound.
4. Ambient water quality criteria developed by the U.S. Environmental Protection Agency (U.S. EPA) or comparable criteria are conservative indices of ecological effects.

The hypothesis tested in Tier 1 is that toxic chemicals in the water body will have significant adverse effects on the survival, growth, or reproduction of aquatic species, or will be significantly bioaccumulated. The hypothesis is tested by comparing expected environmental concentrations (EECs) of chemicals in water and sediment with acute and chronic ecological risk criteria (ERC), or by estimating concentrations of chemicals in aquatic biota using bioconcentration factors.

Risk characterization in Tier 1 is estimated by a quotient, derived by dividing the EEC by the ERC. Values of the quotient greater than 1 indicate a potential risk to the aquatic biota. Using conservative values for the exposure concentration and the risk criterion provides an approach for minimizing the chance that toxic chemicals are falsely determined to pose little or no risk to the aquatic community.

Tier 1 produces a list of the chemicals posing potential risks (the chemicals of potential concern, or COPCs); quotients reporting the magnitude of risks for each chemical and for the entire sample; and a description of the methodology, assumptions, and data. The listed chemicals are ranked in terms of the magnitude of the potential risks they pose.

17.1.2 TIER 2: RISK QUANTIFICATION WITH EXISTING DATA

The Tier 2 AERA methodology provides quantitative, probabilistic risk estimates for each COPC identified in Tier 1. In Tier 2, existing data are evaluated more rigorously so that the ecological risk posed by a given chemical can be more accurately estimated. Tier 2 identifies three classes of COPCs, those that (1) clearly pose no risk, (2) clearly constitute a present threat in the environment and should be subject to risk management and remediation, and (3) may or may not pose a significant ecological risk and may warrant further study and data collection in Tier 3.

Tier 2 differs from Tier 1 in that:

- Fewer assumptions are used.
- Only those COPCs not eliminated in the Tier 1 screening are considered.
- Risks are based on probabilities of adverse effects.
- Aquatic ecological risk is quantified in terms of the percent of species or genera affected by the toxic chemical or chemicals.
- Site-specific ecological receptors and the development of site-specific ecological risk criteria, such as site-specific water quality criteria, may be evaluated.
- The quality of the data used to assess risk better reflects actual field conditions; more sophisticated and comprehensive modeling techniques may be devoted to estimating exposure.
- The uncertainties surrounding estimates of exposure and ecological effects, including the uncertainties in the resulting risk estimates, are rigorously calculated.

The hypothesis tested in Tier 2 is that the EECs of chemicals have adverse effects on the survival, growth, or reproduction of aquatic organisms and the structure of the aquatic community. The hypothesis is tested by comparing probability distributions for EECs with probability distributions developed with ecological effects data.

Species distributions are created in Tier 2 as a means of assessing the risk of toxic chemicals to the environment. The risk assessor can choose specific risk levels and evaluate the expected impact to the aquatic community given the distribution of exposure concentrations.

Once Tier 2 has been completed, many of the COPCs at a site will have been determined to pose either a nonsignificant risk or a risk that is considered to be at an acceptable level. Toxic chemicals in the first category will have been dismissed from further consideration. Those in the second category will have been concluded to pose an established risk, which may require a risk management response. Based on Tier 2 data, however, other chemicals for which a reliable decision cannot be made due to uncertainties created by insufficient site-specific data may have been identified. These COPCs may be examined in Tier 3.

17.1.3 Tier 3: Risk Quantification with New and Existing Data

Although the Tier 3 AERA methodology uses the same basic procedures as Tier 2, it is distinct in that (to the extent deemed appropriate by the risk manager) new, site-specific data are used. These data are collected to reduce uncertainty in the risk estimates.

17.1.4 Overview of SLRA Tiers

As explained in Parkhurst et al. (1996), each risk assessment tier consists of specific approaches to source characterization, exposure assessment, ecological receptor characterization, ecological effects characterization, risk characterization, and risk management. Table 17.1 presents an overview of the tiers and identifies the differences among them for each risk assessment area.

17.2 SPECIES DISTRIBUTIONS IN RISK ASSESSMENT: A TIER 2 APPROACH

The following discussion presents elements of the Tier 2 approach as presented in Parkhurst et al. (1996). An overview of the Tier 2 approach is presented for those major risk analysis elements that involve the use of species distributions for establishing risk levels. A discussion is presented of (1) the development and use of EECs, (2) the development and use of species-derived risk distributions, and (3) risk characterization.

17.2.1 Development and Use of Expected Environmental Concentrations

The Tier 2 exposure assessment uses realistic and bioavailable exposure concentrations. Site-specific exposure pathways are identified and evaluated. Chemical forms considered in the assessment may include the chemical species (e.g., trivalent arsenic, methyl mercury, ionic copper) or their operationally defined phases (e.g., dissolved, particulate, tissue residue), or both. The significance of a pathway depends both on the characteristics of the COPCs and the nature of their releases into the environment. For example, if a lipophilic chemical is released mainly as a particulate in storm water and is deposited in the sediments, the surface water and sediments constitute two pathways. Subsequent chemical behavior (e.g., release from the sediments or uptake by a benthic organism) would define other pathways.

EECs in Tier 2 should be expressed as probability distributions, so that spatial and temporal variation can be evaluated. Data from actual site-specific field measurements are preferable; however, environmental fate-and-transport models can generally be used to forecast future conditions and concentrations in unmeasured matrices. A fate-and-transport model must be selected that is appropriate for the spatial and temporal scale of the problem of interest. The choice of an appropriate model is complex, and is not discussed here in detail. Various guidance documents, however, are available. Simple equilibrium and dilution models, such as quantitative

TABLE 17.1
Comparison of Risk Assessment Methods for Tiers 1 through 3

Step	Tier 1	Tier 2	Tier 3
Source characterization	Using existing data, identify COPCs, sources, and loadings.	Same as Tier 1.	Collect new data on sources of toxic chemicals.
Exposure assessment	Use existing data or simple mass balance models and total concentrations to estimate acute and chronic EECs.	Use existing site measurements or fate-and-transport models and data on spatial and temporal variability to estimate probability distributions for chemicals. Use dissolved concentrations, if possible, to assess bioavailability.	Collect new analytical data on spatial and temporal variability and perform additional fate-and-transport modeling to better estimate probability distributions for chemicals. Assess site-specific bioavailability. Use site-specific data in models.
Ecological receptor characterization	Assume generic, diverse, multitrophic-level community is present and biological integrity is sound.	Based on existing data, characterize site-specific species, populations, and communities.	Based on new and existing data, characterize site-specific species, populations, and communities.
Assessment endpoints	Survival, growth, or reproduction of aquatic populations, and bioconcentration.	Survival, growth, or reproductive rates of site-specific aquatic populations and structure of communities.	Survival, growth, or reproductive rates of site-specific aquatic populations and structure of communities.
Measurement endpoints	Generic: Acute toxicity Chronic toxicity Bioconcentration	Generic or Site-Specific Acute toxicity Chronic toxicity Bioaccumulation Species Populations Communities	Site-Specific: Acute toxicity Chronic toxicity Bioaccumulation Species Populations Communities
Ecological effects characterization	Use the U.S. EPA national water quality criteria, toxicity databases (e.g., AQUIRE), and simple models (e.g., QSARs, BCFs), to estimate toxicity values and bioconcentration. Assume measurement and assessment endpoints are related directly.	For each chemical, develop models of concentration vs. % of species affected. Model can be generic or site specific.	Same as Tier 2.

TABLE 17.1 (continued)
Comparison of Risk Assessment Methods for Tiers 1 through 3

Step	Tier 1	Tier 2	Tier 3
Risk characterization	EEQ = EEC/Ecological risk criterion Acute quotient >0.3 = COPC Chronic quotient >0.3 = COPC Total Risk = Sum of EEQs Uncertainties acknowledged	Test for significant differences between EECs and ecological effects criteria. Integrate results of exposure assessment and ecological effects characterization to estimate % of species affected and to quantify uncertainties.	Same as Tier 2. Use site-specific bioassessments or ambient toxicity tests to validate risk estimates.
Risk management	Compare remediation costs vs. eco-risks vs. assessment costs. To refine risk estimates for COPCs, proceed to Tier 2.	Compare remediation costs vs. eco-risks vs. assessment costs. To refine risk estimates for COPCs, proceed to Tier 3.	Compare remediation costs vs. eco-risks vs. assessment costs.

structure–activity relationships and mass balance models, respectively, are routinely used. More complex models, allowing EECs to be estimated temporally and spatially in various matrices, are widely available (e.g., Mackay and Peterson, 1982; Reinert, 1987; Reuber et al., 1987; Ambrose et al., 1988; Southwood et al., 1989; Mackay, 1991).

In the Tier 2 AERA, both dissolved and total concentrations of COPCs may be considered, but dissolved COPC concentrations should be emphasized, because they are considered (1) most indicative of bioavailability (DiToro et al., 1991; 1992; U.S. EPA, 1994c) and (2) most representative of laboratory toxicity tests.* Equally important, expressing exposure in terms of dissolved or bioavailable chemical concentration may ultimately provide more accurate risk estimates for most chemicals that pose toxic risks to aquatic life. For some chemicals and site-specific situations, however, ingestion of contaminated food may pose a significant exposure pathway for aquatic life and predators in higher trophic levels.

Frequently, data on dissolved metal concentrations in surface waters are lacking and must be predicted. For certain metals, U.S. EPA (1992) presents a generic method for predicting these data based on national average ratios between total and dissolved metals in surface waters (Table 17.2). The uncertainties in the U.S. EPA (1992) ratios for specific metals have not been determined.

---

* Most laboratory toxicity tests are conducted under conditions favoring toxicant bioavailability. For example, most laboratory dilution waters contain low concentrations of suspended solids and organic carbon, conditions that favor bioavailability by removing sorption/complexation surfaces. Standard test methods (e.g., ASTM, 1994) require high-quality dilution water.

TABLE 17.2
Dissolved Metals as the Fraction of Total Metals in Surface Waters, National Averages

Metal	Fresh Water	Salt Water[a]
Aluminum	0.07	0.07
Cadmium	0.04	0.81
Copper	0.40–0.62	0.40–0.50
Lead	0.10	0.08
Nickel	No data	0.60–0.73
Silver	No data	0.11
Zinc	0.20	0.60

[a] Does not include waters near the bottom.

Source: U.S. EPA, 1992.

In Tier 2, EECs are derived from existing water quality data or from water quality models. Thus, the EECs are estimates or predictions, with inherent levels of uncertainty. In Tier 1, the EEC is treated as a single value that is assumed to be accurate. In reality, the Tier 1 EECs represent the highest (or close to the highest) EECs expected. In Tier 2, the EEC is treated as an estimated value for which measures of associated uncertainty can be calculated.

Commonly, the risk assessor is faced with small sample sizes and some measurements that are less than the method detection limits (MDLs). Because Tier 2 involves characterizing risk using more complex statistical and quantitative methods than Tier 1, small sample sizes with many MDLs can be problematic. In particular, a certain amount of data is required for estimating the distribution of EECs at the site. Guidance on strategies for dealing with small sample sizes is available in Parkhurst et al. (1996).

In environmental samples, particularly for toxic chemicals, finding a substantial proportion of the samples reported as MDLs, or nondetects, is common. This does not necessarily mean that the concentration is zero, only that it is between zero and the MDL.

Although MDLs should not be ignored, how they are represented can significantly affect the estimated mean and variance of concentration measurements. In the past, common approaches to handling MDL data have included assuming that all such points were equal to zero, half the MDL, or the MDL. The first option provides the least conservative estimate, and the last option the most conservative. All three approaches, however, produce biased estimates of the mean and standard deviation.

For symmetric distributions, such as the normal distribution, the population median and mean are expected to be equivalent. The median is estimated as the middle value of the sample data and is thus unaffected by the presence of MDL data, as long as at least half of the data are above the MDL. Therefore, a reasonable

TABLE 17.3
Calculation of Winsorized Mean and Standard Deviation

Assume a sample of size n is drawn from a symmetric distribution, arranged in ascending order of magnitude, and that the first k values are nondetects. The Winsorization procedure is as follows:

1. Replace the first k values (the nondetects) by the next largest $(k + 1)$ value.
2. Replace the last k values ($n - k$ to n) by the $n - k - 1$ value.
3. Compute the Winsorized mean as the mean of the resulting set of data.
4. Compute the sample standard deviation, s, of the same data set.
5. Compute the Winsorized standard deviation as $s_w = [s(n - 1)]/[n - 2k - 1]$.

The Winsorized mean and standard deviation may then be used to compute confidence limits on the mean in the usual manner, except that the degrees of freedom for the t-distribution will be $n - 2k - 1$, not $n - 1$.

Source: Gilbert, R. O., *Statistical Methods for Environmental Pollution Monitoring,* Van Nostrand Reinhold, New York, 1987, 313. With permission.

approach to estimating the mean with MDL data for a symmetric distribution is to substitute the median. This approach does not, however, resolve the question of estimating the variance, which is needed for calculating confidence limits and conducting hypothesis tests concerning the mean. Furthermore, if the distribution is actually skewed, the median will be a biased estimate of the mean.

A useful approach for symmetric distributions with a limited proportion (e.g., ≥50%) of MDL data and a single MDL is Winsorization, described by Dixon and Tukey (1968) and by Gilbert (1987). The Winsorized mean and standard deviation calculations are shown in Table 17.3.

Environmental data are often assumed to be lognormally distributed. When this assumption applies, several techniques are available for estimating the mean and standard deviation. The simplest method is log-probit analysis, which can be implemented graphically. When values from a lognormal distribution are plotted against a probit scale, they lie on a straight line. On the probit scale, a probit value of 5.0 is equivalent to the 50th percentile of the data, while ±1 probit value is equivalent to 1 standard deviation (in log-transformed space). If the first k of n values are MDLs, the $k + 1$ value is plotted at the appropriate position (cumulative percentile) within the complete data set, which should produce a linear plot. The point at which the line crosses probit 5.0 is the geometric mean of the distribution (mean of the natural logs), while the slope of the line is the geometric standard deviation (standard deviation of the natural logs). These values may then be used to estimate the arithmetic mean and standard deviation.

The log-probit method and an alternative maximum-likelihood method have been computerized by the U.S. Geological Survey (Helsel and Cohn, 1988) and are also available in various other computerized programs. Both methods can be used for data with single or multiple detection limits.

The same types of summary statistics calculated for water-column EECs in Tier 2 should also be calculated for sediment EECs. The types of COPC data required are sediment interstitial water EECs. These can be based on direct measurements or, for

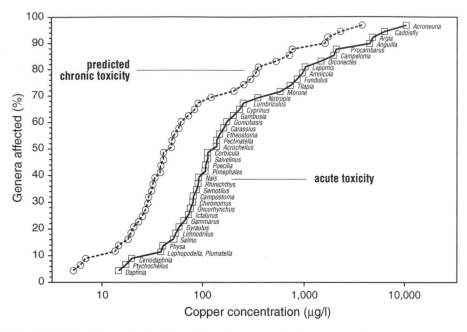

FIGURE 17.1 Acute and chronic toxicity of copper to freshwater aquatic genera.

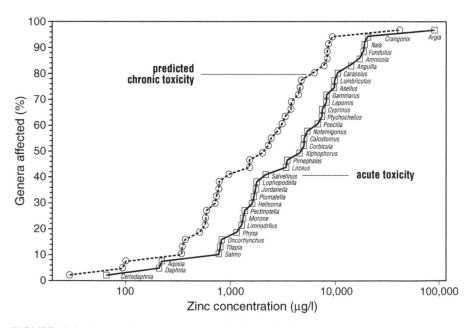

FIGURE 17.2 Acute and chronic toxicity of zinc to freshwater aquatic genera.

neutral hydrophobic organic chemicals, estimated using the equilibrium partitioning method. If such data are unavailable, but are required to perform a quantitative AERA, then Tier 3 is conducted to obtain the required data. An alternate method of estimating exposure and effects for sediments is to conduct sediment toxicity tests.

17.2.2 ECOLOGICAL EFFECTS CHARACTERIZATION: GRAPHICAL REPRESENTATION OF THE ASSOCIATION OF COMMUNITIES AND TOXICITY

The toxicity data used by U.S. EPA to develop its water quality criteria can be used to develop a relationship between concentrations of COPCs and acute and chronic effects to aquatic communities. The relationships can be assumed to be generic (U.S. EPA, 1985a) or made to be site-specific using data on resident species (Carlson et al., 1984; U.S. EPA, 1994b). Figures 17.1 and 17.2 depict the acute and chronic toxicities of copper and zinc to "communities" of species, represented by genera, based on the data in the respective aquatic water quality criteria (AWQC) documents for these metals. The four general assumptions behind these relationships are as follows:

1. As concentrations of COPCs increase, the number of species in the community affected by acute and chronic toxicity increases.
2. The relationships between concentrations of COPCs and effects on the community of species can be estimated from the published data in U.S. EPA water quality criteria documents or similar sources.
3. The relationships defined by these curves are representative of the relationships that would be found in natural aquatic communities exposed to these COPCs.
4. No confounding effects of habitat, water quality, flows, bioavailability, and species, such as competition and predation, are present.

Figures 17.1 and 17.2 plot geometric means of LC_{50} values for genera, called genus mean acute values (U.S. EPA, 1985a). Each genus mean acute value summarizes all acute toxicity data reported for species within a given genus in the water quality criteria document. Because the toxicities of copper and zinc are hardness dependent, the genus mean acute values have been adjusted, for this example, to a total hardness of 50 mg/l $CaCO_3$. Tier 2 risk assessments should use the hardness specific to the water body being assessed. To generate the plot, the genus mean acute values are sorted and ranked, and a cumulative distribution value is created (see Equation 17.24, below).

The U.S. EPA estimates water quality criteria for each chemical by plotting a regression for the four lowest points and estimating the LC_{50} corresponding to the 5th percentile. The corresponding value (i.e., the final acute value), after division by 2, becomes the U.S. EPA-specified water quality criterion for acute toxicity, the criterion maximum concentration (CMC). The U.S. EPA chronic criterion, the criterion continuous concentration (CCC), is generally calculated by dividing the final acute value by the geometric mean acute-to-chronic ratio (ACR) for the chemical.

Thus, the acute and chronic curves have the same configuration and differ in magnitude only by the ACR. This occurs only because the acute and chronic values have been assumed to be in constant proportion for all genera. If chronic toxicity has been measured directly, the two curves may not be parallel. The criteria calculated from these data are community based, because they presume to protect at least 95% of all aquatic genera (species) from acute and chronic toxicity.

Figure 17.1 shows that copper is acutely toxic (at 50 mg/l $CaCO_3$ total hardness) to the community of genera at concentrations between 17 and >5000 µg/l. The daphnids (*Daphnia* and *Ceriodaphnia*) rank at the 2nd and 7th percentiles in sensitivity. Salmonids, in contrast, rank in the 16th percentile (*Salmo*), 28th percentile (*Oncorhynchus*), and 47th percentile (*Salvelinus*). Invertebrates may be either more sensitive than fish (e.g., daphnids, snails, and amphipods) or less sensitive (e.g., crayfish, odonates, caddisflies, and stoneflies). The predicted chronic toxicity of copper ranges from 6 to >3500 µg/l (at 50 mg/l $CaCO_3$ total hardness) for the community.

The acute toxicity of zinc to the species shown in Figure 17.2 ranges from 69 to >88,000 µg/l (at 50 mg/l $CaCO_3$ total hardness), and its corresponding chronic toxicity ranges from 31 to >40,000 µg/l. Two groups of salmonids, trout and salmon, rank in the 11th and 17th percentiles for sensitivity, whereas banded killifish (*Fundulus*) are in the 89th percentile. Fathead minnow (*Pimephales*) approximate the median sensitivity.

17.2.3 STATISTICAL ESTIMATION OF THE DISTRIBUTION OF ECOLOGICAL RISK CRITERIA FOR ACUTE AND CHRONIC DATA

As in the analysis of EECs, the U.S. EPA AWQC for single chemicals are generally calculated and applied without considering their associated uncertainty (Erickson and Stephan, 1988). This section describes one method for estimating the uncertainty associated with AWQC or other ERCs. The method incorporates data used to develop the U.S. EPA AWQC and is based, in principle, on the U.S. EPA methods for developing AWQC (U.S. EPA, 1985a).

The original data U.S. EPA used to develop its AWQC (e.g., see table 1 in U.S. EPA, 1985c and 1987b) are used to estimate the ERC and its associated variance. The data are used in a logistic regression model relating cumulative genus mean acute values to chemical concentrations (Figure 17.3). These same procedures can be applied to species LC_{50} values and chronic toxicity test endpoints such as chronic values (e.g., no-observed-effects concentrations, or NOECs, and lowest-observed-effects concentrations, or LOECs), all of which follow a sigmoidal response function. The function has been shown to fit data for many chemicals, including metals, pesticides, and organics.

From this regression model, a desired risk level for an ERC (e.g., 5%) can be selected and the distribution of concentrations that are predicted to yield this target by the regression model can be determined. Uncertainties for an ERC are thus quantified. At this risk level, an average of 95% of all the genera tested have genus mean acute values greater than the ERC and, therefore, should be protected by the ERC.

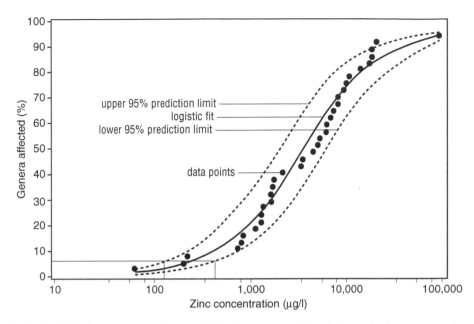

FIGURE 17.3 Logistic regression model for the acute toxicity of zinc to fresh water aquatic genera showing estimation of uncertainty for an acute risk criterion. Solid line is the logistic regression model fit to these data and broken lines are 95% prediction limits for the model.

For some metals where toxicity depends on water hardness (mg/l $CaCO_3$), to prepare the data for use in the logistic regression model, the LC_{50} data must first be adjusted to a consistent hardness. The relationship between the adjusted and original data is as follows:

$$\text{Adjusted } LC_{50} = \left(\frac{50}{\text{hardness}} \right)^{\text{slope}} * LC_{50} \tag{17.1}$$

where the slope is the pooled slope taken from the U.S. EPA AWQC document, and the hardness is that associated with the original test. Geometric mean LC_{50} values for each genus should then be calculated directly from the adjusted LC_{50} data.

For most metals, toxicity is expressed primarily as the dissolved form. The laboratory tests used to develop the U.S. EPA AWQC for these metals had a certain portion of the reported metal concentration present in the particulate form. The U.S. EPA, therefore, issued factors to convert reported concentrations in the laboratory toxicity tests to a dissolved basis (*Federal Register*; 60: 22228–22237, May 4, 1995). These conversion factors should be applied to LC_{50} concentrations for comparison to dissolved metal EECs.

The logistic regression model in our application has the form:

$$\frac{e^{\alpha + \beta_1 * \log(\text{genus mean } LC_{50})}}{1 + e^{\alpha + \beta_1 * \log(\text{genus mean } LC_{50})}} \tag{17.2}$$

A linear regression form of this model would be written:

$$\text{logit}(p) = \alpha + \beta_1 * \log(\text{genus mean LC}_{50}) + \varepsilon \qquad (17.3)$$

where the independent variable is the log of the genus mean acute value (e.g., LC_{50}) and p is the probability of an effect at a specific concentration of the toxicant. The logit transformation is defined as

$$\text{logit}(p) = \ln\left(\frac{p}{1-p}\right) \qquad (17.4)$$

and converts a nonlinear regression into a linear one. The resultant model is some-times called a log-odds model. The logit transformation takes the familiar s-shaped curve associated with LC_{50} data (e.g., Figure 17.3) and changes it into a straight line. This transformation does not alter the relationship between the dependent and independent variables, but simply changes the scale on which the data are viewed. Thus, simple linear regression techniques can be used to develop a model. Parameter estimates for the above model are presented in Table 17.4. For metals whose toxicity is a function of hardness, the data used in constructing this table were adjusted to a total hardness of 50 mg/l as $CaCO_3$. Finally, because the toxicity data of the original metal were based on measurements of total metal, these data were converted to dissolved metal concentrations using the appropriate Fresh Water Conversion Factor (which is consistent with U.S. EPA, 1995d).

A 95% protection level is equivalent to the U.S. EPA AWQC. At this level of protection, 5% of the species will be affected by the toxicant. This corresponds to a logit of –2.9444. To calculate the expected ERC, –2.9444 is inserted into the estimated logistic regression model, and the concentration of the toxicant associated with the ERC is back-calculated from the equation. Associated with the predicted ERC is a 95% prediction interval. Most computer packages will output the prediction limit associated with the linear regression model. On a plot of logit(p) vs. log(LC_{50}), at the point where the risk criterion (i.e., % genera affected) is 5%, a horizontal line can be drawn that intersects the estimated regression line and the upper and lower prediction bounds. The concentration associated with the point of intersection with the estimated regression line is the ERC. The concentrations associated with the points of intersection with the upper and lower prediction bounds form a 95% prediction interval for the ERC. The prediction bounds on the ERC can be obtained from any of the three following methods:

Method 1: The prediction bounds for the ERC can be taken directly from the graph. An example is presented in Figure 17.3.

Method 2: The prediction bounds can be computed using a first-order error analysis. The terms of the equation are the parameters in the linear regression model (α, β_1), the variance of each parameter, and the covariance between the parameters (assume the covariance between the error and each of the parameters is zero, and

TABLE 17.4
Logistic Regression Models for Acute Toxicity

COPC[a]	Parameter	Estimate	Std. Error[b]	Root MSE[c]
Ammonia	α	−1.5759	0.0994	0.4223
	β	2.2519	0.1015	
Cadmium	α	−4.3779	0.2444	0.5296
	β	0.6892	0.0363	
Chlordane	α	−4.2451	0.3970	0.4107
	β	1.2244	0.1093	
Chlorine	α	−10.3013	0.4033	0.3127
	β	2.0636	0.0799	
Chromium III	α	−16.5410	1.4765	0.5086
	β	1.7656	0.1571	
Chromium VI	α	−4.5825	0.4210	0.6471
	β	0.4850	0.0426	
Copper	α	−5.0024	0.2261	0.4301
	β	0.9476	0.0409	
DDT	α	−2.1437	0.1348	0.4178
	β	0.8349	0.0430	
Dieldren	α	−2.7052	0.2139	0.3787
	β	0.8018	0.0564	
Endrin	α	−0.6054	0.1476	0.6631
	β	0.7298	0.0710	
Heptachlor	α	−2.1960	0.1689	0.3369
	β	0.7564	0.0492	
Lead	α	−4.9125	0.3679	0.2987
	β	0.5440	0.0394	
Lindane	α	−5.6742	0.2549	0.2491
	β	1.3242	0.0577	
Nickel	α	−15.7101	0.9635	0.3450
	β	3.8690	0.2363	
PCBs	α	−1.8770	0.3254	0.4098
	β	0.6646	0.0988	
Silver	α	−2.3596	0.2457	0.5282
	β	2.1201	0.1903	
Zinc	α	−8.5337	0.3061	0.3253
	β	1.0539	0.0372	

[a] At a hardness of 50 mg/l as $CaCO_3$ for hardness-dependent metals; metal concentrations expressed on a dissolved basis.

[b] Std. error is the standard error, which is a measure of the uncertainty in the parameter estimate. Two standard errors on either side of the estimate approximate 95% confidence intervals.

[c] Root MSE is the root mean square error, which is a measure of the relative fit of the model; the smaller the root MSE, the better the fit.

the magnitude of the error is insignificant). These values are generated by most linear regression computer programs. The equation is

$$VAR(ERC) = \left(\frac{1}{\beta_1}\right)^2 VAR(\alpha) + \left(\frac{z-\alpha}{\beta_1^2}\right)^2 VAR(\beta_1)$$
$$- 2\left(\frac{1}{\beta_1}\right)\left(\frac{z-\alpha}{\beta_1^2}\right) COV(\alpha, \beta_1)$$

(17.5)

where

$$z = \ln\left(\frac{p}{1-p}\right)$$

(17.6)

Method 3: The prediction bounds and confidence bounds on ERC can be computed by solving quadratic equations. For example, the logit model (Equation 17.3) can be written as

$$Y = a + b * X$$

(17.7)

where

$$Y = \text{logit}(p), \quad X = \log(\text{genus mean } LC_{50})$$

(17.8)

The prediction bounds on ERC (in log-space) at a given risk criterion p can be found by solving the following equation for x at a specified value of y:

$$a + b * x \pm t * s * \sqrt{1 + \frac{1}{n} + \frac{(x - \bar{x})^2}{d}} = y$$

(17.9)

where
  a = parameter estimate of the intercept
  b = parameter estimate of the slope
  t = critical t-value at level $(1 - \alpha/2)$ with $(n - 2)$ degrees of freedom, available in most standard statistics texts; $1 - \alpha$ is the prediction level
  s = root mean square error (MSE) obtained from the regression model
  n = number of observations for fitting the regression model
  $\bar{x}$ = average of all x values, i.e., average of all log(genus mean LC_{50}) values used for fitting the regression model
  $d = \Sigma(x_i - \bar{x})^2$
  $y = \text{logit}(p) = \ln(p/(1 - p))$

After algebraic transformation, the roots for the equation are

$$x_{1,2} = \frac{-B \pm \sqrt{B^2 - 4AC}}{2A} \qquad (17.10)$$

where

$$B = 2*b*n*d*(y-a) - 2*t^2*s^2*n*\bar{x} \qquad (17.11)$$

$$A = t^2*s^2*n - n*d*b^2 \qquad (17.12)$$

$$C = t^2*s^2*(n*d + d + n*\bar{x}^2) - n*d*(y-a)^2 \qquad (17.13)$$

The smaller root x_1 is associated with the lower $(1 - \alpha) * 100\%$ prediction limit on ERC (in log-space), and the larger root x_2 is the upper $(1 - \alpha) * 100\%$ prediction limit on ERC (in log-space).

The confidence bounds on the ERC (in log-space) can be obtained in the same manner, except that the equation to be solved differs slightly:

$$a + b*x \pm t*s* \sqrt{\frac{1}{n} + \frac{(x-\bar{x})^2}{d}} = y \qquad (17.14)$$

Having obtained the 95% confidence limits on the ERC in log-space (i.e., x_1, x_2), the variance of the ERC in log-space can be estimated by $((x_2 - x_1)/4)^2$, because the 95% confidence limits are approximately two standard deviations above or below the mean. Thus,

$$\text{Mean log(ERC)} = \frac{(\text{logit}(p) - a)}{b} \qquad (17.15)$$

and

$$\text{Std Dev log(ERC)} = \frac{(x_2 - x_1)}{4} \qquad (17.16)$$

Calculating the mean and variance of ERC in arithmetic space requires a transformation. The transformation is readily obtained from the properties of the lognormal distribution (see, for example, Benjamin and Cornell, 1970). Consider a natural log transform of data: $X = \ln(Y)$. If X is normally distributed, with mean μ_x and standard deviation σ_x, then Y is said to be lognormally distributed. The arithmetic parameters of Y are calculated from the ln-space parameters of X as follows:

$$\text{Mean of } Y - \mu_y = \exp\left[\mu_x + \frac{\sigma_x^2}{2}\right] \qquad (17.17)$$

$$\text{Std Dev of } Y = \sigma_y = \mu_y \cdot \left[\exp\left(\sigma_x^2\right) - 1\right]^{1/2} \qquad (17.18)$$

The mean and standard deviation of ERC in ln-space can be computed by

$$\mu_x = \text{Mean log(ERC)} * \ln(10) \qquad (17.19)$$

and

$$\sigma_x = \text{Std Dev log(ERC)} * \ln(10) \qquad (17.20)$$

This method was implemented in the WERF Aquatic Ecological Risk Assessment software package (The Cadmus Group, Inc., 1996b). Note that the standard deviation estimates on ERC computed as above are a function of the percent of genera affected; for example, the standard deviation associated with a risk of 2.5% will differ from the standard deviation for a risk of 5%.

17.2.4 ESTIMATION OF CHRONIC ERCs WHEN NO CHRONIC DATA EXIST

For most chemicals, the amount of available chronic toxicity data is often limited. Using ACRs to compute the chronic ERCs from acute data for such chemicals is recommended by the U.S. EPA (1985a). The chronic value is defined as

$$\left(\frac{1}{\text{ACR}}\right) \cdot (\text{Final acute value}) \qquad (17.21)$$

Because the data are analyzed on a log-transformed scale:

$$\log(\text{Chronic value}) = \log(1/\text{ACR}) + \log(\text{Final acute value}) \qquad (17.22)$$

and, therefore,

$$\log(\text{Chronic value}) = \log(\text{Final acute value}) - \log(\text{ACR}) \qquad (17.23)$$

Thus, in log-space, $\text{ERC}_{\text{chronic}} = \text{ERC}_{\text{acute}} - \log(\text{ACR})$ can be used as the chronic ERC. The variance of $\text{ERC}_{\text{chronic}}$ is the same as the variance of the acute ERC. (See Figure 17.2 for a representation of this method using zinc.)

17.2.5 Risk Characterization

Risk characterization involves comparing the results of the exposure assessment and the ecological effects characterization. The following is an illustration of how community-level risks can be estimated using graphical and statistical procedures.

17.2.5.1 Graphical Estimation

Figures 17.4 and 17.5 illustrate two approaches for comparing observed EECs for copper to acute and chronic concentrations from the U.S. EPA water quality criteria databases. In Figure 17.4, the EECs are represented by the minima and maxima, while in Figure 17.5 the same range of EECs is represented by a cumulative probability distribution. In this example, a distributional form is not assumed because data are limited; instead, an empirical distribution of EECs is estimated simply by ranking the data (six observations in this case) and assigning plotting positions as

$$y_j = \frac{j}{n+1} \tag{17.24}$$

where y_j is the cumulative probability of observation j, j is the rank order, and n is the number of observations.

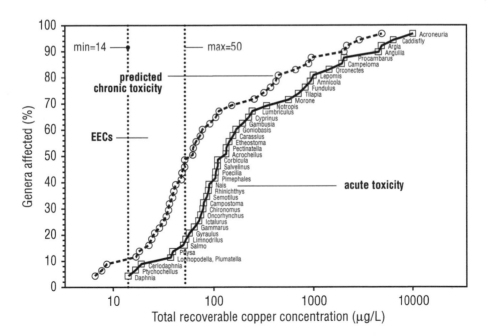

FIGURE 17.4 Risk characterization for copper: comparison of a range (minimum and maximum) of copper EECs and cumulative distributions for acute and chronic toxicity of copper to aquatic genera.

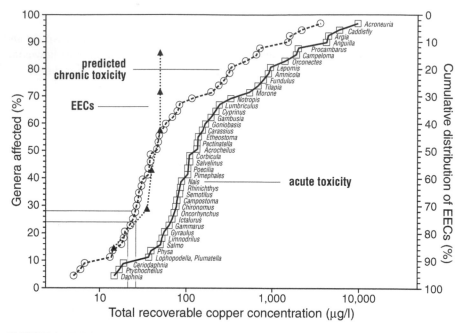

FIGURE 17.5 Risk characterization for copper: comparison of a cumulative distribution of EECs with cumulative distributions for acute and chronic toxicity of copper to aquatic genera.

Figure 17.4 provides a quick, qualitative, visual comparison of toxicity data and EECs. It indicates that aquatic genera are exposed to dissolved copper concentrations ranging from 14 to 50 µg/l with a median of 27 µg/l. To characterize potential risk, the mean concentration should be compared with the chronic toxicity data and the maximum concentration should be compared with the acute toxicity data. A risk of acute and chronic toxicity exists at the mean, median, and maximum concentrations. The median EEC (27 µg/l) is predicted to pose a risk of acute toxicity to 8% of the aquatic genera. Invertebrates related to the daphnids (*Daphnia* and *Ceriodaphnia*) and the squawfish (*Ptycocheilus*), or other equally sensitive species, would be at risk from acute toxicity at the median concentration. Risks of acute toxicity would be even greater at the maximum dissolved copper concentration of 50 µg/l, with 17% of the genera potentially at risk, including the trout *Salmo*, snail *Physa*, and bryozoans *Lophopodella* and *Plumatella*.

The median dissolved copper concentration of 27 µg/l poses a significant risk of chronic toxicity to approximately 25% of the genera (those identified above plus the amphipod *Gammarus*, the brown bullhead *Ictalurus*, and other equally sensitive species), as shown in Figure 17.5. A variety of risk statements can be made using such graphs, permitting simultaneous visualization of the EEC and toxicity distributions. For example, a dissolved copper concentration of 38 µg/l poses a risk of acute toxicity to about 10% of the genera and a risk of chronic toxicity to about 30% of the genera.

17.2.5.2 Estimated Community Risk: Single Chemicals

Procedures have been developed for calculating community risk, which estimates risk as the likelihood that a percentage of genera or species may be affected by acute or chronic toxicity. Two items of information are assumed to be available to the investigator. The first is a model relating chemical concentration to the percent of species (or genera) affected.

In this model, the risk (R, here representing the number of genera affected) is a function of concentration (EEC), so that the risk from a given concentration can be written as $g(R|EEC)$ (this notation indicates that risk is conditional on concentration). The EEC is described by a probability distribution, $f(EEC)$. Risk and concentration then have a joint probability function $h(R,EEC)$ and, from basic probability theory,

$$h(R, \text{EEC}) = g(R|\text{EEC}) f(\text{EEC}) \qquad (17.25)$$

The conditional probability, $g(R|EEC)$, is assumed to be binomial (n,p):

$$g(R|\text{EEC}) = cp^r (1-p)^{n-r} \qquad (17.26)$$

where

$$c = \frac{n!}{n!(n-r)!} \qquad (17.27)$$

The binomial parameter p can be related to toxicity through the logistic regression model (Equation 17.2). The logistic model parameters α and β are estimated by least squares (Equation 17.3).

The parameter p is interpreted to be the probability that the COPC at the given concentration is toxic to a genus or species. At any concentration, the risk (number of genera affected) is $n * p$.

The distribution of the EECs is assumed to be lognormal (i.e., log $C \approx N(\mu,\sigma^2)$). If log-transformed data are denoted by log C, then the distribution function, $f(\log C)$, has the form:

$$f(\log C) = \frac{1}{\sigma\sqrt{2\pi}} \exp\left[\frac{-1}{2\sigma^2}(\log C - \mu)^2\right] \qquad (17.28)$$

Thus, the joint probability of risk and concentration (for a single COPC) has the form:

$$h(R, \log C) = (c) p^R (1-p)^{n-R} \frac{1}{\sigma\sqrt{2\pi}} \exp\left[\frac{-1}{2\sigma^2}(\log C - \mu)^2\right] \qquad (17.29)$$

TABLE 17.5
Expected Community Risk Evaluation Example

EEC	p(EEC)	E(R | EEC)	p(EEC) * E(R | EEC)
50	0.015	0.04	0.0006
175	0.145	0.10	0.0145
325	0.33	0.12	0.0396
625	0.33	0.15	0.0495
875	0.145	0.20	0.0290
1225	0.025	0.50	0.0125

Expected Total Risk (% genera lost, weighted over C)   0.1457

Based on this probability function for a single COPC, the expected value of risk, variance of risk, and marginal probability function of risk can be computed. The Aquatic Ecological Risk Assessment Software, Version 2.0 (The Cadmus Group, Inc., 1996b) computes these quantities by numerically solving three integrals. First, the program computes the expected value of risk, which requires solving

$$E(R) = \int_0^n \int_{-\infty}^{\infty} R \cdot h(R, \log C) d\log C \, dR \tag{17.30}$$

The resulting calculation of community risk is most easily visualized by a simple discrete approximation. Assume that EECs can take one of six discrete values from 50 to 1225 ppm, and that the probabilities of these concentrations, $p(\text{EEC})$, and the associated community risk, $E(R|C)$, are known. The total community risk is then simply:

$$E(R) = \sum_{i=1}^{6} p(\text{EEC}_i) \cdot E(R|\text{EEC}_i) \tag{17.31}$$

The expected community risk is then evaluated, for example, as given in Table 17.5.
Next, the variance of risk is calculated, which involves solving the integral:

$$\int_0^n \int_{-\infty}^{\infty} R^2 \cdot h(R, \log C) d\log C \, dR \tag{17.32}$$

Finally, the marginal probability function for risk is computed, which is given by the equation:

$$f(R) = \int_{-\infty}^{\infty} h(R, \log C) d\log C \tag{17.33}$$

The marginal probability function is then used to compute the cumulative distribution function (CDF) for risk. Plots of (1 − CDF) are useful for assessing the risks of individual chemicals, as well as the total risk from all COPCs at an impacted site.

17.2.5.3 Estimated Community Risk: Multiple Chemicals

Given the probability density function of risk for each individual chemical $f_i(R)$ ($i = 1, n$ and $R = 0, 100$), the distribution of the community risk of multiple chemicals $f_m(R_t)$ can be estimated. This process has two steps: (1) an additive approximation of the risk from all chemicals and (2) normalization of the estimated risk to a 0-to-100% scale.

Step 1. Discrete Approximation of Additive Risk

Assuming the toxic effects of the COPCs are independent and the risks from the COPCs are additive, we can calculate $f_m(R_t)$ by a discrete approximation of the convolution integral. The assumptions of independence and additivity, however, pose a potential problem: The total risk, which is the sum of the risks from all COPCs of interest, might exceed 100%. If the sum of the risks from COPCs is greater than 100%, the total risk can be adjusted by assuming that it is 100%. In this case, a different formula is required for the probability density function for total risk at $R_t = 100\%$. For example, suppose we are interested in two COPCs: C_1 and C_2, and the total risk R_t can take one of the 101 discrete values from 0 to 100; the probability function of total risk R_t from the multiple chemicals (at $R_t = 0$ to 99) can then be obtained by

$$f_m\left(R_t\right) = \sum_{R=0}^{R_t}\left[f_1(R) * f_2\left(R_t - R\right)\right] \tag{17.34}$$

where
   $f_1(R)$ = the marginal probability function of risk for chemical C_1
   $f_2(R)$ = the marginal probability function of risk for chemical C_2

Where $R_1 + R_2 \geq 100$, the probability function for multiple chemicals at total risk of 100% can be estimated as follows:

$$f_m(100) = \sum_{R_1=0}^{100} \sum_{R_2=100-R_1}^{100} \left[f_1\left(R_1\right) * f_2\left(R_2\right)\right]; \quad R_1 + R_2 \geq 100 \tag{17.35}$$

This algorithm can be readily expanded for situations involving toxicity from more than two chemicals. For example, the distribution of the total risk of two chemicals obtained as described above can be considered as a distribution of one "combined" chemical and be iterated back into the above equations.

Step 2. Normalization of Total Risk

The estimate of community risk of multiple chemicals described above is approximate in that it is discrete; moreover, the assumption of additivity inflates the probability of total risk near 100%. Additivity is most appropriate when the sum of risks from individual chemicals is low. In other words, the assumptions are of greater validity for the left-hand portion of the total risk distribution. This leads to the second step: normalization of the total risk, if the probability of total risk at 100%, $f_m(100)$, calculated in the manner described in Step 1, is unreasonably high, for example, greater than 2%.

The risk expressed in log-odds units [denoted by logit(p) in the previous sections] from each individual COPC has a normal distribution, which means the total risk in log-odds units is also normally distributed under the assumption of additivity. The mean and variance of the total risk in log-odds units will fully specify the distribution and can be calculated by using the probability function for total risk (in arithmetic space) obtained in Step 1.

Because the risk (in arithmetic space) from COPCs has a skewed distribution, using the median of the total risk (in arithmetic space) to estimate the mean of total risk in log-odds space is more appropriate. Let R_{50} denote the median of the total risk in arithmetic space, which can be found by taking the 50th percentile of the total risk from $f_m(R_t)$ obtained in Step 1. The mean of the total risk in log-odds space, denoted by μ_{RL}, can be calculated as follows:

$$\mu_{RL} = \text{logit}\left(R_{50}\right) \tag{17.36}$$

The variance of the total risk in log-odds space can be estimated from Step 1 results by the following procedure:

1. Determine the 5th percentile of the total risk in arithmetic space from $f_m(R_t)$, denoted by R_5. The 5th percentile, rather than the 95th percentile, of the total risk is determined, because the probability of the lower end of the total risk is relatively more accurately estimated in Step 1.
2. Calculate the log-odds of R_5 (denoted by P_{R5}) by $P_{R5} = \text{logit}(R_5)$.
3. Because the total risk in log-odds space is normally distributed, the difference between μ_{RL} and P_{R5} will approximate $Z_{0.95} = 1.64$ standard deviations of the total risk in log-odds space. Therefore, the variance of the total risk in log-odds units, denoted by σ_{RL}^2, can be back-calculated as follows:

$$\sigma_{RL}^2 = \left[\left(\mu_{RL} - P_{R5}\right)/Z_{0.95}\right]^2 \tag{17.37}$$

The total risk in log-odds space follows a normal distribution of $N(\mu_{RL}, \sigma_{RL}^2)$, and the distribution of the total risk in arithmetic space can be easily plotted as illustrated in Figure 17.6. As depicted in this figure, cadmium, copper, and zinc are

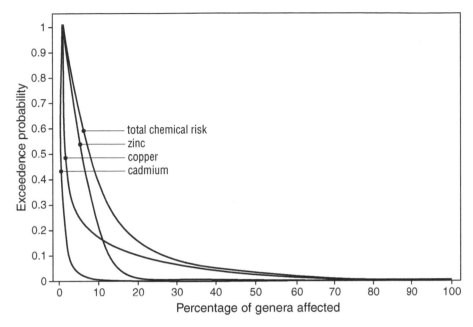

FIGURE 17.6 Clear Creek risk distribution functions for the estimated acute risks of cadmium, copper, and zinc and the total risk of these metals to aquatic life, estimated as the percentage of genera affected.

the three COPCs. For cadmium, only a 1% probability exists that more than 5% of the genera will be affected. For copper, however, there is about a 25% chance that more than 5% of the genera will be affected. About a 41% probability exists that more than 5% of the genera will be affected by zinc. This figure also allows the risk assessor to examine the combined risk due to all COPCs present at the site. The line labeled *total chemical risk* in Figure 17.6 represents the total combined risk from cadmium, copper, and zinc. At this site, the probability that more than 40% of the genera will be affected is approximately 9%, and the probability that more than 5% of the genera will be affected by the combined toxicity of cadmium, copper, and zinc is approximately 72%. This analysis of total risk assumes that the toxic effects of the COPCs are independent (i.e., that the COPCs do not interact) and that the risks from the COPCs are additive. This assumption is probably valid for these divalent, cationic metals that have similar modes of toxicity.

17.3 CASE STUDIES

Two case studies that demonstrate the use of SSDs in the WERF methods are used to evaluate aquatic ecological risks at two locations: (1) the Jordan River near Salt Lake City, Utah, and (2) Salado Creek in San Antonio, Texas. Note that the WERF methods have been used successfully with metals, pesticides, and other chemicals at many sites worldwide.

17.3.1 JORDAN RIVER, UTAH

In this case study, the potential risks of ammonia, mercury, and silver in treated wastewater discharged to the Jordan River during 1996 were evaluated. The WERF methods were used to evaluate whether these chemicals would meet a risk-based effluent limit, which was defined as the concentration of chemicals in the effluent that, after dilution in the Jordan River, would adversely affect 5% or more of the species present. The 5% value was chosen as the risk-based limit, because it is analogous to the level of protection estimated for the U.S. EPA water quality criteria (U.S. EPA, 1985a).

17.3.1.1 Materials and Methods

Existing data on concentrations of contaminants and flows in treated wastewater, tributaries, and the Jordan River were acquired from a wastewater treatment plant and from state and federal agencies and compiled into databases. The monthly mean concentrations and standard deviations of each chemical downstream of the waste-water treatment plant was estimated using daily flow measurements and a mass balance equation:

$$C_D = \frac{(C_E * Q_E) + (C_M * Q_M) + \left(C_{J_{up}} * Q_{J_{up}}\right)}{Q_E + Q_M + Q_{J_{up}}}$$

(17.38)

where
- C = concentration
- Q = flow
- E = effluent
- M = Mill Creek
- J_{up} = Jordan River, upstream
- D = downstream

The estimated concentrations were calculated from measured concentrations of these chemicals in the effluent, Mills Creek — which discharges to the river immediately downstream of the wastewater treatment plant (WWTP) — and upstream river water.

Risk-based effluent limits were estimated with the WERF software using the Single Chemical, Analyses, Risk Evaluation procedure, which fits a logistic regression model to acute or chronic toxicity data and then estimates a risk criterion. The risk criterion in this case was the concentration that would affect 5% or more of the species (e.g., Figure 17.7). The U.S. EPA (1998a) equation was used for predicting the toxicity of ammonia to aquatic life, which treats toxicity as a function of pH, but not temperature. Maximum monthly pH values were used to estimate ammonia. Silver toxicity was assumed to be a function of hardness, and minimum monthly hardness values were used to estimate silver toxicity. Because of the paucity of data on the chronic toxicity of silver to freshwater species, the U.S. EPA (1987a) ACR of

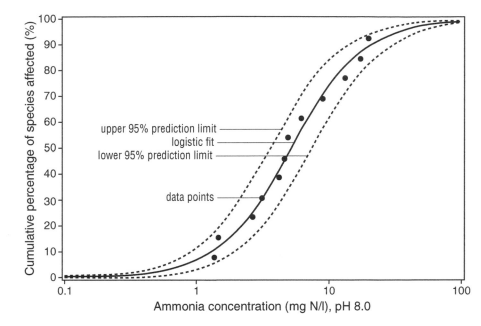

FIGURE 17.7 Logistic regression model for the chronic toxicity of ammonia to fresh water aquatic species at pH 8.0.

15.7 was used to estimate the chronic toxicity. For similar reasons, the U.S. EPA (1996) ACR of 3.731 was used for mercury to predict its chronic toxicity. The Risk Evaluation procedure was used to estimate the risk criterion and its standard deviation.

Based on the estimated monthly mean concentration of each chemical in the Jordan River downstream of the WWTP, the WERF methods were used to estimate the percentage of species adversely affected by each chemical separately and for all three chemicals combined. The estimate was made using the Single Chemical, Analyses, Risk Distribution, and the Multiple Chemicals procedures. If this percentage of species affected was greater than 5%, the risk-based limit was considered to be exceeded.

17.3.1.2 Results and Discussion

Ammonia. In 1996, the effluent would have passed the risk-based effluent limits for ammonia for all but 2 months. During these 2 months (January and September), only 6 to 7% of the aquatic species in the Jordan River were estimated to be adversely affected by chronic ammonia toxicity (Table 17.6). Thus, the risks of these exceedences to aquatic life in the Jordan River were relatively small.

Mercury and Silver. Fewer than 0.1% of the aquatic species would have been at risk from chronic mercury toxicity; consequently, mercury would not have failed the risk-based limits (Table 17.6). Similar results were found for silver. The estimated small risk of mercury largely results because the water quality standard for mercury, 0.012 µg/l, greatly overestimates the direct toxicity of mercury in the water to aquatic

TABLE 17.6
Compliance (expressed as %) with Risk-Based Permit Limits during 1990 and 1996[a]

	Jan	Feb	Mar	Apr	May	Jun	Jul	Aug	Sep	Oct	Nov	Dec
						Ammonia						
Pass		3.6	0.3	<0.1	0.2	0.3	0.3	1		2	<0.1	<0.1
Fail	7								6			
						Mercury						
Pass	<0.1	<0.1	<0.1	<0.1	<0.1	<0.1	<0.1	<0.1	<0.1	<0.1	<0.1	<0.1
Fail												
						Silver						
Pass	<0.1	<0.1	<0.1	<0.1	<0.1	<0.1	<0.1	<0.1	<0.1	<0.1	<0.1	<0.1
Fail												

[a] Values are estimated percentages of species adversely affected by chronic toxicity.

life. This standard is designed to protect human health from mercury that bioaccumulates from the water into fish, which are then consumed by humans. The estimated small risk from chronic silver toxicity results from the high hardness of Jordan River water, 260 to 400 mg/l as $CaCO_3$, which is assumed to significantly decrease the toxicity of silver to aquatic life.

Total Risk of Ammonia, Mercury, and Silver. During 1996, virtually all risk from the three chemicals was caused by ammonia. Mercury and silver presented negligible risks to aquatic life (Figures 17.8 and 17.9).

17.3.2 SALADO CREEK, TEXAS

The objective of this case study was to evaluate the aquatic risks of chemicals present in Salado Creek, a small creek located entirely within San Antonio city limits. No point-source discharges to the creek were present, and all known sources of contaminants were nonpoint.

17.3.2.1 Materials and Methods

Data on the water quality of Salado Creek were compiled from a variety of federal, state, and local agencies and entered into databases. Then, an SLRA was conducted to identify COPCs. Following the SLRA, a Tier 2, probabilistic risk assessment was conducted to quantify the risks of the identified COPCs.

17.3.2.2 Tier 1, Screening-Level Risk Assessment

Exposure Assessment. Following the WERF methods, acute and chronic EECs for each chemical were calculated as the mean plus 2 standard deviations and the mean

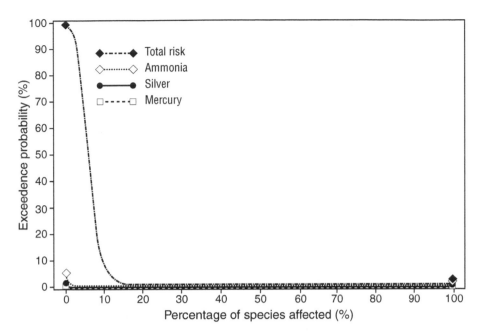

FIGURE 17.8 Risk distribution for the chronic toxicity of ammonia, silver, and mercury to aquatic life in the Jordan River during September 1996.

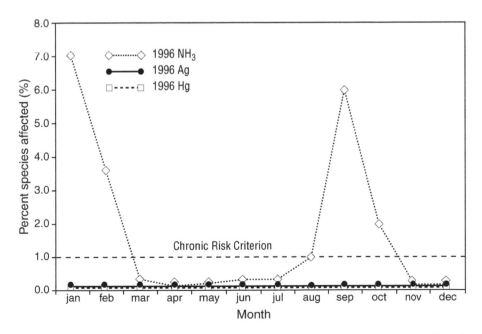

FIGURE 17.9 Total risk of ammonia, silver, and mercury as the percent of species affected downstream of the Central Valley Water Reclamation Facility during 1996.

plus 95% confidence limits, respectively, for each year that data were available. For all chemicals, dissolved measurements were used, because they simulate bioavailable concentrations better than total measurements. We focused on data for 1993 and 1997, because the most complete and extensive analytical chemistry data were available for those years.

Ecological Effects Characterization: Toxicity Benchmarks. For metals and ammonia, current U.S. EPA (1996; 1998a) acute and chronic water quality criteria were used as ecological risk benchmarks to screen the data to identify COPCs. Metals whose standards are hardness dependent were adjusted for Salado Creek hardness, using the hardness values measured at the same time as the metals. The concurrent pH measurements were used to adjust the ammonia criterion.

For diazinon, no acute or chronic water quality standards were available. The U.S. EPA (1998b) has proposed a draft acute criterion of 0.09 µg/l, but was unable to derive a chronic criterion because of the wide range in the geometric mean ACRs between fish (284.1) and invertebrates (1.3).

17.3.2.3 Probabilistic Risk Assessment: Acute and Chronic Risks

The toxicity data from the U.S. EPA (1998b) and the WERF methods and software were used with the following approach to derive acute and chronic benchmarks for diazinon:

1. The WERF software Single Chemical procedure was used to fit a logistic regression model to the species mean acute values (i.e., mean acute 48-h and 96-h LC_{50}) for freshwater species (Table 17.7 and Figure 17.10). Although *Menidia beryllina* (inland silversides) is an estuarine species, its value was included, because *M. beryllina* inhabits Salado Creek. Note the large difference in magnitude of species mean acute LC_{50} values between the species with the lowest LC_{50} values, which are all arthropods, and the other species, which are fish or nonarthropod invertebrates. Diazinon is an insecticide designed to quickly kill arthropods, and thus, arthropods have much higher acute sensitivity to diazinon than nonarthropods. This large difference in sensitivity caused the relatively wide 95% prediction limits for the logistic regression fit to the data (see Figure 17.10). By using the WERF software, an acute ecological risk criterion that would be equal to the LC_{50} for the 5th percentile of the most sensitive species was estimated using the Single Chemical, Analyses, Risk Evaluation procedure. This value was 0.37 µg/l. Because an LC_{50} is estimated to be lethal to 50% of the individuals, the risk criterion was divided by 2 to estimate the threshold acutely toxic concentration (U.S. EPA, 1985a). The final estimated benchmark, which would be protective of all but 5% of the most sensitive species, was 0.19 µg/l. Our benchmark value is about twice the proposed U.S. EPA acute criterion (1998b), which was 0.09 µg/l; however, examination of Figure 17.10 indicates that an acute benchmark of 0.19 µg/l diazinon should be adequately protective of virtually all aquatic species.

TABLE 17.7

Acute Toxicity Database for Diazinon from the WERF Aquatic Ecological Risk Assessment Software

Genus and Species	48–96 h LC_{50} (μg/l)
Gammarus fasciatus	0.20
Ceriodaphnia dubia	0.38
Daphnia pulex	0.78
D. magna	1.05
Simocephalus serrulatus	1.59
Hyalella azteca	6.51
Chironomus tentans	10.70
Pteronarcys californica	25.00
Oncorhynchus mykiss	425.80
Lepomis macrochirus	459.60
Salvelinus namaycush	602.00
S. fontinalis	723.00
Poecilia reticulata	800.00
Menidia beryllina	1170.00
Jordanella floridae	1643.00
Oncorhynchus clarki	2166.00
Pomacea paludosa	3198.00
Lumbricus variegatus	7841.00
Brachydanio rerio	8000.00
Pimephales promelas	8641.00
Carassius auratus	9000.00
Gillia altilis	11000.00
Dugesia tigrina	11640.00

2. Species mean chronic values for diazinon were estimated for each of the species the U.S. EPA (1998b) used to derive its draft acute criterion by dividing the species mean acute values for each fish and invertebrate species by the geometric mean ACRs for fish (284.1) and invertebrates (1.3), respectively (Table 17.8 and Figure 17.11). The ACRs were used to estimate the species mean acute values because of the paucity of chronic diazinon toxicity data. As seen in Figure 17.11, the logistic model does not fit the diazinon chronic toxicity data very well. The chronic data are clustered in two separate groups: (1) four very insensitive species, which are all nonarthropod invertebrates, and (2) the remaining species, which are all arthropods or fish.

3. To account for this difference in chronic sensitivity, the nonarthropods were excluded from the data set and a new logistic regression model was fit to the data for arthropods and fish (Figure 17.12) with the WERF software. This new model fits the data very well. Using this model, the chronic benchmark using the Single Chemical, Analyses, Risk Evaluation procedure was estimated as 0.17 μg/l. Based on the data in Figure 17.12,

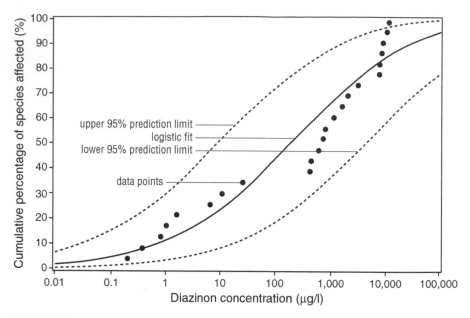

FIGURE 17.10 Acute toxicity aquatic community risk model for diazinon for all freshwater aquatic taxa.

this benchmark should be adequately protective of virtually all freshwater aquatic life, and in particular, arthropods (Figure 17.13) and fish (Figure 17.14). This chronic benchmark could also be considered as a potential, risk-based, chronic water quality criterion for diazinon.

17.3.2.4   Results and Discussion

Tier 1, Screening-Level Risk Assessment. Following the WERF methods, the quotient method was used in the Tier 1 assessment to identify COPCs. At all sites, the acute and chronic EEQs for ammonia and dissolved metals were substantially less than 1; therefore, none of these chemicals is considered to be a COPC. For diazinon, however, both acute and chronic EEQs were >1; therefore, diazinon is considered to be an acute COPC and a chronic COPC in lower Salado Creek.

Tier 2, Probabilistic Risk Assessment. Acute and Chronic Risks. Using the Risk Distribution procedure in the WERF software, the acute and chronic risks of diazinon to aquatic biota in Salado Creek at the Lower Gaging Station were quantified. The most extensive diazinon and flow data were available for 1993 and 1997. Because of the large difference in sensitivity to diazinon between arthropods and fish, risks for arthropods, fish, and all freshwater aquatic taxa were evaluated separately. Separate acute and chronic toxicity databases for diazinon were created with the WERF software for each group. Then, the Risk Distribution procedure was used to enter annual mean concentrations and their standard deviations into the program, and the percentages of taxa affected for each year were estimated for each group of taxa.

TABLE 17.8
Chronic Toxicity Database for Diazinon from the
WERF Aquatic Ecological Risk Assessment Software

Genus and Species	Estimated Chronic Value (μg/l)
Gammarus fasciatus	0.15
Ceriodaphnia dubia	0.29
Daphnia pulex	0.59
D. magna	0.81
Simocephalus serrulatus	1.20
Oncorhynchus mykiss	1.50
Lepomis macrochirus	1.60
Salvelinus namaycush	2.00
Poecilia reticulata	3.00
Salvelinus fontinalis	3.00
Menidia beryllina	4.10
Hyalella azteca	4.90
Jordanella floridae	6.00
Oncorhynchus clarki	8.00
Chironomus tentans	8.10
Pteronarcys californica	19.00
Brachydanio rerio	28.00
Pimephales promelas	30.00
Carassius auratus	32.00
Pomacea paludosa	2408.00
Lumbricus variegatus	5904.00
Gillia altilis	8283.00
Dugesia tigrina	8765.00

Risks from Maximum Concentrations and Storm Waters. Because of rapid acute toxicity of diazinon to arthropods, the acute risks of maximum annual concentrations and storm water concentrations of diazinon were also evaluated using the Single Chemical, Analyses, Acute/Chronic Comparison procedure, which provides a graphical comparison of the maximum concentration with the logistic regression model fit to the data. From these graphs, the percentage of taxa affected by acute toxicity from annual maximum and storm water concentrations was estimated.

Tier 2, Risk Characterization Results. Table 17.9 summarizes the Tier 2 acute and chronic analyses as 50% exceedence probabilities, which are the risks expected on average to occur. In 1993, the year with the highest measured diazinon concentrations, about 13% of arthropod taxa and 11% of all freshwater taxa are estimated to have been affected by chronic diazinon toxicity. This finding indicates that adverse effects on growth, reproduction, and long-term survival rates may have occurred. About 5% of all taxa and 11% of arthropod taxa are estimated to have been adversely affected by acute diazinon toxicity, which indicates that significant mortality of sensitive arthropod species may have occurred. Fish in Salado Creek probably were

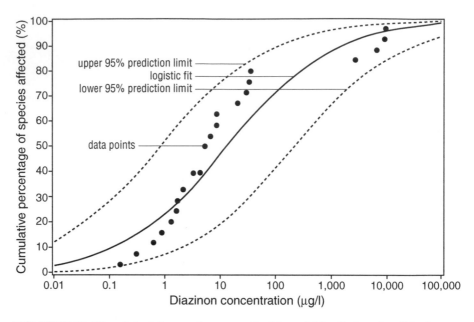

FIGURE 17.11 Chronic toxicity aquatic community risk model for diazinon for all freshwater aquatic taxa.

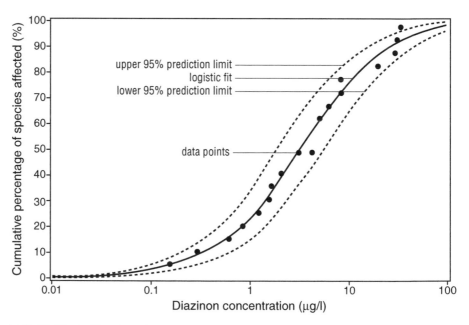

FIGURE 17.12 Chronic toxicity aquatic community risk model for diazinon for all freshwater arthropod and fish taxa.

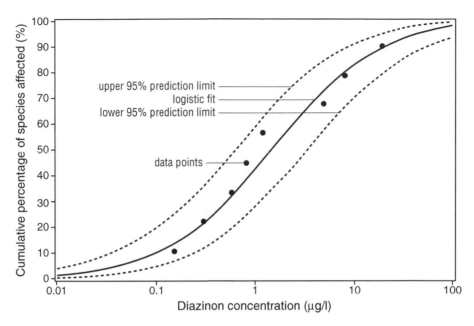

FIGURE 17.13 Chronic toxicity aquatic community risk model for diazinon for all freshwater arthropod taxa.

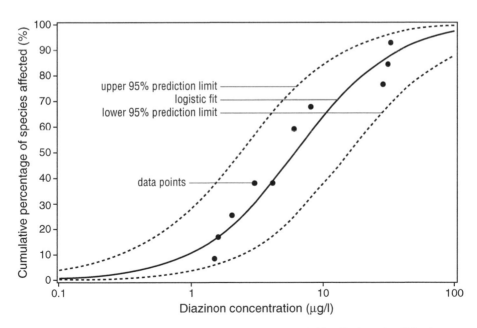

FIGURE 17.14 Chronic toxicity aquatic community risk model for diazinon for all freshwater fish taxa.

TABLE 17.9
Acute and Chronic Risks of Diazinon
in Salado Creek to Aquatic Biota[a]

	1993	1997
Diazinon (μg/l), 1997	0.16	0.065
(SD)	(0.21)	(0.079)
% All taxa: acute	5	4
% All taxa: chronic	11	8
% Arthropod and fish: acute	7	5
% Arthropod and fish: chronic	5	2
% Arthropod taxa: acute	11	6
% Arthropod taxa: chronic	13	7
% Fish taxa: acute and chronic	2	<1

[a] Values are estimated percentages of taxa affected by chronic diazinon toxicity.

not significantly affected by acute or chronic diazinon toxicity. During 1997, estimated risks were somewhat lower. Up to 8% and 4% of all taxa may have been affected by chronic and acute toxicity, respectively.

Figure 17.15 presents the risk distribution for arthropods for the 1993 chronic analyses. On average, an estimated 13% of arthropod species would be affected, but the 95th percentile exceedence probability could be as low as 5% of species affected, whereas the 5th percentile exceedence probability could be greater than 20% of species affected.

Previously, a 0.17 μg/l chronic water quality standard for diazinon that would protect 95% of all freshwater species was estimated. Using the WERF software and the diazinon toxicity databases for arthropods and fish, the diazinon concentration required to protect 95% of all arthropods from chronic toxicity would be 0.02 μg/l (see Figure 17.13). This value could be used as a more protective, risk-based, chronic water quality criterion for diazinon.

Peak diazinon concentration appears to pose even greater risks to aquatic life in Salado Creek. In 1993, for example, the maximum measured diazinon concentration was 0.6 μg/l. Figure 17.16 indicates that this concentration of diazinon could have caused acute mortality to as many as 30% of the more sensitive arthropod taxa in Salado Creek. For the single 1997 storm water measurement, the diazinon concentrations in the storm water that could cause acute toxicity was estimated to be about 20% of the arthropod taxa present (Figure 17.17).

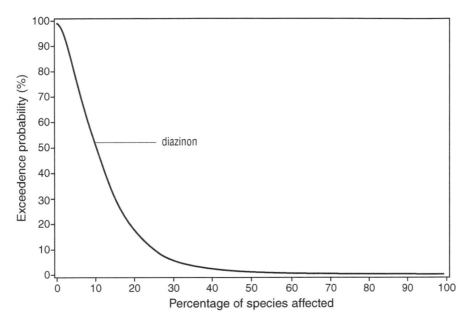

FIGURE 17.15 Risk distribution for the chronic toxicity of diazinon to arthropods in Salado Creek during 1993.

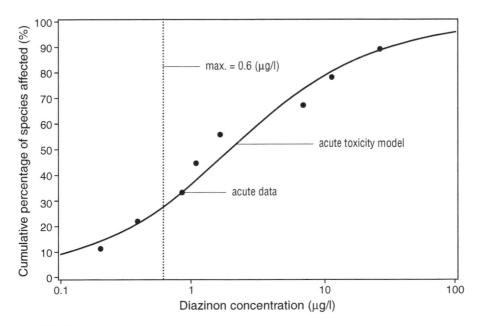

FIGURE 17.16 Estimated percentage of arthropod taxa in Salado Creek affected by acute toxicity from the maximum concentration of diazinon measured during 1993.

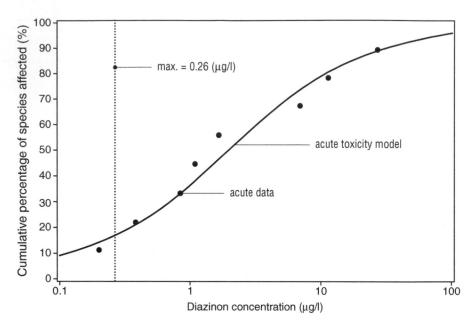

FIGURE 17.17 Estimated percentage of arthropod taxa in Salado Creek affected by acute toxicity from diazinon measured in a storm flow during 1997.

18 Toxicity-Based Assessment of Water Quality

Dick de Zwart and Aart Sterkenburg

CONTENTS

Abstract — This chapter describes the principles and application of a method based on species sensitivity distributions (SSDs) to quantify local toxic potency in aquatic ecosystems. Observed toxic potency is defined by the parameter pT. To derive pT values for a water sample, interspecies variations of acute toxicity observed in organic surface water concentrates with unknown constituents are used to assess potential effects in the naturally exposed aquatic community. The proposed method includes the following steps: (1) Application of an XAD-sorption-based concentration procedure to large volume samples of surface water. (2) Laboratory toxicity testing of the concentrates with five microvolume tests on different species. (3) Acute to chronic effects extrapolation. (4) Calculation of the pT of the original water sample, using the sensitivity distribution of effective concentration factors observed in laboratory testing. The pT value is expressed as the proportion of a generic species assembly that is potentially under toxic stress in the field (species exposed above their chronic no-observed-effects concentration). Using redundancy analysis, the set of applied toxicity tests is demonstrated to reveal different aspects of toxic pollution. With pT values ranging from 0 to about 10% of the species potentially affected, the results suggest that some of the aquatic ecosystems in the Netherlands are stressed by exposure to a variety of organic toxicants. Comparison with measured toxicant concentrations attributes the observed effects mainly to a variety of pesticides and polycyclic aromatic hydrocarbons.

18.1 INTRODUCTION

Several methods are used to obtain information on the "health status" of ecosystems. Chemical monitoring is one of the methods applied on a large scale to obtain information on the toxicological stress put upon ecosystems. One of the major disadvantages of this method is the limited number of substances that can be dealt with analytically. Additionally, ecotoxicological data are generated for only a few substances from the approximately 150,000 which are known in the European Inventory of Existing Chemical Substances (EINECS).

In addition to chemical monitoring, ecological inventories are carried out to obtain detailed information on the health status of ecosystems. This type of study focuses on deviations found when species diversity or ecosystem functioning is compared to reference systems or goals. In most cases, it is difficult to relate observed ecological effects directly to such causal factors as toxicity.

Diagnosis of toxicological stress on ecosystems is still a necessary tool to identify areas of toxicological concern or to evaluate regulatory actions. Bioassays on environmental samples respond to the combined effect of all substances present in the environment. Monitoring of environmental toxicity in the range of subacute or chronic effects is in practice not conceivable because of the long time it may take to find any effect. Alternatives have been developed to measure the total toxicity in environmental samples. One of these is the toxic potency (pT) method introduced by Slooff and De Zwart (1991). By this method, organic substances in the water sample are concentrated and the acute toxicity of the concentrated sample is tested subsequently. RIVM (1997) measured the toxicity of the Rhine River delta in the Netherlands. This method provided a view of the mixture toxicity of the river water for a single bacterial species (Figure 18.1). However, it was recognized that a more integral view of the total toxicity toward a broader set of species was needed.

The present study is aimed at obtaining this integral view. A method is presented that is able to compare the toxicity of samples of surface water taken at different locations and moments in time. For this purpose a species sensitivity distribution

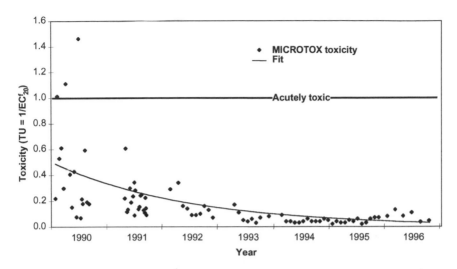

FIGURE 18.1 Microtox toxicity of Rhine River water concentrates at Lobith. The EC_{20}^{f} (20% effective concentration factor) is the number of times a sample of surface water had to be concentrated to induce a 20% reduction of light emission in bacterium *Vibrio fisheri* (*Photobacterium phosphoreum*).

(SSD) model has been developed by which a pT is derived from the results of a variety of bioassays conducted on organic* concentrates of surface water samples.

This chapter presents the toxic potencies found in surface water at 15 locations in the Netherlands during the year 1996. The results will be related to data of chemical analysis of the samples investigated, thus comparing the data found on total toxicity with those obtained from the analysis of individual substances.

18.2 THE MODEL

18.2.1 INDIVIDUAL CHEMICALS

Several procedures have been described that assess the risks associated with the presence of toxic substances in ecosystems. Many of these methods assume that for every compound, the LC_{50} (Kooijman, 1987) or no-observed-effects concentration (NOEC)-values (Van Straalen and Denneman, 1989; Wagner and Løkke, 1991; Aldenberg and Slob, 1993) for single species in a community are stochastically independent variables with a log-logistic or lognormal frequency distribution. To estimate the model parameters a minimum of four to five toxicity values are required, which should be obtained with functionally different species. The minimum requirements depend on the method by which uncertainty due to small sample size is

---

* Note that the environmental mixture of inorganic compounds can be analyzed in a similar manner, given the development of proper concentration procedures.

handled statistically (see, e.g., Chapter 8; Newman et al., 2000). The model developed for individual chemicals is discussed in detail elsewhere in this volume.

18.2.2 COMBINATION TOXICITY OF KNOWN CHEMICALS

When a mixture of toxic substances is present, toxic risk may be calculated for the combined effects of the chemicals. The data required to accomplish the derivation of a combined toxic risk include an estimate of the bioavailable concentration of the individual compounds, their NOEC distributions, and their mode of toxic action. This topic is treated extensively in Chapter 16 of this volume.

18.2.3 UNKNOWN CHEMICALS IN ENVIRONMENTAL SAMPLES

Clearly, with increasing complexity of chemical mixtures, especially when present in low concentrations such as in environmental water samples, the derivation of combined toxic risk from chemical concentrations becomes more difficult. Analogous to the assessment of SSD for individual chemicals, the toxic potency of environmental samples can be determined experimentally by exposing a set of functionally different species to the overall toxicity of the sample. Since the ecosystem may be continuously exposed to the cocktail of toxicants, the determined toxicities from which the toxic potency is inferred should also represent chronic values.

Complication: *No Chronic Toxicity Data Available*

The ecotoxicological evaluation of the water quality in terms of toxic potency, as presented in this chapter, has been designed for application in monitoring programs with a high grid density and a relatively high frequency. In this respect, it is undesirable to conduct chronic exposure assays, because of practical constraints of the labor intensity and high costs involved. For monitoring purposes, acute assays are often preferred. However, the extrapolation of the results of these tests toward chronic toxicity estimates can only be performed using empirically determined values of the acute–chronic ratio (ACR). Empirically derived data on the ACR of complex mixtures is extremely scarce. However, some ACRs can be derived from toxicity data presented in the literature. From a study on the ecotoxicological quality of effluents from oil refineries, chemical industries, and sewage treatment plants (U.S. EPA, 1991), the acute and chronic toxicity values of these complex mixtures for fish, daphnids, and mysids can be compared. More than 90% of the ACR data appear to be less than 10 (Figure 18.2).

Solution

In view of this observation, an arbitrary and rather conservative ACR of 10 is used in the present study to extrapolate acute L(E)C$_{50}$-values to chronic no effect data (NEC).

Complication: *Concentration Necessary*

The toxicity contained in environmental samples as they are taken from the field is generally not sufficient to invoke any effects in standard acute laboratory toxicity

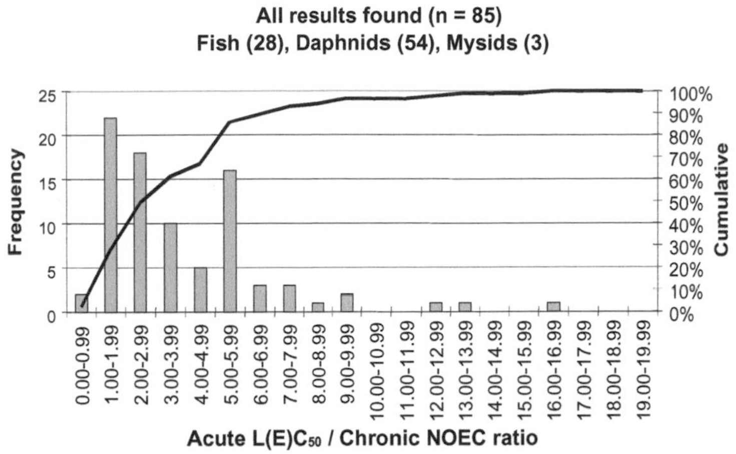

All results found (n = 85)
Fish (28), Daphnids (54), Mysids (3)

Acute L(E)C$_{50}$ / Chronic NOEC ratio

FIGURE 18.2 Acute–chronic toxicity ratio of complex mixtures of chemicals in industrial and municipal effluents.

tests. To be able to detect toxicity in acute tests, it is necessary to extract the toxicity from the sample and to apply a concentration procedure prior to toxicity testing (see Section 18.3). In the concentrated samples with unknown amounts of toxicants, the observed toxicity cannot be expressed in terms of an exposure concentration (mg/l).

Solution

The observed toxicity is given as the factor by which the original sample had to be concentrated to meet the 50% effect criterion in the acute bioassays (LC_{50}^f or EC_{50}^f).

Complication: *Microvolume Toxicity Testing Required*

The application of a concentration procedure prior to toxicity testing requires the use of small-volume toxicity tests to avoid the necessity of processing huge samples with the associated extreme expenditure in labor, facilities, and the chemicals needed for the concentration process.

Solution

Several small-volume acute toxicity tests were applied to test the toxicity of the organic concentrates as explained in the Materials and Methods section on toxicity testing.

18.3 MATERIALS AND METHODS

18.3.1 CHEMICALS

All chemicals used, except the adsorption resins, were analytical grade and supplied by Merck. The mineral water used as a blank and dilution medium was Spa Reine supplied by Spa Monopole N.V. (Spa, Belgium).

18.3.2 Sampling

Samples of surface water (60 l) were taken at 15 sites in the Netherlands, at regular intervals during the year 1996. The samples were immediately transferred to the laboratory by cooled transport.

18.3.3 Concentration of Contaminants in Acetone

Organic substances present in the water were concentrated by adsorption onto a 1:1 mixture of XAD-4 (Rohm & Haas, Antwerp, Belgium) and XAD-8 (DAX-8, Supelco, Sigma-Aldrich, Zwijndrecht, the Netherlands) resins that had been purified prior to use.

The resin purification procedure involved washing with, successively, 4% (w/v) NaOH (repeated 10 times), 4% (w/v) $HClO_4$ (10 times), double distilled deionized water (10 times), and methanol (2 times). After Soxhlet extraction with methanol (24 h) and another washing step with ethanol (3 times) the resins were further Soxhlet-extracted (24 h) with ethanol/cyclohexane (30.5/69.5) azeotropic mixture and finally washed with double-distilled methanol (5 times). The purified resins were stored in methanol, in darkness at room temperature until further use. Immediately prior to the adsorption, a slurry of mixed resins in methanol was transferred to a glass extraction column and reconstituted with, successively, 2 bed volumes of methanol, 2 bed volumes of acetone, 2 bed volumes of methanol, and 12 bed volumes of commercially available mineral water.

Within 48 h after sampling, the nonfiltered water samples were transferred to six 10-l borosilicate bottles, after which the resin mixture was added to a final concentration of 2 ml of resin mixture per liter water. The bottles were rolled for 24 h at 20°C in darkness. Then the resins were sieved out of the water and were allowed to dry in a gentle airstream for 24 h. The XAD mixture was eluted with 1 bed volume of acetone, producing a 500-fold acetone concentrate of 120 ml. The acetone-concentrate was distributed among six 20-ml vials, sealed with a crimp cap, and stored at –20°C until further processing.

18.3.4 Transfer of Contaminants from Acetone to Water

The acetone concentrate was transferred from the 20-ml storage vial into a conical tube. Then, 2 ml of mineral water was added and spherical reflux distillation condensers were placed on top of the conical tube. The acetone present in the concentrate was evaporated at 65°C during approximately 30 min. Distillation was stopped when volume reduction and boiling ceased. The residue was purged for 20 min using a precisely tuned nitrogen stream to reduce the remaining acetone content to an ecotoxicologically acceptable level (less than 0.1 mg/l). After purging, diluent was added to a final volume of 10 ml, resulting in a 1000-fold aqueous concentrate. The diluent was either EPA medium (0.55 mM $CaCl_2$; 0.50 mM $MgSO_4$; 1.14 mM $NaHCO_3$; 0.05 mM KCl; pH 7.6 ± 0.2) or Dutch Standard Water (1.36 mM $CaCl_2$; 0.73 mM $MgSO_4$; 1.19 mM $NaHCO_3$; 0.20 mM $KHCO_3$; pH 8.2 ± 0.2), depending on the toxicity test to be performed. The total of 6 × 10 ml water concentrate was kept at 5°C for no longer than 1 day prior to toxicity testing.

18.3.5 TOXICITY TESTING

Microtox assay (Bulich, 1979; Bulich and Isenberg, 1981): Freeze-dried bacteria (*Photobacterium phosphoreum*, Microbics Inc., Carlsbad, CA) were reconstituted in Microtox reconstitution solution. Approximately 1.5 h after reconstitution the water concentrate was added to the suspended bacteria in a dilution range with concentration factors of 0, 28, 56, 112, 225 with respect to the original sample. Luminescence was measured in a luminometer (Microtox model 500) after 5 and 15 min incubation at 15°C. As EC_{50}^f the lowest concentration factor is taken that decreases light emission by 50%, either measured after 5 or after 15 min of exposure. For both exposure durations, the EC_{50}^f is calculated using the interpolation model supplied by Microbics, Inc.

Algal photosynthesis test: A slightly modified procedure of the algal photosynthesis test of Tubbing et al. (1993) was used. *Selenastrum capricornutum* (approximately 2×10^5 cells/ml), pregrown in Wood's Hole medium in chemostat culture, was exposed for 4 h to the concentrated sample in a dilution series containing concentration factors of 0, 0.2, 0.63, 2, 6.3, 20, 63, 200 with respect to the original sample. The exposure was conducted under continuous illumination (100 $\mu E/m^{-2}/s$) in 50-ml closed glass flasks on a rolling device, at 20°C. Subsequently, a known amount of $^{14}C\text{-}HCO_3^- EC_{50}^{f-}$ (approximately 1 μCi; Amersham, 50 to 60 mCi/mmol) was added. After incubation for 1 h, the assimilation of the labeled bicarbonate was stopped by adding formaldehyde, and the algae suspensions were filtered over 0.45 μm membrane filters. After drying of the filters, the assimilated radioactivity was measured with a liquid scintillation counter (Packard Tricarb, model 1600 TR). The concentration-dependent rate of photosynthesis is expressed as a percentage of the control. The EC_{50}^f is interpolated to represent the concentration factor that causes 50% reduction of photosynthesis by applying log-probit regression.

Rotox test: The Rotoxkit F is a commercially available bioassay kit developed to measure the acute toxicity of toxicants in water (Snell and Persoone, 1989; Snell et al., 1991; Janssen et al., 1993). Cysts of the rotifer *Brachionus calyciflorus* were allowed to hatch in EPA medium during a period of 16 to 18 h under continuous light exposure. Within 2 h after hatching, the rotifers were transferred into disposable multiwell test plates and subsequently exposed to a dilution series of the concentrated sample (0, 32, 64, 125, 250, 500 concentration factor) for 24 h in darkness. Mortality was determined by microscopic observation of mobility. The LC_{50}^f value is taken as the concentration factor leading to 50% mortality when compared with the control, which is calculated using the Spearmann–Kärber method (Hamilton et al., 1977).

Thamnotox test: The Thamnotoxkit F is also a commercially available bioassay kit developed to measure the acute toxicity of toxicants in water (Centeno et al., 1995). Cysts of the crustacean *Thamnocephalus platyurus* were allowed to hatch after 24-h incubation in hatching medium under continuous light. Organisms were allowed to acclimatize to the dilution medium for 4 h after which the concentrated water (0, 32, 64, 125, 250, 500 concentration factor) was added. Instead of the original disposable multiwell test plate supplied with the kit, glass vials with crimp caps were used. After 24 h of exposure in darkness, mortality was determined by microscopic observation of mobility. The LC_{50}^f value is taken as the concentration

---

TABLE 18.1
The Approximate Volume of 1000-fold
Aqueous Concentrate Needed to
Perform the Five Different Toxicity
Tests in Duplicate

Test	Volume in ml
Microtox test	2
Algal photosynthesis test	20
Rotox kit	5
Thamnotox kit	5
Daphnia IQ test	20
Total	**52**

---

factor leading to 50% mortality compared with the control, using the Spearmann–Kärber method (Hamilton et al., 1977).

Daphnia IQ test: In the Daphnia IQ test (Aqua Survey, Inc., 1993) the toxicity is determined by measuring the inhibition of enzymatic cleavage of 4-methylumbelliferyl, β-D-galactoside in living *Daphnia magna*. Starved young daphnids were exposed for 1 h to a dilution range of the water concentrate (0, 8, 16, 32, 64, 125, 250 concentration factor). Thereafter, the galactoside was added to the test vessels. After another 15 min of incubation, fluorescence of the daphnids upon ultraviolet irradiation was observed by eye. The EC_{50}^f value is taken as the concentration factor leading to 50% fluorescing daphnids compared to the control, using the Spearmann–Kärber method (Hamilton et al., 1977). The approximate volume of concentrate needed to perform the five microvolume tests in duplicate is given in Table 18.1.

18.3.6 PROCEDURE FOR THE CALCULATION OF TOXIC POTENCY

The observed acutely effective concentration factors (EC_{50}^f or LC_{50}^f) are extrapolated to represent chronic no-effect concentration factor values (NEC^f) by assuming a constant ACR of 10:

$$NEC^f = \frac{EC_{50}^f}{10} \quad \text{or} \quad NEC^f = \frac{LC_{50}^f}{10} \tag{18.1}$$

The integrated log-logistic distribution function of species sensitivity is analogous to the curve originally described for individual chemicals (Aldenberg and Slob, 1993):

$$F(x) = \frac{1}{1 + e^{-\left(\frac{\log_{10} c^f - \alpha}{\beta}\right)}} \tag{18.2}$$

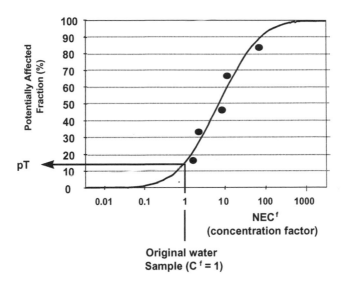

FIGURE 18.3 A graphical representation of the pT calculation and the fitted SSD.

where α is the sample mean of log-transformed NEC^f-values:

$$\alpha = \frac{1}{n} \sum_{i=1}^{n} \log_{10} NEC_i^f \qquad (18.3)$$

β is estimated from the sample standard deviation of log-transformed NEC_f-values:

$$\beta = \frac{\sqrt{3}}{n} \cdot s = \frac{\sqrt{3}}{\pi} \cdot \sqrt{\frac{1}{n-1} \cdot \sum_{i=1}^{n} \left(\log_{10} NEC_i^f - \alpha\right)^2} \qquad (18.4)$$

and C^f is the environmental concentration factor considered.

To estimate the pT of the original untreated water, the value of 1 is substituted for the C^f. Graphically, the above procedure can be depicted as in Figure 18.3.

18.3.7 CONFIDENCE INTERVAL FOR THE TOXIC POTENCY

Because of the lack of statistical tools to determine a confidence interval for the pT calculated according to an assumed log-logistic SSD, uncertainty is treated as if the SSD is lognormal. Applying this procedure can be justified by realizing that both types of distributions are only marginally different in the tails.

For a normal random variable $\log_{10} X_i$, with known mean (μ) and known standard deviation (σ), it is possible to say that exactly a proportion P of the normal population is below $\mu + K_p \sigma$, where K_p is read from a table of the inverse normal probability distribution. When a limited number of samples are taken from the population, the μ and σ are unknown and must be estimated from the samples, which are supposed to be randomly selected from the population

$$\mu = \bar{x} = \frac{1}{n}\sum x_i; \sigma = s = \sqrt{\frac{1}{n-1}\sum_{i=1}^{n}(x_i - \bar{x})^2}$$

A tolerance limit of the form $\bar{x} + ks$ can be used. Since $\bar{x}$ and s will be random variables, the tolerance limit statement can only be made with a given probability attached. The problem then reduces to finding an interval $k_{pl} \le k \le k_{pr}$, such that the probability is γ that at least a proportion P_l and at most a proportion P_r is below $\bar{x} + ks$.

For the evaluation of confidence in the determination of environmental toxicity, the following formulae are required to compute this two-sided confidence interval of the pT value (Owen, 1968: pp. 457–459):

x_i = L(E)C$_{50}$ given as an effective concentration factor with respect to the original sample

n = number of measured data

x^* = toxicant concentration factor for the original environmental sample = 1

f = degrees of freedom = $n - 1$

$$\bar{x} = \frac{1}{n}\sum_{i=1}^{n}\log\frac{x_i}{10} \tag{18.5}$$

$$s = \sqrt{\frac{1}{n-1}\sum_{i=1}^{n}\left(\log\frac{x_i}{10} - \bar{x}\right)^2} \tag{18.6}$$

$$k^* = \frac{\log x^* - \bar{x}}{s} \tag{18.7}$$

$$t = k^* \sqrt{n} \tag{18.8}$$

$$y = \left(1 + \frac{t^2}{2f}\right)^{-1/2} \tag{18.9}$$

$$y' = \frac{t}{\sqrt{2f}}y \tag{18.10}$$

$$\left.\begin{array}{l} \lambda_l(y', f, \gamma = 0.95) \\[2mm] \lambda_r(-y', f, y = 0.95) \end{array}\right\} \text{ Table of noncentrality factors (Owen, 1968)} \tag{18.11}$$

$$\left.\begin{array}{l} \delta_l = t - \dfrac{\lambda_l}{y} \\[2em] \delta_r = -t - \dfrac{\lambda_r}{y} \end{array}\right\} \quad \text{Difference between left- and right-side margin} \quad (18.12)$$

$$\left.\begin{array}{l} k_{pl} = \dfrac{\delta_l}{\sqrt{n}} \\[2em] k_{pr} = \dfrac{\delta_r}{\sqrt{n}} \end{array}\right\} \quad \text{Difference between left- and right-side margin} \quad (18.13)$$

To provide the 90% confidence interval for the pT value, both left and right k values are substituted in the function

$$\frac{1}{\sqrt{2\pi}} e^{-k^2/2}$$

The resulting probabilities can also be found in a table of the inverse standard normal probability distribution.

Because of its binomial nature (proportion of species affected), the probability distribution of toxic potency is skewed right when the pT is considerably below 0.5 (or 50%) and skewed left when the pT is well above 0.5.

18.3.8 CORRECTION FOR PROCEDURAL BLANK pT

To correct for the introduction of any toxicity by the concentration procedure, batches of 60 l of commercially available mineral water were treated in exactly the same way as the surface water samples with respect to the applied concentration procedure and the executed toxicity tests. Blanks were assumed to yield a toxic potency that is solely attributable to the toxicity added by the concentration procedure. Toxic potencies of surface water samples were corrected for toxicity introduced by the sample preparation method following the formula:

$$pT_{corrected} = \frac{pT_{sample} - pT_{blank}}{1 - pT_{blank}} \quad (18.14)$$

18.3.9 CALCULATING COMBINED pT VALUES

The toxicity observed in environmental samples may vary both in time and in place. Monitoring generally implies that a particular water body is sampled regularly or spatially. In reporting the monitoring results, it may be desirable to lump temporal or spatial variability of pT values to produce an overall quality judgment on a particular period in time or a stretch of water body.

Since the probability distribution of the pT is binomial, the distribution can be normalized by applying the following transformation:

$$^{10}\log \frac{pT}{1-pT}$$

Subsequently, a spatially or temporally lumped toxic stress (pT^l) is evaluated by taking the arithmetic average of the transformed individual pT values, followed by exponentiation to derive the geometric average, x, of pT^l/(1 – pT^l), from which the lumped pT^l value can be calculated according to:

$$\frac{pT^l}{1-pT^l} = x \Rightarrow pT^l = \frac{x}{x+1} \qquad (18.15)$$

The 5% (left) and 95% (right) confidence limits of the pT^l are calculated by a likewise transformation of the 5% (CM$_l$) and 95% (CM$_r$) confidence margins of the individual pT observations. Since the probability distribution of the pT is now normalized, which means that both the 5% and the 95% confidence margins are about 2 standard deviation units from the mean, the standard deviation of this distribution can be approximated according to

$$S_l = \frac{\log \dfrac{pT^l}{1-pT^l} - \log \dfrac{CM_l}{1-CM_l}}{2} \qquad (18.16)$$

and

$$S_r = \frac{\log \dfrac{CM_r}{1-CM_r} - \log \dfrac{pT^l}{1-pT^l}}{2}$$

The S_l and S_r values of individual pT observations should be approximately equal. However, in the extreme tails of the distribution (mainly with extremely low pT values or confidence margins), the assumption of transformed normality does not hold. Therefore, only the S_r is used for further processing into an estimate of the confidence interval for combined pT values.

From all S_r values of individual pT observations to be combined, the highest (S_r^{max}) is selected as being a worst-case representative for the second moment of the probability distribution of the pertaining data. This S_r^{max} is divided by $\sqrt{n}$ (the square root of the number of pT values combined) to produce an overall standard deviation (S_r^{max}) for the transformed probability distribution of the lumped pT^l. An estimate for the 5% (CM$_l^l$) and 95% (CM$_r^l$) lumped confidence margins for the pT^l can be calculated by back transformation:

$$\frac{CM_l^1}{1-CM_l^1} = 10^{\left[\log\left(pT^1/\left(1-pT^1\right)\right)-2S_r^{max}\right]} = x \Rightarrow CM_l^l = \frac{x}{x+1} \qquad (18.17)$$

and

$$\frac{CM_r^1}{1-CM_r^1} = 10^{\left[\log\left(pT^1/\left(1-pT^1\right)\right)+2S_r^{max}\right]} = y \Rightarrow CM_r^l = \frac{y}{y+1}$$

18.3.10 CHEMICAL ANALYSIS AND CALCULATED TOXICITY

18.3.10.1 Chemical Analysis

All samples evaluated toxicologically (Table 18.2) were also chemically analyzed to produce the aquatic concentrations of up to 74 organic micropollutants. The organic toxicant concentrations obtained for the six data series were geometrically averaged over the year. Only in case all six concentrations per compound were marked to be below the analytical detection limit, the average concentration has been put to zero. In case one or more observations indicated that a particular compound was actually present, the remaining observations below the detection limit were substituted by the detection limit concentration.

Unfortunately, different subsets of compounds were analyzed at different locations. The most elaborate and comparable chemical analysis was performed for the locations with the codes EYS, LOB, MAA, and SCH, all with the same 74 toxicants analyzed.

18.3.10.2 Toxicity Calculation from Chemical Concentrations

To calculate the combined toxic risk, the multisubstance potentially affected fraction (msPAF) of the mixture, all concentrations are first transformed to toxic units (TU). TU is defined as the ratio of the environmental concentration over the chronic toxicity (NOEC) geometrically averaged over as many species as possible. The chronic toxicities of the compounds and their associated toxic modes of action were obtained as described in Chapter 8. The msPAF was subsequently calculated following the procedure described by Chapter 16, under the assumption that chemicals with the same toxic mode of action can be considered to act concentration additively. Chemicals with different modes of toxic action are assumed to act effect additively.

18.4 RESULTS

18.4.1 OBSERVED TOXICITIES

The untreated toxicity data for six bimonthly series of measurements on concentrates from 15 Dutch surface waters for the year 1996, together with the calculated combined pT, are presented in Table 18.2.

Figure 18.4 shows the toxic potencies, including the 90% confidence intervals lumped over the six data series produced in 1996.

TABLE 18.2
The Measured Toxicity Data

Station–Series	Code	Microtox EC_{50}^f	Algae EC_{50}^f	Rotox LC_{50}^f	Thamno LC_{50}^f	Daphnia EC_{50}^f	pT %
Spa blank–1	CON1	424.50	84.65	104.79	533.93	299.40	0.11
Belfeld–1	BEL1	20.14	68.31	—	105.30	86.40	1.27
Eijsden–1	EYS1	29.18	111.32	63.23	169.57	57.42	0.48
Haringvlietsluis–1	HAR1	10.73	64.92	—	125.02	123.23	6.10
Vrouwenzand–1	VRO1	72.65	113.18	572.40	572.40	190.80	0.25
Keizersveer–1	KEI1	75.02	162.35	68.33	204.98	58.91	0.06
Ketelmeer–1	KET1	94.32	45.31	—	170.10	180.27	0.12
Nieuwegein–1	NIE1	30.02	38.16	—	129.60	109.80	1.04
Lobith–1	LOB1	66.64	169.30	—	270.00	78.00	0.10
Markermeer–1	MAR1	74.83	131.09	194.41	484.12	286.20	0.06
Nieuwe Waterweg–1	MAA1	96.41	78.05	238.52	534.00	166.43	0.12
Noordzeekanaal–1	NOO1	49.36	138.42	128.25	465.50	86.45	0.37
Puttershoek–1	PUT1	15.61	15.23	33.51	126.20	171.12	8.60
Schaar Ouden Doel–1	SCH1	25.18	28.90	30.10	141.90	126.42	2.97
Volkerak-Zoommeer–1	VOL1	59.17	52.83	—	176.11	95.26	0.08
Wolderwijd–1	WOL1	43.7	60.57	206.30	404.88	231.36	0.65
Spa blank–2	CON2	228.57	51.20	107.69	560.20	299.40	0.34
Belfeld–2	BEL2	109.43	52.53	116.24	264.60	68.20	0.10
Eijsden–2	EYS2	65.93	—	219.34	157.53	34.51	0.78
Haringvlietsluis–2	HAR2	72.02	97.94	230.00	264.96	181.36	0.02
Vrouwenzand–2	VRO2	62.94	109.46	126.90	367.20	170.18	0.06
Keizersveer–2	KEI2	—	—	—	—	—	—
Ketelmeer–2	KET2	74.84	124.83	289.88	436.80	256.82	0.05
Nieuwegein–2	NIE2	11.64	78.88	122.16	468.77	190.81	4.54
Lobith–2	LOB2	42.25	94.27	107.64	241.44	198.72	0.16
Markermeer–2	MAR2	56.33	140.62	175.86	403.10	299.40	0.11
Nieuwe Waterweg–2	MAA2	39.32	125.33	269.55	261.62	96.13	0.31
Noordzeekanaal–2	NOO2	66.06	74.81	135.56	333.78	198.57	0.09
Puttershoek–2	PUT2	55.32	163.11	110.97	280.62	106.04	0.05
Schaar Ouden Doel–2	SCH2	28.79	35.12	86.22	220.00	109.12	1.33
Volkerak-Zoommeer–2	VOL2	88.63	67.29	359.31	576.52	274.39	0.27
Wolderwijd–2	WOL2	194.79	186.13	421.47	560.39	301.50	0.00
Spa blank–3	CON3	99.26	80.86	107.38	335.79	301.20	0.06
Belfeld–3	BEL3	37.28	28.90	45.54	242.25	299.40	3.13
Eijsden–3	EYS3	26.45	8.78	69.02	123.25	98.60	7.46
Haringvlietsluis–3	HAR3	21	60.02	55.12	202.10	86.91	1.52
Vrouwenzand–3	VRO3	15.72	84.11	55.92	141.26	62.31	1.95
Keizersveer–3	KEI3	40.24	17.76	135.54	125.50	128.13	2.11
Ketelmeer–3	KET3	38.02	34.35	80.22	106.96	82.00	0.15
Nieuwegein–3	NIE3	45.28	82.32	138.99	268.46	180.36	0.15
Lobith–3	LOB3	60.59	46.30	130.97	266.72	116.00	0.22
Markermeer–3	MAR3	79.74	62.88	213.84	60.26	298.50	0.26

TABLE 18.2 (continued)
The Measured Toxicity Data

Station–Series	Code	Microtox EC_{50}^f	Algae EC_{50}^f	Rotox LC_{50}^f	Thamno LC_{50}^f	Daphnia EC_{50}^f	pT %
Nieuwe Waterweg–3	MAA3	36.5	49.25	70.08	227.28	77.23	0.53
Noordzeekanaal–3	NOO3	23.29	51.62	64.68	229.32	43.82	2.04
Puttershoek–3	PUT3	31.02	48.66	92.54	204.97	178.74	0.80
Schaar Ouden Doel–3	SCH3	17.82	9.14	56.11	110.41	34.72	10.06
Volkerak-Zoommeer–3	VOL3	37.65	59.66	228.69	423.50	122.85	0.97
Wolderwijd–3	WOL3	36.76	78.85	98.56	191.84	138.10	0.15
Spa blank–4	CON4	89.91	120.74	119.76	416.17	221.78	0.03
Belfeld–4	BEL4	29.31	12.05	111.60	100.80	126.76	4.77
Eijsden–4	EYS4	44.98	13.10	136.66	132.86	116.85	3.30
Haringvlietsluis–4	HAR4	51.05	29.46	139.00	131.00	66.00	0.44
Vrouwenzand–4	VRO4	45.44	90.65	160.84	261.74	230.77	0.16
Keizersveer–4	KEI4	40.22	14.76	129.34	116.97	82.74	2.72
Ketelmeer–4	KET4	58.85	30.10	177.18	100.10	126.13	0.40
Nieuwegein–4	NIE4	35.34	32.20	121.24	219.44	242.48	1.51
Lobith–4	LOB4	45.42	66.96	132.00	148.00	115.00	0.03
Markermeer–4	MAR4	69.3	93.69	354.71	255.51	139.28	0.07
Nieuwe Waterweg–4	MAA4	31.7	55.81	127.25	219.44	134.27	0.54
Noordzeekanaal–4	NOO4	44.85	60.13	132.87	153.85	122.12	0.06
Puttershoek–4	PUT4	61.82	56.54	137.41	169.51	110.44	0.02
Schaar Ouden Doel–4	SCH4	40.18	10.10	112.00	223.99	86.29	5.81
Volkerak-Zoommeer–4	VOL4	39.25	24.35	165.67	143.21	148.82	1.41
Wolderwijd–4	WOL4	32.27	76.04	177.18	279.28	122.00	0.54
Spa blank–5	CON5	106.9	176.24	219.02	785.68	174.70	0.05
Belfeld–5	BEL5	80.21	25.99	179.45	187.34	88.74	0.67
Eijsden–5	EYS5	44.38	53.29	123.37	96.29	66.13	0.02
Haringvlietsluis–5	HAR5	65.41	28.80	233.17	152.15	73.70	0.78
Vrouwenzand–5	VRO5	118.76	72.22	132.21	94.57	131.26	0.00
Keizersveer–5	KEI5	125.27	27.45	168.73	125.57	104.48	0.33
Ketelmeer–5	KET5	74.47	59.72	226.32	68.08	63.38	0.09
Nieuwegein–5	NIE5	88.89	57.16	166.91	198.03	161.16	0.02
Lobith–5	LOB5	156.3	62.81	246.26	224.42	179.72	0.01
Markermeer–5	MAR5	435.51	107.65	473.00	264.88	300.30	0.00
Nieuwe Waterweg–5	MAA5	66.03	51.76	157.15	126.82	59.47	0.04
Noordzeekanaal–5	NOO5	33.1	57.20	104.16	129.95	32.67	0.58
Puttershoek–5	PUT5	158.19	66.06	234.70	163.55	301.50	0.01
Schaar Ouden Doel–5	SCH5	47.35	10.26	84.46	63.58	76.38	3.95
Volkerak-Zoommeer–5	VOL5	34.12	43.56	235.52	224.48	43.26	1.92
Wolderwijd–5	WOL5	115.98	61.55	331.11	121.00	58.12	0.20
Spa blank–6	CON6	786.37	162.64	237.75	352.82	246.51	0.00
Belfeld–6	BEL6	180.03	36.60	207.87	118.78	302.70	0.29
Eijsden–6	EYS6	88.78	115.07	149.23	332.37	48.05	0.18

TABLE 18.2 (continued)
The Measured Toxicity Data

Station–Series	Code	Microtox EC_{50}^{f}	Algae EC_{50}^{f}	Rotox LC_{50}^{f}	Thamno LC_{50}^{f}	Daphnia EC_{50}^{f}	pT %
Haringvlietsluis–6	HAR6	218.47	52.92	258.26	204.20	127.71	0.05
Vrouwenzand–6	VRO6	36.43	27.80	68.86	112.98	69.72	0.38
Keizersveer–6	KEI6	151.29	81.43	129.26	123.25	163.02	0.00
Ketelmeer–6	KET6	255.71	98.84	445.37	238.39	66.40	0.11
Nieuwegein–6	NIE6	65.71	60.55	167.44	122.98	108.32	0.01
Lobith–6	LOB6	157.76	75.99	94.35	118.40	126.63	0.00
Markermeer–6	MAR6	104.94	146.75	246.50	241.57	299.40	0.00
Nieuwe Waterweg–6	MAA6	152.85	52.01	33.12	115.92	76.46	0.26
Noordzeekanaal–6	NOO6	128.94	51.66	99.20	68.14	99.10	0.00
Puttershoek–6	PUT6	88.36	139.94	255.42	302.72	133.33	0.00
Schaar Ouden Doel–6	SCH6	47.62	13.58	113.63	41.81	90.63	3.03
Volkerak-Zoommeer–6	VOL6	90.72	48.58	167.33	124.50	79.28	0.02
Wolderwijd–6	WOL6	324.51	94.50	258.66	324.76	82.00	0.04

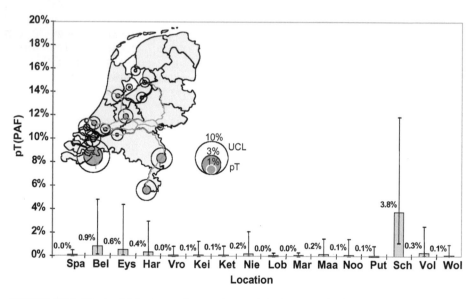

FIGURE 18.4 The toxicities lumped to produce a pT value over six data series over the year 1996.

18.4.2 CHEMICAL CONCENTRATIONS TRANSFORMED TO MSPAF TOXIC RISK

All chemical concentration data analyzed during the year 1996 comprise a set of 3451 observations. Therefore, the scope and size of this chapter does not allow for a presentation of the raw chemical data. Based on the chemical observations,

TABLE 18.3
pT Values Including Their 90% Confidence Intervals
Lumped over 1996 and the Toxic Risk (msPAF) Calculated
from Chemical Observations Averaged over the Year 1996

Station	Toxic Potency, %			Toxic Risk, %	Analyte Number,
	pT	CM_i^c	CM_u^c	msPAF	n
MAA	0.22	0.03	1.51	14.55	74
LOB	0.02	0.00	0.28	13.49	74
EYS	0.58	0.07	4.38	10.83	74
SCH	3.78	1.13	11.85	7.02	74
HAR	0.37	0.04	2.96	5.75	39
NOO	0.13	0.01	1.46	4.87	37
VOL	0.32	0.04	2.54	3.64	37
NIE	0.22	0.02	2.10	2.34	22
PUT	0.09	0.01	0.89	0.77	17
VRO	0.05	0.00	0.84	0.14	17
BEL	0.86	0.15	4.81	0.00	15
KET	0.12	0.02	0.89	0.00	15
MAR	0.02	0.00	0.32	0.00	15
WOL	0.07	0.00	1.00	0.00	15
KEI	0.05	0.00	1.27	0.22	3

Note: Correlation coefficient between pT and msPAF = 0.17.

Table 18.3 presents the calculated msPAF averaged over 1996. The lumped toxic potency, together with its 90% confidence interval, is also shown for comparison.

For the stations most thoroughly analyzed (EYS, LOB, MAA, and SCH), Table 18.4 shows the compounds adding the most to the overall toxic risk assessed as msPAF.

18.5 DISCUSSION

18.5.1 RELATIVE SUSCEPTIBILITY OF THE BIOASSAYS

The median toxicity per test system over all stations and sampling dates reveals a ranking of susceptibility of the different tests. This type of analysis as presented in Table 18.5 demonstrates that the Microtox and the algal productivity tests are by far the most sensitive.

18.5.2 INFORMATION REDUNDANCY

A correlation analysis on the log-transformed and centered toxicity data obtained with the five different test systems (Table 18.6) shows very low correlation coefficients, indicating that the five different toxicity tests provide their share of information, depending on the nature of the toxicants present.

TABLE 18.4
The Average Concentration Expressed in Toxic Units from the Most Prominent Organic Toxicants for the Four Most Elaborately Analyzed Stations

	EYS	LOB	MAA	SCH
Benzo(*a*)pyrene	**0.034**		**0.026**	
Benzo(*a*)anthracene	**0.017**	**0.013**	**0.013**	
Ethyl-Parathion	**0.010**	0.003	0	
Benzo(*b*)fluoranthene	0.002	0.002	0.001	0.002
Fluoranthene	0.002	0.001	0.001	0.001
Chrysene	0.002	0.001	0.001	0.001
α-endosulfan	0.002	0.001	0.001	0.001
Diuron	0.002	0.001	0.001	**0.010**
Fenitrothion	0.001	0.001		0.001
Malathion	0.001	0.001		0.001
Atrazine	0.001			0.002
Diazinon	0.001			0.001
Ethyl-azinphos		**0.026**	**0.030**	
Mevinphos		0.005	0.003	0.002
Methyl-azinphos		0.003	0.003	0.003
Fenthion		0.002	0.001	0.001
MCPP				0.001
245-Trichlorophenol				0.001
ΣTU over above compounds	0.075	0.060	0.081	0.028
ΣTU over all compounds	0.083	0.065	0.091	0.032

Note: Highest contributions per station in **bold** type.

TABLE 18.5
Ranked Susceptibility of the Different Tests

Type of Test	Median $L(E)C_{50}^f$	95% CI	Rank
Microtox	56.33	48–67	1
Algae photosynthesis	59.87	50–71	2
Daphnia-IQ	116.85	102–133	3
Rotoxkit-F	136.66	118–158	4
Thamnotoxkit-F	191.84	168–219	5

A more detailed conclusion can be drawn from a principal component analysis (PCA) on the toxicity data (Figure 18.5). Prior to PCA analysis, the data were centered over observations and tests. The first (horizontal) PC axis explains 42% of the variance and may tentatively be interpreted as the axis representing minimum narcotic toxicity, which is indicated by a comparatively strong influence of the tests with aspecific endpoints such as mortality in the Rotox and Thamnotox tests and

TABLE 18.6
Correlation Matrix of Toxicity Data Obtained
with the Five Test Systems

	ALG50	IQ50	MTX50	RO50	THA50
ALG50	1.0000				
DIQ50	−0.4275	1.0000			
MTX50	−0.3030	−0.2909	1.0000		
RO50	−0.4408	−0.2671	0.1454	1.0000	
THA50	0.0969	−0.0003	−0.6931	−0.2928	1.0000

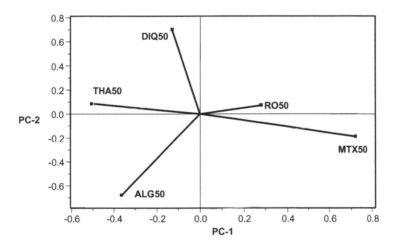

FIGURE 18.5 The loadings of five different toxicity tests performed on 82 sample concentrates on the first two principal axes as calculated with principal component analysis.

Krebs cycle malfunction in the Microtox test. The second axis (with an added explained variance of 30%) may perhaps be interpreted as representing more specific modes of action by the strong influence of more specific endpoints such as algal primary productivity and the enzymatic ability of *Daphnia* to digest a particular substratum.

The opposite orientation of the vectors for the Microtox test and the Thamnotox test on the first axis indicate that these tests are negatively correlated in their response. The same applies to the Algal productivity and the Daphnia IQ tests. It should be realized that negative correlations between the responses of different test systems most probably are attributable to site-specific differences in the relative composition of the cocktail of environmental contaminants.

18.5.3 COMPARISON OF OBSERVED AND CALCULATED TOXICITY

As can be judged from Table 18.3, there are rather large and inconsistent discrepancies between the average observed toxic potencies (pT values) for the 15 stations and the results of the toxic risk calculated from average chemical concentrations (msPAF). For all stations, the observed toxicity is considerably less than the calculated toxicity. In part, this may be explained by the compound selectivity of the concentration procedure applied prior to toxicity testing. This phenomenon may also be attributed to the partial loss of highly volatile toxicants, which is also associated with the concentration procedure. As has been demonstrated by Struijs et al. (1998) the total loss of toxicants may mount up to a range of 40 to 60%. Improper scaling of one or both of the methods would result in a more consistent difference in effect prediction. However, the variance in the msPAF is very poorly explaining the variance in the observed toxic potency as is demonstrated by a low correlation coefficient of 0.17. One of the most probable causes for the lack of correlation between observed and calculated toxicity is that one or more locally important toxicants have been omitted from the chemical analysis. This is clearly the case at the stations that are poorly covered by chemical analysis, but also at station SCH where, according to Table 18.4, the concentrations of measured toxicants are quite low and the measured toxic potency is the highest of all.

It has been observed (this chapter and Slooff, 1983) that the SSDs of bioassays with complex environmental concentrates are generally characterized by a slope that is much steeper than the slopes observed for the individual chemicals in the mixture. In the above study the average β value is 0.17 (SEM = 0.05) vs. a range of 0.28 to 0.71 as found for the β values of individual chemicals (Chapter 8). A β value of 0.17 corresponds to an average maximum difference in toxicity between the tested species of a factor 7. This observation may also cause unexplained differences between observed pT and the results of the toxic risk calculated on the basis of chemical concentrations. A hypothesis to explain the steeper slope for mixture toxicity testing is that the species tested may be more sensitive than average to some of the chemicals in the mixture and less sensitive than average to others. The species tested are unlikely to be extremely sensitive or insensitive to all chemicals in the mixture. As a result, their responses to complex mixtures will tend to be closer to median toxicity than their responses in case of exposure to individual chemicals. Another line of thought to explain the difference between the observed and calculated toxicity is the possibility that the single chemicals in the complex cocktail of environmental contaminants lose their individual properties and start to act as a "single virtual toxicant" with a very narrow action spectrum.

Obviously, the explanation of observed toxicity is urgently in need of further study. The same holds for the validation and calibration of the toxic potency model with the aid of ecological inventories providing information on the actually occurring ecological effects due to toxic compounds.

19 Mapping Risks of Heavy Metals to Birds and Mammals Using Species Sensitivity Distributions

Theo P. Traas, Robert Luttik, and Hans Mensink

CONTENTS

Abstract — A method is presented for calculating and mapping the risk to birds and mammals due to secondary poisoning of heavy metals. Risk is defined as the fraction of species for which the concentration of a certain compound in the environment exceeds their predicted no-effect concentration (PNEC). In case of secondary poisoning, the PNEC is a concentration in the diet. The results are expressed as a risk index between 0 and 1 for a certain geographical ecosystem unit. The method has been applied for cadmium, zinc, and copper in the Netherlands, and yielded maps of the risk index. The risk of cadmium is highest on the sandy soils, but especially along the southern border near a former zinc smelter. The calculated risk for copper is low for all species (<0.1). Risks of zinc exposure to vertebrates are only calculated for the western peat areas. The present model does not estimate population effects that could be expected in the field, only the risk of exceeding the PNEC. It can be expected that significant toxic stress occurs at the highest risk levels, but confirmation of these results with field data on exposure or effects is needed.

19.1 INTRODUCTION

Ecotoxicological risks for heavy metals in soils in the Netherlands are derived from species sensitivity distributions (SSDs) for terrestrial soil fauna (e.g., Chapters 12 and 16). It has been proposed that the 5th percentile of the SSD offers sufficient protection for the entire ecosystem (Van Leeuwen, 1990). The risk of trophic transfer of persistent compounds, however, is not explicitly covered, since the basic toxicological information from which the distribution is derived usually does not include food chain exposure data. In the current Dutch procedure for deriving soil quality criteria, food chain transfer is taken into account by calculating risks for worm-eating birds and mammals (Romijn et al., 1994; Van de Plassche, 1994; Chapter 12). This chapter is an attempt to expand the use of SSDs in risk assessment to more species of birds and mammals at specified locations in the Netherlands.

The mobility of heavy metals in the food chain, especially cadmium, has given rise to concern about the risk posed by food chain transfer of heavy metals (Gorree et al., 1995; Traas et al., 1996). It has been shown that the health of small mammals can be impeded in metal polluted floodplains, mainly by accumulation of heavy metals in kidney and liver (Ma et al., 1991; Hendriks et al., 1995). This chapter aims to identify hot spots where risks for birds and mammals can be expected.

In 1990, the Dutch Ministry of Agriculture, Nature Management and Fisheries announced the establishment of the National Ecological Network (NEN), consisting of core areas (existing nature reserves), natural development areas, and ecological corridors connecting these (Bal et al., 1995). In the near future, the national ecological network will exist of a comprehensive set of ecosystem types. Each of the 132 ecosystem types in this classification specifies an ecological objective in terms of biotic and abiotic components, for which 657 species from ten taxonomic groups were selected as indicator species. The selection is based on an assessment of international significance, knowledge on trends in population size, and on national endangered species lists.

This chapter describes a method to calculate the percentage of the mammalian and avian indicator species of specified ecosystem types in the Dutch NEN, exposed to concentrations higher than the protective level for those species. To calculate the potentially affected fraction (PAF) of species (i.e., the fraction of species potentially at risk), a comparison is made between the potential no-effect concentration of the species (PNEC) and the potential environmental concentration (PEC). Since birds and mammals are mainly exposed through the food chain, a species-specific bioaccumulation factor is calculated based on bioaccumulation factors for different food sources (Jongbloed et al., 1996). Spatially explicit input data such as ecosystem type, associated indicator species, and soil parameters such as pH, organic matter, clay, and heavy metal concentrations were available as digitized maps. The results are presented as spatially explicit risk maps for the Netherlands, depicted by plotting the PAF with a geographic information system (GIS). The method will be illustrated with examples for cadmium, zinc, and copper.

19.2 METHODS

19.2.1 GEOGRAPHICAL INPUT DATA

The area designated for the Dutch NEN was subdivided in units that represent a certain dominant ecosystem, as defined by ecosystem type (Bal et al., 1995). The ecosystem type determines which birds and mammals are considered indicators for achievement of ecological objectives at a specific location. Ecosystem types fall into nine physicogeographical regions: hilly country, higher sandy soils, river clay, peat bogs, marine clay, dunes, former sea-arms, tidal areas, and the North Sea. The map of ecosystem types (Alterra Research Institute, the Netherlands) has a resolution of 1×1 km.

Heavy metal concentrations in the topsoil (0 to 10 cm) were calculated from interpolation of a large database of measured heavy metal concentrations. Known relationships for sorption of heavy metals to soil and historical loading records were used to calibrate a model for extrapolation of locally measured soil concentrations to complete concentration maps (Tiktak et al., 1998). These maps were made available as a 650×560 grid projected over the Netherlands, with a cell size of 500×500 m, which is the output resolution.

19.2.2 PREDICTED ENVIRONMENTAL CONCENTRATION

Heavy metals occur naturally in soil, mainly associated with the mineral fraction of the soil. Regression analysis was used to estimate the background concentrations for the metals Cd, Cu, and Zn in the Netherlands. The lower bound of nonexchangeable metal (HM_{non-ex}) was based on several estimates (Klepper and van de Meent, 1997) and is of the form:

$$HM_{non-ex} = a[\% \text{ clay}] + b[\% \text{ organic matter}] + c[\% \text{ sand} + \text{silt}] \quad (19.1)$$

The upper bound is determined by regressions, based on an analysis of Dutch reference soils and is of the same form (Lexmond and Edelman, 1992). The mean of both estimates is taken to be HM_{non-ex}. Soil properties such as clay and organic matter content were used as input for these regressions, taken from digitized maps with a resolution of 500×500 m (Klepper and van de Meent, 1997; Klepper et al., 1998).

All calculations were programmed in FORTRAN 77 and resulting maps were plotted with MATLAB (v 5.3, MathWorks, Natick, MA) using the IDRISI image map format (Clark Labs, Worcester, MA).

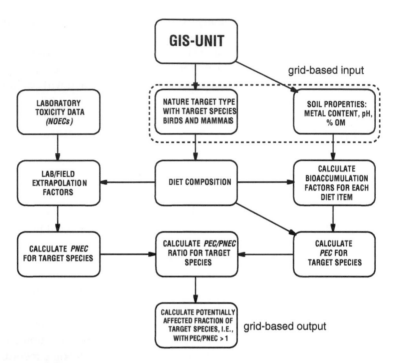

FIGURE 19.1 Flowchart for calculating the PAF for indicator species at a specific location. OM = soil organic matter.

19.2.3 PREDICTED NO-EFFECT CONCENTRATION BASED ON SSDS

To estimate the risk for an indicator species (bird or mammal), the exposure concentration is compared with a PNEC for that particular species.

The toxicity of metals to indicator species might be more readily linked to dose (mg/kg bw/day) than to diet concentrations (mg/kg diet). In chronic toxicity studies with vertebrates, exposure concentrations are usually expressed as diet concentrations (Jongbloed et al., 1994) although this has several shortcomings discussed below. In view of consistency with present methodology for deriving quality criteria for secondary poisoning (Van de Plassche, 1994), risks to indicator species are related to diet concentrations. To calculate the PNEC, several steps were taken (Figure 19.1, left side).

Toxicity data are scarce for wildlife (Posthuma et al., 1995), and therefore laboratory toxicity data on rats, mice, rabbits, quail, etc. are used to estimate wildlife toxicity. When extrapolating toxicity data derived from laboratory tests with bird and mammals to indicator species in the field, several issues concerning differences in exposure and sensitivity should be considered (U.S. EPA, 1993b; Health Council of the Netherlands, 1993b):

1. Metabolic rate, depending on state of activity
2. Caloric content of food, depending on diet composition

3. Food assimilation efficiency, depending on diet composition of indicator species
4. Pollutant assimilation efficiency, depending on the type of pollutant
5. Species sensitivity

The first extrapolation problem is that total dose received should be equal for test species and indicator species of equal weight under equal circumstances. Free-roaming animals use more energy than caged test species (issue 1), eat more food with generally a low caloric content (issue 2), and therefore receive a larger dose given the same diet concentration.

The second extrapolation problem is related to assimilation efficiencies of both food and pollutant (issues 3 and 4), which may differ between test species and indicator species.

The third extrapolation problem is related to the sensitivity of indicator species (issue 5), which could be fundamentally different from the test species, i.e., the SSD for the indicator species is significantly different from the SSD for the test species.

For the first three issues mentioned, the differences between indicator species and test species can be quantified, whereas evidence for the last two issues is scarce and could not be quantified for birds and mammals (Traas et al., 1996).

Since no-observed-effects concentrations (NOECs) for indicator species are generally lacking, NOECs of test species for the sublethal endpoints growth or reproduction and based on food exposure were used as a starting point. From these SSDs, the median NOEC for cadmium (birds and mammals separated), and zinc and copper (birds and mammals combined) was estimated. Birds and mammal NOEC data are separated since the median NOECs for birds seems lower than for mammals (*t*-test, *P* value 0.06). PNECs of indicator species are estimated from the median NOECs using laboratory-to-field extrapolation factors:

$$PNEC = NOEC_{50} \cdot fCC \cdot fFAE \cdot fMR \qquad (19.2)$$

where

PNEC	= PNEC for an indicator species in the field
$NOEC_{50}$	= median NOEC of laboratory test species for Cu, Zn, or Cd
fCC	= extrapolation factor for caloric content (CC)
fFAE	= extrapolation factor for food assimilation efficiency (FAE)
fMR	= extrapolation factor for metabolic rate (MR)

From a review of regressions (Jongbloed et al., 1994) on field metabolic rate (FMR) and existence metabolic rate (EMR), an average FMR/EMR ratio of 0.4 was calculated. This ratio is used as the extrapolation factor (fMR [-])for metabolic rate differences between laboratory and field. Extrapolation factors for food assimilation efficiency (fFAE) and caloric content (fCC) are calculated from a weighted average of the different diet items for the top predator:

$$\text{fFAE} = \frac{\sum_{i=1,n} \text{FAE}_i \cdot \text{fDiet}_i}{\text{FAE}_{\text{ref}}} \quad \text{and} \quad \text{fCC} = \frac{\sum_{i=1,n} \text{CC}_i \cdot \text{fDiet}_i}{\text{CC}_{\text{ref}}} \tag{19.3}$$

where

FAE_i = food assimilation efficiency of diet item i

CC_i = caloric content of diet item i

fDiet_i = fraction of diet of item i

FAE_{ref} = laboratory FAE (equal for birds and mammals)

CC_{ref} = laboratory CC (different for birds and mammals)

Full details were published previously (Traas et al., 1996; Jongbloed et al., 1996).

19.2.4 DIETS

Indicator species for mammals (16 species) and birds (64 species) were selected by Bal et al. (1995), based on an assessment of international significance, knowledge of trends in population size, and on endangered species lists. Diets of birds were taken from Cramp (1977–1994) and diets of mammals from the *Handbuch der Säugetiere Europas* (Niethammer and Krapp, 1978–1994). Details on the species and diets used were described previously (Luttik et al., 1997).

19.2.5 BIOACCUMULATION DATA

Romijn et al. (1993; 1994) and Jongbloed et al. (1994) collected bioaccumulation factors (BAFs) for cadmium, zinc, and copper for the selected diet items as depicted in Figure 19.2. Reported median values taken from field studies were used for invertebrates. BAFs for small vertebrate prey items are based on laboratory feeding studies since relevant field studies are very scarce (see Jongbloed et al., 1994).

Concentrations in prey (Cp_i, mg/kg ww) were calculated for all prey items except earthworms as

$$Cp_i = \text{BAF}_i \cdot \text{metal}_{\text{soil}} \tag{19.4}$$

Concentrations in earthworms were calculated based on total heavy metal concentration in the topsoil and pH (Verhallen and Ma, 1997):

$$\log_{10}\left(\text{metal}_{\text{worm}}\right) = x_0 + a \cdot \log_{10}\left(\text{metal}_{\text{soil}}\right) + b \cdot \text{pH} \tag{19.5}$$

where

$\text{metal}_{\text{worm}}$ = the concentration in earthworms [mg/kg dw]

$\text{metal}_{\text{soil}}$ = the total concentration in the topsoil [mg/kg dw]

x_0, a, b = regression coefficients

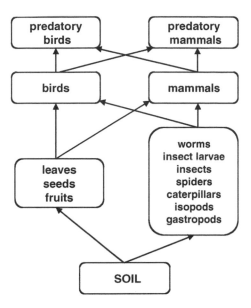

FIGURE 19.2 Diagram of the food web for birds and mammals with the different food sources grouped. Each food source can be selected by the indicator species as a food source. Indicator species are in the top two trophic levels.

Worm concentrations were converted from dry weight to wet weight, assuming a dw/ww ratio of 0.18.

Similar regression equations are available for different worm species and metals (Ma, 1982; Ma et al., 1983), but the influence of pH is not always found to be statistically significant (Sample et al., 1999).

19.2.6 Diet Concentrations

The potential environmental concentration (PEC) for exposure of birds and mammals (see Figure 19.1, right side) is the average concentration in the food. Chronic metal toxicity to birds and mammals is estimated based on exposure concentrations in the food, neglecting exposure by way of drinking water or inhaled air. This underestimates total metal exposure somewhat.

The diet concentration is calculated by weighing the concentrations of the different prey items according to the proportions in the diet (see Figure 19.2). Toxicant concentrations of most soil-dwelling prey items are calculated with median empirical BAFs. The concentration in earthworms is calculated for each grid cell, depending on total heavy metal concentration and pH (see Table 19.1, Equation 19.5).

Water exposure of prey items such as aquatic insects is taken into account by using a single median concentration for all water bodies in the Netherlands and bioconcentration factors for such prey items. The aquatic routes are a small minority of the exposure routes for the selected indicator species, and therefore this is considered to be an acceptable simplification. For those indicator species that forage solely from aquatic ecosystems, e.g., piscivorous wildlife not under consideration

TABLE 19.1
**Regression Coefficients for Accumulation
of Metals in Earthworms**

Coefficients	x_0	a	b	R^2	n
Cadmium	2.60	0.49	−0.2	0.69	88
Copper	1.21	0.43	−0.08	0.65	70
Zinc	3.07	0.27	−0.1	0.45	87

in this study, average exposure is not an acceptable simplification and geographically explicit exposure concentrations should be collected.

PECs were calculated for each grid cell, for those indicator species that are associated with the ecosystem type for that specific location:

$$PEC_{TargetSpecies} = \sum_{i=1,n} fDiet_i \cdot Cp_i \qquad (19.6)$$

where
  $fDiet_i$ = diet fraction of item i (to n items)
  Cp_i = concentration in prey i, calculated from Equation 19.4 or 19.5

19.2.7 RISK CALCULATIONS FOR INDICATOR SPECIES

After calculating the PNEC of indicator species and the PEC for indicator species in each grid cell of the NEN, the PEC/PNEC ratio is calculated for each indicator species j at a specific location p (see Figure 19.1). An indicator species j is considered to be potentially affected if the PEC/PNEC ratio is larger than 1 (i.e., exposure is larger than the PNEC); otherwise, it is considered not to be affected:

$$\frac{PEC_j}{PNEC} > 1 \Rightarrow \text{species } j \text{ potentially affected}$$

$$\frac{PEC_j}{PNEC} < 1 \Rightarrow \text{species } j \text{ not affected} \qquad (19.7)$$

The fraction of the indicator species in an ecosystem type p that is potentially affected by a metal, abbreviated PAF, is the number of indicator species in that ecosystem type with a PNEC/PEC ratio greater than 1, divided by the total number of indicator species in the specific ecosystem type p:

$$PAF_p = \frac{nPot.affected_p}{nTargetSpecies_p} \qquad (19.8)$$

Following Klepper and van de Meent (1997), the fraction of heavy metals associated with the mineral fraction is considered to be the background level. Natural stress by background levels of metals is considered to be an occurrence like other natural stress factors such as nutrients, moisture, or temperature (Klepper et al., 1998). For this reason, the PAF is calculated twice, once for the background concentrations (PAF_{bg}) and once for the total concentrations in the soil (PAF_t). In this way the PAF due to addition of metals by anthropogenic activity (PAF_a) can be distinguished from PAF_{bg} by using the scaling formula (Klepper et al., 1998):

$$PAF_a = \left(PAF_t - PAF_{bg}\right) / \left(1 - PAF_{bg}\right) \tag{19.9}$$

Because of extrapolation error in the regressions for metals associated with the mineral fraction, the estimated PAF_{bg} can be higher than PAF_t in a limited number of grid cells. These phenomena occurred in 1.5% of the grid cells of the NEN for cadmium, 3.9% for zinc, and 0.03% for copper, and these cells are not taken into account when calculating the PAF_a.

Combined effects of several substances can be calculated as well and expressed as a multisubstance PAF (msPAF).

Assuming that there is no interaction of effects, response addition of metals (Klepper et al., 1998; Chapter 16) leads to

$$msPAF = 1 - \left(\left(1 - PAF_{zinc}\right) \cdot \left(1 - PAF_{copper}\right) \cdot \left(1 - PAF_{cadmium}\right)\right) \tag{19.10}$$

The possible values that a PAF can assume for indicator species depends on the number of indicator species per ecosystem type. For an ecosystem type with only ten indicator species, the PAF can only be 0, or multiples of 0.1 with a maximum value of 1.0. Calculation of the combined PAF for several metals can lead to values of PAF that are not exact multiples of 0.1. Because of the use of classes in mapping the PAF, this will not lead to a significant difference in interpretation of the PAF map.

19.3 RESULTS

19.3.1 PREDICTED NO-EFFECT CONCENTRATIONS FOR BIRDS AND MAMMALS

By extrapolating the limited set of toxicity data from laboratory testing, PNECs for indicator species were calculated using extrapolation factors from Equations 19.2 and 19.3. Because two of three extrapolation factors depend on the diet of a particular indicator species, each species has a different PNEC. By plotting the cumulative frequency of PNECs, we can inspect the resulting sensitivity distribution of indicator species (Figure 19.3).

The median calculated cadmium PNEC for birds and mammals is 0.74 mg/kg ww of diet (range 0.17 to 4.0 mg/kg diet ww). The median PNEC for copper is

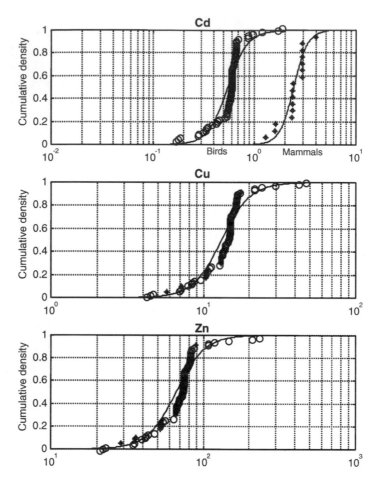

FIGURE 19.3 PNECs for birds and mammals for the heavy metals cadmium, copper, and zinc. Open circles: PNECs for birds. Crosses: PNECs for mammals. Drawn line: theoretical cumulative logistic distribution. For cadmium, SSDs for birds and mammals are separated.

13.2 mg/kg diet ww, and the range is 4.2 to 46 mg/kg diet ww. The median PNEC for zinc is 66 mg/kg diet ww and the range is 21 to 230 mg/kg diet ww. For copper and zinc the most sensitive indicator species is estimated to be a factor of 10 more sensitive than the least-sensitive indicator species; for cadmium this factor is approximately 40.

The cumulative PNEC frequencies for cadmium differ from the curves for zinc and copper, because two different data sets for cadmium toxicity were used, one for birds ($n = 5$) and one for mammals ($n = 5$). Birds appear more sensitive to cadmium than mammals and the SSD for birds, estimated with the diet factors (Equation 19.2 and 19.3), is distinct from the SSD for mammals. For zinc and copper, very few studies were available ($n = 5$) so a combined data set was used as a starting point for PNEC calculations for all indicator species.

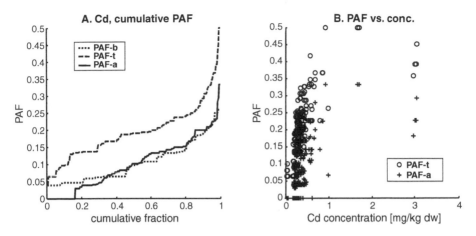

FIGURE 19.4 Random sample from the risk map for cadmium. The cumulative anthropogenic PAF$_a$ frequency is plotted as calculated from total PAF$_t$ and background PAF$_{bg}$ (A) and a scatterplot of concentration vs. PAF$_a$ and total PAF$_t$ (B).

19.3.1.1 Cadmium

In Figure 19.4, results for a random sample over all soil and ecosystem types of the NEN map are presented for cadmium. Figure 19.4A allows an analysis of which proportion of the map (*x*-axis) is associated with low or high PAF values (*y*-axis). PAF values are presented for PAF$_t$, PAF$_{bg}$, and PAF$_a$. Figure 19.4B allows an analysis of the nonlinear relation between concentrations (*x*-axis) and the PAF (*y*-axis).

The cumulative frequencies of the calculated PAF levels for cadmium (Figure 19.4A) show that the maximum fraction affected is about 50% for PAF$_t$ and that the PAF$_a$ value is cut off at low values of total PAF$_t$. Because of the scaling procedure employed (Equation 19.9), the PAF$_a$ value is related to the background PAF$_{bg}$ instead of just being the difference of PAF$_t$ and PAF$_{bg}$. The scatterplot (Figure 19.4B) shows that relatively low cadmium concentrations up to 1 mg/kg dw result in an almost linear increase of PAF$_a$ up to 50%. Above 1 mg/kg, PAF$_a$ values level off, and the highest soil contents do not necessarily lead to the highest PAF$_a$ levels.

The most sensitive areas on the cadmium risk map (see Figure 19.7), i.e., with high PAF$_a$ values, are acidic sandy soils (with pH < 5) poor in organic matter. These soils are found mainly in the central and southeastern part of the country. The highest risks are predicted along the southern border, where a former zinc smelter has emitted both zinc and cadmium in the past. In this area, high cadmium levels are found in earthworms and organs of voles and shrews (Ma et al., 1991). This contamination of intermediate prey serves as input to the model, leading to the highest PAF levels for cadmium.

19.3.1.2 Copper

The cumulative frequencies of calculated copper PAF (Figure 19.5A) show that the difference between risks due to background metal and total metal is almost negligible,

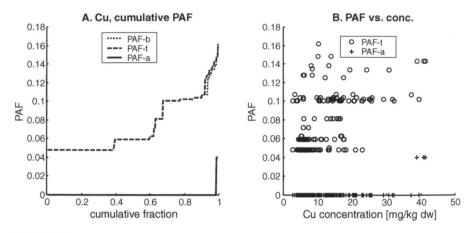

FIGURE 19.5 Random sample from the risk map for copper. The cumulative anthropogenic PAF_a frequency is plotted as calculated from total PAF_t and background PAF_{bg} (A) and a scatterplot of concentration vs. PAF_a and total PAF_t (B).

except for the highest percentiles. From this it can be concluded that the PAF_a is very low.

The background Cu PAFs are relatively high, occurring mainly in the dunes and along the greater rivers (not shown). In sensitive soils with a high copper input from cattle manure, such as agricultural areas on sandy soils, copper levels were about a factor of 9 higher than in forest soils (Van Drecht et al., 1996) but such areas are usually not represented NEN. The scatterplot shows that total PAF_t is generally between 0.04 and 0.18 PAF units, but the anthropogenic PAF_a becomes clearly separated from background PAF_{bg} only above 30 mg/kg dry soil.

The PAF_a for copper is generally low (<0.05) and is highest in the western part of the Netherlands with peat or clay-on-peat soils (see Figure 19.7). This indicates that the predicted risk of the essential metal copper to wildlife is probably not serious since a PAF value of 0.05 is regarded as generally protective of field effects (Versteeg et al., 1999). Because of the uncertainties of the food chain approach, this finding, which is based on direct toxicity in aquatic micro- or mesocosm studies, should be verified for wildlife before being used in risk management.

It is observed that copper and zinc, as essential metals that are regulated within a relevant physiological range, do not bioaccumulate much at intermediate trophic levels compared to mobile elements such as cadmium. Some form of homeostatic regulation was observed for Zn and Cu in shrews and voles and even in invertebrate prey (Ma et al., 1991). If not "overloaded" by high exposure levels causing the homeostasis mechanism to break up, this may prevent strong bioaccumulation by small vertebrates from their prey.

19.3.1.3 Zinc

The analysis of calculated PAF for zinc shows that background levels of zinc dominate PAF_a in Dutch soil (Figure 19.6A), since background and total PAF_t are

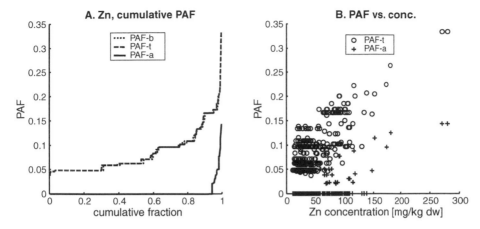

FIGURE 19.6 Random sample from the risk map for zinc. The cumulative anthropogenic PAF_a frequency is plotted as calculated from total PAF_t and background PAF_{bg} (A) and a scatterplot of concentration vs. PAF_a and total PAF_t (B).

very much alike. The total PAF_t for birds and mammals is mainly below 20% and only the highest percentiles of the cumulative frequencies show a difference between the total and the anthropogenic contribution. These spots are only found in areas with high zinc concentrations above 50 mg/kg dw in soil (Figure 19.6B).

The PAF_{bg} is relatively high in the peat areas of the western and northern parts of the Netherlands and in the floodplains of the large rivers (not shown). The anthropogenic PAF_a is mainly found in the western peat areas (see Figure 19.7) with PAF_a levels up to 0.15 units.

19.3.2 COMBINED RISK FOR CADMIUM, COPPER, AND ZINC

In Figure 19.7, the combined PAF (based on response addition, see Chapter 16) for cadmium, copper, and zinc is presented. The map shows a great resemblance to the cadmium map. High PAF_a values are found in the sandy soils of the Netherlands but especially along the southern border where the former zinc smelter is located. In addition, elevated PAF_a levels are found in the western peat areas as mainly determined by zinc.

19.3.3 EXCEEDING SELECTED PAF LEVELS

With the available geographical information, it is possible to calculate the area of the NEN where certain ecotoxicological risk limits are exceeded (Table 19.2). The area where selected PAF_a levels for birds and mammals are exceeded is compared between the metals, with a low risk limit of 0.05 PAF and a serious risk limit of 0.5 PAF units. The low risk limit is associated with ecotoxicological quality criteria in the Netherlands (Chapter 12), the serious risk limit is associated with soil contamination that trigger considerations for soil cleanup measures, the so-called ecotoxicological intervention values (VROM, 1999). Both risk limits are preferably derived from an SSD analysis.

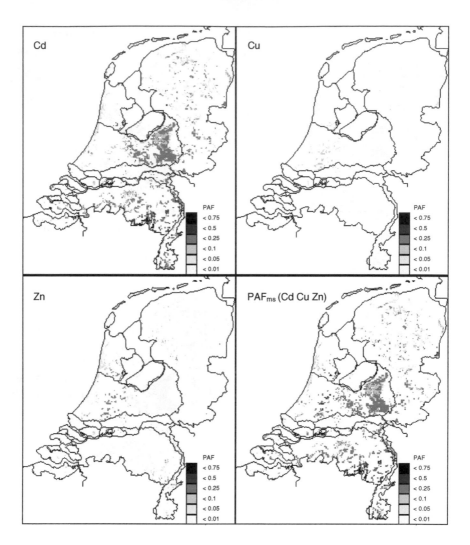

FIGURE 19.7 Calculated risk indices for toxic risk on mammals and birds, plotted as PAF_a values between 0 and 1 for the NEN in the Netherlands. Shown are seperate PAF_a maps for Cd, Cu, and Zn and the combined risk index msPAF.

Cadmium probably poses the largest risk in the NEN (see Figure 19.7), as indicated by the large proportion of the NEN where the 0.05 PAF_a level for cadmium is exceeded (see Table 19.2). In more than half of the NEN, sublethal effects may occur for more than 5% of the indicator species. The serious risk limit of 50% PAF is exceeded in only 0.8% of the NEN. These results are in line with several studies indicating that cadmium is a mobile metal in the food chain (e.g., Ma et al., 1991; Shore and Douben, 1994).

TABLE 19.2
Area of the NEN (%) Where
Specific Risk Limits Are Exceeded

	PAF > 0.05	PAF > 0.5
Cadmium	62.5	0.8
Copper	2.0	0.0
Zinc	6.7	0.0

The low risk limit for zinc is exceeded in 6.7% of the NEN. This is the case for copper in only 2.0% of the NEN, indicating that essential heavy metals are unlikely to pose a threat to birds and mammals in the NEN.

19.4 DISCUSSION

For cadmium, high PAF_a values of about 50% of species are calculated for an area contaminated by mostly historical smelter activities. The question arises what the estimated NOEC exceedence of 50% of indicator species means in terms of field effects. Although the toxicological endpoints in the laboratory tests (e.g., growth, reproduction, and mortality) are relevant for the survival of populations, the translation to indicator species and population-level effects is not straightforward. A decrease in reproduction need not be meaningful if a species population level is determined by food availability (Kooijman, 1993; Ferson et al., 1996). The populations of some species may only be affected at reproduction rates close to zero. In addition, negative effects determined in the laboratory may not show in the field where populations could be adapted genetically or physiologically.

Against these arguments pointing to a conservative nature of the PAF, there are several reasons to expect that the estimation methods used are not overly protective. Chronic laboratory tests have a relatively short duration when compared to the life expectancy of vertebrates. Effects could occur after longer exposure, perhaps even after several generations. Of particular concern in this respect are effects on reproduction that may occur at concentrations below the NOEC for such endpoints as growth or mortality. The eventual effects on population fitness depend on life cycle variables (Kammenga et al., 1996), which should be assessed in a more detailed population analysis.

As a starting point for the PNEC estimate, the median estimate (50th percentile) of the NOEC distribution of the laboratory species is used, instead of a worst-case approach where the lowest NOEC from laboratory testing is used. In addition, for all other parameters used in the food web, median or mean estimates are used (e.g., BAF values and the composition of the diet).

Apart from these arguments about toxicological relevance and uncertainty, spatial aspects of exposure require some consideration. Averaging of exposure concentrations in space and time by the predator, due to random selection of prey over the home range, was ignored in the present study because of limitations of the GIS system used. The implementation of averaging over the home range of a species as shown by Clifford et al. (1995) will probably lead to less extreme PAF values. A matter of attention in fragmented landscapes such as those discussed here is habitat preference. Alternative, but less preferential habitat may be available outside the NEN. Home range averaging should also take habitat preference into account to provide realistic exposure estimates.

Another possible scenario with implications for exposure patterns that are hard to predict is that predators preferentially feed on prey that is weakened by the contaminant and thus are exposed to more than just the average contamination level of the prey. Even with such shortcomings, the present procedure may prove useful for identifying potential hot spots for bird and mammal exposure.

Metals in the environment are partly of natural origin. The background concentration of zinc and copper, as estimated by Klepper and van de Meent (1997), leads to noticeable background PAFs for indicator species (see Figures 19.4 through 19.6), which may be considered a major shortcoming of the present method. This could be caused by using the concept of SSDs (see, e.g., Aldenberg and Slob, 1993) for both essential and nonessential metals. Some assumptions of this method may not hold for essential metals, since very low concentrations of copper or zinc can lead to metal deficiency (Janus et al., 1989; Janus, 1993). Such levels are usually not encountered in the Netherlands. It also possible that the concept is correct in predicting that some (not-adapted) organisms are very sensitive to copper or zinc and that these organisms would be affected at background levels. However, the scaling procedure devised by Klepper and van de Meent (1997) largely rules out the influence of both essentiality and very sensitive species by only looking at the anthropogenic contribution to the risk index PAF_a.

A reason for the relatively high PAF_{bg} values is our imperfect knowledge of the influence of soil properties on partitioning of heavy metals and its influence on bioavailability. For such animals as earthworms, soil properties such as clay, pH, and organic matter clearly influence uptake of metals. Regressions were used to describe the bioconcentration factor as a function of soil properties. For all other invertebrates and plants, these functions are not clear-cut and median BAFs had to be used. This can easily lead to under- or overestimation of the true metal uptake by soil fauna and influence both background levels PAF_{bg} and PAF_r. This influence is ruled out to a certain extent by the present method where the PAF_a levels determine the risk index for indicator species.

19.5 CONCLUSIONS

The percentage of birds and mammals that are exposed to concentrations above their PNECs was calculated based on geographically explicit soil concentrations and conditions. It was found that cadmium is responsible for the largest contribution to PAF_a to birds and mammals but serious risks are only predicted in a natural NEN area close to a former smelter. Zinc and copper generally seem to pose little risk. The PAF in combination with GIS and exposure assessment can be used to identify locations where significant toxic risk is expected, but the model does not give an estimation of the population effects to be expected in the field. Toxic effects at the highest PAF_a levels for cadmium seem probable, but confirmation of these results with field observations on individual organisms or populations is needed.

ACKNOWLEDGMENTS

The authors are indebted to two anonymous reviewers and the section editors for comments that significantly improved the manuscript.

20 Ecotoxicological Impacts in Life Cycle Assessment

Mark A.J. Huijbregts, Dik van de Meent,
Mark Goedkoop, and Renilde Spriensma

CONTENTS

Abstract — Dealing with ecotoxicological impacts is a common part of the environmental life cycle assessment (LCA) of products. Until recently, characterization factors, used for aggregating ecotoxicological impacts in LCA, were based on the hazard index (PEC/PNEC ratio). However, an index of toxic pressure based on the potentially affected fraction of species (PAF), as derived from species sensitivity distributions (SSDs), may allow a more proper aggregation of toxic impacts due to exposure to multiple substances. Two methods for implementation of the PAF concept in LCA are explained. The first method combines noninteractive concentration and response addition rules of calculus, and the second method implicitly starts from noninteractive concentration addition rules of calculus. Outcomes of both methods are compared for 33 substances in freshwater aquatic ecosystems, showing differences as large as 2.2 orders of magnitude. In addition, the prerequisites of a conversion of ecotoxicological impacts to ecosystem damage are explored. Although calculations are technically possible, substantial research is still needed to underpin the validity of both PAF approaches to assess multisubstance damage and guarantee well-validated conversion rules for ecotoxicological impacts to ecosystem damage.

20.1 INTRODUCTION

Life cycle assessment (LCA) is a tool for the assessment of the environmental impact of product systems (Heijungs et al., 1992). It considers the life cycle of a product

1-56670-578-9/02/$0.00+$1.50

from resource extraction to waste disposal. In LCAs, substance-specific weighting factors, also called characterization factors in LCA circles, are used to aggregate a wide variety of emissions causing ecotoxicological impacts.

Characterization factors based on the predicted environmental concentration/predicted no-effect concentration (PEC/PNEC) ratio have been widely used for ecotoxicological impact assessment in LCA (Guinée and Heijungs, 1993; Guinée et al., 1996; Hauschild et al., 1998; Huijbregts et al., 2000). Recently, Goedkoop and Spriensma (1999) have applied a no-observed-effects concentration (NOEC)-based potentially affected fraction of species (PAF), as proposed by Klepper et al. (1998) and Van de Meent (1999), for this purpose. An advantage of the PAF concept is that it may be extended to the potentially disappeared fraction of species (PDF) to allow comparison of ecosystem damage of environmental stressors in general (Klepper et al., 1999).

This chapter starts with a general introduction of the steps involved in LCA. Thereafter, it explains and compares two possible approaches for the calculation of ecotoxicological characterization factors based on PAF rules. Finally, it discusses possible extension of the PAF concept to the PDF.

20.2 LIFE CYCLE ASSESSMENT

According to ISO standardization guidelines (ISO, 1997), an LCA study can be divided into four phases: goal and scope definition, inventory analysis, impact assessment, and interpretation (Figure 20.1).

In the goal and scope definition, the aim and the subject of an LCA study are determined and a "functional unit" is defined. An example of a functional unit is "cleaning 100 kg of colored clothes" with, for example, the aim to compare the environmental impacts of different types of cleaning methods. In the inventory

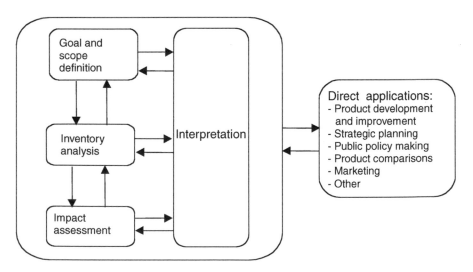

FIGURE 20.1 Phases taken in a LCA. Information on SSDs is applicable to LCIA. (From ISO, 1997.)

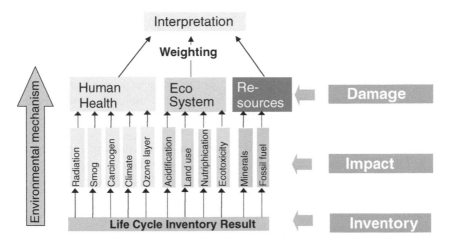

FIGURE 20.2 Identification of environmental problems at the "impact level" and the "damage level." Information on SSD can be used in the calculation of characterization factors at the impact level (ecotoxicity) and the damage level (damage to ecosystem quality). (Derived from Goedkoop and Spriensma, 1999.)

analysis, for each of the product systems considered data are gathered for all the relevant processes involved in the life cycle. A product system can be considered as a combination of processes needed for the functioning of a product or service. The outcome of the inventory analysis is a list of all extractions of resources and emissions of substances caused by the functional unit for every product system considered, generally disregarding place and time of the extractions and releases.

Life cycle impact assessment (LCIA) aims to improve understanding of the inventory result. First, it is determined which extractions and emissions contribute to which impact categories. An impact category can be defined as "a class representing environmental issues of concern into which results from the inventory analysis may be assigned" (ISO, 1998). As shown in Figure 20.2, impact categories may be identified at the "impact level" of environmental problems, such as ecotoxicological impacts (Udo de Haes, 1996), or at "damage level," such as human health damage or damage to ecosystem quality (Goedkoop and Spriensma, 1999).

The next important step in the impact assessment is the characterization. The aim of the characterization is to aggregate the releases of pollutants and the extractions of resources of a product system for a number of environmental impact categories defined at the damage level or the impact level. As LCA applications tend to be change oriented ("what is the additional environmental impact if one extra product is produced?"), the approach of "marginal change" is advocated in LCIA (Udo de Haes et al., 1999). It assumes that an additional amount of a certain stressor introduces very small changes on top of a *ceteris paribus* background situation. The change in impact per unit amount of additional "release" represents the relative importance of the stressor to an impact category (Heijungs et al., 1992). This conversion factor is referred to as the characterization factor Q for the pollutant and impact category considered (Equation 20.1; Heijungs and Hofstetter, 1996).

$$S_j = \sum_i \sum_x Q_{j,x,i} \times M_{x,i} \qquad (20.1)$$

where S_j is the impact score for impact category j per functional unit (e.g., multi-substance PAF/year for ecotoxicity); $Q_{j,x,i}$ is the characterization factor for impact category j of substance x emitted to compartment i (e.g., multisubstance PAF/year/kg for ecotoxicity); and $M_{x,i}$ is the emission of substance x to compartment i per functional unit (kg).

Here, it should be stressed that the summation of emissions over space and time in the life cycle inventory implies that in LCA only *potential* impacts for a "generic environment" can be assessed. As a consequence, it is only possible in LCA to focus on the total *mass* of releases to the environment. This is a major methodological difference compared with risk assessment, which estimates the actual risk of predicted *concentrations* of chemicals in the environment.

Commonly applied sets of characterization factors at the impact level are human toxicity potentials (Hertwich et al., 1998; Huijbregts et al., 2000). At the damage level, human damage factors, based on the concept of disability-adjusted life years (DALYs), are often used in LCIA (Hofstetter, 1998).

In most cases, both fate and effects of a substance are taken into account in the calculation of the characterization factor (Jolliet, 1996; Nichols et al., 1996):

$$Q_{j,x,i} = F_{j,x,i} \times E_{j,x} \qquad (20.2)$$

where $F_{j,x,i}$ is the fate and exposure factor for impact category j of substance x emitted to compartment i (e.g., mg/m^{-3}/year/kg for ecotoxicity); and $E_{j,x}$ is the effect factor for impact category j of substance x (e.g., multisubstance PAF/mg/m^3 for ecotoxicity).

Fate factors for substances causing ecotoxicological impacts are commonly calculated with environmental multimedia fate models, such as Simplebox (Brandes et al., 1996) and Caltox (McKone, 1993). As will be discussed in detail in the next section, information on species sensitivity distributions (SSD) can be very useful in the calculation of ecotoxicological effect factors (impact level) and may serve as input for the calculation of effect factors toward ecosystem quality (damage level).

The last (optional) step of the impact assessment is the calculation of an environmental index by aggregation of the impact categories. This can be done by attributing weighting factors to the different impact categories. The final phase in an LCA study is the interpretation of the results from the previous three steps, to draw conclusions and to formulate recommendations for decision makers.

20.3 ECOTOXICITY IN LCA

20.3.1 INTRODUCTION

As stated before, characterization factors are used for a weighted aggregation of emissions coming from the inventory analysis (Equation 20.1). Various sets of characterization factors have been proposed for the aggregation of pollutants causing ecotoxicological impacts. Until recently, the use of characterization factors for ecotoxicity based on the marginal change of the hazard index (PEC/PNEC ratio) was generally preferred in LCA (Guinée and Heijungs, 1993; Guinée et al., 1996; Hauschild et al., 1998; Huijbregts et al., 2000). In these hazard index calculations, the effect factor (PNEC) for a substance is generally defined as the 5th percentile of its SSD[NOEC]. The problem with this approach is, however, that the hazard index may be a poor basis for the aggregation of pollutants. The same marginal change in the hazard index of two or more substances may still reflect an unequal change in ecotoxicological impacts, because the shape and actual working point on the SSD curves are not considered in the hazard index calculations. Therefore, characterization factors that allow a more proper aggregation of pollutants causing ecotoxicity, would increase the validity of LCA. A candidate is the PAF, reflecting the fraction of all species that is exposed above its NOEC as based on SSD[NOEC] values (Klepper et al., 1998). In this approach the marginal change in a combination of effects of various compounds being present simultaneously can be accommodated.

Two possible approaches for implementation of the PAF concept in LCA will be reviewed and used. The first approach follows the PAF calculation rules of Hamers et al. (1996a), and the second approach was recently used by Goedkoop and Spriensma (1999). The main difference between the two approaches is that following Hamers et al. (1996a) the marginal change in the multisubstance PAF is calculated by combining noninteractive concentration and response addition rules of calculus, whereas Goedkoop and Spriensma (1999) implicitly start from noninteractive concentration addition rules of calculus. For 33 substances, the differences obtained between these two approaches are shown for freshwater aquatic ecotoxicity.

20.3.2 COMBINED CONCENTRATION AND RESPONSE ADDITION

The concept of multisubstance PAF as an approach for calculating the toxic impact of mixtures of chemicals on sets of species was first introduced by Hamers et al. (1996a). They assumed that (1) concentration addition is applicable for hydrophobic, chemically inert organic substances (i.e., narcotics), (2) response addition is applicable for all other compounds, and (3) no correlation in sensitivities or interaction of effects between substances exist, and calculated the combined PAF of all toxic substances (msPAF) from the combined fraction affected of narcotic chemicals, $PAF_{\Sigma narcotics}$, and the fractions affected of individual chemicals, PAF_x, by

$$msPAF_e = 1 - \left(1 - PAF_{\Sigma narcotics,e}\right) \times \prod_x \left(1 - PAF_{x,e}\right) \qquad (20.3)$$

where $msPAF_e$ is the msPAF in ecosystem e (e.g., soil, water, sediment); $PAF_{\Sigma narcotics,e}$ is the PAF in ecosystem e for all narcotics after concentration addition; and $PAF_{x,e}$ is the PAF in ecosystem e of substance x with a specific mode of action. This model relates to the assignment of compounds with similar and dissimilar toxic modes of action (e.g., Verhaar et al., 1992), and the prediction of ecotoxicological impacts on the basis of calculation rules as derived from basic toxicology and pharmacology.

PAF$_{x,e}$ may be calculated with help of the cumulative logistic distribution function at a measured or calculated field concentration $C_{x,e}$ (Aldenberg and Slob, 1993):

$$\text{PAF}_{x,e} = \frac{1}{1+e^{-\left(\log_{10} C_{x,e}-\alpha_{x,e}/\beta_{x,e}\right)}} \tag{20.4}$$

where $\alpha_{x,e}$ is the location parameter, estimated by the sample mean of the $\log_{10}$-transformed species toxicity values, and $\beta_{x,e}$ is the scale parameter, estimated from the standard deviation of the log-transformed species toxicity values of substance x for ecosystem e (see also Chapters 4 and 5).

If the approach of marginal change is applied to ecosystem stress by toxic substances and substance x has a specific mode of action, the characterization factor $Q_{e,x,i}$ of substance x causing ecotoxicological impacts in ecosystem e after release to compartment i is the derivative of Equation 20.3:

$$
\begin{aligned}
Q_{ETe,x,i} &= \frac{\text{dmsPAF}_e}{dM_{x,i}} = \left(1-\text{PAF}_{\text{rest},e}\right) \times \frac{d\text{PAF}_{x,e}}{dM_{x,i}} \\[2mm]
&= \frac{\left(1-\text{PAF}_{\text{rest},e}\right) \times \left(1-\text{PAF}_{x,e}\right)}{\left(1-\text{PAF}_{x,e}\right)} \times \frac{d\text{PAF}_{x,e}}{dM_{x,i}} \\[2mm]
&= \frac{1-\text{msPAF}_e}{1-\text{PAF}_{x,e}} \times \frac{d\text{PAF}_{x,e}}{dM_{x,i}} \\[2mm]
&= \frac{1-\text{msPAF}_e}{1-\text{PAF}_{x,e}} \times \frac{\gamma_{x,e}}{10^{\alpha_{x,e}}} \times \frac{dC_{x,e}}{dM_{x,i}}
\end{aligned}
\tag{20.5}
$$

in which dmsPAF$_e$/$dM_{x,i}$ is the marginal change in the multicompound PAF caused by a marginal change in the emission of substance x to compartment i (e.g., air, water, agricultural soil) (in msPAF/year/kg); PAF$_{\text{rest},e}$ is the PAF caused by all substances other than substance x in ecosystem e; $\gamma_{x,e}$ is the slope factor at the working point of the PAF curve of substance x in ecosystem e (e.g., aquatic, sediment, terrestrial); and $dC_{x,e}$/$dM_{x,i}$ is the marginal change in the concentration of substance x in ecosystem e due to a marginal change in the emission of substance x to compartment i (mg/m^{-3}/year/kg). Note that if substance x is a narcotic compound, the marginal change of the multicompound PAF is calculated by changing PAF$_{x,e}$ in PAF$_{\Sigma\text{narcotics},e}$ in Equation 20.5.

As can be derived from Equation 20.5, application of Hamers' rules for calculating marginal PAF increase requires knowledge of (1) the α_x values of the chemicals, (2) the ambient msPAF, (3) the β_x values and ambient concentrations of substances with a specific mode of action; and (4) the β value and ambient PAF for narcotics.

20.3.3 Concentration Addition Only

An alternative approach for calculating the marginal PAF increase is applied by Goedkoop and Spriensma (1999). Goedkoop and Spriensma started from a slight

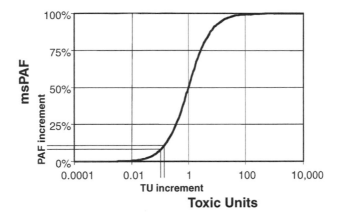

FIGURE 20.3 The dose–response relationship of the ambient mixture of substances assumes a logistic curve. Concentrations of single substances are standardized to toxic units (TU), representing equitoxic concentrations of chemicals with simple similar action (noninteractive concentration addition).

modification of Hamers' approach. They not only applied the rules of concentration addition to the narcotic chemicals, but they also generalized it to all chemicals with similar toxic modes of action. Taking this approach, Goedkoop and Spriensma reasoned that, as a general case, all possible modes of action are represented in the environment already, so that addition of any chemical would imply concentration addition to a mechanism already present. The difficulty remains that one needs to decide which β value to apply in calculating the concentration-additive PAF increase, and how much of this toxic activity (in terms of toxic units) is present already. This information about the ambient toxic activity is usually not available. Goedkoop and Spriensma's way around this is to use one (weighted average) β value to calculate concentration additively the marginal PAF increase resulting from addition of a chemical to the environment. This approach effectively uses one general PAF curve, expressed in toxic units, with one average scale factor β (Figure 20.3). Adopting this concentration-additive calculation procedure implicitly assumes that all chemicals considered, ambient as well as added chemicals, toxicologically act similarly. Consequently, the characterization factor $Q_{ETe,x,i}$ is a function of the α_x values of the chemicals, the chosen standard β value, and the chosen working point on the generalized PAF curve, and Equation 20.5 simplifies to

$$Q_{ETe,x,i} = \frac{\mathrm{dmsPAF}_e}{dM_{x,i}} = \frac{\gamma_e}{10^{\alpha_{x,e}}} \cdot \frac{dC_{x,e}}{dM_{x,i}} \qquad (20.6)$$

in which γ_e is the slope factor at the working point of the overall PAF curve for ecosystem e.

20.3.4 COMPARISON OF BOTH APPROACHES

Empirical support for the hypothesis that effects at low concentrations of hydropho-bic organic substances with specific modes of toxic action and metals can be approx-imated assuming a concentration-additive rule of calculus has indeed been reported (Könemann, 1981; Hermens and Leeuwangh, 1982; De Zwart and Slooff, 1987; Deneer et al., 1988a; Enserink et al., 1991; Calamari and Vighi, 1992; 1993; Bro-derius et al., 1995; Khalil et al., 1996; Weltje, 1998). For organic substances, it has been hypothesized that, in a large mixture of differently acting compounds, the concentrations of the individual compounds may be so low that their specific toxic effects per compound separately will not be apparent, while the unspecific toxicity related to the hydrophobicity and narcotic effects of the mixture may remain (Köne-mann, 1981; Hermens and Leeuwangh, 1982; Deneer et al., 1988a; McCarty and Mackay, 1993). On the other hand, concentration-additive effects are generally obtained in short-term single-species laboratory studies with mortality as endpoint. First of all, experimental results obtained via single-species toxicity testing may not be copied simply to the multiple species situation. In contrast to the single-species tests, concentration addition no longer only refers to the mode of action of the different compounds, but also to the response of the different species. This means that "same mode of action" should be applicable not only between different com-pounds, but also between different species for the same compound (Hamers et al., 1996a). Furthermore, substances for which the acute toxic action is simple similar do not necessarily have simple similar chronic actions (Könemann, 1981), and an increase in the sensitivity of the endpoint chosen may lead to joint toxicity deviating from concentration additivity (Hermens et al., 1984; Deneer et al., 1988b).

The response addition rules proposed by Hamers et al. (1996a) are, however, also not without problems. The practical application for a large number of substances may be difficult, because background concentrations of individual substances in the environment and toxicity data needed in the calculation of substance-specific dose–response curves will be very difficult to obtain. Although the scale parameter β of substances with no (chronic) species toxicity data available may be estimated on the basis of acute toxicity data and/or information of the working mechanism of the substance involved (see Chapter 8), representative data on environmental con-centrations are far more difficult to obtain. This is particularly the case for LCA for which dimensions of time and space in the inventory data are currently lacking.

At present, it appears that there are no data providing full theoretical and empirical support for a final preference of one of the two rules of calculus. As there seems insufficient scientific evidence, the question arises whether the two approaches give approximately the same characterization factors. Therefore, as an example, character-ization factors (Q values) for freshwater ecosystems of 33 substances directly emitted to fresh water were calculated using the calculation rules of Hamers et al. (1996a) and Goedkoop and Spriensma (1999), respectively. The steady-state version of the multi-media fate, exposure, and effects model USES-LCA, developed by Huijbregts et al. (2000), was modified for this purpose. In USES-LCA, both emissions to and impacts in the freshwater compartment are modeled on the West European scale. Table 20.1 shows the toxicity data, environmental concentrations, and background PAFs applied

TABLE 20.1
Toxicity Data, Background Concentrations, and PAFs Needed in the Calculation of Characterization Factors of 33 Substances for the Freshwater Compartment

Nr.	Substance	α (log mg/m³)	β (–)	$C_{dissolved}$ (mg/m³)	$PAF_{x,fw}$ (–)	dC/dM (mg/m³/year/kg)	Ref.
1	1,2,3-Trichlorobenzene[a]	3.11	0.39	n.r.[b]	$1.5 \cdot 10^{-2}$	$2.4 \cdot 10^{-8}$	1
2	1,2,4-Trichlorobenzene[a]	3.13	0.39	n.r.	$1.5 \cdot 10^{-2}$	$2.5 \cdot 10^{-8}$	1
3	1,3,5-Trichlorobenzene[a]	2.93	0.39	n.r.	$1.5 \cdot 10^{-2}$	$2.6 \cdot 10^{-8}$	1
4	2,3,7,8-TCDD	−2.31	0.68	$1.4 \cdot 10^{-7}$	$1.3 \cdot 10^{-3}$	$1.9 \cdot 10^{-8}$	1
5	2,4-D	3.31	0.76	$1.0 \cdot 10^{-2}$	$9.2 \cdot 10^{-4}$	$3.7 \cdot 10^{-7}$	2, 3
6	Atrazine	1.88	0.47	$1.1 \cdot 10^{-2}$	$2.8 \cdot 10^{-4}$	$1.3 \cdot 10^{-6}$	2, 3
7	Azinphos-methyl	0.08	0.64	$6.0 \cdot 10^{-3}$	$2.7 \cdot 10^{-2}$	$5.7 \cdot 10^{-8}$	2, 3
8	Benzene[a]	3.29	0.39	n.r.	$1.5 \cdot 10^{-2}$	$2.0 \cdot 10^{-8}$	1
9	Benzo[a]pyrene[a]	0.91	0.39	n.r.	$1.5 \cdot 10^{-2}$	$1.2 \cdot 10^{-7}$	1
10	Cadmium	1.18	0.55	$4.0 \cdot 10^{-2}$	$9.1 \cdot 10^{-3}$	$4.8 \cdot 10^{-8}$	2, 4
11	Carbendazim[a]	1.94	0.39	n.r.	$1.5 \cdot 10^{-2}$	$7.1 \cdot 10^{-7}$	3, 5
12	Copper	1.26	0.41	2.3	$1.0 \cdot 10^{-1}$	$1.2 \cdot 10^{-7}$	2, 4
13	Dibutylphtalate	2.91	0.46	$2.8 \cdot 10^{-1}$	$5.4 \cdot 10^{-4}$	$7.4 \cdot 10^{-8}$	1
14	Dichlorvos	1.56	0.91	$4.0 \cdot 10^{-3}$	$1.3 \cdot 10^{-2}$	$8.0 \cdot 10^{-9}$	2, 3, 5
15	Diethylhexylphtalate	2.29	0.53	$1.2 \cdot 10^{-1}$	$2.3 \cdot 10^{-3}$	$1.9 \cdot 10^{-8}$	1
16	Diuron	1.42	0.57	$1.0 \cdot 10^{-2}$	$2.5 \cdot 10^{-3}$	$3.7 \cdot 10^{-7}$	2, 3
17	DNOC	2.95	0.56	$4.0 \cdot 10^{-3}$	$7.1 \cdot 10^{-6}$	$2.2 \cdot 10^{-7}$	2, 3
18	Endosulfan	1.45	0.81	$1.0 \cdot 10^{-3}$	$4.1 \cdot 10^{-3}$	$5.1 \cdot 10^{-8}$	2
19	Fentin-acetate	0.37	0.46	$4.0 \cdot 10^{-3}$	$2.4 \cdot 10^{-3}$	$1.2 \cdot 10^{-7}$	2, 3, 5
20	Fluoranthrene[a]	2.10	0.39	n.r.	$1.5 \cdot 10^{-2}$	$2.9 \cdot 10^{-7}$	1
21	Hexachlorobenzene[a]	1.37	0.39	n.r.	$1.5 \cdot 10^{-2}$	$3.4 \cdot 10^{-8}$	1
22	Lead	2.17	0.38	0.1	$2.4 \cdot 10^{-4}$	$9.8 \cdot 10^{-9}$	2, 4
23	Lindane	1.23	0.84	$4.0 \cdot 10^{-3}$	$1.3 \cdot 10^{-2}$	$6.1 \cdot 10^{-7}$	2
24	Mecoprop	3.91	0.57	$4.0 \cdot 10^{-2}$	$9.0 \cdot 10^{-6}$	$1.4 \cdot 10^{-7}$	2, 3, 5
25	Mercury	0.80	0.48	$1.0 \cdot 10^{-2}$	$2.9 \cdot 10^{-3}$	$3.7 \cdot 10^{-8}$	4, 6
26	Metabenzthiazuron	2.47	0.70	$1.1 \cdot 10^{-2}$	$1.8 \cdot 10^{-3}$	$8.6 \cdot 10^{-7}$	2, 3, 5
27	Mevinphos	0.69	0.57	$2.0 \cdot 10^{-3}$	$2.6 \cdot 10^{-3}$	$8.7 \cdot 10^{-8}$	2, 3, 5
28	Monolinuron	1.52	0.51	$3.0 \cdot 10^{-3}$	$3.6 \cdot 10^{-4}$	$7.2 \cdot 10^{-7}$	2, 3, 5
29	Parathion-ethyl	−0.05	0.85	$3.0 \cdot 10^{-3}$	$5.2 \cdot 10^{-2}$	$2.1 \cdot 10^{-7}$	2, 3
30	Pentachlorophenol	0.83	0.46	$2.0 \cdot 10^{-3}$	$4.7 \cdot 10^{-4}$	$2.3 \cdot 10^{-7}$	2
31	Simazine	1.98	0.45	$1.0 \cdot 10^{-2}$	$1.4 \cdot 10^{-4}$	$3.5 \cdot 10^{-7}$	2, 3, 5
32	Toluene[a]	3.28	0.39	n.r.	$1.5 \cdot 10^{-2}$	$2.0 \cdot 10^{-8}$	1
33	Zinc	1.97	0.38	7.1	$5.0 \cdot 10^{-2}$	$5.6 \cdot 10^{-8}$	2, 4

[a] Narcotic substances with $\beta_{fw} = 0.39$ (Chapter 8) and $PAF_{\Sigma narcotics,fw} = 1.5\%$ (derived from Van de Meent, 1999).

[b] n.r. = not relevant.

References: (1) Bakker and Van de Meent (1997); (2) Klepper and Van de Meent (1997); (3) Crommentuijn et al. (1997a); (4) Crommentuijn et al. (1997b); (5) Van de Meent (1999); (6) Slooff et al. (1995).

TABLE 20.2
Calculation of the Freshwater Ecotoxicity Potential of Cadmium Emitted to Fresh Water Using Either Combined Concentration and Response Addition or Concentration Addition Only

Calculation Step	Unit	Concentration and Response Addition	Concentration Addition Only
Emission to fresh water	kg/year	1	1
Concentration increase in fresh water	mg/m^3	$4.79 \cdot 10^{-8}$	$4.79 \cdot 10^{-8}$
Location parameter $\alpha_{fresh\ water}$	$\log_{10}$(mg/m^3)	1.18	1.18
Slope factor $\gamma_{fresh\ water}$ at working point	—	2.71	0.50
$(1 - \text{msPAF})/(1 - \text{PAF})$	—	0.71	n.r.
msPAF increase	msPAF/year/kg	$6.05 \cdot 10^{-9}$	$1.57 \cdot 10^{-9}$

n.r. = not relevant.

in the calculation of the characterization factors (i.e., Equations 20.5 and 20.6). Because of lack of representative multicompound PAF information and measurements of concentrations on the West European scale, an msPAF of 30% and measured concentrations representative for Dutch regional surface waters were used in the calculations (Klepper and Van de Meent, 1997; Bakker and van de Meent, 1997; van de Meent, 1999). Substance-specific parameters needed in the fate analysis were taken from Huijbregts (1999). A value for β of 0.4 for mixture toxicity, needed in the "concentration addition only" calculations, was taken from Goedkoop and Spriensma (1999). As an example, Table 20.2 shows for both methods the calculation steps of the freshwater ecotoxicity potential of cadmium emitted to fresh water.

As can be seen in Figure 20.4 the calculation rules of Goedkoop and Spriensma (1999) give generally lower marginal changes in the msPAF compared with the approach based on calculation rules of Hamers et al. (1996a; up to 2.2 orders of magnitude) and in turn may influence the outcomes of LCA. The reason for the results obtained is that the $\beta_{x,fw}$ of the majority of the 33 substances is considerably higher than 0.4 (i.e., the β assumed by Goedkoop and Spriensma, 1999, for the overall PAF curve). This leads to higher marginal msPAF changes in the range of the working points on the individual PAF curves (PAF$_{x,fw} \leq 0.1$), compared with marginal changes found with the overall PAF curve. In contrast, for substances with a β considerably smaller than 0.4, the marginal msPAF change based on calculation rules of Hamers et al. (1996a) will be smaller in the range of the working points on the individual PAF curves than the marginal multicompound PAF change based on the approach of Goedkoop and Spriensma (1999).

20.4 TOWARD ECOSYSTEM DAMAGE IN LCA

Interpretation of the NOEC-based PAF of species, msPAF^{NOEC}, as a measure of ecological risk (Van de Meent, 1999) has opened possibilities for extension to the

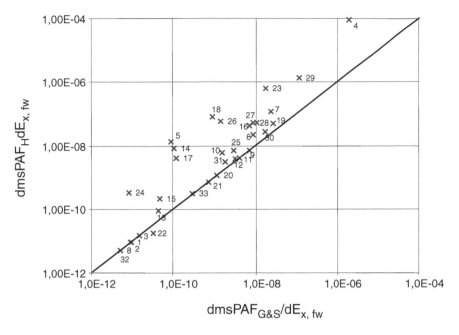

FIGURE 20.4 Comparison of freshwater ecotoxicity potentials after direct emission to fresh water using the approach from Goedkoop and Spriensma (1999; $dmsPAF_{G\&S}/dE_{x,fw}$) and the approach based on Hamers et al. (1996a; $dmsPAF_H/dE_{x,fw}$).

level of ecosystem damage and, hence, to the damage level. This is the most important advantage of the PAF-based characterization factors compared with the PEC/PNEC-based characterization factors. Ultimately, the importance of being able to calculate the marginal change in ecosystem damage, caused by a wide range of chemical and physical stressors, such as land use, desiccation, eutrophication, acidification, and toxicity, is of primary interest in LCA. For this purpose, the PDF would be desirable in LCA as a damage indicator of ecological impact. The Potentially Disappeared Fraction of species (PDF) can be interpreted as the fraction of species that has a high probability of no occurrence in a region due to unfavorable conditions (Goedkoop and Spriensma, 1999). PDF-like damage indicators for different stressors could then be considered as independent effects, and aggregated according to the rules for response addition, so that

$$msPDF_e = 1 - \prod_x \left(1 - PDF_{x,e}\right) \qquad (20.7)$$

and

$$Q_{eco,x,i} = \frac{dmsPDF_e}{dS_{x,i}} = \frac{1 - msPDF_e}{1 - PDF_{x,e}} \cdot \frac{dPDF_{x,e}}{dS_{x,i}} \qquad (20.8)$$

where msPDF_e is the multistress PDF in ecosystem e (e.g., soil, water, sediment); $S_{x,i}$ is the introduction of stressor x to compartment i (e.g., air, fresh water, agricultural soil) in kg/year (e.g., toxic substances), m^3/year (e.g., use of groundwater), or m^2/year (e.g., land use); and $\mathrm{PDF}_{x,e}$ is the PDF in ecosystem e of stressor x.

Empirical information on the probability of species occurrence, as used in the assessment of ecosystems effects due to acidification, eutrophication, and desiccation (Latour and Reiling, 1993; Latour et al., 1994; Kros et al., 1995), could directly be used in the calculation of characterization factors for ecosystem damage. As an alternative, the msPDF for ecotoxicity could be derived from msPAF, using such simple linear relationships as

$$\mathrm{msPDF}_e = K_{\mathrm{PAF} \rightarrow \mathrm{PDF}} \cdot \mathrm{msPAF}_e \qquad (20.9)$$

where $K_{\mathrm{PAF} \rightarrow \mathrm{PDF}}$ is a proportionality constant for converting PAF into PDF.

It should be kept in mind that the NOEC-based PDF provides no information other than that a certain proportion of the species is exposed to concentrations at which some population-relevant toxic effect (increased mortality, decreased reproduction or growth, etc.) may occur. Although it seems reasonable to expect a relationship between the magnitude of this proportion and the impact on ecosystems in terms of presence or absence of species, empirical evidence for the existence of such a direct relationship between toxic pressure and ecosystem damage is, however, scarce. Recently, Vonk et al. (2000) have coupled empirical observations on terrestrial vegetation with model-calculated toxic pressures from heavy metals. They derived a multivariate regression model that explained 30 to 40% of the spatial and temporal variance in occurrence of plant species in the Netherlands from known or calculated variances in soil pH, nitrogen availability, groundwater table, and PAFs. In this study, Vonk et al. found a statistically significant correlation between toxic pressure and occurrence of plant species; they concluded that increased concentrations of copper, cadmium, and zinc in soils have been responsible for a 10% decrease in occurrence of terrestrial plants in the Netherlands between 1950 and 1995. More evidence of this kind is necessary to decide whether this relationship between toxic pressure and species occurrence holds more generally. To our knowledge, no such work has been reported so far.

Empirical relationships as reported by Vonk et al. (2000) could provide a basis for deriving a proportionality constant $K_{\mathrm{PAF} \rightarrow \mathrm{PDF}}$. However, Vonk et al. have not attempted to generalize and simplify their findings into a format that can be applied directly to LCA. Klepper et al. (1999), considering disappearance of sensitive species or genotypes as the negative effect of interest, suggested that $K_{\mathrm{PAF} \rightarrow \mathrm{PDF}}$ may be as great as 1 for persistent toxic substances, such as metals. In the absence of further empirical evidence, Goedkoop and Spriensma (1999) have proposed to use the value of 0.1 for $K_{\mathrm{PAF} \rightarrow \mathrm{PDF}}$. Still, substantial research needs to be carried out before general conversion rules from PAF to PDF can be applied (Traas et al., 1999).

20.5 CONCLUSION

In the calculation of impact scores for ecotoxicity in LCA (impact level), characterization factors based on the marginal change in the msPAF^{NOEC} may be preferred compared with characterization factors based on the marginal change in the hazard index. Marked differences are found between marginal msPAF changes for individual substances based on calculation rules of Hamers et al. (1996a) and Goedkoop and Spriensma (1999). As the marginal change in ecosystem damage due to a marginal change over a wide range of chemical and physical stressors is of primary interest in LCA (damage level), a further conversion of the msPAF^{NOEC} to the msPDF seems necessary. Although this conversion is possible for both calculation procedures, substantial research still is needed to underpin the validity of both PAF approaches and to guarantee well-validated conversion rules for PAF to PDF.

Section IV

Evaluation and Outlook

This final section presents an overview of the current field and of options for future developments. The concepts and data presented in the preceding chapters and in the literature have been analyzed in view of the criticisms of SSDs that have been voiced in the past, and during the Interactive Poster Session that was held in 1999 at the 20th Annual Meeting of the Society of Environmental Toxicology and Chemistry in Philadelphia, Pennsylvania. In the concluding outlook chapter, all preceding chapters have been reconsidered to determine the prospects for resolving the criticisms and problems of SSDs. Some of these issues, those that seem amenable to solution, have been extrapolated to the near future, to stimulate discussion and thought on further SSD evolution.

21 Issues and Practices in the Derivation and Use of Species Sensitivity Distributions

Glenn W. Suter II, Theo P. Traas, and Leo Posthuma

CONTENTS

1-56670-578-9/02/$0.00+$1.50
© 2002 by CRC Press LLC

Abstract — As is clear from the preceding chapters, species sensitivity distributions (SSDs) have come to be commonly used in many countries for setting environmental quality criteria (EQCs) and assessing ecological risks (ERAs). However, SSDs have had their critics, and the critics and users of SSD models have raised conceptual and methodological concerns. This chapter evaluates issues raised in published critiques of SSDs (e.g., Forbes and Forbes, 1993; Hopkin, 1993; Smith and Cairns, 1993; Chapman et al., 1998), in a session at the 1999 SETAC Annual Meeting (Appendix A), and in the course of preparing this book. The issues addressed include conceptual issues, statistical issues, the utility of laboratory data, data selection, treatment of data, selection of protection levels, and the validity of SSDs. When considering these issues, one should be aware that the importance and implications of these issues may depend on the context and use of an SSD. The consequences of this evaluation for further development of SSDs are elaborated in Chapter 22.

21.1 THE USES OF SSDS

Models of species sensitivity distributions (SSDs) with respect to a toxic substance can be used in two conceptually distinct ways (Chapters 1 and 4). The first use is to estimate the concentration that affects a particular proportion of species, the HC_p. This is the older so-called inverse use, and is employed in the derivation of environmental criteria. The second use is the forward use of SSDs, which estimates the potentially affected fraction (PAF) of species, or the probability of effects on a species (PES) at a given concentration.

The PAF or PES can be calculated for single chemicals and these values can be aggregated to a single value for mixtures of chemicals. In any of these uses, it is assumed that protection of species and communities may be assured by considering the distribution of sensitivities of species tested individually. Although some regulatory agencies have embraced the concept of risk embedded in the use of SSDs (Chapters 2 and 3) the assumption that SSD-derived criteria are protective is an open question. The definition and interpretation of risk as defined previously (Suter, 1993; Chapters 15 through 17) play a major part in the interpretation of the outcome of SSD methods, as discussed below.

21.1.1 SSDs FOR DERIVATION OF ENVIRONMENTAL QUALITY CRITERIA

As discussed in the introductory chapters, SSDs were developed to derive criteria for the protection of ecological entities in contaminated media. That is, criteria are set at an HC_p or an HC_p modified by some factor.

Such criteria may be interpreted as, literally, levels that will protect $1 - p\%$ of species or simply as consistent values that provide reasonable protection from unspecified effects. If the criteria are interpreted as protecting $1 - p\%$ of species from some effect with defined confidence, then they are potentially subject to scientific confirmation. Some studies have attempted to confirm SSD-based quality criteria in the last decade by comparing them to contaminant effects in the field (Chapter 9 and Section 21.8.2). However, if criteria derived from SSDs are interpreted simply as reasonable and consistent values, their utility is confirmed in that sense by a record of use that has been politically and legally acceptable. That is, if they were not reasonable and consistent, they would be struck down by the courts or replaced due to pressures from industry or environmental advocacy groups.

The U.S. Environmental Protection Agency (U.S. EPA) National Ambient Water Quality Criteria and the Dutch Environmental Risk Limits for water, soil, and sediment have achieved at least the latter degree of acceptance. A general acceptance of the SSD methodology is not necessarily negated by challenges incidentally posed to individual SSD-based criteria such as the challenge of the environmental quality criterion (EQC) for zinc by European industries (RIVM/TNO, 1999).

The general acceptance of SSD-derived criteria should not suggest a uniformity of methods around the globe. Adopted methods for deriving EQCs vary in many ways among countries, including the choice and treatment of input data, statistical models, and choice of protection level (Chapters 10 through 20; Roux et al., 1996; Tsvetnenko, 1998; Vega et al., 1997; Tong et al., 1996; ANZECC 2000a,b; etc.). One homology is that SSDs defined by unimodal distribution functions are the basis for deriving EQC in several countries. Polymodality of the data may, however, occur for compounds with a taxon-specific toxic mode of action (TMoA) (Section 21.5.5), and Aldenberg and Jaworska (1999) suggested polymodal model for EQC derivation. The HCp values in the protective range of use (e.g., 5th percentile) estimated with this model were shown to be numerically fairly robust toward deviations from unimodality in some selected cases (Aldenberg and Jaworska, 1999). For compounds with a specific TMoA, it can be argued that the variance in species sensitivity as estimated from the total data set is larger and not representative of the variance of the target species. This would lead to overprotective criteria since the HC_p is very sensitive to this variance. On the other hand, it can be argued that the total variance may lead to more protective criteria, providing some safety against unknown or unexpected side effects. Conclusive numerical data remain to be presented in this matter. On non-numerical grounds, but driven by considering the assessment endpoints, the estimate of a specific HC_p for a target taxon may be preferred over an HC_p based on the total data set (Chapter 15).

The diversity of operational details and the invention of new approaches like polymodal statistics suggest that discussions will proceed in the use of SSD for deriving environmental quality standards. The history of SSD use (Chapters 2 and 3) teaches that it is important to distinguish clearly in the discussion between issues related to assessment endpoints, methodological details of SSDs, and choices within the SSD concept related to the policy context.

21.1.2 SSDs for Ecological Risk Assessment

The goal of risk assessment is to estimate the likelihood of specified effects such as death of humans or sinking of a ship. The growing use of SSDs in ecological risk assessments and the diverse terminology used so far (Chapter 4; Chapters 15 through 20) necessitate a sharp definition of the outcome of SSDs in terms of predicted risks for specific ecological endpoints. Also, unlike criteria, risk assessments must deal with real sites, which requires modeling the effects of mixtures. SSDs have been incorporated into formal ecological risk assessment methods developed by the Water Environment Research Foundation (WERF, Parkhurst et al., 1996), the Aquatic Risk Assessment and Mitigation Dialog Group (ARAMDG, Baker et al., 1994), and the Ecological Committee on FIFRA Risk Assessment Methods (ECOFRAM, 1999a,b).

21.1.2.1 Assessment Endpoints and the Definition of Risk

The appropriateness of SSDs in risk assessment depends on the endpoints of the assessment as well as the use of the SSDs in the inferential process. Assessment endpoints are the operational definition of the environmental values to be protected by risk-based environmental management (Suter, 1989; U.S. EPA, 1992). They consist of an ecological entity such as the fish assemblage of a stream and a property of that entity such as the number of species. Assessment endpoints are estimated from numerical summaries of tests (i.e., test endpoints such as LC_{50} values) or of observational studies (e.g., catch per unit effort). The extrapolation from these measures of effect to an assessment endpoint is performed using a model such as an SSD.

If SSDs are used inferentially to estimate risks to ecological communities, it is necessary to define the relationship of the SSD to the assessment endpoint, given the input data (test endpoints). Currently, two types of test endpoints are most often used, acute LC_{50} values* and chronic no-observed-effects concentrations (NOECs) or chronic values (CVs), which yield acute (SSD^{LC50}) and chronic (e.g., SSD^{NOEC}) SSDs with different implications.

The acute LC_{50} values are based on mortality or equivalent effects (i.e., immo-bilization) on half of exposed organisms. Hence, this test endpoint implies mass mortality of individuals. At the population level, it could be interpreted as approx-imately a 50% immediate reduction in abundance of an exposed population. As discussed in Chapter 15, some populations recover rapidly from this loss, but other populations are slow to recover. The immediate consequences of mass mortality are, however, often unacceptable in either case. Hence, if such SSDs are considered to be estimators of the distribution of severe effects among species in the field, then the acute SSDs (SSD^{LC50}) may be considered to predict the proportion of species experiencing severe population reductions following short-term exposures. An example

---

* For brevity, we use LC_{50} to signify both acute LC_{50} and EC_{50}.

of the relationship between SSD and an acute assessment endpoint is shown in Chapter 9, where SSD^{LC50} values for chlorpyrifos are compared with SSDs for arthropod density in experimental ditches. In this specific example, the SSD model seemed to adequately predict the assessment endpoint "arthropod density" in acute exposures. This shows that SSDs based on acute toxicity data for toxicants with a defined TMoA can adequately predict acute changes in appropriate measures of effect. These SSDs likely predict *that* something will happen, and also (approximately) *what* (a degree of mortality).

The situation is more difficult for chronic assessments. As discussed below (Section 21.3.1), the conventional chronic endpoints represent thresholds for statistical significance and have no biological interpretation. Assessors commonly assume that they represent thresholds for significant effects (Cardwell et al., 1999), but that assumption is not supportable. Conventional chronic endpoints correspond to a wide range of effects on populations (Barnthouse et al., 1990). Hence, the relationship of chronic SSDs to measures of effects in the field is less clear than for acute SSDs. Further, ecosystem function and recovery are not embraced in conventional chronic tests or in the SSD models that utilize them. It is important to apply SSDs to endpoints for which they are suited, and not to overinterpret their results. The chronic SSDs may simply predict the proportion of species experiencing population reductions ranging from slight to severe following long-term exposures.

Ecological risk assessors have tended to focus on techniques and to avoid the inferential difficulties of defining and estimating assessment endpoints. For example, the aquatic ECOFRAM (1999a) report provides methods for aquatic ecological risk assessment that rely heavily on SSDs but does not define the assessment endpoints estimated by those methods. Rather, it discusses population and ecosystem function and suggests that they will be protected when 90% of species are protected from effects on survival, development, and reproduction. Similar ambiguities occur in the ARAMDG and WERF risk assessment methods (e.g., Baker et al., 1994; Parkhurst et al., 1996). The ambiguity in the relationship of SSDs to assessment endpoints is due in part to the lack of guidance from the regulatory agencies. The U.S. EPA has not defined the valued environmental attributes that should serve as assessment endpoints (Troyer and Brody, 1994; Barton et al., 1997). The risk managers must identify the target and then risk assessors can design models and select data to hit it. However, the U.S. EPA and other responsible agencies have been reluctant to be more specific than "protect the environment," "abiotic integrity," "ecosystem structure and function," or "ecosystem health." It is not surprising that risk assessors have tended to be equally vague when specifying what is predicted by SSD models.

The lack of a clear relationship of SSDs to assessment endpoints is less problematical if the goal of an assessment is simply comparison or ranking (e.g., Manz et al., 1999). For example, SSDs based on NOECs are used in the Netherlands for mapping regional patterns of relative risks (Chapter 16). In particular, the PAF[NOEC] was hypothesized to be a measure of the relative risk to the clear ecological endpoint, vascular plant diversity.

Risk characterization need not be based solely on SSDs, but on a weighing of multiple lines of evidence. In those cases SSDs may play a supporting role rather than serving as the sole estimator of risk (De Zwart et al., 1998; Hall and Giddings,

2000). In particular, effects may be estimated from biosurveys or field experiments and the laboratory data may indicate the particular chemicals that cause the effect. For example, in an assessment of risks to fish in the Clinch River, Tennessee, effects were estimated using survey data, the toxicological cause of the apparent effects was established from toxicity tests of ambient waters and biomarkers, and SSDs were used simply to establish the plausibility of particular contaminants as contributors to the toxicity (Suter et al., 1999). The assessment endpoint was a "reduction in species richness or abundance or increased frequency of gross pathologies." A 20% or greater change measured in the field or in toxicity tests of site waters was considered significant. The chronic SSDs for individual chemicals were considered reasonably equivalent to this endpoint, because chronic tests include gross pathologies (when they occur) and the chronic test endpoints correspond to at least 20% change in individual response parameters, which in combination, over multiple generations, may result in local population extinction (Suter et al., 1987; Barnthouse et al., 1990).

SSDs have been suggested as a key tool in a proposed formal tiered risk assessment scheme for contaminated soils, where multisubstance PAFs (msPAFs) functions in a "weight of evidence" approach, in which none of the parameters is able to present the whole "truth." In this context, the msPAF is considered along with bioassay and field inventory results (De Zwart et al., 1998), arraying them on a dimensionless 0 to 1 scale. When all results point in a similar direction, the investigations are ended at the lowest possible tier with a conclusion.

A risk-based approach using SSDs as one line of evidence may also be used to derive environmental criteria for specific sites. The guidelines for water quality in Australia and New Zealand recommend the use of bioassessment and toxicity tests of effluents or ambient media along with SSD-based trigger values to derive defensible regulatory values (ANZECC, 2000a).

Risk assessment approaches may also be used in the enforcement of criteria. The interpretation of criteria is usually binary (i.e., the criterion is or is not exceeded) or in terms of an exceedence factor (e.g., the concentration exceeds the criterion by 5 times). However, a more risk-based alternative would use an SSD to determine the increase in the number or proportion of species at risk as a result of exceeding the criterion (Knoben et al., 1998).

21.1.2.2   Ecological Risk Assessment of Mixtures

Because SSDs have historically been based on single-chemical toxicity tests, they have been criticized for not incorporating the combined effects of mixtures of chemicals (Smith and Cairns, 1993). Since mixtures are the rule rather than the exception in field conditions, this subject requires attention.

Since single-chemical test data are the major source of data to construct SSDs, methods have been developed to predict the joint risk of chemicals in a mixture (Chapters 16 and 17). They extend the SSD methodology with concepts from toxicology and pharmacology (Plackett and Hewlett, 1952; Könemann, 1981). This is technically feasible, since the units in which risks are quantified (PAFs, or similar expressions used in this book) are dimensionless. The resulting fraction of species exposed beyond test endpoint concentrations, given exposure from multiple chemicals, can thus (at least theoretically) be defined, and we propose the term "multisubstance-PAF" (msPAF) for this concept.

The ability to calculate msPAFs as measures of mixture risks relates to the classification of pollutants according to their TMoA (e.g., Verhaar et al., 1992; Vaal et al., 1997). For compounds with the same TMoA, concentration addition rules are applied subsequent to SSD analyses in various forms (Chapters 4, 16, and 17). For compounds with different modes of action the rule of response addition has been used (Chapter 16). Conceptually, the transfer of the toxicological models to the risk assessment context may need further investigation. First, the TMoA is defined in relation to specific sites of toxic action within species, but it may not be constant across species. For example, a photosynthesis inhibitor has a clear dominant TMoA in plants and algae, but it may simultaneously be a narcotic agent for species lacking photosynthesis capacities.

The numerical outcome of these approaches is determined by the algorithms to calculate PAFs for nonspecific and specific modes of action and for aggregation into msPAF. The algorithms encountered in this book have not as yet been rigorously tested for their conceptual soundness (e.g., application of toxicological principles to communities rather than to individuals) or for their predictive ability for specific species assemblages.

A drawback of calculating msPAF from measured concentrations of compounds is that often many compounds go unnoticed, since they are not in the standard measurement array, or their concentrations are below technical detection limits. Alternatively, an msPAF can be derived experimentally. An effluent, complex material, or contaminated ambient medium is tested at different dilutions (or concentration steps) with a sufficient number of species to derive an SSD for that mixture, so that nonidentified chemicals are also taken into account (Chapter 18). For example, an acute criterion was calculated for aqueous dilutions of petroleum, expressed as total petroleum hydrocarbons, using the U.S. EPA methodology (Tsvetnenko, 1998). Trends across time or space in risks from mixtures can be analyzed in this way, again most likely as a relative scaling of toxic stress.

In this experimental context, it has been observed (Slooff, 1983; Chapter 18) that SSDs from tests of complex mixtures generally have steeper slopes than the SSDs of the individual chemicals in the mixture (Figure 21.1). A probable cause is that the single chemicals in a complex cocktail of contaminants not only act as chemicals with a specific toxicity but also contribute to joint additive toxicity, when they are present below their threshold concentration (Hermens and Leeuwangh, 1982; Verhaar et al., 1995). This is often referred to as baseline toxicity. The results of the experimental study by Pedersen and Petersen (1996) seems to be in accordance with this theory. They observed that the standard deviation of a set of toxicity data for a set of five laboratory test species tended to decrease (i.e, the slope of the SSD, plotted as a cumulative distribution function, or CDF, would increase) with an increasing number of chemicals in the mixture, although the number of species in these experiments was small compared to many SSDs or species in field communities.

The relationships between the calculated and measured msPAFs and between these msPAFs and measures of community responses in the field are complicated

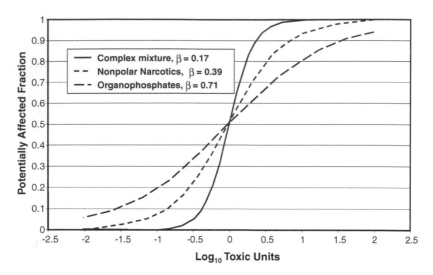

FIGURE 21.1 SSDs for single compounds and a large mixture, showing the steepness (β) of the CDF for the large mixture as compared to individual compounds. (Based on data from De Zwart, Chapters 8 and 18.)

and have not as yet been demonstrated clearly. Variance in the composition of the mixture may lead to varying effects on communities, depending on the dominant modes of action and the taxa present. Obviously, the relation between observed toxicity and the toxicity of mixtures predicted with SSDs requires further development of concepts and technical approaches, to yield outcomes beyond the level of relative measures of risks (Chapter 22).

21.1.3 Probability of Effects from SSDs

The criteria generated from SSDs and the risks estimated from SSDs (PAFs or PESs) are often described as probabilistic without defining an endpoint that is a probability (Suter, 1998a,b). This issue relates to the problem discussed above that the users of SSDs often do not clearly define what they are estimating when they use SSDs. The issue becomes important when communicating SSD-based results to risk managers or other interested parties.

When SSDs are used as models of the PES for an individual species, the sensitivity of the species is treated as a random variable. The species that is the assessment endpoint is assumed to be a random draw from the same population of species as the test species used to estimate the distribution (Van Straalen, 1990; Suter, 1993). The output of the model is evidently probabilistic, namely, an estimate of the PES on the endpoint species. For example, the probability of toxic effects on rainbow dace given an ambient concentration in a water body may be estimated from the distribution of the sensitivity of tested fish. As with the use of SSDs as models of communities (i.e., to calculate PAFs), uncertainties and variability are associated with estimating a PES. Given the parameter uncertainty due to sampling and sample size, a confidence interval for the PES can be calculated (Chapters 5 and 17; Aldenberg and Jaworska, 2000). That is, one could calculate the probability that the PES is as high as P_z. However, at present, none of the standard SSD-based assessment methods claims to estimate risks to individual species.

More commonly, SSDs are used to generate output that is not a probability. That is, when calculating HC_p, p is the proportion of the community that is affected, not a probability. Similarly, when calculating a PAF, the F is a fraction (or equivalently, a proportion) of the community affected, not a probability. If we estimate the distributions of these proportions, then we can estimate the probability of a prescribed proportion. Hence, one could estimate the probability that the PAF is a high as F_x or the HC_p is as low as C_y given variance among biotic communities, uncertainty due to model fitting, or any other source of variability or uncertainty. Parkhurst et al. (1996) describe a method to calculate the probability that the PAF is as large as F_x at a specified concentration given the uncertainty due to model fitting. The calculation of confidence intervals on HC_p to calculate conservative criteria is conceptually equivalent (Van Straalen and Denneman, 1989; Aldenberg and Slob, 1993).

The practical implications of this become apparent when considering the need to explain clearly the results of risk assessments to decision makers and interested parties (Suter, 1998b). One must explain that the probabilities resulting from various SSD-based methods are probabilities of some event with respect to some source of variance or uncertainty. In the explanation of SSD results, it should be clear that there are various ways by which the SSD approach may analyze sources of uncertainty and variability (see Chapters 4 and 5), and many sources that may be included or excluded. Hence, risk assessors should be clear in their own minds and in their writings concerning the endpoint that they intend to convey.

21.2 STATISTICAL MODEL ISSUES

21.2.1 SELECTION OF DISTRIBUTION FUNCTIONS AND GOODNESS-OF-FIT

The choice of distribution functions has been the subject of much debate in published critiques of the use of SSDs. Smith and Cairns (1993) objected to the fact that there is no good basis for selecting a distribution function when, as is often the case, the number of observations is small. Many users of SSDs simply employ a standard distribution that has been chosen earlier by a regulatory agency or by the founders of their preferred assessment method. This can lead to SSDs that badly fit the data. See, for example, Figure 21.2, or Aldenberg and Jaworska (1999). Although the use of a standard model can be defended as easy, consistent, and equitable, poor fits cast doubt on the appropriateness of the method. There are various alternatives for selecting distribution functions.

First, a chosen function may be considered acceptable based on failure to reject the null hypothesis that the distribution of the data is the same as the distribution defined by the function. Fox (1999) correctly raised the objection to this criterion that failure to reject the null hypothesis does not mean that the function is a good fit to the data. Statistical inference does not allow one to accept a null hypothesis based on failure to reject.

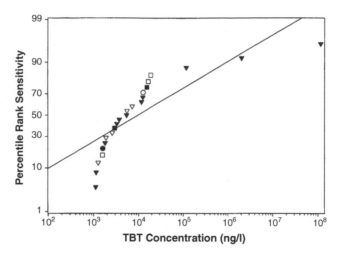

FIGURE 21.2 A probit function (linearized lognormal) fit to freshwater acute toxicity data for tributyltin. (From Hall, L. W., Jr. et al., *Human and Ecological Risk Assessment,* 6(1), 141, 2000. With permission.)

Second, it is preferable to choose functions based on goodness-of-fit or other statistical comparisons of alternative functions, rather than by testing hypotheses concerning a chosen function. Versteeg et al. (1999) used this approach, fitting the uniform, normal, logistic, extreme value, and exponential distributions to 14 data sets. Hoekstra et al. (1994) compared lognormal and log-logistic fits to data for 26 substances and found that the lognormal was consistently preferable. However, Van Leeuwen (1990) pointed out that the demonstrations of good fits of the log logistic are based on relatively large sets of acute LC_{50} and EC_{50} values. The much more heterogeneous chronic NOEC data sets may not have the same distribution and usually do not provide enough observations to evaluate the fit rigorously. The method for calculating water quality guidelines for trigger values in Australia and New Zealand specifies selecting a distribution function from the Burr family based on goodness-of-fit analyses (ANZECC, 2000).

Third, functions may be selected based on their inherent properties rather than their fit to the data. In this respect, statistical arguments have been used more frequently than ecological arguments. Aldenberg and Slob (1993) chose the logistic because it is more conservative than the normal distribution (generates lower HC_5 values), and because it is more computationally tractable. Fox (1999) objected that mathematical tractability is not an appropriate basis for choosing a function. Aldenberg and Jaworska (1999) suggested a bimodal function to address misfits caused by bimodality of the data set, which are in turn caused by the inclusion of subgroups of sensitive and insensitive species. Fox (1999) and Shao (2000) argued for the three-parameter Burr type III function, of which the logistic is a special case, because the additional parameter provides greater flexibility. However, for both approaches, the estimation of additional parameters enhances concerns with small sample sizes. Wagner and Løkke (1991) preferred the normal distribution based on its central position in statistics, promising wide applicability. Aldenberg and Jaworska (2000) supported that argument. However, it was recognized early in the development of SSDs that many data sets are not fit well by normal or lognormal distributions (Erickson and Stephan, 1985). The U.S. EPA used the log-triangular distribution because of its good fit (particularly with its truncated data sets) and its form, which is consistent with the biological fact that there are no infinitely sensitive or insensitive species (U.S. EPA, 1985a). Some use empirical distributions because they do not require assumptions about the true distribution of the data (Jagoe and Newman, 1997; Giesy et al., 1999; Newman et al., 2000; Van der Hoeven, 2001). Others have used empirical distributions as a way to display the observed distribution of species sensitivities when neither PAFs nor HC_p values are calculated (Suter et al., 1999), when a simple method is desired for early tiers of assessments (Parkhurst et al., 1996), or when none of the parametric distributions is appropriate (Newman et al., 2000). The use of linear interpolation to calculate HC_p values is equivalent to the use of an empirical distribution.

Finally, knowledge of the chemical may guide the choice of model. For example, specifically acting chemicals will tend to have large variances and asymmetry due to extremely sensitive or insensitive species (Vaal et al., 1997). If it is not possible to partition the data sets for such chemicals (Section 21.5.5), it may be wise to use empirical distributions rather than symmetrical functions.

Some have argued that the choice of function makes little difference, because the numerical results are similar in many cases. An OECD workshop compared the lognormal, log-logistic, and log-triangular distributions and concluded that the differences in the HC_5 values were insignificant (OECD, 1992). Smith and Cairns (1993) also stated that those distributions give relatively similar results. Fox (1999) argued that the choice of function matters, based on his demonstration that adding a parameter can make "up to a 3-fold difference." Newman et al. (2000) argued that the use of parametric functions is a mistake, because they often fail to fit real data sets. Therefore, they stated that empirical distributions fit to data by bootstrapping should be preferred to avoid indefensible assumptions. Others have argued that this practice is defensible only for large sample sizes (Van der Hoeven, 2001). Also, the use of parametric models is more suited to extensions of the extrapolation model, such as the addition of variation in bioavailability or biochemical parameters related to partitioning (Aldenberg and Jaworska, 2000; Van Wezel et al., 2000).

An issue that is likely to be more important for the numerical outcome than the choice of model is the related issue of data pretreatment discussed in Section 21.5. The choices made in data treatment, often related to ecological issues, can influence model choice and output precision; therefore, the debate should not focus solely on statistical concerns.

21.2.2 CONFIDENCE LEVELS

EQC may be based on protecting a percentage of species or protecting a percentage with prescribed confidence. An example of the former practice is that the U.S. EPA has used the HC_5 without uncertainty estimates to calculate criteria (U.S. EPA, 1985a; Chapter 11). Examples of the latter are Kooijman (1987), who developed

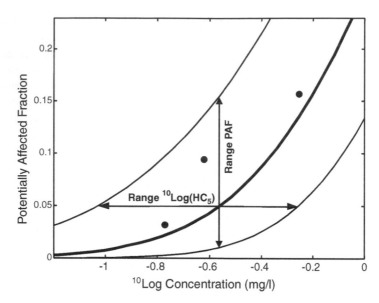

FIGURE 21.3 A graphical representation of confidence bounds for the HC_5 and the PAF. The figure shows the 5 and 95% confidence limits of $^{10}log(HC_5)$ and the 5th, 50th, and 95th percentiles of the PAF. Dots represent toxicity test results. (Courtesy of Tom Aldenberg.)

factors to protect all members of a community with 95% confidence, and Van Straalen and Denneman (1989) and Aldenberg and Slob (1993), who developed factors to protect 95% of species with 95% confidence. Wagner and Løkke (1991) also developed a method for protecting 95% of species with 95% confidence and showed that the confidence intervals of HC_p values are similar to what are called "tolerance limits" in distribution-based techniques for quality control of industrial products.

Suter (1993) pointed out that the those calculations are incomplete analyses of uncertainty concerning the HC_p. They account for uncertainty due to fitting a function to a sample but not due to uncertainties in the individual observations including extrapolations from the test endpoints to the values to be protected and systematic biases in the test data sets.

Van Leeuwen (1990) argued that the use of lower 95% confidence bounds, particularly when n is low, leads to unrealistically low values (Figure 21.3). However, the use of confidence bounds on the HC_p is still advocated as a prudent response to the uncertainties in the method (Newman et al., 2000), and confidence bounds are now routinely reported when calculating HC_p values in the Netherlands (Verbruggen et al., 2001).

The issue of whether to use confidence intervals is also important in the context of risk assessment (see Figure 21.3). The use of confidence intervals may be limited by the presentation method, as in the case of spatial mapping of PAF values (Chapters 16 and 19). There is also a theoretical objection. Solomon and Takacs (Chapter 15) argue that the use of confidence intervals on SSDs is inadvisable unless important species can be weighted, because the use of confidence intervals assumes that all species are equal in the sense of their roles and functions in the ecosystem and that they can be treated in a purely numerical fashion. That objection is applicable to any use of SSDs, with or without uncertainty analysis. In practice, accounting for uncertainties concerning predicted effects is desirable, both to improve the basis for decision making and for the sake of transparency concerning the reliability of results. Thus, the context of the application and the preferences of the decision maker may limit or promote the reporting of confidence intervals or probabilities of pre-scribed PAF or PES levels. In any case, it is important to specify what sources of uncertainty are included in the calculation.

21.2.3 CENSORING AND TRUNCATION

Because of the symmetry of most of the distribution functions used in SSDs, asym-metries in the data can affect the results in unintended ways. In particular, even after log conversion, many ecotoxicological data sets contain long upper tails due to highly resistant species (see, e.g., Figure 21.2). If these data are used in fitting the distri-bution, the fitted 5th percentile can be well below the empirical 5th percentile.

One approach to eliminating this bias is to censor the values for the highly resistant species, as recommended by Twining and Cameron (1997). To avoid both the bias and the apparent arbitrariness of censoring, the U.S. EPA simply truncates the distribution when calculating risk limits (U.S. EPA, 1985a; Chapter 11). That is, all data are retained, but only the lower end of the distribution is fit. This, however, can lead to a misfit to the total data set, as shown by Roman et al. (1999). Hence, its use is limited to the calculation of criteria, as in U.S. EPA (1985a), or to risk assessments with low PAFs.

Another approach is to analyze the data set by fitting mixed (i.e., polymodal) models to generate risk limits. Aldenberg and Jaworska (1999) applied a bimodal-normal model to the (log) toxicity data to this end. The parameter estimates were generated through Bayesian statistics and provide estimates for the HC_p for the most sensitive group of species, independent of prior knowledge about sensitive species. This practice can eliminate the need for censoring or truncating but is computation-ally intensive (Aldenberg and Jaworska, 1999). Shao (2000) used a mixed Burr type III function for the same purpose.

Pretreatment of data may reduce the need for censoring or truncation by reducing biases in data sets due to differences in bioavailability or other confounding factors (Section 21.5). Fitting alternative models may also remove the need for censoring and truncation.

21.2.4 VARIANCE STRUCTURE

Smith and Cairns (1993) point out that the data sets used in SSDs are likely to violate the assumption of homogeneity of variance. That is, test results from different laboratories using different test protocols are likely to have different variances. They recommend the use of weighting to achieve approximate homogeneity.

21.3 THE USE OF LABORATORY TOXICITY DATA

SSDs are derived from single-species laboratory toxicity data. Some of the criticisms of SSDs are actually criticisms of any use of those data, and pertain also to other approaches, such as the application of safety factors. These issues will be discussed only briefly here, because they are not peculiar to SSDs.

21.3.1 TEST ENDPOINTS

SSDs are most often distributions of conventional single-species toxicity test end-points, and the HC_p values and other values derived from them can be no better than those input values. All of the conventional test endpoints have some undesirable properties (Smith and Cairns, 1993), but whether these are serious depends on the context of SSD application. Furthermore, the appropriateness of test endpoints cannot be fully judged until their relationships to the assessment endpoints are clarified.

LC_{50} values represent severe effects that are unlikely to be acceptable in regulatory applications of SSDs to derive quality criteria for routine exposures. However, they may be properly applied in assessments of short-term exposures, as in spills or upsets in treatment operations.

No-observed-effect concentrations (NOECs) and lowest-observed-effect concentrations (LOECs) have all of the problems of test endpoints that represent statistically significant rather than biologically or societally significant effects. In particular, they do not represent any particular type or level of effect, so distributions of NOECs or LOECs are distributions of no specific effect (Van Leeuwen, 1990; Van der Hoeven, 1994; Laskowski, 1995; Suter, 1996). NOECs are particularly problematical because they may be far below an actual effects level or may correspond to relatively large effects, which are not statistically significant because of high variance and low replication (Van der Hoeven, 1998; Fox, 1999). Wagner and Løkke (1991) recognized these problems, but used NOECs anyway as the best available option to derive EQCs. Van Straalen and Denneman (1989) argued that NOECs are reasonably representative of effects thresholds in the field. They recommend using only NOECs for reproductive effects to derive criteria, both to increase consistency and because of the importance and sensitivity of reproduction.

The relationship between SSDs and ecosystem processes has been an issue of debate. Smith and Cairns (1993) argued that criteria based on SSDs do not protect ecosystem functional responses, implying that such responses are likely to be more sensitive than organismal responses. Hopkin (1993) argued that SSDs are unlikely to protect ecosystem processes, because key processes may be dominated by a few species, such as large earthworms, and those species may be more sensitive than 95% of species. Forbes and Forbes (1993) also suggested that SSDs do not address ecosystem function, but they argued that ecosystem function is likely to be less sensitive than structure, and therefore SSDs will be overprotective. Various authors have stated in the context of pesticide risk assessment that ecosystem function is likely to be less sensitive than organismal responses, because of functional redundancy (Solomon, 1996; Solomon et al., 1996; Giesy et al., 1999). Neither these

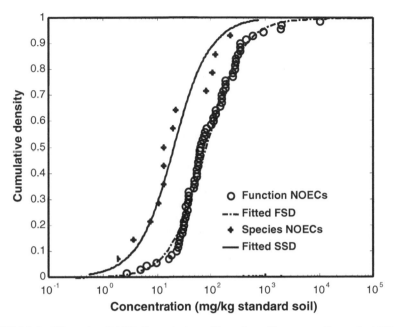

FIGURE 21.4 SSD and soil FSD for cadmium. (Data from Crommentuijn et al., 1997.)

arguments from theory nor the attempts to confirm SSDs using mesocosm data (Section 21.8.2) have resolved this issue. The appropriate resolution in particular cases should depend on the assessment endpoints chosen.

One might respond to both sides by pointing out that SSDs, which are based primarily, if not entirely, on tests of vertebrate and invertebrate animals, should not be expected to estimate responses of ecosystem functions which, in aquatic systems, are dominated by algae, bacteria, and other microbes. As a pragmatic solution, Van Straalen and Denneman (1989) argue that, if ecosystem functions are of concern, criteria should be derived using appropriate test endpoints. This pragmatic solution was adopted by using terrestrial microbial functions for derivation of soil quality criteria in the Netherlands. Distribution functions for data sets of microbial and fungal processes and enzyme activities (Chapter 12), are used to derive FSDs (function sensitivity distributions) and the lowest FSD or SSD is chosen to derive the EQC (Figure 21.4). The Canadian approach in deriving EQCs applies another pragmatic approach, using test endpoints that relate to the assessment problem directly (Chapter 13).

21.3.2 LABORATORY TO FIELD EXTRAPOLATION

From the beginning of the use of SSDs, the importance and difficulty of laboratory-to-field extrapolation has been discussed (U.S. EPA, 1985a; Van Straalen and Denneman, 1989). Differences believed to be important include a range of phenomena (see Chapter 9, Table 9.1), including bioavailability, spatial and temporal variance in

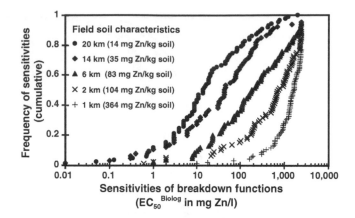

FIGURE 21.5 FSDs for microbial communities showing the reduced variance (increased steepness of CDFs) for metal-tolerant communities. Tolerance was measured using activity measurements of sampled microbial communities on Biolog™ plates. Microbial tolerance increases with decreasing distance from a former zinc smelter and with increasing soil zinc concentrations. (Courtesy of M. Rutgers.)

field exposures, and genetic or phenotypic adaptation. However, the issue of laboratory-to-field extrapolation is another problem that is generic to laboratory toxicology and not peculiar to SSDs. If the use of laboratory data cannot be avoided, due to the lack of field data or problems with field–field extrapolation, the laboratory data can be adjusted or pretreated with the aim to improve field relevance.

For example, concerning bioavailability, Smith and Cairns (1993) argued that SSDs are inappropriate because environmental conditions, particularly water chemistry, do not necessarily match test conditions. However, test endpoints may be adjusted for environmental chemistry, or exposure models may be used to estimate bioavailable concentrations rather than total concentrations.

Data treatment cannot solve all extrapolation problems, because of the complex nature of ecological phenomena. For example, genetic adaptation or pollution-induced community tolerance may occur when populations or communities are chronically exposed to contaminants. Field populations or communities may become less sensitive due to evolved capabilities to physiologically exclude or sequester contaminants or to compensate for effects, a phenomenon not addressed in laboratory toxicity tests. Strong evidence has shown the existence of such responses upon contaminant exposure (Posthuma et al., 1993; Rutgers et al., 1998). The occurrence of genetic adaptation by sensitive species may cause reduced variance of sensitivities in a community, which may lead to a "narrowed" SSD in the field, as observed by Rutgers et al. (1998) (Figure 21.5).

The selection and treatment of input data for use in SSDs can address some discrepancies between the laboratory and field, and various options are treated in Sections 21.4 and 21.5. However, other discrepancies must be treated as sources of uncertainty until they are resolved by additional research.

21.4 SELECTION OF INPUT DATA

The dependence of SSDs on the amount and quality of available data has been particularly obvious to critics. This section discusses issues of data selection and adequacy.

21.4.1 SSDs FOR DIFFERENT MEDIA

EQC are set for specific compartments: water, air, soil or sediment in different countries (Chapters 10 through 14). To this end, specific SSDs are constructed from data of terrestrial, aquatic, or benthic species. However, complications arise when associating SSDs with media.

The U.S. EPA and other environmental agencies have routinely derived separate freshwater and saltwater criteria (U.S. EPA, 1985a). Because the toxicity of some chemicals is not significantly influenced by salinity (Van Wezel, 1998), that distinction is not always necessary (Chapter 15). In particular, the U.S. EPA combines freshwater with saltwater species for neutral organic chemicals (U.S. EPA, 1993). The Dutch RIVM combines saltwater and freshwater data if no statistical significant difference can be demonstrated. Solomon et al. (2001) combined freshwater and saltwater data, unless the intercept or slope of the probit SSD models was different.

The assignment of species to a medium may be unrealistic, since some species are exposed through various environmental compartments, either during their whole life cycle (e.g., a mammal that drinks water and feeds from terrestrial food webs) or during parts thereof (amphibians). This may pose specific problems related to combining different species in an SSD and the use of one species in SSDs for more than one compartment. When species are significantly exposed through several compartments, SSDs can be based on total dose received by those species rather than ambient concentrations. Subsequently, given the SSD, criteria for the different environmental compartments can be calculated using a multimedia model (e.g., Mackay, 1991) to calculate critical concentrations in the relevant environmental compartments. When both direct and food chain exposure exists in a species assemblage, they can be combined by relating the food chain exposure to concentrations in a common exposure compartment such as water or sediment by using bioconcentration factors (BCFs) or biota-to-sediment accumulation factors (BSAFs) (Chapter 12). When species with multiple exposure routes are omitted or when exposure routes are ignored, results may be biased.

An alternative solution to the problem of complex exposures is to use body burdens as exposure metrics (McCarty and Mackay, 1993). That is, the SSD would be distributed relative to concentration of the chemical in organisms rather than in an ambient medium. This approach would be expected to yield lower variances among species. It would be particularly useful for risk assessments of contaminated sites.

Problems also arise when media have multiple distinct phases. In particular, sediment contains aqueous and solid phases, and soil contains aqueous, solid, and gaseous phases. This problem is addressed by assuming a single dominant exposure medium such as sediment pore water, so the exposure axis of the SSD is simply taken as the concentration in water (e.g., Chapter 12). However, equilibrium assumptions may not hold, and these cases may also need to be treated as multimedia exposures resulting in a combined dose.

21.4.2 Types of Data

A fundamental problem of SSDs is defining the range of test data that is appropriate to a model and to an environmental problem. If SSDs are interpreted as models of variance in species sensitivity, it is necessary to minimize other sources of variance. These sources of extraneous variance potentially include variance in test methods, test performance, properties of test media, and test endpoints. This consideration has led to specification of acceptable types of input data as in the U.S. EPA procedure for deriving EQC (U.S. EPA, 1985a; Chapter 11).

Rather than eliminating or minimizing extraneous variance, sources of variance may be explicitly acknowledged as part of the SSD methodology. For example, in deriving soil screening benchmarks, Efroymson et al. (1997a,b) recognized that variance in test soils was significant, so they considered their distributions to be distributions of species/soil combinations (see also Section 21.5.1). Such inclusiveness can quickly carry us beyond the topic of SSDs. For example, in setting benchmark values for sediments, various laboratory and field tests and field observations of organisms, populations, and communities have been combined into common distributions that are difficult to characterize (Long et al., 1995; MacDonald et al., 1996). It may well be that SSDs for soil will almost always have other sources of variance that are large relative to the variance among species, with or without provisional correction for bioavailability. In that sense, SSD results become part of a multivariate description, in which the species sensitivities are one of the descriptor variables and pH, etc. are others. This multivariate approach has been taken in modeling effects of multiple stressors on plants (Chapter 16).

The selection of data with consistent test endpoints may be difficult. As discussed above (Section 21.3.1) test endpoints based on statistical hypothesis testing are inherently heterogeneous. Hence, SSDs based on NOECs, LOECs, or CVs contain variance due to differences in the response parameter and the level of effect. Conceivably, one could select data to minimize this variance. For example, one could use only NOECs that are based on reproductive effects and that do not cause more than a 10% reduction in fecundity. However, that is not part of current practice.

SSDs are models of the variance among species, so species should be selected that are of concern individually or as members of communities. For example, algae and microbes are usually valued for the functions they perform and not as species. Therefore, the exclusion of algae and microbes from SSDs, as in the U.S. EPA method, may be appropriate (U.S. EPA, 1985a).

In contrast to these concerns, Niederlehner et al. (1986) suggested that the selection of species may not matter. Based on a study of cadmium, they argue that the loss of 5% of protozoan species in a test of protozoan communities on foam substrates is equivalent to the U.S. EPA HC_5, which is derived from tests of diverse fish and invertebrates. However, it seems advisable to choose species based on their susceptibility to the chemical, particularly when assessing compounds with specific modes of action such as herbicides or insecticides, and on whether they represent the endpoint of concern.

21.4.3 DATA QUALITY

The issue of data quality has received considerable attention in frameworks to derive quality criteria. This is because the use of data sets that have not been quality assured can introduce extraneous variance into SSDs, and can introduce biases into SSD models. The U.S. EPA has specified quality criteria for the data used to derive water quality criteria (U.S. EPA, 1985a). Emans et al. (1993) used OECD guidelines for toxicity tests to qualify data for their study. The aquatic ECOFRAM used quality criteria from the Great Lakes Initiative (U.S. EPA, 1995). In the Netherlands, all data used for derivation of quality criteria for water, soil, or sediment with SSDs are evaluated according to a quality management test protocol that is continuously updated (Traas, 2001).

Some SSD studies apparently accept all available data, with unknown effects on their results. It has been argued that all available data should be used because variance among species is large relative to variance among tests (Klapow and Lewis, 1979). It should be noted in this respect that the readily accessible databases usually have some degree of quality control on the input data, and this quality control applies indirectly to the SSDs derived from them. However, quality control is needed even when using generally accepted databases. After merging data from various databases, De Zwart (Chapter 8) applied an extensive quality control prior to using the merged data set. This was not based on quality checks to *all* original references (>100,000), but on removal of double entries and a check for false entries based on pattern recognition. Whatever the application, explicit and well-described data quality procedures improve transparency and repeatability of an analysis as well as the reliability of the results.

21.4.4 ADEQUATE NUMBER OF OBSERVATIONS

In the derivation of environmental quality criteria, various requirements have been suggested regarding the adequate number of observations based on differing tolerances for uncertainty concerning the HC_p (Figure 21.6). The smallest data requirement ($n > 3$) was specified by early Dutch methods (Van de Meent and Toet, 1992; Aldenberg and Slob, 1993). Van Leeuwen (1990) indicated that five species would be adequate based on uncertainty and ethical and financial considerations. Danish soil quality criteria also require a minimum of five species (Chapter 14). The U.S. EPA method requires at least eight species from different families and a prescribed distribution across taxa (U.S. EPA, 1985a; Chapter 11).

Various suggestions for adequate numbers have been given for SSDs used in ecological risk assessments, based on statistical and ecological grounds. The method applied by the Water Environmental Research Foundation in the United States does not specify a minimum n, but the authors indicate that the eight chronic values for zinc were too few, while the 14 values for cadmium were sufficient (Parkhurst et al.,

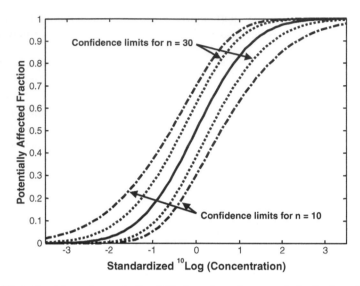

FIGURE 21.6 Confidence intervals for SSDs based on the normal distribution, depending on the number of data only. The lines show the median and 5th to 95th percentiles of the PAF for $n = 10$ and $n = 30$. (The figure was kindly prepared by T. Aldenberg according to Aldenberg and Jaworska, 2000.)

1996). Four chronic or eight acute values were required by the Aquatic Risk Assessment and Mitigation Dialog Group (Baker et al., 1994). Cowan et al. (1995) stated that SSDs may be useful when more than 20 species have been tested, because that number is required to verify the form of the distribution. Newman et al. (2000) estimated that the optimal number of values in an SSD is 30, the median number needed to approach the point of minimal variation in the HC_5. Vega et al. (1999) and Roman et al. (1999) conclude that, for logistically distributed data, this point is approached when ten or more values are available.

De Zwart (Chapter 8) presented evidence that the shape of the SSD (the slope) was associated with the TMoA of the compound. Given the idea of such an intrinsic (mode-of-action related) shape parameters for SSDs, he found that the number of test data needed to obtain the required value of the shape parameter for a certain compound would range from 25 to 50. However, due to the observed mode-of-action-related patterns among shape parameters for different compounds, it was suggested that the use of surrogate shape values, derived from data of compounds with the same TMoA, could solve the problem of data limitation. Estimation of the position parameter requires far fewer data than estimation of the slope.

Apparently, there are numbers, beyond which the SSD does not change considerably in shape or estimated parameters. Aldenberg and Jaworska (2000) gave relationships for confidence limits related only to the number of input data. For example, at $n = 4$, the estimated HC_5 is rather imprecise, with a 90% confidence interval between 0.07 and 37%. This means that the median HC_5 derived from this low number of data is very often not protective of the fraction of species specified as being protected in as many as 37% of cases (secondary risk). If decision makers want to be more certain that 95% of the species are protected, upper confidence limits can be calculated from the known patterns (e.g., $n = 8$), where the upper confidence limit of the HC_5 falls below 25%.

When data are in short supply, as is the case for many substances, an optimum number of observations will not be reached. With such limitations, decision makers can either ask for estimated HC_p values with specified confidence, or for specification of the uncertainty in the ecological risk assessment.

21.4.5 BIAS IN DATA SELECTION

The species used for toxicity testing are not a random sample from the community of species to be protected (Cairns and Niederlehner, 1987; Forbes and Forbes, 1993). This nonrandomness can potentially bias the SSDs. However, the magnitude and direction of the bias are not clear.

One argument is that species are selected for their sensitivity, and therefore SSDs have a conservative bias (Smith and Cairns, 1993; Twining and Cameron, 1997). In the United States, the validity of this argument is supported by the fact that the U.S. EPA defined its base data set to ensure the inclusion of taxa that were believed to be sensitive (U.S. EPA, 1985a). The ARAMDG recommended choosing species at "the sensitive region of the distribution" (Baker et al., 1994). Further, for biocides, testing may be focused on species that are closely related to the target species and therefore likely to be sensitive. An OECD workshop concluded that, in general, this bias is not unreasonable (OECD, 1992). However, there is little basis for this conclusion beyond expert judgment.

Another source of bias in data selection is the narrow range of taxa tested relative to the range of taxa potentially exposed (Smith and Cairns, 1993). For example, aquatic toxicity testing has focused on fish and arthropods and has tended to neglect other vertebrate and invertebrate taxa. Algal and microbial taxa are almost inevitably underrepresented when SSDs are intended to include them. Toxicity testing of birds has focused on the economically important Galliformes and Anseriformes rather than the much more abundant Passeriformes. In addition, species from some communities such as deserts tend to be underrepresented (Van der Valk, 1997). These biases will tend to reduce the variance of the SSD, which could cause an anticonservative bias in estimates of low percentiles.

21.4.6 USE OF ESTIMATED VALUES

Toxicity data estimated from models have been used in cases where the available test data are not sufficiently numerous to derive an SSD. Models have been developed for compound-to-compound, SSD-to-SSD, or species-to-species extrapolation.

Models may be used for compound-to-compound extrapolation. Van Leeuwen et al. (1992) assembled a set of 19 quantitative structure–activity relationships (QSARs) that estimate NOECs for chemicals with a baseline narcotic TMoA. Because all of their QSARs used the octanol–water partitioning coefficient (K_{ow}) as the independent variable, it was possible for the authors to derive a formula for estimating the HC_5 for any narcotic chemical from its K_{ow}. If any test data were available, they could be used along with the QSAR-derived values in an SSD. The same approach could be used to supplement any data set when an appropriate QSAR is available. However, the use of QSARs adds additional uncertainties due to imprecision of the model and the potential for misclassification of the chemical. The QSAR model is also the average fit to the individual toxicity data, thereby possibly reducing the variance of the SSDs compared to the original data. DiToro et al. (2000) used a similar approach to estimate the HC_5 for polycyclic aromatic hydrocarbon (PAHs) from K_{ow} and a QSAR for narcosis.

SSDs have also been approximated for small numbers of species by using properties of SSDs derived from large data sets. One approach is to assume that the mean and standard deviation are independent (Chapter 8; Luttik and Aldenberg, 1997). One may then derive a SSD from a mean estimated from the small set of test endpoints for the chemical of concern and a pooled variance derived from several SSDs for the same class of chemicals. Alternatively, one may use resampling of small data sets from large sets used to derive HC_p values, calculate the quotients of the lowest endpoint in the samples to the HC_p for that chemical, and derive a distribution of that ratio for each small n (Host et al., 1995). These approximations introduce another source of uncertainty to the use of SSDs, so the authors of both methods recommend conservative estimates of the HC_p. Alternatively, when few chronic data are available, one may approximate the chronic SSD by applying an acute-to-chronic ratio to the acute SSD (Section 8.5.2).

SSDs have been supplemented by using models to extrapolate from a small set of test species to a larger set of species of interest. Traas et al. (Chapter 19) used this approach to derive SSDs for avian and mammalian wildlife from small sets of avian and mammalian toxicity data. Their extrapolation models were based on differences in dietary composition and intake rates among species. For animals exposed through their diet, the exposure component of sensitivity is difficult to separate from the intrinsic toxicity of chemicals, unless based on concentrations in target sites. These models are thus more based on exposure distributions than on intrinsic toxicity distributions. However, other interspecies extrapolation models could also serve that purpose.

Extrapolation models used in the construction of an SSD introduce additional uncertainties. In particular, if they do not incorporate all of the relevant sources of variance among species, they will tend to overestimate HC_p values (i.e., be less protective) when p is small.

21.5 TREATMENT OF INPUT DATA

Some of the limitations of, and objections to, SSDs may be addressed by processing the data prior to calculation of the SSD. There are two possibilities for this. First, processing changes the data with the same factor(s) for all species, for example, by using a fixed factor correcting for differences in exposure between laboratory and field (e.g., Traas et al., 1996). This shifts the distribution on the log-concentration axis. Second, processing may consist of applying different factors for different species, so that the variance of the data changes. Preprocessing of data is usually done to improve the association of the SSD with the assessment endpoint. This may lead to better predictions of effects and may reduce one of the main problems with the verification studies of SSDs, the dissimilarity of variances between SSDs in the laboratory and in the field.

21.5.1 HETEROGENEITY OF MEDIA

The most common preprocessing of input data is normalization to reduce variance among tests due to the physicochemical properties of the test medium. Those properties may influence the biological availability or toxicity of the compound.

Preprocessing of data is done for several metals, ammonia, and phenols in the U.S. EPA water quality criteria (U.S. EPA, 1985a). In the Netherlands, metal concentrations in soil are adjusted for soil chemistry (Van Straalen and Denneman, 1989). Variables used have included pH, hardness, temperature, clay content, and organic matter content. In the Netherlands, empirical formulae have been derived by fitting a statistical function to sets of data that include the range of chemical conditions encountered in the field. For example, normalization of metal concentrations to a standard soil (with a fixed percentage organic matter and clay) is applied, by using regression equations that relate metal contents to soil properties for a range of relevant areas (Lexmond and Edelman, 1986). These equations are routinely used in normalizing laboratory toxicity data to a so-called standard soil. After normalization, an SSD is made for the standard soil, and the EQCs for metals are derived accordingly (Chapter 12). In applying these EQCs to field soils, the regression equations are used inversely, so that a certain degree of site specificity is created in the EQCs. As a secondary use of the empirical formulae, one can calculate whether exceedence of EQCs will occur in case of changing substrate characteristics. For example, Van Straalen and Bergema (1995) tested the expectation that soils will become more acidic when normal agricultural practice ceases and therefore more toxic due to the heavy metal load already present.

The proposed formulae are intended to normalize data to a standard chemistry, so that the SSD does not contain extraneous variance. U.S. water quality criteria and Dutch EQCs are adapted to local conditions by adjusting for the chemistry of local media. They may be used to adjust the HC_p or the entire distribution for local conditions when performing a site-specific risk assessment (Hall et al., 1998). Standardization algorithms are important but are a source of debate in the use of SSDs. They should at least be verified for their intended purpose, reducing the extraneous variance in SSDs or improving the accuracy of site-specific assessments.

21.5.2 ACUTE–CHRONIC EXTRAPOLATIONS

For many chemicals, the available data are primarily acute, whereas regulators and assessors are primarily concerned with chronic effects. Usually, there are insufficient chronic test data to derive a chronic SSD.

The simplest option to fill this gap is to use a generic acute–chronic ratio to convert the acute values to estimated chronic values. For example, De Zwart and Sterkenburg (Chapter 18) used a factor of 10 and the Danish method uses a factor of 3 (Chapter 14).

Alternatively, the extrapolation factor may be chemical specific. When there are not enough data to derive a chronic SSD, the U.S. EPA derives a chemical-specific acute–chronic ratio (U.S. EPA, 1985a). The Water Environment Research Foundation method uses acute–chronic ratios to estimate chronic SSDs, even when, as for copper and zinc, a relatively large number of chronic values are available to derive a chronic distribution directly (Parkhurst et al., 1996). Although the latter method allows for direct derivation of chronic SSDs, in practice the authors prefer the SSDs obtained from the larger number of acute values and assume that the use of acute SSDs and acute–chronic ratios to estimate chronic SSDs does not increase uncertainty.

A further alternative, based on a different assessment of the acute–chronic pattern, makes use of whole SSDs. The chapter on SSD regularities (Chapter 8) has shown that the uncertainty of acute–chronic ratios for specific TMoAs can be calculated from SSD parameters for chemicals with the same TMoA. This uncertainty can then be combined analytically with that of the SSD itself. This uncertainty is, however, of a quite different magnitude than acute–chronic ratios derived directly from acute–chronic pairs for species, which also has consequences for further statistical properties (e.g., calculation of confidence intervals).

21.5.3 COMBINING DATA FOR A SPECIES

Often, more than one value will be available for a particular chemical, species, and test endpoint. In such cases, it is generally desirable to reduce those multiple values to a single observation.

For effects-based test endpoints where the test endpoint is clear (such as with LC_{50} values) and test conditions do not significantly differ, one may choose one of the tests or average them. Selection might be based on the quality of the tests, the magnitude of the result (e.g., choose the lowest value), or their relevance to the situation being assessed (e.g., similarity of test water chemistry to the site; Suter et al., 1999). If all values are acceptable, they may be averaged. The U.S. EPA uses geometric means (U.S. EPA, 1985a).

For test endpoints based on hypothesis testing (i.e., most chronic toxicity data), there is the additional problem that the endpoints are usually not for the same response, the same level of response, or the same test conditions, so that averaging is generally not appropriate. In such cases, the lowest value is commonly used (Okkerman et al., 1991; Aldenberg and Slob, 1993). Van de Meent et al. (1990) proposed using the lowest value when the test endpoint responses are different and the geometric mean when they are the same. Such decisions can bias SSDs.

21.5.4 COMBINING DATA ACROSS SPECIES

In the use of SSDs, it is implicitly assumed that a set of tested species represents independent observations from a random distribution. However, the variance in sensitivity among species is not random. In particular, the responses of species that are closely related taxonomically are more highly correlated than those distantly related (Suter et al., 1983; LeBlanc, 1984; Slooff et al., 1986; Suter and Rosen, 1988;

Fletcher et al., 1990). This correlation structure is to be expected given the evolutionary underpinnings of taxonomy and the dependence of toxicological sensitivity on the physiological and morphological traits that differentiate taxa.

In recognition of this problem, the U.S. EPA combines test endpoints for all congeneric species, so that the input data for its SSDs are genus mean values (U.S. EPA, 1985a; Chapter 11). Danish methods combine "closely related species," including, for example, all earthworms (Chapter 14). The latter example implies aggregation at least across families. These aggregation approaches eliminate the worst cases of lack of independence. However, some lack of randomness is inevitable due to higher taxonomic relationships as well as other factors such as the similarity of results from a particular laboratory using a particular water source. The U.S. EPA has been criticized for aggregating species within genera, because aggregation provides less resolution than using all species (Giesy et al., 1999). However, the critics do not suggest a solution for the problem of lack of independence.

Versteeg et al. (1999) presented other arguments for aggregating species. First, they point out that species identification for some genera is questionable, and aggregation eliminates that concern. Second, aggregation of species within genera eliminates the common problem of predominance of certain taxa, particularly the genus *Daphnia*, in ecotoxicological data sets.

Aggregation of species inevitably results in a reduction of the number of input data. This can be counteracted by requiring a minimum number of taxa. However, combining data across species to reduce overrepresentation should always be considered in view of the community under consideration and the assessment endpoint. If, for example, the assessment is particularly concerned with effects on a rare salmonid fish or on a community dominated by salmonids, it might be undesirable to combine species within genera of that family.

21.5.5 COMBINING TAXA IN A DISTRIBUTION

SSDs may contain observations from all taxa exposed to a compound in an environmental medium, or separate SSDs may be derived for major taxa or functional groups. For example, one might derive an SSD for all aquatic species, derive one for animals and another for plants, or derive one each for fish, invertebrates, and algae.

Wagner and Løkke (1991) and the aquatic ECOFRAM argued for the derivation of SSDs for separate taxa to the extent that the available data allowed (see also Chapter 15). This approach has several advantages:

1. Distributions that lump distantly related species are likely to be polymodal due to systematic differences in sensitivity. In particular, target and non-target taxa are likely to have different sensitivity distributions for selective biocides. See examples in Hall et al. (1999) and Chapter 15.
2. The loss of species from different higher taxa or functional groups has different ecological implications, so separate distributions provide a better basis for ecological assessments. For example, in an assessment of aquatic risks of diazinon, aquatic insects were fitted separately not only because

they are highly sensitive, but also because they were treated as food species for the endpoint fish species (Giddings et al., 2000).

3. Different taxa and functional groups have different values to society, so decision makers may choose different percentiles as protection levels for each (Twining and Cameron, 1997).

4. Different taxa may have different modes of exposure, so, because they do not have the same exposure scale, it may be inappropriate to lump them in a common distribution.

One response to the calls for increased taxon specificity is the derivation of SSDs for rotifers (McDaniel and Snell, 1999).

In contrast, lumping taxa within an SSD has two advantages. First, it permits the use of SSDs when data sets are too small to partition. Second, it simplifies the process of setting criteria by yielding only one value for a medium.

There are various options to split taxa over different distributions. A system for grouping species into SSDs was developed by the aquatic ECOFRAM (Chapter 15, Figure 15.12). Versteeg et al. (1999) placed taxa in different distributions if their means were statistically significantly different. Solomon et al. (2001) placed vertebrates and arthropods in different SSDs because they are not valued in the same way, and because the difference in their HC_{10} values was large (Figure 21.7).

Criteria other than taxonomy may be used to separate species into different SSDs. Based on ecological characteristics of the species, Solomon and Tacaks (Chapter 15) suggested separating species that recover rapidly from those that

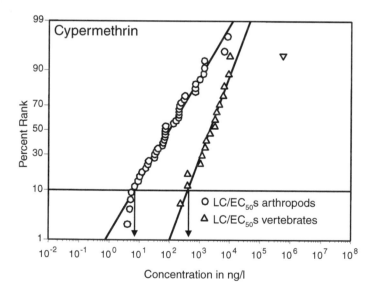

FIGURE 21.7 SSDs for aqueous acute exposures to the pesticide cypermethrin. The two taxa are separated based on differences in their slope and position as reflected in the large difference in their HC_{10} values. (From Solomon, K. R. et al., *Environmental Toxicology and Chemistry,* 2001. With permission.)

recover slowly, based on life history characteristics. This procedure is particularly relevant to compounds with episodic exposures, such as pesticides.

The issue of combining heterogeneous sets of taxa in an SSD may be addressed by fitting a polymodal function. Aldenberg and Jaworska (1999) suggested a bimodal normal distribution to capture taxa with apparently nonuniformly distributed sensitivities, to derive environmental quality criteria. On the one hand, they concluded that unimodal distributions did not poorly estimate HC_5 values from distributions that were in fact bimodal, so that HC_5 estimates appeared robust. However, they also pointed out that analysis of polymodality can help to identify sensitive taxa and estimate risks to them. In addition, unimodal functions fitted to multimodal data would not be expected to produce good estimates of risk of members of particular taxa, since these have different sensitivity profiles. In the more recent use of SSDs in ecological risk assessment (e.g., Chapter 15), and considering the increasing availability of data (e.g., Chapter 8), the splitting of taxa is gaining support.

These observations suggest that use of SSDs to assess specific assessment endpoints may be considerably improved by splitting taxa or other groups among SSDs. However, lumping SSDs may still be desirable for deriving EQCs.

21.5.6 Combining Data across Environments

When there are insufficient data for terrestrial species to calculate a soil standard, the Danish method allows use of data for aquatic species (Chapter 14). This is done by assuming that exposures to soil contaminants are in actuality exposures to the aqueous phase (Løkke, 1994). Given that assumption, the aqueous exposure concentration for soils is estimated using partitioning coefficients. Then, the soil-normalized aquatic data are combined with soil test data. A similar strategy is followed in the Netherlands when there are insufficient data for soil or sediments (Chapter 12).

21.5.7 Combining Data across Durations

To obtain SSDs that reflect variance in species sensitivity, assessors have attempted to keep the exposure durations of tests relatively constant, by selecting appropriate test durations. When exposure durations are variable, as in acute episodic exposures such as spills or failures of treatment equipment, it is desirable to derive separate SSDs for different durations that may be most relevant to the exposure scenarios (e.g., 24, 48, and 96 h, Campbell et al., 2000).

It is common practice to lump tests into categories of acute and chronic. For example, Solomon et al. (2001) used 24- to 96-h test data in their assessment of cotton pyrethroids. Defining exposure intervals of chronic tests is more difficult. The most common strategy when deriving chronic SSDs is to use all nominally chronic data. That is, include all tests that are described as chronic tests or as equivalent to chronic tests. An ecologically more defensible strategy is to use all data that include critical life stages such as production of young or, ideally, a full life cycle. An alternative approach based on toxicological insights is to use all tests that are of sufficient duration for the test organisms to reach equilibrium with the test medium. A model for estimating this time for fish is presented by Cowan et al. (1995).

When setting criteria, it is probably sufficient to use broadly defined categories of duration such as acute and chronic. However, for risk assessments, it is important to try to match the time to response in the tests to the duration of exposure.

21.5.8 COMBINING CHEMICALS IN DISTRIBUTIONS

If the available data set is too small to derive an SSD for a chemical, it may be appropriate to lump similar chemicals (Smith and Cairns, 1993). This recommendation may be appropriate if the variance in toxicity of the set of chemicals is small relative to the variance in sensitivity of species. This assumption may be fulfilled by normalizing the toxicity of the chemicals. For example, it is possible to normalize the toxicity of halogenated dicyclic aromatic hydrocarbons to the toxicity of a 2,3,7,8-TCDD using toxicity equivalency factors (Safe, 1998; Van den Berg, M. et al., 1998). Versteeg et al. (1999) applied this approach to linear alkyl sulfonates by normalizing them to a dodecyl chain length. The lumping of chemicals in an SSD may be appropriate also for sets of chemicals such as PAHs that typically occur in the environment as complex and poorly defined mixtures (Chapter 14). In that case, the SSD represents the distribution of sensitivity of species to the range of PAH mixtures found in the environment. In either case, this practice may raise concerns about the independence of observations and underlying polymodality. It is also important to determine whether the toxicity equivalency factors are applicable to all species, as was done in the review of dioxin-like toxicity by Van den Berg, M. et al. (1998). The practice of combining chemicals in an SSD is uncommon and should be employed with caution.

21.6   SELECTION OF PROTECTION LEVELS

When using SSDs for derivation of EQC, it is necessary to select one or more proportions p of the community or taxon as trigger values. The value of p varies among countries and may differ among assessment endpoints. The most common value in the lower range is 5%, which is used by regulators in the United States, the Netherlands, and Denmark. The selection of this value in the Netherlands has been described as arbitrary and a result of political compromise (Van Leeuwen, 1990; Emans et al., 1993).

It is necessary to define criteria that are sufficiently protective, yet not so conservative to be unachievable. An early proposal to protect all species with 95% confidence based on the method of Kooijman (1987) resulted in criteria that were far below background levels for most naturally occurring chemicals (Chapter 3).

The value of p may vary among taxa or functional groups. In particular, protection of 95% of species in groups like algae and bacteria, which are valued for their function, are highly diverse, and have low public appeal, may not be necessary (Twining and Cameron, 1997). Protection of function may be achieved by means other than SSDs such as functional sensitivity distributions (FSD, see Section 21.3) based on microbial processes such as specific enzyme inhibitions.

For readily degradable compounds, the value of p may be the same as for nondegradable compounds, but the half-life of the compound may be used as an associated criterion to calculate a maximum acceptable half-life (given p). In that case, it is assumed that the ecological risk is related to the time required for the compound concentration too fall below the level of p (Van Straalen et al., 1992).

The p value may also vary among ecosystems. The ANZECC (2000b) method employs an HC_5 for slightly moderately disturbed ecosystems and the HC_1 for high-conservation-value ecosystems.

When SSDs include considerable variance among tests and media as well as among species, it may not be necessary to use a p as low as 5%. For example, screening values for sediment (the effects range–low, ER-L) and soil (screening benchmarks) were based on p values of 10% (Long and Morgan, 1991; Efroymson et al., 1997a,b).

High p values may be chosen to designate concentration levels triggering investigations into remediation options and urgency, rather than to designate low concentration thresholds for concern. In the Netherlands, risk limits for classification of soil as "seriously contaminated" are calculated from the HC_{50} of the SSD^{NOEC} (Chapter 12).

Values of p might also be chosen for statistical or ecological reasons rather than a desire to provide a certain level of protection. The ARAMDG and aquatic ECOFRAM chose a p value of 10% for fish and invertebrates, because it was approximately the inflection point for many distributions (Baker et al., 1994; ECOFRAM, 1999a). They wished, thereby, to avoid the portion of the SSD in which a unit reduction in concentration results in the protection of relatively few additional species (Giesy et al., 1999). It has also been argued that the assessment endpoints should be functional properties of communities and ecosystems, which implies that numerous species may be lost without significant effects (Solomon, 1996; Giesy et al., 1999). Therefore, those authors argue that a relatively large p value is justified. In response to the argument by Lee and Jones-Lee (1999) that effects on 10% of species are unacceptable, Solomon and Giddings (2000) argued that significant effects rarely occur at the HC_{10} because of factors such as variance in exposure and recolonization that are not included in their assessment models.

The 5% protection level was apparently chosen independently in the United States and the Netherlands. More recently, a range of p values have been chosen for specific uses, and debates continue on the choice of generic p values with or without safety factors superimposed on them based on the number of input data (e.g., for European Union Technical Guidance Documents, January 2001).

21.7 RISK ASSESSMENT ISSUES

Several authors have been concerned with the lack of site specificity of SSD predictions (Forbes and Forbes, 1993; Smith and Cairns, 1993; Chapter 9) and the difficulty of linking SSD predictions to ecological endpoints that can be observed in the field. Some issues are treated here regarding exposure, ecology, and risk interpretation. More general aspects of the concept of SSDs for risk assessment have been discussed in previous sections and are omitted here.

21.7.1 EXPOSURE

Although this book is concerned with a toxic effects model, it is important to remember that risk is a function of exposure as well as toxicity. The success of an effects model depends on the quality of exposure estimates and the concordance of exposure metrics with the exposure–response metrics. This is true for application of environmental criteria as well as risk assessment, but criteria are less often adapted to local conditions, because they are intended to be generically applicable.

The duration of the toxicity studies used for SSDs relative to exposure durations is important for the predictive ability of SSDs (Steen et al., 1999; Chapter 9). One would not expect SSDs based on acute LC_{50} values to predict chronic effects in the field. The input data of SSDs that are used in ecological risk assessment should be compared to a field-relevant exposure scenario, to avoid misinterpretations (see Section 21.5.7).

Although chemical concentrations in the laboratory tests used to derive SSDs are highly uniform and often considered readily available for uptake in organisms, exposures in the field may be highly variable in space, time, and chemical form. The general assumption of homogeneous compartments is often not correct for field applications of pesticides due to various loss processes and redistribution over the various environmental compartments.

Sorption differences may explain observed discrepancies between predicted and observed effects (Chapter 9), and can be considered as a necessary extension of the SSD model when calculating risks for soils or sediments (Klepper et al., 1998; Chapters 16 and 19). The algorithms currently used for correcting bioavailability are, however, mainly based on soil sorption studies. The biological aspects of bioavailability such as habitat, routes of exposure behavior, and life history characteristics are important as well and should be part of further study to unravel the environmental influences from the intrinsic species sensitivities (Peijnenburg et al., 1997). In any case, sufficient attention should be paid to exposure quantification, especially in solid media (soil, sediment), where availability differences may range over an order of magnitude.

21.7.2 Ecological Issues

The ability of SSDs to predict effects in the field is of prime concern. The principal goal of the confirmation study in this volume (Chapter 9) has been to establish whether any relation exists between SSD-based predictions and field effects. Now that this seems to be the case in certain conditions, an effort should be made to establish the most favorable test endpoints related to relevant properties of communities such as extinction probabilities or densities.

SSD-based predictions are derived from sets of tested species that are very often not representative of the species at a specific site (Smith and Cairns, 1993). If we deal with a well-researched toxicant, we may make taxon-specific predictions (as in Chapters 9 and 15), which can lead to a considerable improvement in the accuracy of SSD predictions. The validity of SSDs for prediction of effects on specific taxonomic groups seems to increase when the TMoA is known and taxa are selected accordingly (Chapter 9).

Even if species are well represented in SSDs, the question remains whether all important ecological properties are represented. One reason to doubt this is the lack of species interactions in the tests used or in the SSD model itself (Forbes and

Forbes, 1993; Smith and Cairns, 1993; De Snoo et al., 1994; Chapter 4). Another reason for concern is the lack of test data for functionally important microbes. Chapter 9 presents evidence that indirect effects of the herbicide linuron and the fungicide carbendazim can occur either below or above threshold concentrations for direct effects. This suggests that ecological interactions are relevant to ecological risk assessment and need to be addressed, likely with tools other than SSDs (Hommen et al., 1993). The same probably applies to ecosystem functions (Forbes and Forbes, 1993), since there is no simple relationship between the number of species affected and changes in ecosystem function.

Sensitive species that are affected by a toxicant are often replaced by less-sensitive species (Rutgers et al., 1998). This allows ecosystem functions, such as leaf litter mineralization, to proceed at almost the same rate even though sensitive species are strongly affected (Van Beelen and Fleuren-Kamilä, 1999). A combined modeling–field experimentation study on nematode species composition in copper- and zinc-contaminated plots indicates that species composition can change while system function stays relatively unharmed (Klepper et al., 1998; Smit et al., in press).

Acclimation of individuals or adaptation of populations may change the relative sensitivity of species and increase mean sensitivity over time, as noted in micro- and mesocosm studies with herbicides and fungicides (Chapter 9). These changes can reduce or obscure any relationship between effects on functions and effects on species in the laboratory.

Based on these considerations and the case studies mentioned here and in Chapter 9, it is evident that SSDs are not models of all ecological attributes (see, e.g., Figure 18.4). Hence, their utility depends on the type of ecological endpoints and the issue to be addressed.

21.7.3 JOINT DISTRIBUTIONS OF EXPOSURE AND SPECIES SENSITIVITY

If exposure is estimated as a distribution due to variability over time or space, then the relationship between that distribution and the SSD must be addressed (Van Straalen, 1990; Parkhurst et al., 1996; Chapters 4, 5, 15, and 17). The distribution of exposure levels may be used simply to calculate the probability of exceeding an HC_x (Twining and Cameron, 1997), or both distributions may be used to calculate risk. The WERF and ARAMDG methods for aquatic risk assessment calculate what is called the "total risk," which is actually the expectation of the PAF, given that the environmental concentration is distributed. This method also allows for the ranking of sites, based on the exceedence probability for a certain percentile of the SSD. In this way, a ranking of the hazard of different toxicants is possible, given measurements over the same set of locations, or a ranking of locations for the same chemical. A very similar approach was used to estimate the distribution in space of the risks to a plant species or microorganism due to metals in soils (Manz et al., 1999). Chapter 5 discusses that the distribution of the PAF (or the ecological risk delta as formulated in Chapter 4) depends very much on the standard deviation of the exposure. Two compounds or two scenarios for the same compound can have the same expectation of PAF (the same "total risk") but different PAF distributions due to exposure uncertainty.

The implications for risk management depend on the source of variation in the estimated environmental concentration. The estimated risk depends on whether the exposure is distributed over time (e.g., annual risk), over space (risk of the effect per km^2 treated), over episodes (risk per pesticide application), over instances (risk of the effect per pulp mill), or some other variable (Suter, 1998a). Clear analysis of these exposure distributions allows identification in time or space of ecological risk, based on SSDs.

21.8 THE CREDIBILITY OF SSDs

The utility of SSDs in practice depends on whether the results are reasonable and consistent, and whether results are confirmed by responses in exposed field communities. It also depends on the availability of alternative approaches that may do better in these respects.

21.8.1 REASONABLE RESULTS

SSDs that are used to derive EQC are not necessarily intended to estimate particular types or levels of effects. However, they should at least produce reasonable results. Two criteria for reasonableness have been cited in critiques of SSDs: protection of important species and generation of criteria that are higher than background levels.

Environmental criteria should provide a high confidence of protecting species that are considered important. The use of criteria based on HC_p values, where $p > 0$, implies that some species may be lost, or at least may experience toxic effects. This implies in turn that in some cases ecologically or societally important species will not be protected (Forbes and Forbes, 1993; Hopkin, 1993; Smith and Cairns, 1993). The U.S. EPA, the Dutch RIVM, and others have addressed this issue by examining observations of toxic effects on species below the HC_p for any particularly important species, and adjusting the standard if any are found (U.S. EPA, 1985a; Giesy et al., 1999). Van Straalen (1993) argued that he did not know of a case in which an important species was not protected by the HC_5. However, the strategy of adjusting quality criteria when important species are at risk does not account for sensitive or important species that have not been tested.

Criteria should not be set below background concentrations or concentrations that are nutritionally essential (Hopkin, 1993). The frequent occurrence of criteria within the range of natural background concentrations contributed to the abandonment of the lower 95% confidence limit on HC_p as a basis for criteria in the Netherlands in favor of the median HC_p. In the Netherlands, the problem of concentrations below background is pragmatically solved by adding the HC_p estimated by an SSD to the natural background (Struijs et al., 1997; Crommentuijn et al., 2000). For this procedure, it is assumed that background chemicals do not participate in toxicity due to their form, bioavailability, or some other factor. The resulting EQC value thus differs among sites with different background concentrations. However, background concentrations may be irrelevant to contaminant concentrations because of differences in speciation or bioavailability between the naturally occurring and anthropogenic forms. Moreover, background levels of some chemicals in some locations are toxic to some species (Van Straalen and Denneman, 1989; Van Straalen, 1993). The Danish method for calculating soil criteria also results in values within background or nutritional ranges (Chapter 14). These values are corrected using expert judgment. Although the Dutch and Danish solutions are pragmatic, they do not resolve the conceptual problem. Similar issues apply when criteria are compared to nutritionally essential levels, with the additional problem that nutritional requirements of nonvertebrate species are poorly characterized.

21.8.2 CONFIRMATION STUDIES

Since the introduction of SSDs for derivation of EQC, there has been interest in the issue of predictive accuracy of the SSD concept (Van Straalen and Denneman, 1987). Originally, in view of the earliest SSD applications, attention focused on the level of protection associated with quality criteria, whereas the most recent studies focus on whole SSDs (Chapter 9). The studies all make use of a study design, in which SSD-based results are compared to micro- and mesocosm data or to field studies. These efforts have, over time, become more formal and sophisticated.

A group at the Dutch RIVM attempted to confirm aquatic quality criteria derived from SSDs by comparing them to results of toxicity tests that were believed to be inherently valid because the test systems contain multiple species (Emans et al., 1993; Okkerman et al., 1993). They found that HC_5 values based on single-species NOECs were approximately equal to the multispecies NOECs (Emans et al., 1993). However, the basis for the result was weak, with only seven multispecies results formally judged to be reliable (Emans et al., 1993) and considerable variance among multispecies NOECs for individual chemicals.

These studies were followed by studies of Dutch terrestrial quality criteria, which emphasized the role of modifying factors, such as compound availability. These studies focused on the field relevance of both the input data (Smit, 1997) and the quality criteria (Posthuma et al., 1998). The dualistic conclusion was that the field relevance of the input data could be considerably improved, but, nonetheless, the soil quality criteria derived from them appeared to relate grossly to the policy targets for which they were designed. When risk limits were compared to mesocosm data (Posthuma et al., 1998a; 2001), the HC_5, according to the added risk approach, appeared to predict no community effects, or only microbial responses (Klepper et al., 1999). When attention was also paid to the middle region of the SSD^{NOEC}, it appeared that the HC_{50} (the trigger for remediation investigations) distinguished the concentration where large effects appear. Finally, a relationship was shown between sensitivity distributions of functional endpoints (FSDs) and microbial responses in mesocosms exhibited as pollution-induced community tolerance (PICT; Van Beelen et al., 2001).

Versteeg et al. (1999) compared distributions of single-species NOECs (or equivalent test endpoints) for 11 chemicals to distributions of NOECs from model ecosystems ranging from 2-l flasks to 15-m^2 ponds. They concluded that the HC_5 values were typically lower than the geometric means of NOECs and were reasonably good predictors of the lower 95% confidence bounds on those NOECs.

In addition, some authors of SSD-based assessments have qualitatively compared their results to published results of multispecies tests (e.g., Giesy et al., 1999; Hall and Giddings, 2000). Those authors concluded that single-species data are conservative estimators of ecosystem responses. Other authors have found a good agreement between single-species data and mecososm responses (Van den Brink et al., 1997; Cuppen et al., 2000).

There are a number of difficulties with these confirmation studies. One important problem is that differences in exposure between the laboratory data and the field mesocosm data and microcosm data may be greater than the inaccuracy of SSDs as effects models. Single-species laboratory tests generally maintain constant concentrations of test chemicals in highly bioavailable forms. In contrast, field tests may have highly variable concentrations including rapid declines in concentration and low-exposure "refugia," the chemical may partition to media that do not occur in the laboratory tests, and the bioavailable concentrations may be much lower than the nominal concentrations. The opposite may also occur (e.g., high availability of metals in acid field soils, while laboratory tests operate in the range of moderate pH values). This disjunction in exposure has been used to justify choosing high p values as thresholds for the aquatic compartment (Giesy et al., 1999). However, SSDs, which are effects models, should not be blamed for inaccurate exposure modeling. The appropriate response is to better estimate effective exposures in laboratory and field tests, not to reject or adjust the effects model to account for poor exposure estimates.

A more fundamental problem is the assumption that the results of tests conducted in microcosms or mesocosms are good standards against which to compare SSD results (Dobson, 1993). Although microcosms and mesocosms are useful for revealing aspects of the fate and effects of chemicals that are not apparent in ordinary laboratory tests, their ability to reveal the true threshold for ecologically or societally significant effects in the field is questionable. First, conventional mesocosm test results do not correspond to any particular type or level of effect. This is in part because their test endpoints are, in general, based on hypothesis testing. In this sense they are even worse than the chronic laboratory tests, because of the larger number of responses that are typically measured. It is also because the set of responses measured in these tests is highly inconsistent, ranging from fish mortality to protozoan community structure and algal production. Second, mesocosm tests are often poorly replicated, few exposure levels are tested, and the intervals between exposure levels are relatively large, so the effects thresholds are ill-defined and often high (Graney et al., 1989). This is not necessary given recent improvements in test design with more exposure levels (Campbell et al., 1998) and the strongly improved statistical evaluation of community composition changes based on Principal Response Curve analyses (Van den Brink and ter Braak, 1999). Third, most microcosm and mesocosm tests are relatively short term, so they may not provide information on effects of long-term exposures. Fourth, the enclosure units used for micro- and mesoscosms and the small artificial channels do not represent all the ecosystems that we wish to protect. In particular, they have small volumes so they often do not include vertebrate predators, they almost never include top predators, and they are more highly influenced by benthic processes than most aquatic systems of concern. At least one confirmation study focused solely on periphyton in the absence of invertebrates (Versteeg et al., 1999). Finally, aquatic mesocosm studies have been conducted primarily to support pesticide registration, so they may not represent an unbiased sample of the range of chemicals to be assessed and regulated.

Although micro- and mesocoms studies have been the most important source of data for SSD confirmation, the above-mentioned issues pose questions about the use of controlled experimental units in SSD confirmation studies. In summary, this is because estimated thresholds are often questionable, and the thresholds may not be representative for all chemicals to be managed and for all real ecosystems to be protected. Hence, mesocosm results and SSDs are independent estimators of assessment endpoints for populations, communities, or ecosystems; each has particular strengths and weaknesses.

An alternative approach, which eliminates the concern about the representativeness of microcosms and mesocosms for the field, is to compare SSD results to results of studies of contaminated ecosystems (Dobson, 1993). Niederlehner et al. (1986) compared a chronic HC_5 value and a concentration causing loss of 5% of protozoan species to cadmium concentrations causing effects in the field. They found that both laboratory-derived benchmarks fell at approximately the boundary between unaffected and damaged ecosystems. The same conclusion was drawn by Posthuma et al. (1998a; 2001) and Smit et al. (in press) who compared HC_5 and HC_{50} values and the occurrence of adverse community effects from contaminated soils. In the study of Posthuma et al. (1998), the confirmation attempts also addressed the issue of mixtures, as mixtures are commonplace in the field. By assuming concentration additivity, concentrations of the occurring compounds were expressed as contamination units, which were equivalent to the HC_5 and HC_{50}, for the two adopted levels of comparisons. However, the field studies used highly diverse measurement endpoints and field exposures are often highly variable and poorly defined (Posthuma, 1997), so that usually very few useful data can be generated or traced from the literature. Like micro- and mesocosm data, field studies from the literature are an imprecise standard for SSD confirmation, although they also, in incidental cases, yielded important insights into the meaning of exposure of communities at or beyond the level of the quality criteria. The primary insight is that microcosm, mesocosm, and field data have not as yet provided evidence suggesting a change in p values for generic use from those currently used, because the studies are insufficiently imprecise to enable this (Posthuma et al., 1998a).

Field studies are realistic but present practical problems as a means of confirming the utility of SSDs. For example, a study of the effects of pesticide runoff into estuaries could not determine the accuracy of the PAFs, because uncertainties concerning episodic exposures in these complex systems dominated the results (Morton et al., 2000).

The criticisms of confirmation studies that relate to the dissimilarity of the endpoints of the SSDs and the microcosm, mesocosm, or field studies may be negated by applying SSD models to all studies. The authors of Chapter 9 determined the SSDs of both laboratory and field communities, and compared the models along the whole concentration range. This showed that the laboratory-based SSD was similar to the field-based SSD for one particular case: an acute exposure scenario for a specifically acting compound. In contrast, dissimilarity occurred in the other case, a chronic exposure scenario for a nonspecifically acting compound. In the latter case, the misfit may be related to a range of causes, including a complete nonoverlap of species incorporated in both SSDs.

Finally, SSDs may be partially confirmed by comparing their results to those of population or ecosystem models. This approach was followed by Forbes et al. (2001), who compared HC_5 values based on chronic NOECs to concentrations causing a 10% reduction in the population multiplication rate for mathematically simulated populations with a range of different life histories. They concluded that the HC_5 values are protective for the investigated populations in most cases, although some populations appeared not to be protected. They acknowledge that their results are dependent on a number of assumptions that should be explored through additional research. This approach was also followed by Klepper et al. (1999), who studied the responses of nematode communities in soil using a mixed approach of observations and food web modeling. This study clearly showed that soil contamination induces changes in biotic communities at almost any degree of contamination, be it micro- or macroscale responses (Posthuma, 1997). This in turn reinforces the argument that assessment endpoints should be chosen prior to SSD modeling, since confirmation or nonconfirmation of a certain use of SSDs would otherwise be determined by the ecological investigators bias toward micro- or macroscale phenomena.

21.8.3 SSD vs. Alternative Extrapolation Models

Most of the debates about the adequacy of SSDs have not carefully or clearly addressed alternative extrapolation models. The major alternative types of extrapolation models, factors and regression models, are discussed here. Other approaches should be considered for development, including those based on lower-level processes (i.e., toxicokinetics and toxicodynamics) or higher-level processes (i.e., population, community, or ecosystem dynamics).

The use of safety factors or assessment factors is commonly dismissed by advocates of SSDs as arbitrary and unsupported by theory. In contrast, Forbes and Forbes (1993) argued that SSDs are no better than "arbitrary assessment factors" because of several conceptual and technical issues, discussed earlier in this chapter. It has also been argued that SSDs are more difficult and uncertain than factors, and they are overly ambitious in their use of limited laboratory data (Calow, 1996; Calow and Forbes, 1997; Calow et al., 1997). Further, factors may be used with very small and inconsistent data sets (Fawell and Hedgecott, 1996).

The other alternative is the regression of one species, taxon, life stage, or other test result against another (e.g., Suter et al., 1983; Slooff et al., 1986). For example, one might regress rainbow trout LC_{50} values, all fish LC_{50} values, or all fish NOECs against fathead minnow LC_{50} values. The use of regression-based extrapolation models was dismissed without explanation by Wagner and Løkke (1991). Van Straalen and Denneman (1989) dismissed regression-derived extrapolation methods as lacking theoretical support. However, it is not apparent that any ecological theory supports the use of SSDs (Calow, 1998). Any set of data will have some distribution from which percentiles may be calculated. In contrast, the regression-based method of Suter et al. (1983), Suter and Rosen (1988), and Suter (1993) is based on the theory that species that are taxonomically similar are likely to have similar sensitivity to chemicals.

All these models (factors, regression equations, and SSDs) are more empirical than theoretical. The relative utility of each of the possible approaches should be demonstrated by comparisons of their predictive performance of the alternative models. This exercise has not been done.

Approaches that are seriously considered for routine use should be practical and acceptable for risk managers. The SSD concept has reached a certain status of acceptance in some countries due to its record of practical use in various conditions and insights into its conceptual adequacy. We believe that the SSD concept can be further developed, to address major points of criticism, including new insights in the availability of data and data patterns (Chapter 8). However, as ecotoxicology and risk assessment advance, it will be important to continue to consider alternatives to SSDs.

21.9 CONCLUSIONS

SSDs are now commonly used both to set environmental quality criteria and to estimate risks, but, as this chapter makes clear, many conceptual and technical issues have a context-dependent solution, or are as yet unresolved. Acceptance or abandonment of SSDs as a tool in the risk assessors' toolbox appeared to be strongly dependent on interactions between risk assessors and regulators at the national or international level. These interactions are influenced by historical decisions and precedents associated with the earliest applications of SSDs for derivation of EQC. Various technical solutions to criticisms have been put forward, including better use of existing data (Chapters 6 and 8), differentiating risks to taxonomic groups (Chapters 9 and 15), and using knowledge of TMoAs (Chapters 8 and 16). These solutions may or may not play a role in future application of SSDs, depending on the context. Analysis of SSDs may even improve the use of safety factors by making them more data driven.

It has become clear that SSDs are no panacea for risk assessment problems, since extrapolation is inherently uncertain (Chapman et al., 1998). Efforts should be directed at further confirmation of SSDs to achieve sufficient predictive capability for adequate risk-based criteria and ecosystem risk assessment for all types of compounds and situations. However, SSDs cannot be further confirmed or improved until we are clear concerning what endpoints we are attempting to predict. Clearly, the utility of SSDs is different for predicting responses of specific important species, of community structures, or of ecosystem functions.

Depending on the test endpoints used as input, species- or function-based sensitivity distributions (SSDs and FSDs, respectively) may be good models for the prediction of episodes of acute lethality, loss of species, diminished functions, or other population or community properties. Distribution-based methods may serve well as predictive models, or may serve to organize conventional toxicity data in a way that supports weight-of-evidence analyses by suggesting what responses are likely in biological surveys or tests of ambient media (De Zwart et al., 1998; Suter et al., 1999; Hall and Giddings, 2000). Because of these diverse interpretations, SSD models have been introduced into various frameworks for risk management such as life cycle assessment (Chapter 20).

Further demonstration of the potential of SSDs to predict ecological effects or to provide insight into the nature of ecotoxicological effects will depend on the development of the basic features of the SSD model itself and on strengthening the use of basic biochemistry, toxicology, and statistics in ecotoxicology. These developments are addressed in Chapter 22.

22 Conceptual and Technical Outlook on Species Sensitivity Distributions

Leo Posthuma, Theo P. Traas, Dick de Zwart, and Glenn W. Suter II

CONTENTS

1-56670-578-9/02/$0.00+$1.50
© 2002 by CRC Press LLC

Abstract — Species sensitivity distributions (SSDs) have been adopted in various risk
assessment frameworks, despite fundamental and operational criticisms. In this chapter,
several fundamental and technical options are suggested for further development and
implementation of the SSD concept. These suggestions have evolved from existing
lines of SSD evolution, which are extrapolated into the near future. A schematic table
for a "complete" SSD analysis is presented and illustrated with data from a hypothetical
ecosystem loaded with a mixture of contaminants. Various fundamental issues are
addressed by adopting concepts and data from toxicology and ecology, giving special
attention to the toxic mode of action in relation to species' features and the ecological
structure of the exposed ecosystem. The approach is compatible with the types of
community effects encountered in exposed field communities and the distinct measures
of effects that are used to quantify such responses at the level of different ecological
groups. It is a plausible general concept, but also provides the basis for further improved
SSD interpretations that may evolve when technical solutions offered by different
disciplines (statistics, ecology, toxicology, and environmental chemistry) are explicitly
linked to the SSDs. In our opinion, there is considerable latitude for improvement of
the SSD concept as a useful component of ecological risk assessment, especially when
it evolves from an approach in which statistical features dominate into one that is still
statistical in nature, but driven by a clear definition of the assessment endpoints and
by basic concepts from other disciplines.

22.1   USE AND OPTIONS OF SSDs

The review of issues and practices in the derivation and use of species sensitivity
distributions (SSDs) (Suter et al., Chapter 21) shows that two facts are relevant to
the developments of SSDs. First, many debates on the advantages and disadvantages
of SSDs have not yet been resolved; a selection of issues is given in Table 22.1.
Second, despite these unresolved issues, SSDs are currently used in practice, as
shown in Table 22.2 with details in the preceding chapters. The use of SSDs in
derivation of environmental quality criteria, risk assessment, and even life cycle
assessment of industrial products (Chapters 15 through 20) shows the breadth of
subdisciplines in which SSD-based results can be used, and underlines the importance of further development of SSD-based risk assessment methods.

TABLE 22.1
Key Advantages and Disadvantages of SSDs Mentioned in the Scientific and Regulatory Debates since the SSDs Were Introduced for Regulation of Toxic Compounds in the Environment

Advantages	Disadvantages
Concept and Use	
Conceptually transparent to decision makers and stakeholders	
More defensible than safety factors	Not proven more reliable than alternatives
Widely accepted by regulators and practitioners	
Easily understood	
Choice of percentiles and confidence limits may be based on the risk manager's preferences	
Input Data and Ecological Relevance	
Uses common toxicological data	Requires relatively large data sets
Toxicology	
No advantages	No disadvantages
Statistics	
Simple and familiar functions	No mechanistic components, purely empirical
	Fits of standard functions may be poor
	Diverse species sets result in polymodal distributions
Output Data and Ecological Relevance	
Include methods of individual chemicals and mixtures	Criteria based on confidence limits on low percentiles are overprotective
Represent potential effects on biodiversity	Test species are not a random sample of receptor species
May represent effects on species (probability of effects) or communities (proportion affected)	Species importance not incorporated
Clear graphical summary of the results of tests of multiple species	Sensitive species may be overrepresented
	Important species may be in the unprotected range
	Does not represent ecosystem functions

Note: The list is not exhaustive.

Source: Poster presentation of Suter et al. from the Interactive Poster session at the SETAC-U.S. conference in Philadelphia, 1999 (see Section A.2.1).

To further the development of SSD-based risk assessment, a conceptual and technical outlook on SSDs is presented in this chapter. The issues that are treated share the following characteristics:

TABLE 22.2
Adoption of SSDs by National and International Authorities

Adoption of SSDs	EQCs	ERA
National Authorities		
Australia	x	
Canada	x	x
Denmark	x	
The Netherlands	x	x
New Zealand	x	
United States	x	x
International Authorities		
OECD	x	
EU	x	

Note: The list contains the authorities where SSDs have some legal status; SSDs have also been mentioned by researchers in other countries (e.g., Germany, the Basque country; see Chapter 21).

1. They are likely to be of quantitative importance.
2. The outlines for potential improvements can be given based on current insights.
3. They are of particular importance to SSDs; that is, we do not address criticisms of the use of laboratory toxicity data that pertain to *all* methods based on such data.

The current insights have resulted from the works of the authors of the preceding chapters and from the debates that have taken place up to now. The identification of main issues for the outlook was further shaped by oral discussions between the authors of this chapter and various experts consulted in the final stages of preparing this book, regarding the question what they would see as main outlook issues. The lines of evolution in the SSD concept summarized in the preceding chapters are extrapolated into the near future, to sketch the characteristics of SSDs if the main conceptual problems can be solved.

The aims of this chapter are to stimulate discussion and thought on the further evolution of the SSD concept itself, and on what SSDs can offer to ecological risk assessment. The chapter is not intended to provide final answers. Research on the important issues should do that.

22.2 SCOPE OF THE OUTLOOK ISSUES

22.2.1 FUNDAMENTAL ISSUES

Suggested improvements to SSDs are various in nature, ranging from fundamental to the concept of SSDs to improvements of technical choices and procedures. For example, the lack of species interactions is a fundamental problem for SSDs, while the choice of distribution function is a technical issue. We recognize that the technical issues are important to the outcome of SSD-based assessment and regulation and that they require proper solutions. In this chapter, attention is focused on fundamental issues rather than on technical issues, although the distinction is not always clear.

The bottom line for any model including SSDs is the question: Does it accurately predict real phenomena? There are two fundamental questions addressed in this chapter:

1. Can we envision effective and feasible options to modify SSDs so that they more accurately predict the toxic effects of contaminants in a range of environmental compartments, for different communities, for a range of compounds (alone or in mixtures), and for various assessment purposes?
2. Can we determine the range of potential validity of estimates of effects derived from SSDs?

According to the analyses of Van den Brink et al. (Chapter 9), there are cases in which phenomena in the field are accurately predicted by an SSD, despite the ecological simplicity of the SSD concept. However, there are also cases of misfit between field responses and SSD prediction, and these might be of help to identify the key issues that need to be addressed to improve predictive accuracy of SSDs. This might change the idea that SSDs are a black box that, when fed with toxicity data, yields some profitable outcome, into a model that requires systematic information from various ecotoxicological subdisciplines — ecology, toxicology, and environmental chemistry — and transparent choices (Figure 22.1).

22.2.2 OPERATIONAL ISSUES

Some criticisms on the type of data used in SSDs cannot be solved because of limitations of the available data. It is evident that laboratory toxicity data may have limited relevance to the field; nonetheless, various adopted assessment methods (e.g., risk quotients) have been based on them. The reason for choosing laboratory toxicity data is an operational one. For the majority of compounds, there are no alternative data for any assessment method. The criticisms put forward in this respect apply to all assessment methods based on them. Further, all alternative types of ecotoxicological data have limitations when used to predict effects. Although some issues will be discussed, criticisms that apply to the state of development of ecotoxicology as a discipline are beyond the scope of this outlook.

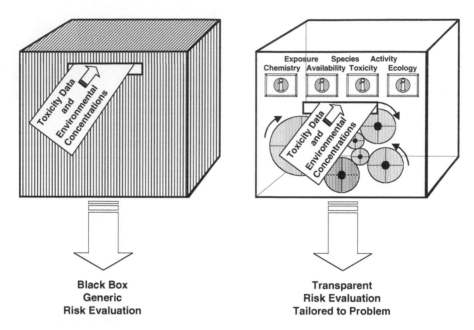

Black Box   **Transparent**
Generic   **Risk Evaluation**
Risk Evaluation   **Tailored to Problem**

FIGURE 22.1 Opposing extreme views on SSDs: (left) SSDs as a black box that yields the desired answers automatically vs. (right) SSDs as a transparent model that requires specific input of information from various disciplines to function properly.

22.2.3 INTRINSIC SSD CHARACTERISTICS

SSDs have clear characteristics and limitations intrinsic to statistical–empirical models, and they will therefore not yield the type of results that can be obtained with other models or studies such as bioassays in substrate samples from contaminated environments. Statistical–empirical approaches have not been designed to *explain* phenomena, and SSDs are thus, for example, inadequate models for demonstrating cause–effect chains. It is of no use to criticize SSDs for that if other methods that have the desired intrinsic property can be chosen in pertinent cases. It should be explicitly acknowledged that, like many other empirical relationships that are used in ecotoxicology (e.g., quantitative structure–activity relationships, QSARs), SSDs may be *of use* for certain assessment problems, just because they describe a phenomenon of interest, the relative sensitivity of species, or because they predict certain phenomena within acceptable limits of accuracy.

22.2.4 COMPARISON TO OTHER METHODS

SSDs are considered adopted methods that can be positioned in a range of assessment methods (Table 22.3). Each method has its own typical requirements and advantages and disadvantages, including technical requirements, data requirements, and output adequacy. Although the methods can simply be classified as in Table 22.3, we are not aware of systematic studies that would suggest SSDs to be *generally* better or

TABLE 22.3
Risk Assessment Techniques (middle column) Have Key Requirements Regarding Biological Input Data (left column) and Typical Characteristics (right column)

Key Requirement Regarding Biological Data	Risk Assessment Technique	Typical Characteristics
Laboratory toxicity data	Risk or hazard quotients	Often conservative choices
Laboratory toxicity data	Species sensitivity distributions	Statistical–empirical treatment of test data
Laboratory toxicity data and demographics	Population modeling	Simulation of important species
Laboratory toxicity data and ecosystem structure	Ecosystem/food web modeling	Simulation of trophic and competitive interactions
Toxicity test in environmental sample	Bioassays	Effects of environmental mixture in substrate of concern
pT	Bioassays + species sensitivity distribution	Effects of environmental mixture on range of species
Field observation	Field inventories	Problems with demonstrating effect and demonstrating cause–effect

Note: The combination of requirements and characteristics determines whether a technique can be considered "best available" for a given assessment problem. Risk or hazard quotients may be good techniques for Tier 1 assessments, where emphasis is on avoiding false negatives; field observations may suit best if certainty is required on local effects.

worse in adequately predicting community effects of toxicants in the environment than other methods. SSDs are widely used to derive environmental quality criteria (EQC) (Chapters 11 through 15) and to identify ecological risk at sites of concern, risk of compounds of concern, or sometimes risk for taxa of concern (Chapters 15 through 19). They can function as the sole instrument in a certain assessment, or in combination with other methods (e.g., a proposed triad-like approach, see De Zwart et al., 1998; Suter et al. 1999; Chapter 21) in order to enhance "weight of evidence." Because of a lack of studies that address the prediction accuracy of all available methods, the relative utility of the various approaches is beyond the scope of this chapter. It is, however, an important topic for future research.

22.3 FROM ASSESSMENT ENDPOINT TO ASSESSMENT APPROACH

The SSD approach has been adopted in various legal frameworks, and for a range of assessment types. Comparison of the adopted methods based on the contributed chapters shows that many operational aspects differ between the methods. This type of use evidently has many appearances and versions, which are similar only with respect to the central role of SSDs. For example, despite homology of protection concerns, and despite the common choice of the 5th percentile of an SSD, as in the United States and the Netherlands, the minimum number of data required, pretreatment of data, averaging in case of multiple data for the same species and endpoint, etc. all differ among the two countries. Part of the differences relate to local policy settings, part to expert judgment for a particular assessment. The use of SSDs in risk assessment of contaminated sites is less prescribed by routine methods, although formal protocols are operational (Warren-Hicks et al., Chapter 17). This lack of standardization is caused by the diversity of assessment problems that can be addressed using SSDs. As shown, for example, in Chapters 15 and 16, the questions posed to the risk assessor range from determining the relative local risks of contamination for a range of contaminated sites, a range of species assemblages, a range of compounds, and a range of time and spatial scales. The diversity of current SSD applications both in the derivation of EQCs and for ecological risk assessment demonstrates that the SSD concept is usually strongly adapted ("tailored") to the local context or to the assessment problem.

This has various implications. First, the definition of the assessment endpoint is of fundamental importance for shaping the SSD. The assessment problem should be clearly defined, and only after that can one judge whether SSDs are an appropriate option and, if so, which SSD. For example, if the protection target should be defined generically, on a national scale, without specific concerns for particular species, and with low chance of false negatives (using a conservative cutoff percentile), one SSD fitted to all data might suffice. One should not expect results, however, that provide a scientific proof that particular species groups or ecological functions are also protected. In those cases, the particular species groups or the ecological functions should have led to the definition of other SSDs, or other approaches. The approaches might differ in degree (SSD with other operational choices), like distinction of sensitive groups in the toxicity database, or of kind. Other kinds of approaches are, for example, functional sensitivity distributions (see Sijm et al., Chapter 12; Suter et al., Chapter 21), food web modeling (see Klepper et al., 1999), or assessments based on triad analyses (Chapman, 1986; De Zwart et al., 1998).

Second, the conclusions of the validation efforts concern primarily the *tailored* SSD that is being validated, and conclusions are only applicable to other SSD approaches when both are numerically indistinguishable. For example, the (U.S.) HC_5 that was subject of a validation study by Versteeg et al. (1999) has been derived with operational choices other than the (Dutch) HC_5 of Okkerman et al. (1991) and Emans et al. (1993). The different quantitative operational choices between different SSD analyses, e.g., how data are treated, require further attention to avoid misinterpretation of validation studies.

In conclusion, tailoring of the SSD to the assessment problem and its context has yielded a range of technical options to work with SSDs. Tailoring in future assessments seems especially useful in cases that are less simple than assessments of acute effects of specifically acting compounds, although, even then, the SSD tailored by selecting sensitive groups might better fit the field data than the overall SSD (Van den Brink et al., Chapter 9, Figure 9.2). In our opinion, the assessment endpoint, and the operational choices made to tailor SSDs to that assessment problem, should be explicitly listed for any assessment. This will strongly improve the possibilities to eventually choose the SSD (including related models, see below) that performs best regarding prediction accuracy.

22.4 FUNDAMENTAL TAILORING OF SSD DESIGN

22.4.1 FROM COMMUNITY RESPONSES IN THE FIELD TO FUNDAMENTAL SSD DESIGN

One of the fundamental limitations of SSDs that has often been mentioned is the lack of ecology. Many issues listed in Table 22.1 relate to this. Insights from ecology can be of use in two ways: in the outline of the SSD approach that is chosen (this section) and in the choice of input data (Section 22.5.1).

An analysis of ecological responses in contaminated environments may help to identify how an SSD approach could be optimized regarding ecology, assuming that sufficient toxicity data are available.

An analysis of existing field data suggests that species and species groups in contaminated ecosystems respond differently to exposure. At moderate contamination levels, there are opportunistic species or groups that show increased densities, indifferent species or groups that show no density change, and sensitive species or groups that show density reduction. Further, attention of scientists usually focuses on separate groups, for which different measurement techniques apply, for example, density counts for arthropods vs. activity measurements for microbial functions. These measures of effect often cannot be aggregated to a single response value. Moreover, according to the criticisms (Table 22.1), protection focuses often on specific types of species groups rather than on "an average community."

Examples of studies on species-level responses are Smit et al. (in press, terrestrial) and Van den Brink and co-workers (aquatic) cited in Chapter 9. One of the examples illustrating the range of responses of a broad range of species groups and functions has been provided in the work of Tyler (1984). Figure 22.2 illustrates the different typical response patterns as observed for a range of species groups and functions in a field pollution gradient.

The extension of the response typology from species to groups of species relates in part to the toxic mode of action of the compound. For example, the sensitive response of microarthropod species to organophosphate compounds relates to the presence of a receptor for these compounds in this whole group.

This analysis of field responses has implications for the way SSDs might need to be shaped. The generally recognizable response pattern is not reflected in all SSDs. Often, current SSDs yield a single estimate for a hazardous concentration, or a single estimate of risk for "a community," while no distinction is made between groups. Apparently, there is a difference between what is predicted (often one value) and the true phenomena to which the prediction should apply (more values).

According to the analysis, and when no limits are imposed upon the SSD methodology by the assessment context, it may be profitable to improve prediction accuracy by using several SSDs. In some current SSD applications, a distinction

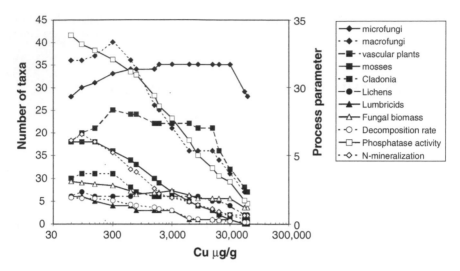

FIGURE 22.2 An illustration of the difference in response patterns of various species groups (black markers) and functions (white markers) in a field gradient with metal contamination of copper, zinc, and cadmium. The *x*-axis marks only the copper concentrations, but the response is also determined by the other metals, with concentrations correlated to the copper concentrations. The pattern has been reconstructed from the original figures of Tyler (1984), and serves to illustrate the diversity of response patterns in field observations.

between species groups is already made; see, for example, Solomon and Takacs (Chapter 15) and Traas et al. (Chapter 16). For certain assessment problems, various SSDs are fitted to subselections of the available data set, yielding multiple SSDs and estimates of risk. This is of major importance for validation studies. A single risk estimate of, say, 0.7 based on an SSD of all available toxicity data is difficult to interpret if some species groups show an indifferent response (effect: zero), while large density reductions (say, 80 to 100%) occur in other groups. Validation efforts using an overall SSD would fail, while efforts based on SSDs for various groups could show that SSDs are accurate, using the same field and toxicity data sets. The issue is illustrated in Figure 22.3.

22.4.2 FROM FUNDAMENTAL SSD DESIGN TO IMPROVED PREDICTION ACCURACY OF SSDS

The fundamental improvements and technical consequences of the SSD concept following from the analysis of community responses in the field have in part already been formulated (e.g., Solomon and Takacs, Chapter 15, and references therein). They relate to combining knowledge from the ecology of species with knowledge on the toxic mode of action of toxic compounds. A well-known example of this line of reasoning is the ecological risk assessment of atrazine, a photosynthesis inhibitor, for which the data are divided (among other distinctions) according to the presence or absence of the capacity for photosynthesis. Technically, the approach can become feasible for communities exposed to a mixture of pollutants, when database analyses

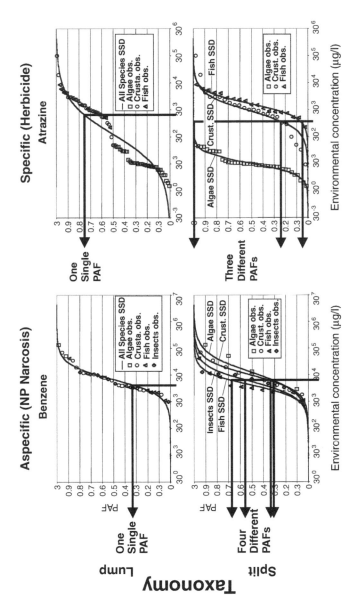

FIGURE 22.3   An illustration of assignment of one SSD to all compiled aquatic toxicity data for a compound (top panels) vs. the assignment of various SSDs to subselections of data (lower panels) for a compound with nonspecific and specific toxic modes of action, respectively (benzene, left panels; atrazine, right panels), for a case where SSDs are used to quantify risk of environmental contamination. The approach with multiple SSDs relates conceptually better to the phenomena occurring in exposed field communities (see Figure 22.2), allows for better discrimination among assessment endpoints, and is compatible with concepts and data from ecology and toxicology. Note that, although statistically the single SSD fit to the benzene data seems an obvious choice with sufficient goodness-of-fit, ecological considerations might suggest subdividing the data is better with respect to field responses and assessment endpoints. The risk estimates for benzene vary between 0.3 and 0.7.

TABLE 22.4
Extending the Evolutionary Line of SSD Development by Distinguishing Risks for Separate Groups, by Taking into Account the Toxic Mode of Action of Compounds (horizontal) and the Presence or Absence of Specific Target Sites of Toxic Action in Species Groups (vertical)

Taxonomical Group	Compound(s): TMoA: Exceedence HC$_5$:	Benzene + Naphthalene NP Narcosis No	Zinc Zn-action No	Cadmium Cd-action Yes	Atrazine Photos. inh. Yes	Malathion AChE inh. No
Algae		0.010	0.013	**0.058**	**0.143**	ND
Cyanobacteria		ND	ND	ND	**0.381**	ND
Crustaceans		0.005	**0.062**	0.014	0.001	**0.283**
Insect larvae		0.010	0.026	0.000	0.000	0.009
Mollusks		0.006	**0.104**	0.007	0.000	0.000
Worms and leeches		0.003	0.012	0.001	ND	0.000
Amphibia		0.004	ND	0.000	0.000	0.000
Fish		0.007	0.045	0.017	0.000	0.000

Note: The values are estimates of risk from SSD analyses expressed as msPAF (see Chapter 16), scaled between 0 and 1, but can pertain to any risk estimate presented in the previous chapters. The example pertains to a hypothetical freshwater pond, for which a hypothetical but realistic exposure was assumed. The calculations are made on the basis of the database of De Zwart (Chapter 8). The environmental quality criterion for atrazine and cadmium, defined as the 5th percentile of the SSD (HC$_5$) of all toxicity data for a compound was exceeded. Note that this is not the case for the other compounds, even though the HC$_5$ values for specific species groups (in **bold**) are exceeded. For explanation, see text. TMoA = distinguished toxic modes of action; NP narcosis = nonpolar narcosis; Photos. inh. = photosynthesis inhibition; AChE inh. = acetylcholinesterase inhibition. ND = not determined.

as done by e.g., De Zwart (Chapter 8) and theory on mixture assessments (Traas et al., Chapter 16) improve data availability and interpretation.

When this line of reasoning is followed a bit further for a hypothetical community that is exposed to a mixture of contaminants with a range of toxic modes of action, the result will look like the example of Table 22.4. The example builds further on the illustration given in Figure 22.3.

Table 22.4 shows estimates of risk for a hypothetical but realistic aquatic community that is exposed to a hypothetical but realistic mixture of contaminants. The estimates of risk in the central part of the table are direct results from 40 separate SSDs constructed from the available toxicity data from which 40 estimates of risk were calculated. In this example, risks are expressed as potentially affected fractions of species (PAF) (see Traas et al., Chapter 16), but the same principle can be followed for other estimates of risk, that is, those involving both SSDs and exposure concentration distributions (see Aldenberg et al., Chapter 5; Solomon and Takacs, Chapter 15; Warren-Hicks et al., Chapter 17, and references therein).

Many will object to this approach because of the lack of toxicity data. In case of lack of data, the confidence interval for each estimated value might increase as a trade-off of "data dilution," and "empty cells" might appear. However, several methods exist to tackle the problem of small data sets (Aldenberg and Luttik, Chapter 6). The example is representative of some metals and other well-studied and important pollutants, and it may provide some relevant insights that are worth further study.

The groups have in part been distinguished based on the toxic modes of action in the mixture. For example, photosynthesis inhibition led to the distinction of species groups with and without photosynthesis capabilities. Further, acetylcholinesterase inhibition led to the distinction of species groups with and without acetylcholine as a neurotransmitter. Both distinctions require coupling of toxicological insights with ecological knowledge of organisms. The result will likely be unimodal SSDs rather than polymodal SSDs fitted to all the data. Further, the distinction in this example was driven by interest for specific groups of species (e.g., fish), causing the eventual subdivision of taxa as shown in the table. Evidently, the final exercise can depend on both the assessment endpoints and ecological and toxicological knowledge. An earlier formalization on the distinction of groups in data sets is presented by Solomon and Takacs (Chapter 15, Figure 15.12), and is driven by toxic mode of action, taxonomy, habitat, and autecology of species. A recent application is provided by Brix et al. (2001).

A limit for concern could be set at the 5th percentile (HC_5) of the SSD based on all data. In this case, the results shown in Table 22.4 suggest that groups likely experience different risks due to different compounds. For example, algae and cyanobacteria, both possessing the capacity for photosynthesis, are the only groups moderately at risk from an exposure concentration of atrazine that only slightly exceeds the limit of concern. The other groups have no specific receptor for a herbicide, and are unlikely to suffer from atrazine. An important phenomenon is shown for the toxicity of malathion to crustaceans. For this group, a risk value of 0.28 was calculated, whereas on the basis of the HC_5 calculated from all toxicity data no risk larger than 5% was expected (Table 22.5). The value found for mollusks in relation to cadmium exposure was, however, not expected from knowledge of specific toxic mode of action of cadmium, or from knowledge that mollusks would have a specific "cadmium receptor." The identification of chemicals of negligible concern in an initial (lower-tier) risk assessment should be handled with care. Compounds of concern should not only be identified in case the limit of concern based on *all* toxicity data is exceeded, but also when a similar limit for certain groups is exceeded. Unexpected sensitivities may show up, as in the case of mollusks and cadmium.

Further inspection of the meaning of the SSD results in Table 22.5 reinforces a conclusion drawn earlier in Chapter 15 by Solomon and Takacs (Figure 15.3), that responses in the field need not be similar to the expectations of SSD analyses. There will likely be concentration-dependent biases, with fewer effects at low concentrations and more effects at high concentrations (and risk estimates). This relates to ecological interactions. Ecological interactions are *not* addressed by SSDs as a model (they never have and they likely never shall be from SSDs only) but they can be deduced with some common ecological reasoning, as in Table 22.4. If, for example, the risk for algae and cyanobacteria for atrazine were >0.8, then it would not be difficult to imagine the ecological consequences at a range of trophic levels, even without formal food web modeling. Such reasoning is impossible when a single risk

TABLE 22.5
Extension of the SSD Approach with Concepts from the Disciplines of Toxicology and Ecology

Taxonomical Group	Compound(s): Benzene + Naphthalene TMoA: NP Narcosis	Zinc Zn-action	Cadmium Cd-action	Atrazine Photos. inh.	Malathion AChE inh.	Aggregated Risk/ Joint Toxicity (over compounds, per group)
Algae	0.010	0.013	**0.058**	**0.143**	ND	**0.211**
Cyanobacteria	ND	ND	ND	**0.381**	ND	**0.381**
Crustaceans	0.005	**0.062**	0.014	0.001	**0.283**	**0.341**
Insect larvae	0.010	0.026	0.000	0.000	0.009	0.045
Mollusks	0.006	**0.104**	0.007	0.000	0.000	**0.116**
Worms and leeches	0.003	0.012	0.001	ND	0.000	0.016
Amphibia	0.004	ND	0.000	0.000	0.000	0.004
Fish	0.007	0.045	0.017	0.000	0.000	**0.068**
						msPAF: **0.077**[a]
All Species	0.015	0.029	0.095	0.072	0.008	msPAF: **0.204**[b]

Note: The estimates of risk obtained from SSDs (data columns 1–5, same data as Table 22.4) can be aggregated to overall risk (expressed as msPAF) of all compounds for a species group (rightmost column). Different protocols can be followed. For the hypothetical case, msPAF values per species group were calculated assuming concentration addition within toxic modes of action, and response addition between toxic modes of action. msPAF values higher than the HC_5 are printed in **bold**. TMoA = distinguished toxic modes of action; NP narcosis = nonpolar narcosis; Photos. inh. = photosynthesis inhibitor; AChE inh. = acetylcholinesterase inhibitor. ND = not determined.

[a] The overall aggregation value (msPAF) shown here is calculated as the average of normalized values in the last column. Various aggregation protocols can be envisaged (see text).

[b] The msPAF value for all species is calculated assuming concentration addition within toxic modes of action, and response addition between toxic modes of action.

estimate is produced from an overall SSD for each compound. For the current example, there are only a few "Not Determined" entries. This relates to the choice of compounds. When less frequently tested compounds are assessed, one can make use of small data set statistics (as elaborated by Aldenberg and Luttik, Chapter 6, which in turn relates to regularities between SSDs as addressed by De Zwart, Chapter 8), or use the fallback option of constructing overall SSDs from all toxicity data.

The subdivision of toxicity data can be driven by a variety of reasons. As an extension to the reasons presented when introducing Table 22.4, one may, for example, wish to assess the local risk for separate functional groups in a contaminant-exposed community. Redundancy and resiliency can thus be introduced in an assessment. This would address various criticisms, such as the lack of attention for important species or functional endpoints (see Table 22.1). A maximum acceptable risk might be set for each functional group, to avoid unacceptable threat to any ecosystem function. Although the functional groups might not have a unimodal sensitivity pattern, this would allow assessment of food web resiliency subsequent to more generic SSD application. First results from incorporation of SSDs in a tritrophic food web model indicated that overall ecosystem functions are to a certain extent resilient to an increased risk of toxic compounds, but values-sensitive species that are already at low risk are replaced by more tolerant ones (Klepper et al., 1999; Figure 22.4). This effect also seemed to occur in chronically exposed nematode communities (Klepper et al., 1999; Smit et al., in press). We think that there is considerable room for further improvement of SSDs by explicitly addressing toxic mode of action and ecosystem function and structure.

22.4.3 FURTHER USE OF TOXICOLOGICAL INSIGHTS: ASSESSMENTS FOR MIXTURES

One of the major perceived advantages of SSD-based risk assessment over risk quotients (see Table 22.1, Solomon and Takacs, Chapter 15; Traas et al., Chapter 16; Warren-Hicks et al., Chapter 17, and consulted experts) is the improved estimate of the cumulative risk of mixtures, as dealt with in several assessments (see, e.g., Traas et al., Chapter 16; Warren-Hicks et al., Chapter 17). They generally make use of classical toxicological principles, such as concentration or response addition. Validation of the proposed risk assessment methods for mixtures is, however, necessary in view of the almost complete lack of validation in literature (see Suter et al., Chapter 21). This is especially important in the context of the not necessarily justified "linear" transfer of toxicological concepts (built on molecular concepts of receptor interaction in organism tissues) to communities of organisms.

The "design" of Table 22.4 is compatible with the idea of mixture assessment using concentration and response addition, at least when the assignment of subgroups has been (partly) based on toxic modes of action present in the mixture. Aggregation of the simultaneous action of the compounds for a species group likely makes toxicologically more sense than aggregation of risk for "whole communities," since the issue of toxic mode of action is addressed specifically.

The following background information is important. Given a mixture, members of a species group can be exposed to compounds for which they have specific

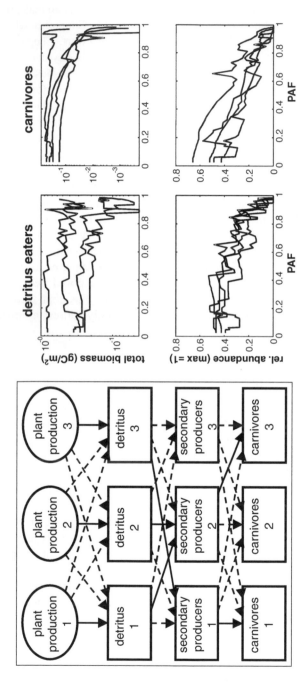

FIGURE 22.4 (Left) A tritrophic food web, to analyze the ways ecological interactions may influence community responses to toxicants: if predators and/or competitors are more sensitive than a particular species, density of a species may increase, whereas toxicant exposure in combination with food shortage and predation may drive a species to extinction before an acutely lethal concentration is reached. The most energy-efficient food chains in the web are indicated with solid arrows. (Right) The responses of functional groups (detritus eaters and carnivores) to increased exposure, quantified as PAF^{NOEC}. Considerable redundancy exists for total biomass in functional groups (top panels), but at the same time, a steady loss of densities per species and reduction of species diversity occurs (lower panels). The lines are four different model runs with ten species for each group. (Modified from Klepper et al., 1999.)

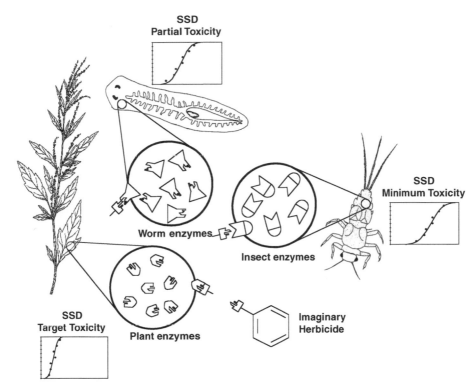

FIGURE 22.5 Target, partial, and minimum toxicity, as illustrated here, will be distinguishable when SSDs are worked out as in Table 22.4. Target toxicity implies specific interactions between contaminant molecules and molecular receptors at the site of toxic action. Minimum toxicity implies no specific interaction. Partial toxicity implies that the contaminant molecules apparently have an interaction with molecular receptors with partial inhibition of function.

receptors (what can be called "target toxicity") and to other compounds for which the interaction may be described in two ways. The first way concerns minimum or baseline toxicity (e.g., Verhaar et al., 1992), the toxicity that any compound exerts on organisms due to nonspecific interactions with membranes. The second way is what can be called "partial toxicity," that is, the toxicity that exceeds minimum or baseline toxicity, but that is attributable to partial binding to a specific receptor. The three types of toxicity are illustrated in Figure 22.5. The different mechanisms need not be assigned in the analysis, but the three mechanisms are of importance since they show up in the shapes of the SSDs that are basic to the risk estimates in Table 22.4. The shape of the SSDs will probably change from sensitive to moderately sensitive to insensitive (SSDs shift to the right) and from steep to semiflat and flat (declining slope), for target toxicity, partial toxicity, and baseline toxicity, respectively. For example, the slope of the SSD for atrazine for algae is relatively steep, whereas the slope for fish is flatter (for a logistic SSD, $\beta = 0.22$ vs. $\beta = 0.34$, respectively).

The aggregation of the SSD-based risk data in Table 22.4 can (technically) proceed according to different assumptions, and will yield estimated risks of the mixture for each species group (Table 22.5, right-hand column). One can make use of (relative) concentration addition for all aggregations, since various reviews suggest that this rule of calculus (although not mechanistically justified) appears to predict joint toxicity of chemicals adequately at the species level (see Traas et al., Chapter 16, for references). One can also make use of both concentration addition and response addition, as done by Traas et al. (Chapter 16).

In Table 22.5, the aggregation protocol was as follows: concentration addition for compounds with the same toxic mode of action (benzene and naphthalene), and response addition (assuming $r = 0$) for compounds with dissimilar toxic modes of action. The results show that for five of eight groups the mixture risk exceeds the level of concern of 0.05. Some groups are likely to be affected by this mixture; others are unlikely. The estimated value of aggregated risk (multisubstance PAF, or msPAF) over all compounds based on SSDs of the total toxicity data in this example is 0.204, which is clearly less informative regarding the species groups most at risk. Exaggerated, the question may be posed 0.204 *for what*?

Exploratory calculations (D. van de Meent, personal communication) have shown that the slopes of the SSDs of the compounds in the mixture can be important for the aggregated estimate of risk. When the slopes are all moderate (with a logistic SSD β value of approximately 0.4 as often found for nonpolar narcosis), the aggregated estimate of risk is similar for both rules of calculus (assuming $r = 0$ for the response addition approach). However, the aggregated estimates become different if slope differences exist as attributable to mechanistic reasons (as illustrated in Figure 22.5). The correct protocol to address mixture risks should therefore be established by further theoretical and practical confirmation of the aggregated risks of Table 22.5.

22.4.4 FURTHER AGGREGATION

From a technical viewpoint, the aggregated risk estimates in the right-hand column of Table 22.5 can be further aggregated to yield one value of risk for the contaminated site. Such an overall value may be of use when the assessment endpoint is to rank sites in decreasing order of ecological risk, e.g., for deciding remediation urgency. Various aggregation protocols may be chosen. On the one hand, a mathematical solution can be chosen, such as taking the normalized average of the separate estimates, which yields the msPAF value shown in Table 22.5 (0.077). On the other hand, the aggregation may be directed by the assessment endpoint. For example, for ecological reasons, one may value primary producers more than primary consumers. Alternatively, driven by assessment targets, if fish are ten times more valued than algae because of game fishing, the protocol can be adapted to that endpoint definition. The wish to execute such aggregations may, however, be less frequent than toxicological aggregation. Assessment of cumulative risks on species groups seems a more logical question in many assessment frameworks than assessment of the overall effects of one compound.

In whatever direction these issues may evolve, the proposed analysis would eventually allow for ranking sites (using the highest data aggregation), ranking species groups within sites (using the joint-effect aggregation), or ranking the most potent compounds (using the basic risk values).

22.4.5 CONCLUSIONS ON FUNDAMENTAL TAILORING OF SSD DESIGN

SSD methodologies can evolve further into approaches that encompass subselection of data to derive various SSDs for a single assessment (see Table 22.5). This evolution is compatible with the range of response types found in field communities in contaminant-exposed ecosystems, with toxicological theories on molecule–receptor interactions, with toxicological theories on mixture effects, and with ideas on autecological features of species. Implementation requires further theoretical work on the issues identified above, especially the methodology to distinguish relevant groups in relation to the toxicological modes of action, ecological features of organism, and the joint effects of chemicals at the community level. Further, experimental evidence that those SSD approaches yield more adequate predictions of field effects is needed. Depending on the assessment endpoint, the design of Table 22.5 can be modified, or only part of the table is required. The number of data needed for a complete analysis is higher than for the classical single-SSD approach, but the regularities among SSDs and the techniques to handle small data sets imply better practicality than might be expected.

In estimating risk for separate groups for each compound, clear linkage of the statistics of SSDs to current insights in ecology and toxicology requires further thought. If this can be established, the most fundamental criticisms on SSDs, namely, that they lack or neglect insights from toxicology and ecology, are addressed. We think that the fundamental improvements that can be made when incorporating ecological and toxicological concepts into the SSD concept bring us to the borders of what would be logical applications of SSDs.

22.5 FURTHER TAILORING OF SSDS USING ECOLOGICAL, TOXICOLOGICAL, AND ENVIRONMENTAL CHEMICAL INFORMATION

In the previous paragraph, the issue of field relevance was addressed by extending the lines of SSD evolution regarding ecology and toxicology. Many additional issues must still be addressed before risk estimates can be printed in the central cells of Table 22.5, such as biological availability differences between laboratory and field conditions. These issues are treated as they relate to input data selection and pretreatment (see Table 22.1). They are relevant for all SSDs derived in the process of risk assessment, whether using "lumped" or "split" toxicity data sets (see Figure 22.3).

22.5.1 DATA COLLECTION: ECOLOGICAL ASPECTS REGARDING TEST ENDPOINT

The endpoint of the assessment currently determines the selection of toxicity data types that are summarized in SSDs. Classical choices for toxicity data to construct

SSDs are no-observed-effects concentrations (NOECs) (Chapters 11, 12, and 14) or low-effect concentrations (Chapter 13) for derivation of protective quality criteria, and the use of other test endpoints in risk assessment (Chapters 15 through 19). No theoretical limitation exists regarding any test endpoint to be used for SSD construction. There might even be preference for acute data over chronic data, because of higher data availability. In view of the assessment endpoint, SSDs could also be based on test endpoints other than the classical ones, such as LC_{20} values to judge acute effects of spills, or EC_5 values if the NOEC is being replaced by statistically sounder test endpoints. A practical objection can be raised against reconsideration of original toxicity data. That is, this could be interpreted as a need to reconstruct existing databases, by extending them with all original dose–response data or with parameters of full dose–response curves. To redesign the existing databases (>100,000 entries) into the desired format would require a major effort. On the other hand, for assessment problems under investigation, the existing databases could be used for the purpose of identifying the original literature sources, which could be reinterpreted with respect to the desired test endpoint. The latter approach is considerably more pragmatic, since for many compounds the number of tests to be interpreted is small. Stepwise, existing databases could be extended along with the reconsideration of compounds.

The test endpoint choice is also relevant in view of the slopes of original concentration–effect curves, which may differ between species (as in Figure 22.5 for SSDs). There is a homology between these slopes and the slopes of SSDs. Risk quotient calculations fail to produce sensible results where they are used to identify the compound posing the highest risk when for both compounds, say, a five times exceedence of the quality criterion is observed. Because of the dissimilarity of the slopes of the SSDs, the risk of a five-times exceedence may be much higher for specifically acting compounds than for compounds with minimum toxicity (see Solomon and Takacs, Chapter 15). Similarly, risk predictions based on SSD^{NOEC} values might yield conclusions different from the ones based on SSD^{EC20} values, simply because of neglect of slope divergence of the original concentration–effect curves of the tested species. The effect of slope divergence between toxicity tests on risk assessment output is illustrated in Figure 22.6.

Figure 22.6 (top panel) shows the SSDs resulting from acute toxicity tests with chlorpyrifos. Acute concentration–effect data for this specifically acting compound were collected for three crustaceans and four aquatic insects (larval stages). For comparisons with the derivation of HC_p values from chronic data, a typical SSD^{NOEC} is provided, which summarizes the available chronic toxicity data on target and nontarget organisms. The SSDs based on acute data ($SSD^{Acute,EC5}$, $SSD^{Acute,EC50}$, $SSD^{Acute,EC95}$) are positioned (far) right to the $SSD^{chronic}$, and are steeper. This shows that target organisms are on average far more susceptible than the "average" aquatic organism, reinforcing the conclusion that separate SSDs for target and nontarget organisms should be interpreted for appropriate ecological resolution and interpretation (see also Figure 22.3). An environmental concentration of 1 µg/l will expose only about 30% of the collection of "generic aquatic organisms" above their chronic NOEC, but will most probably eradicate the more susceptible species since the risk associated with this concentration (PAF^{EC95}) is about 50%. Further, the acute SSDs

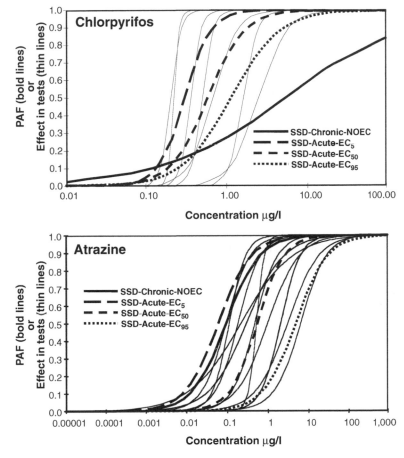

FIGURE 22.6 (Top) Concentration–effect curves for acute exposure of seven species to chlorpyrifos (thin lines), and the SSDs derived from them for different effect levels (SSD[EC5], SSD[EC50], SSD[EC95]), including an SSD[NOEC] of all available chronic toxicity data. (Bottom) Concentration–effect curves for chronic exposure of 11 species to atrazine (thin lines), and the SSDs derived from them (SSD[EC5], SSD[EC50], SSD[EC95], and SSD[NOEC] for comparison). (Chlorpyrifos data courtesy of R.P.A Van Wijngaarden, Alterra, the Netherlands.)

have slightly different slopes. Hence, a single environmental concentration can generate various estimates of risk with shifting relative risk ratios. For example, at 0.3 µg/l, the EC_5, EC_{50}, and EC_{95} are exceeded for approximately 80, 50, and 10% of the species (8:5:1); this ratio shifts to (e.g.) 1.3:1.15:1 at approximately 0.8 µg/l. Evidently, when there is slope divergence for the original test data, the conclusions of risk assessment procedures will differ when different test endpoints are chosen.

The other example pertains to atrazine toxicity (Figure 22.6, lower panel). The concentration–response relationships of this photosynthesis-inhibiting compound were collected for a range of aquatic organisms. The species are cyanophyta (1), chlorophyta (5), crustaceans (1), and fishes (4). Despite that algae are the target group, the SSDs were based on the overall data set. First, this example shows that the SSD^{NOEC} is almost identical to the SSD^{EC5}. An environmental concentration of 0.01 mg/l will only expose about 10% of the aquatic species above their chronic NOEC or EC$_5$. An environmental concentration of 10 mg/l will pose risk to approximately 100% of the species to some extent when judged by SSD^{NOEC} or SSD^{EC5}. The latter risk estimates are hardly informative, and indistinguishable from the risks of halved or doubled concentrations. However, when considering the SSD^{EC50} or the SSD^{EC95}, the analyses suggest that environmental concentration of 10 mg/l will likely imply 95% risk of EC$_{50}$ exceedence and 70% of EC$_{95}$ exceedence. Results can be badly interpretable when highly sensitive test endpoints are used to make risk assessments at high exposure levels (e.g., NOEC exceedence of 100%).

One of the most important issues regarding the choice of test endpoints to construct SSDs is the relationship between the test endpoint and population performance. For NOECs, Suter et al. (Chapter 21) already demonstrated that the relationship between test endpoint and population performance is unclear for especially chronic test endpoints, partly for statistical, partly for ecological reasons. Van Straalen et al. (1989) and Kammenga et al. (1996) showed that population performance under exposure was not necessarily determined by toxic effects on the most sensitive life cycle characteristic. Barnthouse et al. (1990) demonstrated that for some fish populations, typical chronic effects thresholds correspond to population extinction if exposure continued for multiple generations. There is, however, no fundamental theory on the basis of which one can *a priori* predict how a population would respond to toxic stress. For selection of the most appropriate test data, research into modeling the population-level implications of chronic exposure for a range of species and life cylce types is needed to uncover relationships between life history types, toxicant sensitivity of test endpoints, and population sensitivity. An overview of the current status of theories and data is provided by Kammenga and Laskowski (2000).

22.5.2 DATA COLLECTION: IDENTIFYING AND CORRECTING FOR NONRANDOM SPECIES SELECTION

One of the serious failings of SSDs is the assumption that a nonrandom sample of test species represents the receiving communities. The obvious solution, testing random sets of species, is not feasible. Full use of the available toxicity data is the usual basis for executing assessments, instead of a scrupulously selected data set that would mimic the exposed community.

An examination of the sets of test data used in SSDs reveals that some taxa such as daphnids in water and oligochaetes in soil are overrepresented and others are underrepresented. This may or may not have implications for the assessment, depending on the relative range of sensitivities in the test species and in the field. An option to identify whether nonrandomness of species selection occurs is comparison of the test data for the compound under consideration with test data from related compounds, for which a broader range of test data is available. In case the comparisons yield similar slopes for the SSD, the implication of nonrandom sampling seems more limited than for highly dissimilar slopes. However, even large data sets are not unbiased.

When there is serious doubt of nonrandom species selection, the first step to correct for bias introduced by this is the taxonomic characterization of the target community. Only after that, corrections can be made for bias, by weighting or by filling in the missing species. It is evident that this characterization is difficult when SSD results are generically used, that is: What is the composition of the receiving community when national quality criteria are to be set? For specific assessments, community composition is more clearly defined by an identifiable receiving set of species.

A relatively simple approach might be to weight the observations based on the proportion of the community that they represent (Suter, 1993). For example, if there are an equal number of fish and invertebrate tests for an SSD for aquatic animals, and fish constitute only 10% of species in an aquatic community, then each invertebrate observation could be given nine times the weight of the fish. Clearly, more detailed taxonomic information could be used in the weighting. Alternatively, weighting could be based on abundance or biomass rather than the number of species in a taxon, or characteristics other than taxonomy such as trophic level might be used in weighting. The weighting procedure is thus strongly influenced by the assessment endpoint.

An alternative to weighting might be to use interspecies extrapolation models to fill out the community. One may define a set of representative species as a surrogate for the community (see, e.g., Traas et al., Chapter 19) or characterize the community. For vertebrates and vascular plants, it is reasonably simple to list the resident species for a site or region. For invertebrates it may be more appropriate to list the genera or families. One may then estimate the sensitivity of each species or higher taxon using some property that is believed to control sensitivity. For terrestrial vertebrates, this is commonly assumed to be body mass or a fractional exponent of mass, and models are developed by allometric regression (Sample and Arenal, 1999). Allometry has also been applied to aquatic species (Newman and Heagler, 1991). Taxonomy provides another basis for extrapolation from test species to individual receptor species (Suter, 1993; 1998c). One may use regression models or factors derived from regression models to extrapolate across species, genera, families, orders, or classes depending on the relationship between a receptor species and the most closely related test species (Calabrese and Baldwin, 1994). Quantitative structure sensitivity relationships (QSSRs) have been derived to identify sensitivity patterns among species for a range of compounds (Vaal et al., 1997a,b; 2000), and database pattern analyses (De Zwart, Chapter 8) may provide relevant data. Alternatively, extrapolation could be performed using toxicokinetic or toxicodynamic models. Traas et al. (Chapter 19) go some distance in this direction by using exposure models to estimate intake for particular species as a function of soil concentration. This approach could be extended to include modeling of delivered dose (toxicokinetics) and response to delivered doses (toxicodynamics).

The approaches discussed all have potentially important advantages. If the sensitivity of all species in a community is measured or estimated, then we have not only eliminated the problem of sampling bias, but also the problem of nonindependence of sampled observations because we are no longer sampling. Instead, we are describing the distribution of the population of species (i.e., the community). However, uncertainties associated with the extrapolation models are introduced. All of these suggestions would need to be the subjects of research and methods development before they could be recommended. Because of the importance of the sampling biases in SSDs, they or some alternative for correcting for bias must be developed.

22.5.3 DATA COLLECTION: CORRECTING FOR LACK OF INDEPENDENCE BETWEEN TOXICITY DATA

The phylogenetic relationships among the species violate the assumption that sensitivities of toxicity data used to construct SSDs are independent. The U.S. EPA practice of combining data for congeneric species greatly reduces the problem but does not entirely eliminate it. An alternative is to correct for the correlation structure inherent in taxonomy, as is commonly done in comparative biology (Nagy et al., 1999). Techniques to generate statistically independent observations of species properties have been published (Garland et al., 1992; 1993). However, these techniques have not been applied to SSDs. If, like Nagy, one is trying to define a taxonomy-independent relationship between body size and metabolism, then it is necessary to eliminate taxonomic correlation. However, if one is interested in describing communities of real species, it seems to these authors that a taxonomy-independent model is not desired. Someone should look carefully into the independence issue to determine its importance in practice and to explore potential solutions.

22.5.4 DATA COLLECTION: ECOLOGICAL ASPECTS REGARDING AUTECOLOGICAL FEATURES OF SPECIES AND BIOGEOGRAPHY

There are various current SSD approaches that consider autecological features of species in the phase of data collection. The most evident one is the subdivision of aquatic toxicity data into a data set for freshwater species and a data set for saltwater species. For certain compounds, the SSDs may be highly different, motivating the use of this distinction in making the assessment, whereas for other compounds there is no statistical difference between sensitivity in both water types. The issue relates to the criticism that test species are not a random choice for the receptor species.

A pragmatic option to address this criticism is that toxicity data should only be used for an assessment when the species do or can occur in the area under consideration. This suggests an evident solution: if one knows the biogeographical distribution of species, one can simply select the appropriate species for an assessment. This would ask for explicit linkage of biogeographical databases to the SSD concept.

Strict application of this principle would, however, likely cause a serious reduction in the number of appropriate toxicity data, with associated loss of statistical accuracy. Alternatively, the autecological features of species could be taken into account, by asking the question "*can* a species occur in the environmental compartment under consideration," replacing *actual* occurrence with an estimate of *potential* occurrence. Systematic collection of autecological data could be useful to assess potential occurrence of species in SSD-based assessments. For many species used in toxicity testing, some autecological information has been gathered

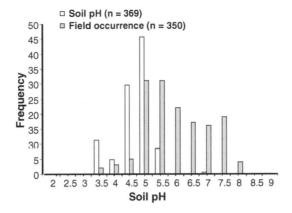

Soil type	Occurrence			
	pH	OM	...	Overall
Sand	no	yes	yes	no
Clay	yes	yes	yes	yes
Peat	no	yes	no	no

FIGURE 22.7 A comparison of autecological features of organisms with substrate characteristics. (Top) Frequency distribution for the oligochaete worm *Allolobophora caliginosa* regarding occurrence in field soils in relation to soil pH (150 observations) are compared to the distribution of soil pH values of calcium-poor sandy soils in the Netherlands (169 data points). (Bottom) Comparisons can be made for various ecologically relevant characteristics, and interpretation of "probability of occurrence" over all relevant factors can be used to identify the soil types in which the species can be regarded as potentially present, calibrated on biogeographical data. (Data courtesy of E.M. Dirven-van Breemen and P.L.A. van Vlaardingen.)

in the phase of designing tests in order to optimize the test. An example of such an analysis, from current work in the Netherlands, is provided in Figure 22.7. Evidently, the result of selecting species on biogeographical distribution or autecological preferences causes a reduction of data availability (since irrelevant data are factored out), but to a lesser extent than biogeographical information only.

22.5.5 ENVIRONMENTAL CHEMISTRY OF INPUT DATA AND FIELD DATA

A fundamental problem that applies when laboratory toxicity data are used for risk assessment purposes is the influence of substrate characteristics on contaminant sorption. In water, but especially in soil and sediment, the influence of substrate characteristics on compound sorption is of high quantitative importance. A low-quality assessment of exposure concentrations may be one clue to the observation that SSDs for the terrestrial compartment fit less well to field data than aquatic SSDs (Chapter 9). Clearly, the exposure element in an SSD analysis needs explicit consideration of appropriate concepts and data from environmental chemistry.

In most toxicity tests, toxic effects are expressed on the basis of total concentrations, but these are not always the relevant exposure concentrations. This means that the concept of sensitivity, which is central in the SSD concept, requires specific attention. In all chapters, the sensitivity is apparently quantified as an NOEC or EC_{50} that is composed of inherent sensitivity mixed with environmental influences upon bioavailability, likely because toxicity data are unclear in this respect. Although several corrections for environmental chemistry (influencing bioavailability) have been proposed, effects are currently a function of all these characteristics rather than of inherent toxicity alone.

The concept of "sensitivity" needs a conceptually clear definition and operationalization in the context of SSDs, especially if one pretends to describe SSDs rather than multivariate relationships between apparent sensitivity and environmental characteristics (like pH, organic carbon content, and co-occurring contaminants). Although multivariate statistics could be used to summarize a data set, the current strategy is to quantify exposure in laboratory and field conditions based on the bioavailable fraction in both conditions. Approaches can be based on mechanistic concepts (e.g., Hare and Tessier, 1996; Klepper and Van de Meent, 1997), but also on statistical–empirical approaches. In the latter case, the relationship between uptake and substrate characteristics is described as multivariate formulae based on uptake studies in a range of field-sampled contaminated substrates.

This line of work has received much attention for soil organisms, both with respect to a conceptual framework (e.g., Peijnenburg et al., 1997) and data collection for soil invertebrates and plants (Janssen et al., 1997b; Posthuma et al., 1998b; Peijnenburg et al., 1999a,b; Vijver et al., 2001). These analyses have suggested that uptake is highly variable among species, since it is driven by characteristics of the compound, substrate, and species. However, a pragmatic subdivision of species over groups with grossly similar uptake characteristics may be feasible.

The distribution of species over subgroups depends on differences in morphology, habitat, food choice, and behavior, leading to different relative importance of exposure pathways. Eventually, this manifests as different transfer functions that relate substrate characteristics to exposure. Two examples are given in Figure 22.8: (1) a worm is assumed mainly exposed through the skin, suggesting that partitioning over solid and liquid phase is important, and (2) a beetle is assumed mainly exposed through the alimentary canal, where conditions are regulated and may negate the effects of soil properties such as pH. Note that the statistical approach does not exclude pathways: for each organism, all exposure pathways may be relevant, but only the quantitatively dominant ones are recognized; that is, earthworms certainly will experience uptake of metals through the walls of the alimentary channel, but apparently this route is not dominant, possibly due to neutral gut pH. For practical reasons, the number of subgroups with different exposure pathway dominance needs to be limited.

The distribution of species over different subgroups is relevant for the designation of subgroups for different SSDs (see Table 22.4). Usually, one does not know the transfer functions for the tested species, but only for the members of some groups. Optimally, the distribution of species over subgroups is compatible with the exposure types, so that for each "cell" in Table 22.5 only one generic transfer function can be applied. For example, the transfer function to calculate atrazine exposure for algae from total concentrations and water chemistries may be applicable to all algae, but it may be different from the function applicable to fish.

Chemical Transfer Functions

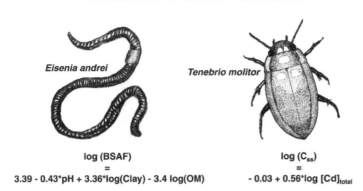

Eisenia andrei

Tenebrio molitor

log (BSAF)
=
3.39 - 0.43*pH + 3.36*log(Clay) - 3.4 log(OM)

$\log (C_{ss})$
=
- 0.03 + 0.56*log [Cd]$_{total}$

FIGURE 22.8 Two types of species with different autecological features and different exposure patterns, as illustrated by different empiricial relationships between soil characteristics
and uptake. The formulae have been derived in toxicokinetic studies with cadmium in field
soils with different soil characteristics. The formula for the biota-to-soil acumulation factor
(BSAF) of the oligochaete *Eisenia andrei* typically suggests pore water–mediated uptake of
cadmium due to the important role of pH, whereas the concentration in beetles (*Tenebrio
molitor* larvae in the pertinent study) was related to total substrate concentrations. (Data
courtesy of W.J.G.M. Peijnenburg and M. Vijver.)

A second issue that is important in relation to environmental chemistry is
background concentrations of naturally occurring compounds. For metals, background concentrations differ between ecosystems on the basis of the geological
origin of a substrate. This issue relates to one of the criticisms of early SSD results,
namely, that environmental quality criteria were determined at concentrations that
could be lower than local natural background concentrations and below physiological
requirements of species. This problem has been addressed by correcting for an
assumed nonavailable fraction in both laboratory tests and in field substrates (see
Struijs et al., 1997). A more fundamental approach would be to select only data for
those species that can inhabit the site for which the assessment is made (see also
Section 22.5.4). The motive for this is the same as for constructing separate SSDs
for saltwater and freshwater species; the ecological relevance of the species used to
construct SSDs for the site under consideration. By definition, background concentrations in tests and in the field can be considered to be sufficient for the nutritional
requirements of all species selected for such SSDs, and it could be highly unlikely
that this would cause environmental quality criteria below background concentrations, provided the chosen p is not too low. Solving this fundamental problem for
toxicity data would ask for investigations regarding geological origin and natural
backgrounds of a site and nutritional requirements of species, and it would be in
line with the approach to use autecological features of species to construct field-
relevant SSDs. Preliminary investigations into the role of autecological features and
the biogeographical distribution of species in constructing SSDs are in progress in
the Netherlands (see Figure 22.7).

A third issue of importance relating to environmental chemistry is the rate of breakdown of compounds in the environment. Breakdown obviously influences exposure in field conditions, and focuses on the importance of the issue "time" in risk assessments. Time may be important to assess ecological responses in the light of the life cycle characteristics of exposed species (see, e.g., Solomon and Takacs, Chapter 15), and may imply that short-term exposures to pesticide spraying may be viewed acceptable if recovery can occur before the next spraying. Van Straalen et al. (1992) proposed a risk assessment formula that is based on SSDs, but also on half-lives of the compound and size of the sprayed area. This type of approach can be further worked out, both operationally and scientifically. This may require setting limits on various parameters based on the ecological characteristics of the receiving species with respect to recolonization ability, etc.

For the insights on environmental chemistry to be of use in practice, the growing understanding of the issues that control exposure should allow for quantitative (empirical–statistical) exposure modeling to calculate the most likely exposure concentrations from total concentrations on the basis of compounds, substrate characteristics, and species' features. It is likely that building such models requires pragmatic distinctions between types of species with similar exposure characteristics. Further, guidance on how to assign test species to exposure types should be provided.

22.6 STATISTICAL ASPECTS OF SSDS

22.6.1 STATISTICS OF TOXICITY DATA SETS AND MODEL CHOICE

When an SSD analysis is started, choices must be made regarding statistical techniques to describe the data in the face of the assessment endpoint. Cumulative distribution functions (CDFs) can be a parametric or nonparametric estimate, they can be fit as straight lines after probit transformation, or they can be based on nonlinear fitting techniques. To date, the debate has focused on which approach should be generally favored over others, such as general preference for the lognormal, log-logistic, or log-triangular parametric approaches, for bootstrapping (Newman et al., Chapter 7) or a hybrid of bootstrapping and parametric techniques (Grist et al., in preparation). There are no convincing arguments that the distribution of species sensitivities should follow any particular statistical function. Therefore, the choice of approach or function must be based on pragmatic considerations rather than first principles. Pragmatic considerations of statistical kind should be weighted while considering the assessment endpoint, to determine which part of the data set is of interest, whether to model the data set as a whole or subsets of the data. For example, a bimodal curve (Aldenberg and Jaworska, 1999) could be a pragmatically good choice to fit a single SSD from all data to yield a robust estimate of the HC_5, but for a different assessment with the same data set, such as a risk assessment of contaminated media, it can be a poor choice (see Figure 22.3).

Prior to choosing any function or approach, a statistical analysis could be made of the available sets of input data. This insight has been put forward both for ecological and human risk assessment (Newman et al., Chapter 7; Aldenberg, personal communication). What are the statistical characteristics of the data sets? A start can be made by presenting some basic descriptive statistics to summarize the data: minimum, maximum, mean, standard deviation, quartiles (0.25th quantile, median, and 0.75th quantile), and so forth. Quartiles are often considered robust indicators of spread and skewness. These estimates, and combinations of them, could point to deviations from symmetrical distributions, or other features of the distribution of the data set.

Similarly, basic statistical techniques can be employed to derive the probability density function (PDF) from the data. For example, kernel density estimation may be useful to detect possible shapes of SSDs, without actually assuming certain parametric form, or requiring symmetry (Aldenberg, personal communication). As suggested by Aldenberg et al. (Chapter 5), a range of further techniques is available to assess the goodness-of-fit, or to take various levels of uncertainty into account, including Bayesian approaches. Uncertainty may concern the input data (both sensitivity in the SSD and exposure concentrations in the exposure concentration distribution) and the model. Both basic and advanced statistical techniques for summarizing data sets offer ways to assist in choosing a best-fit approach for any data set, be it uni- or multimodal, with parametric or nonparametric methods. According to this approach, the data set defines the proper model, not the assessment context. When new statistical techniques are introduced, guidance on the statistical techniques that can be applied before choosing among the options for a given data set should be made available to avoid misinterpretation. For example, when choosing the "best-fitting" statistical procedure, one should consider to which area of the curve the "best-fit" criterion should be sensitive in view of the assessment endpoint. Should it focus on the lower tails or on the whole curve. Do we want a model to be rejected when there is proper fit in the lower tail and misfit in the higher concentration range? This idea is implemented by the U.S. EPA by truncating the data for derivation of environmental quality criteria (Stephan, Chapter 11). Further, the statistical power of goodness-of-fit tests should be considered, to avoid overinterpretation of the test statistic in cases with small data sets. In such cases, the test might fail to detect misfit of any model, that is, several models would pass the test.

The availability of a variety of functions and approaches might lead to "shopping" for a particularly conservative or anticonservative choice if the statistical criteria for selecting a function are not clearly specified. For any assessment, it seems obvious that for a statistics-based technique (like the SSD concept) state-of-the-art statistical techniques should be employed to describe the data after selecting them. Currently, for the sake of procedural consistency or because of policy considerations, fixed functions are prescribed, sometimes yielding good and sometimes yielding bad fits to the data sets considered. Fixed functions seem defensible only from a policy viewpoint (transparency, consistency), not from an ecological or statistical perspective. The situation can be improved by replacing fixed functions by guidance on a range of data analysis steps that define the appropriate approach on statistical grounds.

22.6.2 Confidence Intervals

Table 22.1 mentions that the presentation of confidence limits is an advantage of SSD-based analyses. Confidence limits seem to offer information on the accuracy of the assessment, as they are wider for smaller data sets and when uncertainty or variability is large. When confidence limits are reported, the following issues should be considered.

First, reported intervals on an estimate of risk need not have an interpretable ecological meaning, even when the confidence interval is small (as for atrazine in Figure 22.3, whole data set SSD), when the estimate of risk itself is poor or undefined in the face of the assessment.

Second, reported intervals of different statistical methods on the same data set with the same assessment target but a different statistical procedure need not be numerically similar. For example, confidence intervals from nonparametric techniques are usually wider than from parametric techniques, as a result of the assumption of some shape of the model.

Third, SSD-based output may not be the final result of an assessment. Propagation of confidence through further assessment steps may need further development, both technically and regarding interpretation. Further consideration is required for the propagation of uncertainty information when aggregating risk estimates for mixtures from the risks of separate compounds (see Table 22.5) and when addressing all levels of uncertainty in data or models that have been identified by Aldenberg et al. (Chapter 5).

Confidence intervals that are presented together with estimates of HC_p values can be useful, to maintain insight into the data quality and to meet the desired level of confidence of an environmental quality criterion or of estimated risk in a risk assessment. The decision to maintain or drop confidence intervals is context dependent, and may be useful only for well-informed risk managers. Maintaining insight during the assessment process may avoid overestimating the confidence in the results, especially when various subsequent assessment steps must be taken. In any case, the meaning of a confidence interval should be presented explicitly, given the exposure data (are they uncertain or variable in time or space?) and the SSD analyses and subsequent aggregations. This is of prime importance for risk communication. Do the estimated confidence intervals of the various risk values relate to statistical uncertainty, to ecological variability or uncertainty, or to both, and what do the intervals imply regarding the assessment endpoint? Because of the hierarchy of methods to address uncertainty and variability, various fundamental questions still need to be addressed (see Aldenberg et al., Chapter 5).

22.6.3 SMALL DATA SETS AND SSD PRACTICALITY

An often-mentioned disadvantage of SSDs is that they require relatively large data sets for each compound (see Table 22.1). Discussions of numbers of data that are required continue to this date. For example, in a recent discussion on EU-Technical Guidance Documents, a minimum of ten toxicity data representing at least eight taxa was proposed. For comparison, the minimum data set in the Netherlands is currently four data representing four taxa. From a scientific perspective, and in relation to the analysis of fundamental statistical features of the data set (Section 22.6), the number of data that is required for a profitable assessment is not fixed. The number of required data depends both on the properties of the data set that is already available and on the assessment endpoints. In some cases, four data for the compound itself may be sufficient, providing an SSD model that does not further improve when adding toxicity data for that compound. In other cases, especially when many data are available from a congeneric compound, even fewer data may be sufficient, by using the statistical techniques of Aldenberg and Luttik (Chapter 6) together with the regularities in toxicity data to estimate surrogate SSD shape parameters (De Zwart, Chapter 8). Acute-to-chronic extrapolation might provide another route to tackle apparent data needs (Parkhurst et al., 1996a; De Zwart, Chapter 8). The results may take the format of normal SSD output (that is, an environment quality criterion or a risk estimate for the compound), or of a safety factor. In the latter case, the magnitude of the safety factor is not fixed, but it depends on, among other considerations, the number of data available (Aldenberg and Luttik, Chapter 6; Elmegaard and Jagers op Akkerhuis, 2000; Mineau et al., 2001).

A recent European workshop recommended developing methods to execute risk assessment for small data sets and to develop guidance on how to implement them (Hart, 2001). When these techniques cannot offer a solution for the assessment problem because of lack of data or data from similarly acting compounds, the number of toxicity tests can be increased, to the level that is sufficient to describe the SSD with the desired confidence. The number of data needed is naturally determined by the compound properties and the species sets in the receiving communities.

22.7 OUTPUT ISSUES

The ultimate issue of concern is whether SSDs yield output that is meaningful in the face of the assessment targets. What remains when the fundamental and technical issues are addressed is an output in terms of risk to species, for each relevant cell of Table 22.5. What is the meaning of these values to ecological entities exposed in the field? In our opinion, there are various issues that require further attention, all related to the issue of "validation."

If we assume that both exposure and sensitivity data are distributed, which is the most general statistical definition of the problem (see Aldenberg et al., Chapter 5), then the SSD result requiring validation is the joint probability curve (JPC). Validition studies should address two different issues, namely, the statistical validity of the model, and the ecological validity of the model output. Van den Brink et al. (Chapter 9) have made both types of comparisons (e.g., Figure 9.2), and used the term *confirmation* for the latter validation efforts, in line with Oreskes et al. (1994).

The statistical validity of the model should be addressed with fundamental statistical techniques, by comparing the statistical features of laboratory-based PDFs or CDFs with those of their field-based equivalents, that is, field-based PDFs and CDFs. Evidently, the data sets should be commensurate before statistical comparisons are made; that is, they should be selected because they make ecological sense, regarding issues like exposure duration and frequency. Statistical techniques for PDF and CDF comparisons are well developed (Aldenberg et al., Chapter 5), and might show that laboratory and field PDFs or CDFs are of similar shape and position for both the exposure and the sensitivity distributions (such as Figure 9.2). Such analyses may also show differences that should be further addressed, such as a shift of the field distribution to the left or right from the laboratory distribution. These techniques should be applied to test the validity of criticisms on fundamental misfits in the data set, or to calibrate the laboratory CDF to the field CDF.

Subsequently, the ecological meaning of the CDF or PDF should be addressed. Unlike the statistical step, there are no fundamentally transparent methodologies to perform this confirmation step. This relates in the first place to the fact that both the assessment endpoint and the measures of effect are often unclearly defined. For example, which response curve of Figure 22.2 should be taken to "validate" the SSD of metals for soil organisms? Further, the meaning of the JPC itself needs consideration, both regarding its statistical interpretation and its "validity."

Regarding the statistical interpretation, Aldenberg et al. (Chapter 5) have already mentioned the importance of further study into the transfer and meaning of uncertainties and variability through the assessment steps taken to construct the JPC. In current approaches, various uncertainties (model, data) are not taken into account, and it is unclear what these mean for the appearance and interpretation of JPCs. For example, what does a distributed PAF estimate tell us regarding field effects to be expected?

Regarding validity, the basis for the JPC is that the exposure and sensitivity distributions are independent. This is an appropriate assumption when the exposure distribution relates to a data set of exposure concentrations that are randomly distributed over an area, and a species data set regarding species that are randomly distributed over this area. In the field, the spatial distribution of exposure concentrations may, however, be not random (hot spots), which influences the distribution of species at the studied area. Depending on species mobility and sensitivity, hot spots within the area will likely contain more tolerant species. In that case, the field SSDs for the community at the hot spot will be shifted to the right on the x-axis compared to the "generic" SSD (pollution-induced community tolerance; see also Figure 21.5). The assessment of ecological validity is further complicated by indirect effects that are not addressed by the SSD approach at all.

Validation studies are, in our opinion, the crucial way to improve the prediction accuracy of SSDs. Field studies may identify the main issues that should be adopted in SSD approaches. In an iterative process of comparisons, the SSD model and its technical features will likely be stepwise improved.

22.8 DATABASE NEEDS

Like any statistical method, SSDs require input data, and for SSDs these are usually extracted from existing toxicity databases that have been compiled for a range of purposes, not necessarily SSDs alone. An example of a large toxicity database is AQUIRE of the U.S. EPA (U.S. EPA, 1984a), which is accessible worldwide, or the database compiled by De Zwart from various sources (Chapter 8). In other cases, all data or additional data are collected for an assessment problem by reviewing the literature on a case-by-case basis.

The outlook on future developments of SSDs suggests prediction accuracy of SSDs might improve when taking into account a broader range of toxicological data (toxic mode of action), ecological data (biogeography, taxonomy, autecological features of species, presence of target sites of toxic action for different compounds), and physicochemical data (those determining bioavailability). A fundamental redesign of toxicity databases might be needed, providing the keys to select the data that are appropriate for an assessment. The assessment eventually determines whether the classical ecotoxicity databases suffice, or additional data sources are needed. It is evident that the practicality of SSD-based assessments is promoted by the compilation of worldwide accessible and high-quality databases (Hart, 2001).

22.9 TOXICOLOGICAL RISKS AND MULTIPLE STRESS CONDITIONS

Calculation of risks in either of the applications described in this book usually addresses only the accumulation of risk over compounds in an environmental mixture. Predictions pertain to "risks" for communities or subcommunities with negligence of additional stress factors that might be present in the field. However, just as single compounds are the exception and mixtures the rule, mixtures of compounds are seldom the sole stress factor in a contaminated ecosystem. Contamination is often associated with other environmental problems.

Improving the utility of SSD-based results within a multiple-stress context is a difficult task, not yet broadly supported by accurate data or theories. The concept of SSDs might be extended by describing the sensitivity patterns to multiple stressors in multivariate models. Alternatively, one can try to address all other stress factors with appropriate methods or models, thereby expressing all effects or risks in the same measures of effect. Covariation of stresses and effects may need to be addressed, either mechanistically or empirically. Eventually, the multiple-stress analysis could provide insights in the total risk of environmental deterioration.

As described by Traas et al. (Chapter 16), some multiple-stress analyses have been made using SSDs and multivariate empirical techniques. By using Geographical Information System (GIS, grid-based, spatially explicit), SSD-based information on risks of metal mixtures for vascular plants was analyzed in conjunction with information of other soil factors that are of recognized importance for plant distribution (such as pH and groundwater table elevation; De Heer et al., 2000). These analyses showed that various soil factors, but also metal msPAF (or an unmeasured factor covarying with it), explained significant parts of the observed variance in plant species distribution in the field. This suggests that the reduced biodiversity of vascular plants in the Netherlands since 1950, the chosen reference year, can be attributed to various soil factors, including msPAF of metals.

The ecological interpretation of SSD-based results in the context of multiple environmental stressors is an interesting new field for statistical–empirical approaches. Efforts are currently extended to other species groups and other types of compounds such as pesticides. For pesticides, the analysis may provide insight in the measures to be taken to reduce side effects, e.g., by making scenario studies for pesticides with the same agricultural role but different half-lives (see Section 22.5.5). Further, the context of multiple-stress analysis is highly relevant for the interpretation of SSD confirmation studies in contaminated ecosystems.

22.10   CONCLUSIONS

This outlook chapter presents several possibilities to transform an originally almost purely statistically formulated approach with many weakly underpinned assumptions into a statistical model that explicitly adopts concepts and data from related disciplines. Like many other models in ecotoxicology, SSDs are subject to criticism, but despite this, the concept is currently used in an array of decision-making processes (Chapter 21). Improvements in SSD design and use can be traced to three important topics. First, the *specificity* of SSDs can be improved by tailoring the fundamental design of the analysis to the problem. Second, SSD techniques and associated techniques that are applied in the assessment can yield *improved assessment accuracy*, by using concepts from environmental chemistry, ecology, toxicology, biogeography, and taxonomy. Third, the *fundamental statistical techniques* can be improved, and regularities in toxicity databases can be used to address problems of small sample size. The evolution of SSDs along these lines may limit the relevance of criticisms for certain SSD approaches.

The current and future applications will likely continue in many ways: as stand-alone applications (of a generic kind, such as the derivation of national quality criteria) to SSD applications along with other methods ("weight of evidence"), as elements of multi-criteria analyses (like life cycle assessment, or LCA), or as a starting point for other models such as food web analyses. From a scientific point of view, we believe that there is considerable latitude for fundamental statistical, ecological, and technical improvements in the SSD concept itself, as demonstrated in this volume.

ACKNOWLEDGMENTS

This chapter is the result of a range of interactions among the authors of this chapter and contributors to this book and other experts, either during correspondence on earlier versions of draft manuscripts, or when interviewing some experts on important outlook issues. All discussions and visions were highly appreciated.

Appendices

APPENDIX A

The Interactive Poster Session "Use of Species Sensitivity Distributions in Ecotoxicology"

Timo Hamers, Theo P. Traas, and Leo Posthuma

Abstract — This is a report of the presentations and plenary discussions at the Interactive Poster Session "Use of Species Sensitivity Distributions in Ecotoxicology" held at SETAC-USA at Philadelphia, PA, November 18, 1999. The session was organized by G. W. Suter II (U.S. EPA), D. van de Meent (RIVM), L. Posthuma (RIVM), D. de Zwart (RIVM), T. P. Traas (RIVM), and chaired by D. van de Meent and G.W. Suter II.

A.1 INTRODUCTION

On November 18, 1999 an interactive poster session was organized called "Use of Species Sensitivity Distributions in Ecotoxicology." The session was held at the 20th Annual Meeting of the Society of Environmental Toxicology and Chemistry (SETAC) at Philadelphia, PA and was organized by Dr. D. van de Meent, Dr. L. Posthuma, Dr. D. de Zwart, and Dr. T. P. Traas (all RIVM–the Netherlands) and Dr. G. W. Suter II (U.S. EPA).

A.1.1 SPECIES SENSITIVITY DISTRIBUTIONS

Species sensitivity distributions (SSDs) are calculated statistical distribution curves based on differences between species in sensitivity to toxicant exposure. Over the past two decades, SSDs have been used in ecotoxicology for two purposes:

1. Environmental quality criteria (EQC) are derived by interpolating a maximum field concentration from a defined maximum percentage of species that may be exposed above critical concentration levels (NOEC, $E(L)C_{50}$). This application is also known as the "inverse application": an x-value of the SSD (i.e., a critical concentration) is calculated from a defined y-value (i.e., an accepted percentage of species exposed to toxic concentrations).

2. Ecological risk assessment (ERA) is performed by calculating the potentially affected fraction (PAF) of species being critically exposed at a given environmental concentration. This application is also known as the "forward application": a y-value of the SSD (i.e., a percentage of species) is calculated from a defined x-value (i.e., a given concentration).

A.1.2 Aims and Approach

The interactive poster session formed part of a project by RIVM–the Netherlands and U.S. EPA aiming at:

1. Bringing together all scientific and regulatory issues related to SSDs,
2. Collecting all relevant data,
3. Reviewing and evaluating the state of the art,
4. Identifying prospects for future developments.

To meet these goals, the knowledge of the principles and uses of SSDs in ecotoxicology was at the time of the meeting to be compiled in this book, from the contributions of an international group of experts working on or with SSDs. Special effort was made to improve the fundamental theory of derivation and use of SSDs and to promote appropriate use of SSDs in decision making.

The aim of the interactive poster session was to discuss the use of SSDs in ecotoxicology with respect to the fundamental theory, the practical problems encountered, and the interpretability of the outcome. During a viewing period, seven posters were presented on recent applications of the SSD concept followed by a short oral explication by the author, which was chaired by Dik van de Meent. Next, Glenn Suter chaired a plenary discussion with the audience to evaluate the support and criticism of SSD usage, both from scientific and regulatory points of view.

A.1.3 Presentations

The seven posters presented fit into three of the four sections, which have been defined for the book on the principles and uses of SSDs in ecotoxicology:

1. Section I: General Introduction and History of SSDs
   • Suter, G. W. II, Posthuma, L. and Traas, T. P.: *The Use of Species Sensitivity Distributions in Ecotoxicology*
2. Section II: Scientific Principles and Characteristics of SSDs
   • Van de Meent, D. and Van Straalen, N. M.: *Two-Way Use of Species Sensitivity Distributions in Ecotoxicological Risk Assessment*
   • Van den Brink, P. J., Posthuma, L., and Brock, T. C. M.: *Confirmation of the SSD-Concept: Field Relevance and Implications for Ecological Risk Assessment*
   • De Zwart, D.: *Regularities Observed in Sensitivity Distributions for Aquatic Species*

- Ownby, D. R., Newman, M. C., Mézin, L. C., Powell, D. C., Christensen, T. R. L., Kerberg, S. B., Padma, T. V., and Anderson, B. A.: *Applying Species Sensitivity Distributions in Ecological Risk Assessment: Assumptions of Distribution Type and Sufficient Number of Species*
3. Section III: Applications of SSDs
   - De Zwart, D. and Sterkenburg, A.: *Toxicity Based Assessment of Water Quality*
   - Hart, A.: *The Use of Species Sensitivity Distributions in Avian Risk Assessments for Pesticide Registration*
4. Section IV: Evaluation and Outlook
   - This section was covered by the final plenary discussion, rather than by any of the posters.

A.1.4 ATTENDANCE

The posters were attended continuously by an interested audience. Many people from the audience asked for further explanation of the posters and consequently had lively discussions with the attending authors by commenting on the SSD work shown and by displaying interest in the huge number of toxicity data collected and the principle of aggregating these data. The plenary discussion lasted 40 min and was attended by about 150 people, of whom 11 actively took part in the discussion.

A.2 POSTER PRESENTATIONS

In this section, a short synopsis is given of the contents of each poster and the subsequent oral presentation by the first author.

A.2.1 THE USE OF SPECIES SENSITIVITY DISTRIBUTIONS IN ECOTOXICOLOGY, BY GLENN SUTER

This poster presented a general introduction to SSDs, and to the two-way (inverse–forward) application in ecotoxicology. An overview was given of typical steps in SSD derivation, such as compiling input data, statistical treatment, and interpretation of output. Finally, advantages, disadvantages, and discussion topics on SSDs were summarized. Advantages referred mainly to the wide applicability of the SSD concept with respect to the exposure (individual compounds or mixtures of toxicants) and to the input data (sensitivity of species, functions, or processes), and the transparent approach requiring an explicit percentage in standard setting rather than a safety factor. Disadvantages referred mainly to the lack of ecology in the SSD concept: ecosystem interactions between species, differences in sensitivity between wildlife and a set of laboratory test organisms, the possible sensitivity of a key species, laboratory–field extrapolation, site-specific data, and background concentrations are not taken into account. Furthermore, SSDs have no mechanistic background and are empirical results, which have not yet proved to be more reliable than safety factors. The points of discussion raised by the author were later used as a guideline for the plenary discussion (see Section A.3, Plenary Discussion).

A.2.2　TWO-WAY USE OF SPECIES SENSITIVITY DISTRIBUTIONS IN ECOTOXICOLOGICAL RISK ASSESSMENT, BY DIK VAN DE MEENT

An overview was given of the assumptions made when applying the SSD concept, and the strong and weak parts of the concept were discussed. In both inverse and forward use, the NOECs used as input parameters are assumed to be population relevant and log-logistically distributed. The inverse use of SSDs to derive environmental quality criteria (EQC) further assumes that ecosystems are protected if the individual species are protected, with a protection of 95% of the species sufficient to protect the ecosystem. The forward use of SSDs for the purpose of ecological risk assessment further assumes that the potentially affected fraction (PAF) is an indicator for toxic risk in ecosystems that is related to field effects on an ecosystem level. Two of these assumptions were considered as weak points of the concept, i.e., the oversimplification of the ecosystem as a set of equally important species and the lack of a relation between PAF and field effects. Another weak aspect that was addressed was the great uncertainty in calculations of EQC or PAF, which was attributed to the low number of input data, rather than to the concept itself. Three strong points were raised: the scientific rationale of the SSD concept, the quantitative use of all available data, and the possibility to aggregate chemicals by calculating the multisubstance PAF of chemical mixtures, using the same methods as in single-species mixture toxicity, namely, concentration addition in case of the same mode of action and response addition across different modes of action. However, questions were raised by the author on the reliability of this aggregation method, because in practice the cumulative SSD curve for an environmental mixture was shown to be about a factor 50 steeper than expected based on a calculated curve for a hypothetical mixture of 25 compounds.

A.2.3　CONFIRMATION OF THE SSD-CONCEPT: FIELD RELEVANCE AND IMPLICATIONS FOR ECOLOGICAL RISK ASSESSMENT, BY PAUL VAN DEN BRINK

The objectives of this poster were to investigate the validity per se of the SSD concept to estimate field effects and to express limitations and/or points of care on the use of SSD results in decision making. To reach these objectives, weight of evidence was presented to confirm that both inverse and forward use of SSD yields results that are a good basis for decision making. For this purpose, data from terrestrial and aquatic semifield experiments were used. Great similarity was found between laboratory and field SSD curves when the investigated taxonomic groups and exposure conditions were similar. Less or no similarity was found in cases when these conditions were not fulfilled. Field effects could be predicted from the forward use of SSDs, when taking into account the major uncertainties related to the mode of action (e.g., species groups tested, exposure regime), whereas the 5th percentile of the SSD (inverse use) seemed to be applicable to protect acutely exposed communities. Finally, the author stressed that location- and community-specific information (e.g., on the mode of action, target species, etc.) should be quantitatively addressed when forwardly using the general SSD concept to predict site-specific ecological damage.

A.2.4 *REGULARITIES OBSERVED IN SENSITIVITY DISTRIBUTIONS FOR AQUATIC SPECIES,* BY DICK DE ZWART

In this poster, statistical methods were applied to relate the parameters of log-logistic SSDs for acute and chronic toxicity data. Aquatic toxicity data together with the toxic mode of action (TMoA) were collected for as many chemicals as possible, yielding 58,929 data of the toxicity of 3462 substances to 1683 test species. Out of the 3462 substances, acute and chronic toxicity was tested for 3420 substances and 707 substances, respectively. The TMoA was only known for about 900 substances.

The log-logistic SSD is defined by two model parameters, i.e., the first moment α, which is the average of the log-transformed toxicity and thus determines the location of the curve relative to the x-axis, and the second moment β, which is 0.55* the standard deviation of the log-transformed toxicity and thus determines the slope of the curve. Comparing acute and chronic SSDs, it was concluded that acute SSDs can be calculated into chronic SSDs by the following rules of thumb: $\alpha_{chronic} = \alpha_{acute} - 1$ and $\beta_{chronic} = \beta_{acute}$. Furthermore, the calculations showed that compounds with similar TMoA had parallel SSD curves when sufficient numbers of species were tested. From this, it was concluded that the β value is strongly biased by the number of species tested and that each TMoA has a specific β value. These findings may help to define the SSD in case insufficient chronic toxicity data are available. In that case, the author suggested to recalculate α_{acute} into $\alpha_{chronic}$ and to use the β value of a compound with a similar TMoA for which sufficient species were tested.

A.2.5 *APPLYING SPECIES SENSITIVITY DISTRIBUTIONS IN ECOLOGICAL RISK ASSESSMENT: ASSUMPTIONS OF DISTRIBUTION TYPE AND SUFFICIENT NUMBER OF SPECIES,* BY DAVID OWNBY

In this poster, three assumptions were tested that are generally accepted in SSD methods, i.e., the data fit the model distribution, the sample size is sufficient, and the species are adequately protected at HC_5. Of 51 toxicity data sets, 27 failed to fit the lognormal distribution, which is generally assumed in North American SSD methods. For percentiles <10%, effect concentrations (HC_p) predicted from the improper SSD curve were often more than tenfold too high or low when compared with the actual concentrations. A bootstrap-based alternative, which does not require a specific distribution, was then adopted for further HC_5-calculations. Adequate sample size (i.e., number of input data) was determined using a formal, yet arbitrary, stopping rule of 10%. This means that convergence is satisfied when the decrease in the magnitude of the 95% confidence interval of the HC_5 with the increase in observation number is less than 10% of the interval estimated for the previous sample size (increments of five species). For pentachlorophenol, fenitrothion, malathion, and hexachloro-cyclohexane, accurate prediction of HC_5 was reached for sets of 40 to 60 input data. Finally, SSDs were calculated for taxonomic subsets of species. In general, HC_5 values, the connected 95% confidence interval and the required number of input data decreased with increased subsetting. The author stressed that this advantage must be balanced against the increased total number of species data, to generate the most meaningful and precise estimates of effect.

A.2.6   TOXICITY BASED ASSESSMENT OF WATER QUALITY, BY DICK DE ZWART

This poster showed an application of SSDs for the purpose of testing the integrated toxic potency of mixtures of environmental pollutants at field concentrations, which are often below chemical analytical detection limits. The toxicity of organic surface water concentrates was tested in a set of five toxicity tests with different species, i.e., an algae productivity test (PAM), the Daphnia IQ test, the Microtox test, the Thamnotox F kit, and the Rotox F kit. For all five bioassays acute $E(L)C_{50}$ values were determined in terms of concentration factors, which were recalculated into chronic NEC values using an arbitrary and rather conservative acute-to-chronic ratio (ACR) of 10. A log-logistic SSD was calculated for the five no-effect concentration (NEC) values, after which the PAF value for the field situation was interpolated at concentration factor = 1. SSDs and related PAF values were determined in six bimonthly series of samples in 1996 for 15 locations. PAF values per location differed between sampling periods, and average (year-round) PAF values differed between the locations from PAF = 0.0 to PAF = 3.8%. A principal component analysis on the five different toxicity tests revealed that there was hardly any redundancy in the results provided by the selected tests.

A.2.7   THE USE OF SPECIES SENSITIVITY DISTRIBUTIONS IN AVIAN RISK ASSESSMENTS FOR PESTICIDE REGISTRATION, BY ANDY HART

This poster illustrated the use of SSDs in assessing risks to birds (blue tit *Parus caeruleus*) from chlorpyrifos use in U.K. orchards. Based on literature data, a lognormal SSD curve was fit to LD_{50} values reported for 17 birds, and a normal distribution curve was calculated for the dose–response slopes, which were reported for only three species. Distributions were also fitted for available exposure data, i.e., insecticide residues on insects (lognormal) and the percentage of time spent in the orchard (truncated β). Risk assessment was performed by a graphical method by overlaying exceedence curves for two parameters, e.g., median sensitivity (SSD curve) and the time spent in the orchard. This approach was simple only when two variables are allowed to vary. The author suggested another method, using Monte Carlo simulations to simulate the uncertainty (due to interspecies differences) and the variability (affecting sensitivity and exposure). The outcome of this second approach was easy to interpret as a probability that a certain percentage of effect is exceeded. Sensitivity analysis showed that the four variables used had similar degrees of influence on the risk, indicating that both the median toxicity and the slope of the dose–response are important. From his work, the author concluded that uncertainty and variability in risk assessment need to be distinguished and can be handled separately and objectively by the Monte Carlo method proposed. The method provides very interpretable outputs, in case the assessment endpoints are clearly defined. Great care should be given to the following:

- The potential for bias in the original data, e.g., residue data were obtained from insects collected in pitfall traps, which is not the way birds collect insects.
- The sampling variation in the original data, e.g., time spent in the orchard was monitored for only 23 birds.
- The systematic variation, e.g., residue data came from 46 sites, which had been treated with different pesticides.
- The unit of sampling, e.g., the timescale of pooled pitfall samples differs from actual exposure.
- The choice of the distributions, i.e., parametric or bootstrap.
- The quality of the information, especially at the tails of the distribution where the poor data are available have a high influence on the shape of the curve.
- Preventing correlation between the input variables.
- Uncertainties that are not included and their possible effect on the result.

A.2.8 COMMENTS ON THE POSTERS

After the viewing period, the authors of the posters were asked for the main comments (positive, negative), suggestions, or questions they received in their one-to-one interaction with the audience.

The most important issues discussed are given below in random order.

- It was suggested that the principle of aggregating mixtures of chemicals into a multisubstance potentially affected fraction (msPAF) might be a very useful tool in life cycle assessment (LCA).
- A question was raised for how many chemicals msPAF can be calculated. msPAF can be calculated for all chemicals for which α and β values can be produced. The method presented by Dick de Zwart (see Section A.2.4) can be applied to include compounds for which insufficient toxicity data are available. The method is presented in Chapter 8, and related information in Chapter 6.
- Experimental work should be performed to confirm and validate the concept of aggregating toxic potencies of mixtures into a single msPAF value.
- Aggregated toxic potencies of mixtures expressed as msPAF may again be recalculated backward into an equivalent concentration of a standard compound with a TMoA that is relevant for the compound or the situation at issue.
- Additional explanation was requested particularly about the forward vs. inverse use of the SSD concept. Many people anticipate a software program:
  1. To calculate SSDs based on a (minimum) set of input data;
  2. To allow them to perform both inverse and forward use of the SSD concept.
- The kind of distribution function and the minimum number of input data raised considerable discussion.

- The fact that between-species variation in slope of dose–response was as important as variation in the median (LD_{50}) in the example of Andy Hart was considered to be very important. This is interesting given:
  1. People generally restrict their interest to variation in the median.
  2. Slopes are required from regulatory studies but almost never used in the risk assessment.

A.3   PLENARY DISCUSSION

Before the discussion started, chairman **Glenn Suter** pointed out five topics for discussion. In the minutes given below, the discussion has been reported in order of these topics, rather than in chronological order. The topics themselves have been put in declining order of the discussion time they consumed, which is considered to be a relative measure of their importance.

1. *How should SSDs be interpreted? As SSDs are based on the results from single species toxicity tests their use may be not applicable for the field situation because properties of communities (e.g., recovery) are disregarded and the probability (sensitivity) for a single species is assumed to be a constant.*

Mike Newman addressed the use of LC_{50} values as input data to construct SSDs. If an LC_{50}-based curve were used in standard-setting or risk assessment, how much of the community would actually be protected when the field concentration already accesses the LC_{50} of a certain percentile, which may even include target species? He thought that the use of chronic NOECs might improve the field relevance of SSDs. **Glenn Suter** stressed that in European risk assessment procedures SSDs are already based on chronic NOECs, whereas in the United States acute toxicity data (LC_{50} values or EC_{50} values) are extrapolated into chronic NOECs. **Mike Newman** responded that he is looking forward to the application of the SSD concept to other input data with a higher relevance for populations. He proposed the use of population-level endpoints such as Malthusian parameters, which implicitly state the response of a whole population in terms of population persistence. **Jeff Giddings** replied that **Paul van den Brink**'s poster (see Section A.2.3) showed that even though "crummy parameters" as NOECs were used of endpoints that are not evidently related to population maintenance (e.g., body growth), no adverse responses were found at the ecosystem level at concentrations smaller than the 5th percentile of the SSD. From this, he suggested that the predictive value of SSDs is rather conservative and that the 5th percentile of the SSD curve (HC_5, used for standard setting) is a protective concentration for effects in the field.

Dwayne Moore supported the issue raised by **Mike Newman**: how to compare a prediction that, for example, "10% of the species is exposed to concentrations higher than their laboratory LC_{50}" to the community responses in actual field situations? Direct lethality is difficult to measure under field conditions. In view of the confirmation issue on the field relevance of SSD-based results raised in the posters of **Dik van de Meent** (see Section A.2.2) and **Paul van den Brink** (see Section A.2.3), SSDs should therefore be based on endpoints that are metric in the field. **Paul van den Brink** replied that the first example on his poster shows that the laboratory SSD of LC_{50} values of arthropods is comparable to the field SSD of arthropod abundance, indicating that abundance seems to be a sufficient metric endpoint for survival of this taxonomic group.

Mark Goodrich commented on the relevance of SSDs with respect to field exposure, which fluctuates both in space and in time. For example, the relevance of the input data (e.g., NOECs) is questionable for bivalves that are very sensitive to a constant exposure under laboratory conditions. However, if true exposure in the field is pulsed, this species can escape from exposure by refraining from ventilation.

Jules Loos wondered how the SSD is affected by the fact that some taxonomic groups are poorly tested. He proposed that SSDs should address key species rather than typical laboratory species. **Glenn Suter** admitted that the composition of taxonomic groups that can be found in the field is unequal to the frequency of taxa that are regularly tested in the laboratory. Regulatory risk assessment often requires toxicity data on a set of different prescribed species. For example, the U.S. EPA guidelines require a set of eight different species to be tested, of which only one must be an insect. This proportion is much lower than that found in the field, and therefore may bias the field relevance of SSDs. In practice, SSDs are often based on whatever toxicity data are available, so the potential for bias is great. **Mark Goodrich** addressed the point of concern of poorly tested taxonomic groups. As an example, he referred to the fact that although both rats (rodents) and ruminants are mammals, they have very different sensitivities when orally exposed to some compounds, because of the complex digesting system of ruminants. It is impossible to make a risk assessment for a taxonomic group when no toxicity data are available.

2. *Should single-species toxicity data be aggregated to a higher taxonomic level (e.g., from multiple $LC_{50\text{-species}}$ into a $LC_{50\text{-genus}}$ or $LC_{50\text{-family}}$) and how should these higher taxa be integrated in a distribution? Or should separate SSDs be derived per taxonomic group as suggested by* **Paul van den Brink** *(see Section A.2.3)?*

Keith Solomon stated that aggregation of toxicity data should not be based on a taxonomic basis, because the use of separate SSD curves for separate taxa may exclude (target) species from risk assessment. Therefore, he argued that aggregation should be based on the toxic mode of action of the compounds. Thus, taxa are classified according to their sensitivity to a certain mode of action, which is a biologically more relevant way of aggregating species. Separate SSDs could then be constructed for those species, which are sensitive to the mode of action at issue. **Glenn Suter** pointed out that for many compounds the mode of action is not known.

Niels Nyholm criticized the fact that "taxonomy drives the assessment" when applying SSDs. He strongly favored a mechanistic approach, based on biochemical thresholds, rather than an assessment based on an old-fashioned classification of organisms. **Glenn Suter** commented that taxonomy is not wrong or obsolete, just because it is an old discipline. He pointed out that similar species are similarly sensitive to toxicants: an increase in taxonomic distance between species corresponds to an increase in toxicological difference between species. Most variability in sensitivity is found at higher taxonomic levels, whereas variability within one species is comparable to variability within congeric species.

3. *What kind of distribution function should be used for SSDs? Log-logistic, lognormal, and log-triangular distribution functions are commonly used, but recently bootstrap techniques were suggested by several authors from the poster session, e.g.,* **David Ownby** *(see Section A.2.5).*

Niels Nyholm argued that nonsymmetrical SSD distributions, rather than symmetrical, should be used. Because of the symmetrical shape of the SSD curves currently used, the 5th percentile of the distribution (HC_5) decreases when a relative high NOEC or LC_{50} value is added to the data set. This implies that the addition of a relative insensitive species to the data set actually decreases the critical field concentration that is considered to be safe. **Glenn Suter** responded that fitting a truncated distribution curve through the lowest 50% of the available toxicity data can solve this problem. The U.S. EPA has already adopted this procedure. **Tom Aldenberg** replied that toxicity data around the 5th percentile of the distribution curve are far more important for the location of the HC_5 than toxicity data from the right ("insensitive") part of the curve. He concluded this from his own analyses performed on the shift of the distribution curve by manipulating NOECs from a real data set. Despite the symmetry of the SSD curve, the HC_5 appeared to increase even when individual NOECs higher than the HC_{50} were increased. The HC_5 finally decreased when very high NOECs ($>HC_{90}$) were further increased. Instead of omitting the highest 50% of the toxicity data from the data set, **Tom Aldenberg** further suggested using a bimodal distribution curve, in case one wants to make a distinction between sensitive and insensitive species groups. Nevertheless, he mentioned that his analyses had shown that in practice the unimodal approach is already very robust for the derivation of environmental quality objectives in the 5% range.

4. *What criteria should be set for the input data of SSDs, with respect to their quality, quantity, and bias?*

Sverker Molander addressed the issue of the quantity and quality of input data for SSDs: **David Ownby**'s poster (see Section A.2.5) showed that >40 species should be tested under laboratory conditions to minimize the uncertainty in SSD methods, whereas **Dick de Zwart**'s database (see Section A.2.4) contained toxicity data on >1500 species! To increase the set of laboratory species to be tested for SSD use, **Sverker Molander** proposed to do this in a clever way, based on the information that can be extracted from **Dick de Zwart**'s database (see Section A.2.4). Possible criteria are as follows:

- Representativeness of the species for the field situation,
- Predictive power of the species for the field situation,
- Sensitivity to different toxic modes of action,
- Robustness in the SSD curve.

Sverker Molander also has considerable data from multispecies test systems (>30 species), and proposed comparison of his outcomes with SSD results.

5. *How should uncertainty be dealt with, both of the input data and the final SSD curve?*

Andy Hart addressed the point that >40 species are suggested by **David Ownby** to reduce uncertainty in SSDs. He raises two questions:

1. What ways do we have for expressing overall uncertainty?
2. How do we decide how much certainty we need?

Glenn Suter responded that uncertainty can be expressed as a tolerance interval for each percentile of the SSD curve. The amount of accepted uncertainty depends on the situation. To make a risk assessment, one should first estimate which is the best model available. This model is not necessarily SSD. Alternative risk assessment models are recommended, especially when SSDs are unsuitable because of the availability of only one or two items of toxicity data.

A.4   CONCLUDING REMARKS

From the Interactive Poster Session the following conclusions may be drawn:

1. SSDs are a hot topic in environmental risk assessment. The concept attracts a wide variety of ecotoxicologists and policy makers, but its applicability and methodology are still under construction and under discussion.
2. The criticism of the inverse application of SSDs for setting environmental criteria is the lack of ecology in the concept:
   a. An ecosystem is expected to be protected when the majority of species (e.g., >95%) is not affected, neglecting all possible interactions among species and between species and the environment.
   b. The community of species is regarded to be sufficiently represented by a much smaller subset of species tested in the laboratory, which is often selected on grounds other than representativeness.
   c. Laboratory tests are performed under optimal conditions with constant exposures, which are not comparable to fluctuating exposure conditions (in time and space) in the field.
   Nevertheless, the predictive value of SSDs seems to be rather conservative because the HC_5, used for standard setting, seems to be a protective concentration for effects in the field.

3. The criticism of the forward use of SSDs for environmental risk assessment is the difficulty of interpreting the outcome of the PAF, especially when the PAF value is based on acute $E(L)C_{50}$ values. With respect to the input parameters of SSDs, the general consensus is that chronic toxicity data should be used, preferably with relevant endpoints for the population persistence. Alternative methods have been proposed to extrapolate acute toxicity data into chronic toxicity values, in the case that (population-relevant) NOEC values are not available. With respect to the outcome of the forward use of SSDs (PAF values), experimental confirmation and validation under field conditions are needed to increase the interpretability for risk assessment procedures.

4. The concept of aggregating the toxic potency of mixtures into multisubstance PAF values is very attractive, but needs even more confirmation, as the underlying principle of aggregating compounds, which was originally developed for single-species dose–response relationships (concentration addition and effect addition), needs to be validated for the SSD concept.

5. According to many contributors, knowledge about the TMoA is poor, needs to be increased, and should further be incorporated into several aspects of the SSD concept:

   a. It was suggested to classify different species according to their sensitivity to a certain TMoA, which is a biologically more relevant basis for aggregation than taxonomic similarity, because the use of separate SSD curves for separate taxa may exclude (target) species from risk assessment.

   b. In case the TMoA for a compound is known, but insufficient toxicity data are available to fit an SSD, the slope parameter β of the log-logistic SSD can be estimated at the β value of a compound with the same TMoA for which sufficient species are tested.

   c. The principle of aggregating the toxic potency of a mixture of compounds into a single multisubstance PAF value is based on the TMoA.

6. No consensus was obtained on the choice of a specific fitted distribution curve, or on the adoption of an alternative bootstrapping method. It was stressed, however, that fitted SSD curves should be calculated not only for the median toxicity, but also for the slope of the species-specific dose–response curves. Those in favor of the use of bootstrap techniques in SSDs argued that the outcome of bootstrapping is easier to interpret in the case of forward use of SSDs and less uncertain in the case of inverse use of SSDs. Those in favor of fitted distributions argue that HC_5 values obtained in the inverse use of SSDs are robust and seem to be protective.

7. Most contributors agreed that because of the lack of sufficient data, in practice often as many input data as available should be used for SSD calculations. However, no consensus was reached on the minimum number of input data required. To reduce the uncertainty to acceptable proportions, minimum numbers of input data were suggested ranging from 20 to 60, depending on the accuracy required and the goal of the use of SSDs. It

was suggested to increase the set of laboratory species to be tested for SSD use in a clever way, based on well-defined criteria as representativeness, predictive power, sensitivity to different TMoA, and robustness.

8. Aggregating toxicity data of different species is a matter of discussion. Separate SSDs for different taxonomic groups have been proposed. For similar taxonomic groups, less input data were required to achieve acceptable uncertainty and better correspondence was found between laboratory and field SSDs. Aggregation of species based on similar TMoA also has been proposed, to reduce the possibility of excluding certain (target) organisms from protection and risk assessment. Possibly, application of both principles will yield similar results, as it was argued that species that are taxonomically closely related are similarly sensitive to a certain TMoA.

APPENDIX B

List of Computer Software Programs

Various chapters in the book refer to one or more existing computer programs, to execute the calculations that are explained in the text. Software is available as follows.

For Derivation of HC$_5$ or Environmental Quality Criteria from a Set of Toxicity Data

Chapters 5 and 12:

- A program is available to calculate HC$_5$ values according to various statistical models (log-logistic, lognormal, triangular) from a set of toxicity data.
- The program is called E$_T$X, and is implemented as MS-DOS executable.
- Reference: Aldenberg (1993).
- To obtain the program, contact T.Aldenberg@RIVM.NL or TP.Traas@RIVM.NL.
- An update is scheduled for the end of 2001.

Chapter 7:

- A program is available to calculate HC$_5$ values using bootstrapping from a set of toxicity data.
- The program is written in FORTRAN.
- Reference: Newman et al., this book.
- To obtain the program, contact Newman@VIMS.EDU.

For Environmental Risk Assessment of Contaminated Ecosystems

Related to Chapter 12:

- A program is available to calculate ecological risks of single compounds and mixtures on the basis of PAF and msPAF calculations, using sets of toxicity data selected according to the methods outlined in Chapter 12.
- The program is called OMEGA (Optimal Modelling for Ecotoxicological Assessment) and is implemented as Microsoft Excel™ worksheet.

- Reference: Beek and Knoben (1997), Durand-Huiting (1999).
- To obtain the program, contact m.beek@riza.rws.minvenw.nl.

Chapter 15:

- A program is available to calculate ecological risks from data sets in which both exposure and effects are distributed, using the idea of joint probability curves.
- The program is implemented as Microsoft Excel™ or QuattroPro™ worksheets.
- Reference: Solomon and Takacs, this book.
- To obtain the program, contact Ksolomon@tox.uoguelph.ca or Jgiesy@aol.com.

Chapter 17:

- A program is available to calculate ecological risks from data sets in which both exposure and effects are distributed, using the idea of joint probability curves in a formal, tiered system, operational in the framework of U.S. Water Quality Guidelines.
- The program is implemented as Windows-based software.
- References: (1) Warren-Hicks et al., this book; (2) Parkhurst et al. (1996).
- To obtain the program, contact WERF (Water Environment Research Foundation) at its Washington, D.C., office.

References

Ahlborg, U.G., G.C. Becking, L.S. Birnbaum, A. Brouwer, H.J.G. Derks, M. Feeley, G. Golor, A. Hanberg, J.C. Larsen, A.K.D. Liem, S.H. Safe, C. Schlatter, F. Waern, M. Younes, and E. Yrjänheikki. 1994. Toxic equivalence factors for dioxin-like PCBs. Report on a WHO-ECEH and IPCS consultation, December 1993. *Chemosphere,* 28: 1049–1067.

AIS. 1989. Principles for environmental risk assessment of detergent chemicals, in *Proceedings of AIS Workshop,* Limelette, 1989. Association Internationale de la Savonnerie et de la Detergence, Brussels, Belgium.

AIS. 1992. Practical aspects of environmental hazard assessment of detergent chemicals in Europe, in *Proceedings of AIS 2nd Workshop,* Limelette, 1992. Association Internationale de la Savonnerie et de la Detergence, Brussels, Belgium.

AIS. 1995. Environmental risk assessment of detergent chemicals, in *Proceedings of the A.I.S.E./CESIO Limelette 3rd Workshop,* 28–29 November 1995. AISE/CESIO, Brussels, Belgium.

Aldenberg, T. 1993. $E_T X$ 1.3a, A program to calculate confidence limits for hazardous concentrations based on small samples of toxicity data. RIVM report 719102 015. National Institute of Public Health and Environment, Bilthoven, the Netherlands.

Aldenberg, T. and J.S. Jaworska. 1999. Bayesian statistical analysis of bimodality in species sensitivity distributions. *SETAC News,* 19: 19–20.

Aldenberg, T. and J.S. Jaworska. 2000. Uncertainty of the hazardous concentration and fraction affected for normal species sensitivity distributions. *Ecotoxicology and Environmental Safety,* 46: 1–18.

Aldenberg, T. and W. Slob. 1993. Confidence limits for hazardous concentrations based on logistically distributed NOEC toxicity data. *Ecotoxicology and Environmental Safety,* 25: 48–63.

Allen, H.E., G. Fu, and B. Deng. 1993. Analysis of acid-volatile sulfide (AVS) and simultaneously extracted metals (SEM) for the estimation of potential toxicity in aquatic sediments. *Environmental Toxicology and Chemistry,* 12: 1441–1453.

Altenburger, R., T. Backhaus, W. Boedeker, M. Faust, M. Scholze, and L.H. Grimme. 2000. Predictability of the toxicity of multiple chemical mixtures to *Vibrio fischeri:* mixtures composed of similarly acting chemicals. *Environmental Toxicology and Chemistry,* 19: 2341–2347.

Ambrose, R.B., T.A. Wool, J.P. Connolly, and R.W. Schanz. 1988. WASP4, a Hydrodynamic and Water Quality Model. Model Theory, User's Manual, and Programmer's Guide. EPA 600/3-87/039. U.S. Environmental Protection Agency, Environmental Research Laboratory, Athens, GA, USA: 317 pp.

Anderson, P.S. and A.L. Yuhas. 1996. Improving risk management by characterizing reality: a benefit of probabilistic risk assessment. *Human and Ecological Risk Assessment,* 2: 55–58.

Ang, A.H.S. and W.H. Tang. 1984. *Probability Concepts in Engineering Planning and Design.* Volume II: *Decision, Risk, and Reliability,* Wiley, New York, USA: 562 pp.

Ankley, G.T. 1996. Evaluation of metal/acid-volatile sulfide relationships in the prediction of metal bioaccumulation by benthic macroinvertebrates. *Environmental Toxicology and Chemistry,* 15: 2138–2146.

Ankley, G.T., G.L. Phipps, E.N. Leonard, D.A. Benoit, V.R. Mattson, P.A. Kosian, A.M. Cotter, J.R. Dierkes, D.J. Hansen, and J.D. Mahony. 1991. Acid volatile sulphide as a factor mediating cadmium and nickel bioavailability in contaminated sediments. *Environmental Toxicology and Chemistry,* 10: 1299–1307.

Ankley, G.T., V.R. Mattson, E.N. Leonard, C.W. West, and J.L. Bennett. 1993. Predicting the acute toxicity of copper in freshwater sediments: evaluation of the role of acid-volatile sulphide. *Environmental Toxicology and Chemistry,* 12: 315–320.

Ankley, G.T., D.M. DiToro, D.J. Hansen, and W.J. Berry. 1996. Technical basis and proposal for deriving sediment quality criteria for metals. *Environmental Toxicology and Chemistry,* 15: 2056–2066.

Anonymous. 1991. Guidelines for the Submission of Applicants for Registration of Pesticides. Part H. Toxicity to Organisms in the Environment. Pesticides Bureau, Wageningen, the Netherlands.

ANZECC. 2000a. Australian and New Zealand Guidelines for Fresh and Marine Water Quality. Vol. 1: The Guidelines. Paper 4. Australia and New Zealand Environment and Conservation Council and Agriculture and Resource Management Council of Australia and New Zealand, Canberra, Australia.

ANZECC. 2000b. Australian and New Zealand Guidelines for Fresh and Marine Water Quality. Vol. 2: Aquatic Ecosystems — Rationale and Background Information. Paper 4. Australia and New Zealand Environment and Conservation Council and Agriculture and Resource Management Council of Australia and New Zealand, Canberra, Australia.

Aqua Survey, Inc. 1993. *Daphnia magna* IQ toxicity test, technical information update. Aqua Survey, Inc., 499 Point Breeze Rd., Flemington, NJ, USA.

Armitage, P. and T. Colton. 1998. *Encyclopedia of BioStatistics*, John Wiley, Chichester, U.K.: 4898 pp.

ASTM. 1980. *Standard Practice for Conducting Toxicity Tests with Fishes, Macroinvertebrates and Amphibians,* American Society for Testing and Materials, Philadelphia, PA, USA, ASTM E 729-80.

ASTM. 1981. *Proposed Standard Practice for Conducting Toxicity Tests with Freshwater and Saltwater Algae,* American Society for Testing and Materials, Philadelphia, PA, USA, ASTM E 47.01.

ASTM. 1994. *Standard Practice for Conducting Acute Toxicity Tests with Fish, Macroinvertebrates and Amphibians.* In *1994 Annual Book of ASTM Standards,* Volume 11.04, *Water and Environmental Technology,* American Society for Testing and Materials, Philadelphia, PA, USA.

Bacci, E. 1994. *Ecotoxicology of Organic Contaminants*, CRC Press, Boca Raton, FL, USA.

Backhaus, T, R. Altenburger, W. Boedeker, M. Faust, M. Scholze, and L.H. Grimme. 2000. Predictability of the toxicity of a multiple mixtures of dissimilarly acting chemicals to *Vibrio fischeri*. *Environmental Toxicology and Chemistry,* 19: 2348–2356.

Bak, J. and J. Jensen. 1999. Critical loads for lead, cadmium and mercury in Denmark. A first attempt for soils based on preliminary guidelines. Working report 96, The National Environmental Research Institute, Silkeborg, Denmark.

Baker, J. L., A.C. Barefoot, L.E. Beasley, L. Burns, P. Caulkins, J. Clark, R.L. Feulner, J.P. Giesy, R.L. Graney, R. Griggs, H. Jacoby, D. Laskowski, A. Maciorowski, E. Mihaich, H. Nelson, R. Parrish, R.E. Siefert, K.R. Solomon, and W. van der Schalie. 1994. *Final Report: Aquatic Risk Assessment and Mitigation Dialog Group*, SETAC Press, Pensacola, FL, USA.

Bakker, J. and D. van de Meent. 1997. Prescription of the calculation of the indicator effects toxic substances (I_{tox}). RIVM report 607504 003. National Institute of Public Health and the Environment, Bilthoven, the Netherlands [in Dutch].

Bal, D., H.M. Beije, Y.R. Hoogeveen, S.R.J. Jansen, and P.J. van der Reest. 1995. *Handbook of Nature Target Types in the Netherlands*, Informatie — en KennisCentrum (IKC) Natuurbeheer, Wageningen, the Netherlands [in Dutch].

Balk, F., P.C. Okkerman, and J.W. Dogger. 1995. Guidance Document for Aquatic Effects Assessment. Report 20. Organization for Economic Co-operation and Development (OECD), Paris, France.

Barnthouse, L.W. and G.W. Suter II, Eds. 1986. User's Manual for Ecological Risk Assessment. ORNL-6251. Oak Ridge National Laboratory, Oak Ridge, TN, USA.

Barnthouse, L.W., D.L. DeAngelis, R.H. Gardner, R.V. O'Neill, G.W. Suter II, and D.S. Vaughan. 1982. Methodology for Environmental Risk Analysis. ORNL/TM-8167. Oak Ridge National Laboratory, Oak Ridge, TN, USA.

Barnthouse, L.W., G.W. Suter II, A.E. Rosen, and J.J. Beauchamp. 1987. Estimating responses of fish populations to toxic contaminants. *Environmental Toxicology and Chemistry,* 6: 811–824.

Barnthouse, L.W., G.W. Suter II, and A.E. Rosen. 1990. Risks of toxic contaminants to exploited fish populations: influence of life history, data uncertainty, and exploitation intensity. *Environmental Toxicology and Chemistry,* 9: 297–311.

Barton, A., C. Berish, B. Daniel, S. Ells, T. Marshall, J. Messer, M. Powell, M. Rice, A. Sergeant, V. Serveiss, I. Sunzenauer, and M. Whitworth. 1997. Priorities for Ecological Protection: An Initial List and Discussion Document for EPA. EPA/600/S-97/002. U.S. Environmental Protection Agency, Washington, D.C., USA.

Baskin, Y. 1994. Ecosystem function of biodiversity. *Bioscience,* 44: 657–660.

Beek, M.A. and R.A.E. Knoben. 1997. Ecotoxicological risks of compounds for aquatic ecosystems. RIZA report 97.064. Rijkswaterstaat, Lelystad, the Netherlands [in Dutch].

Belanger, S.E. 1994. Review of experimental microcosm, mesocosm, and field tests used to evaluate the potential hazard of surfactants to aquatic life and the relation to single species data. In Hill, I.R., F. Heimbach, P. Leeuwangh, and P. Matthiessen, Eds., *Freshwater Field Tests for Hazard Assessment of Chemicals*, Lewis Publishers, Boca Raton, FL, USA: pp. 287–314.

Benard, A. and E.C. Bos-Levenbach. 1953. Plotting observations on probability paper. *Statistica,* 7: 163–173 [in Dutch].

Benjamin, J.R. and C.A. Cornell. 1970. *Probability and Statistics for Engineers,* McGraw-Hill, New York, USA: 684 pp.

Berthouex, P.M. and L.C. Brown. 1974. *Statistics for Environmental Engineers*, Lewis Publishers, Boca Raton, FL, USA: 335 pp.

Bettger, W.J. and B.L. O'Dell. 1981. A critical role of zinc in the structure and function of biomembranes. *Life Sciences,* 28: 1425–1438.

Bier, V.M. 1999. Challenges to the acceptance of probabilistic risk analysis. *Risk Analysis,* 19: 703–709.

Blanck, H. 1984. Species dependent variation among aquatic organisms in their sensitivity to chemicals. *Ecological Bulletin,* 36: 107–119.

Blanck, H., G. Wallin, and S.-Å Wängberg. 1984. Species dependent variation in algal sensitivity to chemical compounds. *Ecotoxicology and Environmental Safety,* 8: 339–351.

Bliss, C.I. 1939. The toxicity of poisons applied jointly. *Annals of Applied Biology,* 26: 585–615.

Blom, G. 1958. *Statistical Estimates and Transformed Beta-Variables*, John Wiley & Sons, New York, USA: 176 pp.

Bockting, G.J.M., E.J. van de Plassche, J. Struijs, and J.H. Canton. 1992. Soil-water partitioning coefficients for some trace metals. RIVM report 679101 003. National Institute of Public Health and the Environment, Bilthoven, the Netherlands.

Bockting, G.J.M., E.J. van de Plassche, J. Struijs, and J.H. Canton. 1993. Soil-water partition coefficients for organic compounds. RIVM report 679101 013. National Institute of Public Health and the Environment, Bilthoven, the Netherlands.

Boeije G., J.-O. Wagner, F. Koormann, P.A. Vanrolleghem, D. Schowanek, and T. Feijtel. 2000. New PEC definitions for river basins applicable to GIS-based environmental exposure assessment. *Chemosphere,* 40: 255–265.

Brandes, L.J., H.A. den Hollander, and D. van de Meent. 1996. SimpleBox 2.0, a nested multimedia fate model for evaluating the environmental fate of chemicals. RIVM report 719101 029. National Institute of Public Health and the Environment, Bilthoven, the Netherlands.

Brattsten, L.B. 1979. Ecological significance of mixed-function oxidations. *Drug Metabolism Review,* 10: 35–58.

Bricker, S.B., D.A. Wolfe, K.J. Scott, G. Thursby, E.R. Long, and A. Robertson. 1993. Sediment toxicity in Long Island Sound embayments. In *Proc. Long Island Sound Research Conference,* October 23–24, 1992, Southern Connecticut State University, New Haven, CT, USA.

Bringmann, G. and R. Kühn. 1977. Befunde der Schadwirkung wassergefährdender Stoffe gegen *Daphnia magna* [Hazardous substances in water toward *Daphnia magna*]. *Zeitschrift für Wasser- und Abwasserforschung,* 10: 161–166.

Brix, K.V., D.K. DeForest, and W.J. Adams. 2001. Assessing acute and chronic copper risks to freshwater aquatic life using species sensitivity distributions for different taxonomic groups. *Environmental Toxicology and Chemistry,* 20: 1846–1856.

Brock, T.C.M., J. Lahr, and P.J. van den Brink. 2000a. Ecological Risk Assessment of Pesticides in Freshwater Ecosystems. Part 1: Herbicides. Alterra-Report 088, Wageningen, the Netherlands.

Brock, T.C.M., R.P.A. van Wijngaarden, and G.J. van Geest. 2000b. Ecological Risk Assessment of Pesticides in Freshwater Ecosystems. Part 2: Insecticides. Alterra-Report 089, Wageningen, the Netherlands.

Broderius, S.J., M.D. Kahl, and M.D. Hoglund. 1995. Use of joint toxic response to define the primary mode of toxic action for diverse industrial organic chemicals. *Environmental Toxicology and Chemistry,* 14: 1591–1605.

Bros, W.E. and B.C. Cowell. 1987. A technique for optimizing sample size (replication). *Journal of Experimental Marine Biology and Ecology,* 114: 63–71.

Buikema, A.L., Jr., B.R. Niederlehner, and J. Cairns, Jr. 1982. Biological monitoring. IV. Toxicity testing. *Water Research,* 16: 239–292.

Bulich, A.A. 1979. Use of luminescent bacteria for determining toxicity in aquatic environments. In *Aquatic Toxicology.* ASTM 667, Markings, L.L. and R.A. Kimerle, Eds., American Society for Testing and Materials, Philadelphia, PA, USA: pp. 98–106.

Bulich, A.A. and D.L. Isenberg. 1981. Use of the luminescent bacterial system for the rapid assessment of aquatic toxicity. *ISA Transactions,* 20: 29–33.

Burmaster, D.E. 1996. Benefits and costs of using probabilistic techniques in human health risk assessments — with emphasis on site-specific risk assessments. *Human and Ecological Risk Assessment,* 2: 35–43.

Burmaster, D.E. and D.A. Hull. 1997. Using lognormal distributions and lognormal probability plots in probabilistic risk assessments. *Human and Ecological Risk Assessment,* 3: 235–255.

Burns, L.A. 1997. Exposure Analysis Modeling System: User's Guide for EXAMS II, Version 2.97.5. U.S. EPA, Office of Research and Development. Washington, D.C., USA. Report No. EPA/600/R-97/047. March 1997.

Burton, G.A. 1991. Assessing the toxicity of freshwater sediments. *Environmental Toxicology and Chemistry,* 10: 1585–1627.

Burton, G.A. 1992. Plankton, macrophyte, fish, and amphibian toxicity testing of freshwater sediments. In Burton, G.A., Ed., *Sediment Toxicity Assessment,* Lewis Publishers, Boca Raton, FL, USA: pp. 167–182.

Burton, G.A., M.K. Nelson, and C.G. Ingersoll. 1992. Freshwater benthic toxicity tests. In Burton, G.A., Ed., *Sediment Toxicity Assessment,* Lewis Publishers, Boca Raton, FL, USA: pp. 213–240.

Cairns, J.J. 1998. Endpoints and thresholds in ecotoxicology. In Schüürmann, G. and B. Markert, Eds. *Ecotoxicology, Ecological Fundamentals, Chemical Exposure and Biological Effects,* John Wiley & Sons/Spektrum Akademischer Verlag, New York, USA: pp. 751–768.

Cairns, J.J. and B.R. Niederlehner. 1987. Problems associated with selecting the most sensitive species for toxicity testing. *Hydrobiology,* 153: 87–94.

Calabrese, E.J. and L.A. Baldwin. 1993. *Performing Ecological Risk Assessments,* Lewis Publishers, Boca Raton, FL, USA.

Calabrese, E.J. and L.A. Baldwin. 1994. A toxicological basis to derive a generic interspecies uncertainty factor. *Environmental Health Perspectives,* 102: 14–17.

Calamari, D. and M. Vighi. 1992. A proposal to define quality objectives for aquatic life for mixtures of chemical substances. *Chemosphere,* 25: 531–542.

Calamari, D. and M. Vighi. 1993. Scientific bases for the assessment of toxic potential of several chemical substances in combination at low level. Office for Official Publications of the European Communities, Luxembourg.

Calow, P. 1996. Variability, noise or information in ecotoxicology? *Environmental Toxicology and Pharmacology,* 2: 121–123.

Calow, P. 1998. Ecological risk assessment: risk for what? How do we decide? *Ecotoxicology and Environmental Safety,* 40: 15–18.

Calow, P. and V.E. Forbes. 1997. Science and subjectivity in the practice of ecological risk assessment. *Environmental Management,* 21: 805–809.

Calow, P., R.M. Sibly, and V. Forbes. 1997. Risk assessment on the basis of simplified life-history scenarios. *Environmental Toxicology and Chemistry,* 16: 1983–1989.

Campbell, K.R., S.M. Bartell and J.L. Shaw. 2000. Characterizing aquatic ecological risks from pesticides using a diquat dibromide case study. II. Approaches using quotients and distributions. *Environmental Toxicology and Chemistry,* 19: 760–774.

Campbell, P.J., D.J.S. Arnold, T.C.M. Brock, N.J. Grandy, W. Heger, F. Heimbach, S.J. Maund, and M. Streloke. 1999. *Guidance Document on Higher Tier Aquatic Risk Assessment for Pesticides (HARAP).* SETAC-Europe, Brussels, Belgium: 179 pp.

Cardwell, R.D., B.R. Parkhurst, W. Warren-Hicks, and J.S. Volosin. 1993. Aquatic ecological risk. *Water Environment and Technology,* 5: 47–51.

Cardwell, R.D., M.S. Brancato, J. Toll, D. DeForest and L. Tear. 1999. Aquatic ecological risks posed by tributyltin in United States surface waters, pre-1989–1996 data. *Environmental Toxicology and Chemistry,* 18: 567–577.

Carlson, A.R., W.A. Brungs, G.A. Chapman, and D.J. Hansen. 1984. Guidelines for Deriving Numerical Aquatic Site-Specific Water Quality Criteria by Modifying National Criteria. U.S. Environmental Protection Agency, Environmental Research Laboratory, Duluth, MN, USA.

Carlson, A.R., G.L. Phipps, V.R. Mattson, P.A. Kosian, and A.M. Cotter. 1991. The role of acid volatile sulfide in determining cadmium bioavailability and toxicity in freshwater sediments. *Environmental Toxicology and Chemistry,* 10: 1309–1319.

CCME. 1996. A Protocol for the Derivation of Environmental Health and Human Health Soil Quality Guidelines. Canadian Council of Ministers of the Environment, Winnipeg, Manitoba, Canada, CCME-EPC-101E. En 108-4/8-1996E.

CCME. 1999. Canadian Environmental Quality Guidelines. Canadian Council of Ministers of the Environment, Winnipeg, Canada.

CCREM. 1987. Canadian Water Quality Guidelines. Prepared by the Task Force on Water Quality Guidelines, Canadian Council of Resource and Environment Ministers, Winnipeg, Canada.

CEC. 1990. Workshop on Environmental Hazard and Risk Assessment in the Context of Directive 79/831/EEC, 15–16 October 1990, Commission of the European Communities, Joint Research Centre, Ispra, Italy.

CEC. 1996. Technical Guidance Document in Support of Commission Directive 93/67/EEC on Risk Assessment for New Notified Substances and Commission Regulation (EC) 1488/94 on Risk Assessment for Existing Substances, EC Catalog Numbers CR-48-96-001, 002, 003, 004-EN-C. Office for Official Publications of the European Community, Luxembourg.

Centeno, M.D., G. Persoone, and M.P. Goyvaerts. 1995. Cyst-based toxicity tests IX: the potential of *Thamnocephalus platyurus* as test species in comparison with *Streptocephalus proboscideus* (Crustacea, Branchiopoda, Anostraca). *Environmental Toxicology and Water Quality,* 10: 275–282.

CEPA. 1997. Environmental Assessments of Priority Substances under the Canadian Environmental Protection Act. Guidance Manual Version 1.0. Chemical Evaluation Division, Commercial Chemicals Evaluation Branch, Environment Canada, Ottawa, Ontario, Canada. Report EPS/2/CC/3E. March 1997.

Chandrashekar, K.R. and K.M. Kaveriappa. 1994. Effect of pesticides on sporulation and germination of conidia of aquatic hyphomycetes. *Journal of Environmental Biology,* 15: 315–324.

Chapman, P.M. 1986. Sediment quality criteria from the sediment quality TRIAD: an example. *Environmental Toxicology and Chemistry,.* 5: 957–964.

Chapman, P.M. 1995. How should numerical criteria be used? *Human and Ecological Risk Assessment,* 1: 1–4.

Chapman, P.M., R.C. Barrick, J.M. Neff, and R.C. Swartz. 1987. Four independent approaches to developing sediment quality criteria yield similar values for model contaminants. *Environmental Toxicology and Chemistry,* 6: 723–725.

Chapman, P.M., A. Fairbrother, and D. Brown. 1998. A critical evaluation of safety (uncertainty) factors for ecological risk assessment. *Environmental Toxicology and Chemistry,* 17: 99–108.

Clifford, P.A., D.E. Barchers, D.F. Ludwig, R.L. Sielken, J.S. Klingensmith, R.V. Graham, and M.I. Banton. 1995. An approach to quantifying spatial components of exposure for ecological risk assessment. *Environmental Toxicology and Chemistry,* 14: 895–906.

Cooper, D.C. and J.W. Morse. 1998. Extractability of metal sulfide minerals in acidic solutions, application to environmental studies of trace metal contamination within anoxic sediments. *Environmental Science and Technology,* 32: 1076–1078.

Couch, J.A. and J.C. Harshbarger. 1985. Effects of carcinogenic agents on aquatic animals: an environmental and experimental overview. *Environmental Carcinogenesis Reviews,* 3: 63–105.

Cowan, C.E., D.J. Versteeg, R.J. Larson, and P.J. Kloepper-Sams. 1995. Integrated approach for environmental assessment of new and existing substances. *Regulatory Toxicology and Pharmacology,* 21: 3–31.

Cramp, S., Ed. 1977 – 1994. *Handbook of the Birds of Europe, the Middle East and North Africa: The Birds of the Western Palearctic*, Oxford University Press, Oxford, U.K.

Crommentuijn, T., D.F. Kalf, M.D. Polder, R. Posthumus, and E.J. Van de Plassche. 1997a. Maximum Permissible Concentrations and Negligible Concentrations for Pesticides. RIVM Report 601501 002 and annexes. National Institute of Public Health and the Environment, Bilthoven, the Netherlands.

Crommentuijn, T., M.D. Polder, and E.J. Van de Plassche. 1997b. Maximum Permissible Concentrations and Negligible Concentrations for Metals, Taking Background Concentrations into Account. RIVM Report 601501 001. National Institute of Public Health and the Environment, Bilthoven, the Netherlands.

Crommentuijn, T., M.D. Polder, D. Sijm, J. de Bruijn, and E.J. van de Plassche. 2000a. Evaluation of the Dutch environmental risk limits for metals by application of the added risk approach. *Environmental Toxicology and Chemistry,* 19: 1692–1701.

Crommentuijn, T., D. Sijm, J. de Bruijn, M. van den Hoop, C.J. van Leeuwen, and E. van de Plassche. 2000b. Maximum permissible and negligible concentrations for some organic substances and pesticides. *Journal of Environmental Management,* 58: 297–312.

Crommentuijn, T., D. Sijm, J. de Bruijn, M. van den Hoop, C.J. van Leeuwen, and E. van de Plassche. 2000c. Maximum permissible and negligible concentrations for metals and metalloids in the Netherlands, taking into account background concentrations. *Journal of Environmental Management,* 60: 121–143.

Crossland, N.O. 1990. The role of mesocosm studies in pesticide registration. Brighton Crop Protection Conference. *Pests and Diseases,* 6B-1: 449–508.

Crum, S.J.H. and T.C.M. Brock. 1994. Fate of chlorpyrifos in indoor microcosms and outdoor experimental ditches. In Hill I.A., F. Heimbach, P. Leeuwangh, and P. Matthiesen, Eds., *Freshwater Field Tests for Hazard Assessment of Chemicals,* Lewis Publishers, Boca Raton, FL, USA: pp. 217–248.

Crum, S.J.H., G.H. Aalderink, and T.C.M. Brock. 1998. Fate of the herbicide linuron in outdoor experimental ditches. *Chemosphere,* 36: 2175–2190.

CSTE/EEC. 1994. EEC water quality objectives for chemicals dangerous to the aquatic environment (List 1). *Review of Environmental Contamination and Toxicology,* 137: 83–110.

Cunnane, C. 1978. Unbiased plotting positions. A review. *Journal of Hydrology,* 37: 205–222.

Cuppen, J.G.M., P.J. van den Brink, H. van der Woude, N. Zwaardemaker, and T.C.M. Brock. 1997. Sensitivity of macrophyte-dominated freshwater microcosms to chronic levels of the herbicide linuron. II. Invertebrates and community metabolism. *Ecotoxicology and Environmental Safety,* 38: 25–35.

Cuppen, J.G.M., P.J. van den Brink, K.F. Uil, E. Camps and T.C.M. Brock. 2000. Impact of the fungicide carbendazim in freshwater microcosms. II. Water quality, breakdown of particulate organic matter and responses of macro-invertebrates. *Aquatic Toxicology,* 48: 233–250.

CWQG. 1999. Canadian Water Quality Guidelines (and updates). Task Force on Water Quality Guidelines of the Canadian Council of Resource and Environment Ministers, Ottawa, Ontario, Canada.

D'Agostino, R.B. 1986a. Graphical analysis. In D'Agostino, R.B. and M.A. Stephens, Eds., *Goodness-of-Fit Techniques,* Marcel Dekker, New York, USA: pp. 7–62.

D'Agostino, R.B. 1986b. Tests for the normal distribution. In D'Agostino, R.B. and M.A. Stephens, Eds., *Goodness-of-Fit Techniques,* Marcel Dekker, New York, USA: pp. 367–419.

534 Species Sensitivity Distributions in Ecotoxicology

D'Agostino, R.B. 1998. Tests for departures from normality. In Armitage, P. and T. Colton, Eds., *Encyclopedia of BioStatistics,* John Wiley, Chichester, U.K.: pp. 315–324.

D'Agostino, R.B. and M.A. Stephens. 1986. *Goodness-of-Fit Techniques,* Marcel Dekker, New York, USA: 560 pp.

Daniels, R.E. and J.D. Allen. 1981. Life table evaluation of chronic exposure to a pesticide. *Canadian Journal of Fisheries and Aquatic Science,* 38: 485–494.

Danish Environmental Protection Agency. 1998a. Remediation of Contaminated Sites. Guidance Report 6, Miljøstyrelsen, Copenhagen, Denmark.

Danish Environmental Protection Agency. 1998b. Remediation of Contaminated Sites. Appendixes, Guidance Report 7, Miljøstyrelsen, Copenhagen, Denmark.

Danish Ministry of Energy and Environment. 1996. Ordinance 823 of 16 September 1996. Ordinance Concerning the Use of Waste Products for Agricultural Purposes. Miljøstyrelsen, Copenhagen, Denmark.

Danish Ministry of Energy and Environment. 1999. Law Act 370 of 2 June 1999. Soil Act, Copenhagen, Denmark.

Davis, J.C. 1977. Standardization and protocols of bioassays — their role and significance for monitoring, research and regulatory usage. In Parker, W.R. et al., Eds., *Proceedings of the 3rd Aquatic Toxicity Workshop,* Halifax, Nova Scotia, Canada, November 2–3, 1976, EPS-5-AR-77-1: 1–14.

Davison, A.C. 1998. Normal scores. In Armitage, P. and T. Colton, Eds., *Encyclopedia of BioStatistics,* John Wiley, Chichester, U.K.: pp. 3067–3069.

Dawson, D.A., E.F. Stebler, S.L. Burks, and J.A. Bantle. 1988. Evaluation of the developmental toxicity of metal-contaminated sediments using short-term fathead minnow and frog embryo–larval assays. *Environmental Toxicology and Chemistry,* 7: 27–34.

Day, K.E. and N.K. Kaushik. 1987. An assessment of the chronic toxicity of the synthetic pyrethroid, fenvalerate, to *Daphnia galeata mendotae*, using life tables. *Environmental Pollution,* 44.

DCEGDE. 1988. Environmental Aspects of the Use of Cationics as Fabric Softeners. Dutch Consultative Expert Group Detergents-Environment, Utrecht, the Netherlands: 16 pp. [in Dutch].

De Bruijn, J.H.M. and C.A.J. Denneman. 1992. Background concentrations of nine trace metals in surface water, groundwater and soil in the Netherlands. VROM-publikatiereeks Bodembescherming 1992/1, The Hague, the Netherlands [in Dutch].

De Haan, F.A.M. 1996. Soil quality evaluation. In De Haan, F.A.M. and M.I. Visser-Reyneveld, Eds., *Soil Pollution and Soil Protection,* International Training Centre (PHLO), Wageningen University, Wageningen, the Netherlands: pp. 1–17.

De Heer, M., R. Alkemade, M. Bakkenes, M. van Esbroek, A. van Hinsberg, and D. de Zwart. 2000. The incidence of about 900 plant species as a function of 7 environmental variables. RIVM report 408657 002. National Institute of Public Health and the Environment, Bilthoven, the Netherlands.

De March, B.G.E. 1987. Simple similar action and independent joint action — two similar models for the joint effects of toxicants applied as mixtures. *Aquatic Toxicology,* 9: 291–304.

De Nijs, T. and J. de Greef. 1992. Ecotoxicological risk evaluation of the cationic fabric softener DHDTMAC II. Exposure modeling. *Chemosphere,* 24: 611–627.

De Snoo, G.R., K.J. Canters, F.M.W. de Jong, and R. Cuperus. 1994. Integral hazard assessment of side effects of pesticides in the Netherlands — a proposal. *Environmental Toxicology and Chemistry,* 13: 1331–1340.

De Zwart, D. and W. Slooff, 1987. Toxicity of mixtures of heavy metals and petrochemicals to *Xenopus laevis. Bulletin of Environmental Contamination and Toxicology,* 38: 345–351.

De Zwart, D., M. Rutgers, and J. Notenboom. 1998. Assessment of site-specific ecological risks of soil contamination: a design of an assessment methodology. RIVM-Report 711701 011. National Institute of Public Health and the Environment, Bilthoven, the Netherlands [in Dutch].

Delp, C.J. 1987. Benzimidazole and related fungicides. In Lyr, H., Ed., *Modern Selective Fungicides — Properties, Applications, Mechanisms of Action,* Longman Group, London, England: pp. 233–244.

Deneer, J.W. 2000. Toxicity of mixtures of pesticides in aquatic systems. *Pesticide Management Science,* 56: 516–520.

Deneer, J.W., T.L. Sinnige, W. Seinen, and J.L.M. Hermens. 1988a. The joint acute toxicity to *Daphnia magna* of industrial organic chemicals at low concentrations. *Aquatic Toxicology,* 12: 33–38.

Deneer, J.W., W. Seinen, and J.L.M. Hermens. 1988b. Growth of *Daphnia magna* exposed to mixtures of chemicals with diverse modes of actions. *Ecotoxicology and Environmental Safety,* 15: 72–77.

Denneman, C.A.J. and C.A.M. van Gestel. 1990. Soil pollution and soil ecosystems, proposal for C-(reference) values on the basis of ecotoxicological risks. RIVM report 725201 001. National Institute of Public Health and the Environment, Bilthoven, the Netherlands [in Dutch].

DeWitt, T.H., G.R. Ditsworth, and R.C. Swartz. 1988. Effects of natural sediment features on survival of the phoxocephalid amphipod, *Rhepoxynius abronius. Marine Environmental Research,* 25: 99–124.

Diggle, P.J. 1990. *Time Series. A Biostatistical Introduction,* Oxford University Press, Oxford, U.K.

DiToro, D.M., J.D. Mahony, D.J. Hansen, K.J. Scott, M.B. Hicks, and M.S. Redmond. 1990. Toxicity of cadmium in sediments, the role of acid volatile sulfide. *Environmental Toxicology and Chemistry,* 9: 1489–1504.

DiToro, D.M., C.S. Zarba, D.J. Hansen, W.J. Berry, R.C. Swartz, C.E. Cowan, S.P. Pavlou, H.E. Allen, N.A. Thomas, and P.R. Paquin. 1991. Technical basis for establishing sediment quality criteria for nonionic organic chemicals using equilibrium partitioning. *Environmental Toxicology and Chemistry,* 10: 1541–1583.

DiToro, D.M., J.D. Mahony, D.J. Hansen, K.J. Scott, A.R. Carlson, and G.T. Ankley. 1992. Acid volatile sulfide predicts the acute toxicity of cadmium and nickel in sediment. *Environmental Science and Technology,* 26: 96–101.

DiToro, D.M., J.A. McGrath, and D.J. Hansen. 2000. Technical basis for narcotic chemicals and polycyclic aromatic hydrocarbon criteria. I. Water and tissue. *Environmental Toxicology and Chemistry,* 19: 1951–1970.

Dixon, W.J., and J.W. Tukey. 1968. Approximate behavior of the distribution of Winsorized t (trimming/Winsorization 2). *Technometrics,* 10: 83–98.

Dobson, S. 1993. Why different regulatory decisions when the scientific information base is similar? — Environmental risk assessment. *Regulatory Toxicology and Pharmacology,* 17: 333–345.

Durand-Huiting, A.M. 1999. Manual and background information on OMEGA123 and OMEGA45. Optimal Modelling for Ecotoxicological Assessment. RIZA-report 99-173X. Rijkswaterstaat, Lelystad, the Netherlands [in Dutch].

ECETOC. 1993a. DHDTMAC, Aquatic and terrestrial hazard assessment. Technical Report 53, CAS 61789-80-8. European Centre for Ecotoxicology and Toxicology of Chemicals (ECETOC), Brussels, Belgium.

ECETOC. 1993b. Aquatic Toxicity Data Evaluation. Technical Report No. 56. European Centre for Ecotoxicology and Toxicology of Chemicals (ECETOC), Brussels, Belgium.

ECETOC. 1997. The value of aquatic model ecosystem studies in ecotoxicology. Technical report 73. European Centre for Ecotoxicology and Toxicology of Chemicals (ECE-TOC), Brussels, Belgium.

ECOFRAM. 1999a. Ecological Committee on FIFRA Risk Assessment Methods Aquatic Report, peer review draft. Available at http://www.epa.gov/oppefed1/ecorisk/index.htm. U.S. Environmental Protection Agency, Washington, D.C., USA.

ECOFRAM. 1999b. Ecological Committee on FIFRA Risk Assessment Methods Terrestrial Draft Report. Available at http://www.epa.gov/oppefed1/ecorisk/index.htm. U.S. Environmental Protection Agency, Washington, D.C., USA.

Edelman, Th. 1984. Background concentrations of substances in soil. VROM-publicatiereeks Bodembescherming 34, Staatsuitgeverij, The Hague, the Netherlands [in Dutch].

EEC. 1996. Commission Directive 96/12/EC of 8 March 1996 amending Council Directive 19/414/EEC concerning the placing of plant protection products on the market. *Official Journal of the European Communities.* L 65/20.

Efroymson, R.E., M.E. Will, and G.W. Suter II. 1997a. Toxicological benchmarks for con- taminants of potential concern for effects on soil and litter invertebrates and het-erotrophic processes, 1997 revision. ES/ER/TM-126/R2. Oak Ridge National Laboratory, Oak Ridge, TN, USA.

Efroymson, R.E., M.E. Will, and G.W. Suter II. 1997b. Toxicological benchmarks for screen-ing contaminants of potential concern for effects on terrestrial plants. ES/ER/TM-85/R3. Oak Ridge National Laboratory, Oak Ridge, TN, USA.

EG&G, Bionomics. 1980. Laboratory evaluation of the toxicity of material to be dredged from Pensacola Harbor and Bay, Pensacola, Florida. EG&G, Bionomics, Marine Research Laboratory, Pensacola, FL, USA: 104 pp.

Eijsackers, H. 1997. Natuurbeheer voor èn door milieubeheer. Inaugural speech, Vrije Uni-versiteit, Amsterdam.

Elmegaard, N. and G.A.J.M. Jagers op Akkerhuis. 2000. Safety factors in pesticide risk assessment. Differences in species sensitivity and acute-chronic relations. NERI Tech-nical report 325, National Environmental Research Institute, Silkeborg, Denmark.

Emans, H.J.B., E.J. van de Plassche, J.H. Canton, P.C. Okkerman, and P.M. Sparenburg. 1993. Validation of some extrapolation methods used for effect assessment. *Environmental Toxicology and Chemistry,* 12: 2139–2154.

Enserink, E.L., J.L. Maas-Diepeveen, and C.J. van Leeuwen. 1991. Combined effects of metals, an ecotoxicological evaluation. *Water Research,* 25: 679–687.

Environment Canada. 1994. Towards a Toxic Substances Management Policy for Canada: A Discussion Document for Consultation Purposes. Environment Canada, Ottawa, Ontario, Canada, September 1994.

EPRI. 1987. Reliability-Based Design of Transmission Line Structures: Final Report. Volume 2: Appendixes. EPRI EL-4793, Electric Power Research Institute, Fort Col-lins, CO, USA.

Erickson, R.J. and C.E. Stephan. 1985. Calculation of the final acute value for water quality criteria for aquatic organisms. National Technical Information Service, Springfield, VA, USA.

Erickson, R.J. and C.E. Stephan. 1988. Calculation of the Final Acute Value for Water Quality Criteria for Aquatic Organisms. PB88-214994. National Technical Information Serv-ice, Springfield, VA, USA.

Estes, J.A., M.T. Tinker, D. Williams, and D.F. Doak. 1998. Killer whale predation on sea otters linking oceanic and neashore ecosystems. *Science,* 282: 473–476.

Evans, M., N. Hastings, and B. Peacock, B. 2000. *Statistical Distributions.* 3rd ed. Wiley, New York, USA: 221 pp.

Faber, M.J., L.M.J. Smith, H.J. Boermans, G.R. Stephenson, D.G. Thompson, and K.R. Solomon. 1997. Cryopreservation of fluorescent marker-labeled algae (*Selenastrum capricornutum*) for toxicity testing using flow cytometry. *Environmental Toxicology and Chemistry*, 16: 1059–1067.

Faust, M., R. Altenburger, T. Backhaus, W. Boedeker, M. Scholze, and L.H. Grimme. 2000. Predictive Assessment of the Aquatic Toxicity of Multiple Chemical Mixtures. *Journal of Environmental Quality*, 29: 1063–1068.

Fawell, J.K. and S. Hedgecott. 1996. Derivation of acceptable concentrations for the protection of aquatic organisms. *Environmental Toxicology and Pharmacology*, 2: 115–120.

Feijtel, T., G. Boeije, M. Matthies, A. Young, G. Morris, C. Gandolfi, B. Hansen, K. Fox, M. Holt, V. Koch, R. Schroder, G. Cassanin, D. Schowanek, J. Rosenblom, and H. Niessen. 1997. Development of a Geography-Referenced Regional Exposure Assessment Tool for Rivers — GREAT-ER — Contribution to GREAT-ER 1. *Chemosphere*, 11: 2351–2373.

Ferson, S., L.R. Ginzberg, and R.A. Goldstein. 1996. Inferring ecological risk from toxicity bioassays. *Water, Air, and Soil Pollution*, 90: 71–82.

Filliben, J.J. 1975. The probability plot correlation coefficient test for normality. *Technometrics*, 17: 111–117.

Fletcher, J.S., F.L. Johnson, and J.C. McFarlane. 1990. Influence of greenhouse versus field testing and taxonomic differences on plant sensitivity to chemical treatment. *Environmental Toxicology and Chemistry*, 9: 769–776.

Forbes, T.L. and V.E. Forbes. 1993. A critique of the use of distribution-based extrapolation models in ecotoxicology. *Functional Ecology*, 7: 249–254.

Forbes, V.E. and T.L. Forbes. 1994. *Ecotoxicology in Theory and Practice*. Chapman & Hall, London, U.K.

Forbes, V.E., P. Calow, and R.M. Sibley. 2001. Are current species extrapolation models a good basis for ecological risk assessment. *Environmental Toxicology and Chemistry*, 20: 442–447.

Foster, G.D., S.M. Baksi, and J.C. Means. 1987. Bioaccumulation of trace organic contaminants from sediments by Baltic clams (*Macoma balthica*) and soft-shell clams (*Mya arenaria*). *Environmental Toxicology and Chemistry*, 6: 969–976.

Fox, D.R. 1999. Setting water quality guidelines — a statistician's perspective. *SETAC News*, 19: 17–18.

Frampton, G.K. 1999. Spatial variation in non-target effects of the insecticides chlorpyrifos, cypermethrin and primicarb on Collembola in winter wheat. *Pesticide Science*, 55: 875–886.

Frampton, G.K. 2000. SCARAB — effects of pesticide regimes on arthropods. In Young, J.E.B., D.V. Alford, and S.E. Ogilvy, Eds., Reducing Agrochemical Use on the Arable Farm, U.K. Ministry of Agriculture Fisheries and Food (MAFF), London, U.K.

Garland, T.J., P.H. Harvey, and A.R. Ives. 1992. Procedures for the analysis of comparative data using phylogenetically independent contrasts. *Systematic Biology*, 41: 18–32.

Garland, T.J., A.W. Dickerman, C.M. Janis, and J.A. Jones. 1993. Phylogenetic analysis of covariance by computer simulation. *Systematic Biology*, 42: 265–292.

Gibbons, J.D. 1971. *Nonparametric Statistical Inference*. 2nd ed. Marcel Dekker, New York, USA: 408 pp.

Giddings, J.M., R.C. Biever, M.F. Annunziato, and A.J. Hosmer. 1996. Effects of diazinon on large outdoor pond microcosms. *Environmental Toxicology and Chemistry*, 15: 8–629.

Giddings, J.M., L.W. Hall, Jr., and K.R. Solomon. 2000. Ecological risks of Diazinon from agricultural use in the Sacramento–San Joaquin river basins, California. *Risk Analysis*, 20: 545–572.

Giddings, J.M., K.R. Solomon, and S.J. Maund. 2001. Probabilistic risk assessment of cotton pyrethroids: II. Aquatic mesocosm and field studies. *Environmental Toxicology and Chemistry,* 20: 660–668.

Giesy, J.P., K.R. Solomon, J.R. Coats, K.R. Dixon, J.M. Giddings, and E.E. Kenaga. 1999. Chlorpyrifos, ecological risk assessment in North American aquatic environments. *Reviews in Environmental Contamination and Toxicology,* 160: 1–129.

Gilbert, R.O. 1987. *Statistical Methods for Environmental Pollution Monitoring.* Van Nostrand Reinhold, New York, USA: 313 pp.

Giolando, S.T., R. Rapaport, R.J. Larson, T.W. Federle, M. Stalmans, and P. Masscheleyn. 1995. Environmental fate and effects of DEEDMAC, a new rapidly biodegradable cationic surfactants for use in fabric softeners. *Chemosphere,* 30: 1067–1083.

Goedkoop, M. and R. Spriensma, 1999. The Eco-indicator 99. A damage oriented method for life cycle impact assessment. Methodology report and annex. Pré Consultants, Amersfoort, the Netherlands. Available at http://www.pre.nl/eco-indicator99/.

Gorree, M., W.L.M. Tamis, T.P. Traas, and M.A. Elbers. 1995. BIOMAG: a model for biomagnification in terrestrial food chains. The case of cadmium in the Kempen, the Netherlands. *Science of the Total Environment,* 168: 215–223.

Government of Canada. 1991a. Toxic chemicals in the Great Lakes and associated effects, synopsis. Report prepared by Environment Canada, Department of Fisheries and Oceans, and Health and Welfare Canada, Toronto, Ontario, Canada: 51 pp.

Government of Canada. 1991b. Toxic chemicals in the Great Lakes and associated effects. Vol. II, Effects. Environment Canada, Department of Fisheries and Oceans, and Health and Welfare Canada, Ottawa, Canada.

Goyette, D., D. Brand, and M. Thomas. 1988. Prevalence of idiopathic liver lesions in English sole and epidermal abnormalities in flatfish from Vancouver Harbour, British Columbia, 1986. Regional Program Report 87-09. Environment Canada, Vancouver: 48 pp.

Graney, R.L., J.P. Giesy, Jr., and D. DiToro. 1989. Mesocosm experimental design strategies: advantages and disadvantages in ecological risk assessment. In Voshell, J.R., Ed., *Using Mesocosms to Assess the Aquatic Ecological Risk of Pesticides: Theory and Practice,* Misc. Publ. 75, Entomological Society of America, Lanham, MD, USA: pp. 74–88.

Greco, W.R., G. Bravo, and J.C. Parsons. 1995. The search for synergy: a critical review from a response surface perspective. *Pharmacological Reviews,* 47: 331–385.

Griscom, S.B., N.S. Fisher, and S.N. Luoma. 2000. Geochemical influences on assimilation of sediment-bound metals in clams and mussels. *Environmental Science and Technology,* 34: 91–99.

Grist, E.P.M., K.M.Y. Leung, J.R. Wheeler, and M. Crane. In preparation. Better bootstrap estimation of Hazardous Concentration thresholds to protect biological assemblages.

Grothe, D.R., K.L. Dickson, and D.K. Reed-Judkins, Eds. 1996. *Whole Effluent Toxicity Testing, An Evaluation of Methods and Prediction of Receiving System Impacts,* SETAC Press, Pensacola, FL, USA.

Guinée, J.B. and R. Heijungs. 1993. A proposal for the classification of toxic substances within the framework of life cycle assessment of products. *Chemosphere,* 26: 1925–1944.

Guinée, J.B., R. Heijungs, L.E.C.M. van Oers, A. Wegener Sleeswijk, D. van de Meent, T. Vermeire, and M. Rikken, 1996. Uniform system for the evaluation of substances. Inclusion of fate in LCA characterisation of toxic release. Applying USES 1.0. *International Journal of Life Cycle Analysis,* 1: 133–138.

Hågvar, S. and G. Abrahamsen. 1990. Microarthropoda and enchytraeidae (Oligochaeta) in naturally lead-contaminated soil, a gradient study. *Environmental Entomology,* 19: 1263–1277.

Hahn, G.J. and S.S. Shapiro. 1969. *Statistical Models in Engineering,* John Wiley & Sons, New York, USA: 355 pp.

Hall, L.W., Jr. and R.D. Anderson. 1995. The influence of salinity on the toxicity of various classes of chemicals to aquatic biota. *Critical Reviews in Toxicology,* 25: 281–346.

Hall, L.W., Jr. and J.M. Giddings. 2000. The need for multiple lines of evidence for predicting site-specific ecological effects. *Human and Ecological Risk Assessment,* 6: 679–710.

Hall, L.W., Jr., M.C. Scott, and W.D. Killen. 1998. Ecological risk assessment of copper and cadmium in surface waters of Chesapeake Bay watershed. *Environmental Toxicology and Chemistry,* 17: 172–189.

Hall, L.W., Jr., J.M. Giddings, K.R. Solomon, and R. Balcomb. 1999. An ecological risk assessment for the use of Irgarol 1051 as an algaecide for antifoulant paints. *Critical Reviews in Toxicology,* 29: 367–437.

Hall, L.W., Jr., M.C. Scott, W.D. Killen, and M.A. Unger. 2000. A probabilistic ecological risk assessment of tributyltin in surface waters of the Chesapeake Bay watershed. *Human and Ecological Risk Assessment,* 6: 141–179.

Hamers, T., T. Aldenberg, and D. van de Meent. 1996a. Definition report — indicator effects toxic substances (I_{tox}). RIVM-Report 607128 001. National Institute of Public Health and the Environment, Bilthoven, the Netherlands.

Hamers, T., J. Notenboom, and H.J.P. Eijsackers. 1996b. Validation of laboratory toxicity data on pesticides for the field situation. RIVM report 719102 046. National Institute of Public Health and the Environment, Bilthoven, the Netherlands.

Hamilton, M.A., R.C. Russo, and R.V. Thurston. 1977/1978. Trimmed Spearman-Kärber method for estimating median lethal concentrations in toxicity bioassays. *Environmental Science and Technology,* 11: 714–719; Correction, *Environmental Science and Technology,* 12: 417.

Hansen, P.-D. 1989. Ecological requirements for the quality objectives in the aquatic environment. *Tenside Surfactants Detergents,* 26: 80–84.

Hare, L. and A. Tessier. 1996. Predicting animal cadmium concentrations in lakes. *Nature,* 380: 430–432.

Hart, A., Ed. 2001. EUPRA: Report of the European Workshop on Probabilistic Risk Assessment for the environmental impacts of plant protection products [provisional title]. Central Science Laboratory, York, U.K.

Hauschild, M., H. Wenzel, A. Damborg, and J. Tørsløv, 1998. Ecotoxocity as a criterion in the environmental assessment of products. In Hauschild, M. and H. Wenzel, *Environmental Assessment of Products, Scientific Background,* Vol. 2, Chapman & Hall, London, U.K.

Health Council of the Netherlands. 1989. Analysing the risk of toxic chemicals for ecosystems. Report submitted by a committee of the Health Council 28E, The Hague, the Netherlands: 173 pp.

Health Council of the Netherlands. 1993a. Ecotoxicological risk assessment and policy-making in the Netherlands — dealing with uncertainties. *Network,* 6(3)/7(1): 8–11.

Health Council of the Netherlands. 1993b. Secondary poisoning. Toxicants in Foodchains. Health Council (Gezondheidsraad) report GR 1993/04, The Hague, the Netherlands [in Dutch].

Heijungs, R. and P. Hofstetter. 1996. Definitions of terms and symbols. In Udo de Haes, H.A., Ed., *Towards a Methodology for Life Cycle Impact Assessment.* Society of Environmental Toxicology and Chemistry — Europe, Brussels, Belgium.

Heijungs, R., J.B. Guinée, G. Huppes, R.M. Lankreijer, H.A. Udo de Haes, A. Wegener Sleeswijk, A.M.M. Ansems, P.G. Eggels, R. van Duin, and H.P. de Goede. 1992. Environmental life cycle assessment of products. Guidelines and backgrounds. Centre of Environmental Sciences, Leiden, the Netherlands.

Helsel, D.R., and T.A. Cohn. 1988. Estimation of descriptive statistics for multiply censored water quality data. *Water Resources Research,* 24: 1997–2004.

Helsel, D.R. and R.M. Hirsch. 1992. *Statistical Methods in Water Resources.* Elsevier, Amsterdam: 522 pp.

Hendley, P., C. Holmes, S. Kay, S.J. Maund, K.Z. Travis, and M. Zhang. 2001. Probabilistic risk assessment of cotton pyrethroids: III. A spatial analysis of the Mississippi cotton landscape. *Environmental Toxicology and Chemistry,* 20: 669–678.

Hendriks, A.J., W.C. Ma, J.J. Brouns, E.M. de Ruiter-Dijkman, and R. Gast. 1995. Modelling and monitoring organochlorine and heavy metal accumulation in soils, earthworms, and shrews in Rhine-delta floodplains. *Archives of Environmental Contamination and Toxicology,* 29: 115–127.

Hermens, J. and P. Leeuwangh. 1982. Joint toxicity of mixtures of 8 and 24 chemicals to the guppy (*Poecilia reticulata*). *Ecotoxicology and Environmental Safety,* 6: 302–310.

Hermens, J., H. Canton, N. Steyger, and R. Wegman. 1984. Joint effects of a mixture of 14 chemicals on mortality and inhibition of reproduction of *Daphnia magna. Aquatic Toxicology,* 5: 315–322.

Hertwich, E.G., W.S. Pease, and T.E. McKone. 1998. Evaluating toxic impact assessment methods, what works best? *Environmental Science and Technology,* 32: A138–A144.

Hewlett, P.S. and R.L. Plackett. 1979. *An Introduction to the Interpretation of Quantal Responses in Biology,* Edward Arnold Ltd., London, U.K.

Hill, I.R., F. Heimbach, P. Leeuwangh, and P. Matthiessen, Eds. 1994. *Freshwater Field Tests for Hazard Assessment of Chemicals,* CRC Press, Boca Raton, FL, USA.

Hoekstra, J.A. and P.H. van Ewijk. 1993. Alternatives for the no-observed-effect level. *Environmental Toxicology and Chemistry,* 12: 187–194.

Hoekstra, J.A., M.A. Vaal, J. Notenboom, and W. Slooff. 1994. Variation in the sensitivity of aquatic species to toxicants. *Bulletin of Environmental Contamination and Toxicology,* 53: 98–105.

Hofstetter, P. 1998. *Perspectives in Life Cycle Impact Assessment, a Structured Approach to Combine Models of the Technosphere, Ecosphere and Valuesphere.* Kluwer, Dordrecht, the Netherlands.

Hogg, R.V. and A.T. Craig. 1995. *Introduction to Mathematical Statistics.* 5th ed. Prentice-Hall, Upper Saddle River, NJ, USA: 564 pp.

Hommen, U., H.-J. Poethke, U. Dulmer, and H.T. Ratte. 1993. Simulation models to predict ecological risk of toxins in freshwater systems. *ICES Journal of Marine Science,* 50: 337–347.

Hopkin, S.P. 1993. Ecological implications of "95% protection levels" for metals in soil. *Oikos,* 66: 137–141.

Host, G.E., R.R. Regal, and C.E. Stephan. 1995. Analyses of Acute and Chronic Data for Aquatic Life. U.S. Environmental Protection Agency, Duluth, MN, USA. Draft dated 3-16-95.

Hsu, H.P. 1997. *Probability, Random Variables, and Random Processes.* Schaum's Outline Series, McGraw-Hill, New York, USA: 306 pp.

Huggett, R.J., R.A. Kimerle, P.M. Mehrle, and H. Bergman. 1992. Introduction. In Huggett, R.J., R.A. Kimerle, P.M. Mehrle, and H. Bergman, Eds., *Biomarkers: Biochemical, Physiological and Histological Markers of Anthropogenic Stress,* Lewis Publishers, Boca Raton, FL, USA: pp. 1–7.

Huijbregts, M.A.J. 1999. Priority Assessment of Toxic Substances in the Frame of LCA. Development and application of the multi-media fate, exposure and effect model USES-LCA. Interfaculty Department of Environmental Science, Faculty of Environmental Sciences, University of Amsterdam, Amsterdam, the Netherlands. Available at http://www.leidenuniv.nl/interfac/cml/lca2/.

Huijbregts, M.A.J., U. Thissen, J.B. Guinée, T. Jager, D. van de Meent, A.M.J. Ragas, A. Wegener Sleeswijk, and L. Reijnders. 2000. Priority assessment of toxic substances in life cycle assessment, I, Calculation of toxicity potentials for 181 substances with the nested multi-media fate, exposure and effects model USES-LCA. *Chemosphere,* 41: 541–573.

IJC. 1993. A Strategy for the Virtual Elimination of Persistent Toxic Substances. Vol. 1, Report of the Virtual Elimination Task Force to the IJC. International Joint Commission, Windsor, Ontario, Canada.

ISO. 1997. *Environmental Management — Life Cycle Assessment — Principles and Framework.* International Organization for Standardization, Geneva, Switzerland.

ISO. 1998. *Environmental Management — Life Cycle Assessment — Life Cycle Impact Assessment.* International Organization for Standardization, Geneva, Switzerland.

IWINS. 1997. Integrated Environmental Quality Standards for Soil, Water and Air. Interdepartementale Werkgroep Integrale Normstelling Stoffen, VROM 97759/h/12-97, The Hague, the Netherlands [in Dutch].

Jacobs, T.L. 1992. Probabilistic environmental risk of hazardous materials. *Journal of Environmental Engineering,* 118: 878–889.

Jagoe, R.H. and M.C. Newman. 1997. Bootstrap estimation of community NOEC values. *Ecotoxicology,* 6: 293–306.

Janssen, C.R., M.D. Ferrando Rodrigo, and G. Persoone. 1993. Ecotoxicological studies with the freshwater rotifer *Brachionus calyciflorus.* I. Conceptual framework and application, *Hydrobiologia,* 255/256: 21–32.

Janssen, P.J.C.M. and G.J.A. Speijers. 1997. Guidance on the derivation of maximum permissible risk levels for human intake of soil contaminants. RIVM report 711701 006. National Institute of Public Health and the Environment, Bilthoven, the Netherlands.

Janssen, R.P.T., W.J.G.M Peijnenburg, L. Posthuma, and M.A.G.T. van den Hoop. 1997a. Equilibrium partitioning of heavy metals in Dutch field soils. I. Relationships between metal partition coefficients and soil characteristics. *Environmental Toxicology and Chemistry,* 16: 2470–2478.

Janssen, R.P.T., L. Posthuma, R. Baerselman, H.A. den Hollander, R.P.M. van Veen, and W.J.G.M. Peijnenburg. 1997b. Equilibrium partitioning of heavy metals in Dutch field soils. II. Prediction of metal accumulation in earthworms. *Environmental Toxicology and Chemistry,* 16: 2479–2488.

Janus, J.A. 1993. Integrated criteria document zinc: ecotoxicity. Annex to RIVM report 710401 028. National Institute of Public Health and the Environment, Bilthoven, the Netherlands.

Janus, J.A., J.H. Canton, C.A.M. van Gestel, and E. Heijna-Merkus. 1989. Integrated criteria document copper; effects. Annex to RIVM report 758474 009. National Institute of Public Health and the Environment, Bilthoven, the Netherlands.

Jaworska, J., V. Koch, T. Feijtel, J. Rosenblom, J.C. Boutonnet, S. Marshall, S. Webb, and H. Vrijhof. 1999. Probabilistic Environmental Risk Assessment of DODMAC/DHTDMAC for Use as a Fabric Softener. *Abstracts,* SETAC Annual Meeting, Philadelphia, PA, USA.

Jensen, J. and P. Folker-Hansen. 1995. Soil quality criteria for selected organic compounds. Working report 49, Danish Environmental Protection Agency, Copenhagen, Denmark.

Jensen, J., H.L. Lakkenborg, M. Bruus Pedersen, and J.J. Scott-Fordsmand. 1997. Soil quality criteria of selected compounds. Working report 87. Danish Environmental Protection Agency, Copenhagen, Denmark.

Jensen, K.I.N., G.R. Stephenson, and L.A. Hunt. 1977. Detoxification of atrazine in three gramineae subfamilies. *Weed Science,* 25: 212–220.

Jolliet, O. 1996. Impact assessment of human and eco-toxicity in life cycle assessment. In Udo de Haes, H.A., Ed., *Towards a Methodology for Life Cycle Impact Assessment,* Society of Environmental Toxicology and Chemistry–Europe, Brussels, Belgium.

Jongbloed, R.H., J. Pijnenburg, B.J.W.G. Mensink, T.P. Traas, and R. Luttik. 1994. A model for environmental risk assessment and standard setting based on biomagnification. Top predators in terrestrial ecosystems. RIVM report 719101.012 with annex. National Institute of Public Health and the Environment, Bilthoven, the Netherlands.

Jongbloed, R.H., T.P. Traas, and R. Luttik. 1996. A probabilistic model for deriving soil quality criteria based on secondary poisoning of top predators. II. Calculations for dichlorodiphenyltrichloroethane (DDT) and cadmium. *Ecotoxicology and Environmental Safety,* 34: 297–306.

Kalf, D.F., T. Crommentuijn, and E.J. van de Plassche. 1997. Environmental quality objectives for 10 polycyclic aromatic hydrocarbons (PAHs). *Ecotoxicology and Environmental Safety,* 36: 89–97.

Kalf, D.F., B.J.W.G. Mensink, and M.H.M.M. Montforts, 1999. Protocol for derivation of harmonised maximum permissible concentrations (MPCs). RIVM report 601506 001. National Institute of Public Health and the Environment, Bilthoven, the Netherlands.

Kammenga, J. and R. Laskowski, Eds. 2000. *Demography in Ecotoxicology.* John Wiley, Chichester, U.K.

Kammenga, J.E., M. Busschers, N.M. van Straalen, P.C. Jepson, and J. Bakker. 1996. Stress induced fitness reduction is not determined by the most sensitive life-cycle trait. *Functional Ecology,* 10: 106–111

Kammenga, J.E., P.H.G. van Koert, J.H. Koeman, and J. Bakker. 1997. Fitness consequences of toxic stress evaluated within the context of phenotypic plasticity. *Ecological Applications,* 7: 726–734.

Kaplan, S. and B.J. Garrick. 1981. On the quantitative definition of risk. *Risk Analysis,* 1: 11–27.

Kater, B.J. 1995. ERASES — Ecotoxicological risk analysis of the Scheldt Estuary: parameters. Werkdocument RIKZ/AB-95.834x. Rijkswaterstaat, Rijksinstituut voor Kust en Zee, Middelburg, the Netherlands [in Dutch].

Kater, B.J. and F.O.B. Lefèvre. 1996. Ecotoxicologische risico analyse in de Westerschelde. RIKZ report 96.007, Rijksinstituut voor Kust en Zee, Middelburg, the Netherlands [in Dutch].

Kaushik, N.K., G.L. Stephenson, K.R. Solomon, and K.E. Day. 1985. Impact of permethrin on zooplankton communities using limnocorrals. *Canadian Journal of Fisheries and Aquatic Science,* 42: 77–85.

Kedwards, T.J., S.J. Maund, and P.F. Chapman. 1999. Community level analysis of ecotoxicological field studies: II. Replicated-design studies. *Environmental Toxicology and Chemistry,* 18: 158–166.

Kersting, K. and R.P.A. van Wijngaarden. 1999. Effects of a pulsed treatment with the herbicide afalon (active ingredient linuron) on macrophyte-dominated mesocosms. I. Responses of ecosystem metabolism. *Environmental Toxicology and Chemistry,* 18: 2859–2865.

Khalil, M.A., H.M. Abdel-Lateif, B.M. Bayoumi, and N.M. van Straalen, 1996. Analysis of separate and combined effects of heavy metals on the growth of *Aporrectodea caliginosa* (Oligocheata; Annelida), using the toxic unit approach. *Applied Soil Ecology,* 4: 213–219.

Klaine, S.J., G.P. Cobb, R.L. Dickerson, K.R. Dixon, R.J. Kendal, E.E. Smith, and K.R. Solomon. 1996a. An ecological risk assessment for the use of the biocide dibromonitrilopropionamide (DBNPA) in industrial cooling systems. *Environmental Toxicology and Chemistry,* 15: 21–30.

Klaine, S.J., K.R. Dixon, and J.D. Florian. 1996b. Characterization of *Selanastrum capricornutum* response to episodic atrazine exposure. Report 09524. Novartis Crop Protection, Greensboro, NC, USA, January 24.

Klapow, L.A. and R.H. Lewis. 1979. Analysis of toxicity data for California marine water quality standards. *Journal of the Water Pollution Control Federation,* 51: 2054–2070.

Klepper, O. and D. van de Meent, 1997. Mapping the potentially affected Fraction (PAF) of species as an indicator of generic toxic stress. RIVM report 607504 001. National Institute of Public Health and the Environment, Bilthoven, the Netherlands.

Klepper, O., J. Bakker, T.P. Traas, and D. van de Meent. 1998. Mapping the potentially affected fraction (PAF) of species as a basis for comparison of ecotoxicological risks between substances and regions. *Journal of Hazardous Materials,* 61: 337–344.

Klepper, O., T.P. Traas, A. Schouten, G.W. Korthals, and D. de Zwart, 1999. Estimating the effect on soil organisms of exceeding no-observed effect concentrations (NOECs) of persistent toxicants. *Ecotoxicology,* 8: 9–21.

Knezovich, J.P., F.L. Harrison, and R.G. Wilhelm. 1987. The bioavailability of sediment-sorbed organic chemicals, a review. *Water, Air, and Soil Pollution,* 32: 233–245.

Knoben, R.A.E., M.A. Beek, and A.M. Durand. 1998. Application of Species Sensitivity Distributions as ecological assessment tool for water management. *Journal of Hazardous Materials,* 61: 203–207.

Könemann, H. 1981. Fish toxicity tests with mixtures of more than two chemicals, a proposal for a quantitative approach and experimental results. *Toxicology,* 19: 229–238.

Kooijman, S.A.L.M. 1987. A safety factor for LC_{50} values allowing for differences in sensitivity among species. *Water Research,* 21: 269–276.

Kooijman, S.A.L.M. 1993. Dynamic energy budgets in biological systems. In *Theory and Applications in Ecotoxicology,* University Press, Cambridge, U.K.

Kooijman, S.A.L.M. 1996. An alternative for NOEC exists, but the standard model has to be abandoned first. *Oikos,* 75: 310–316.

Kooijman, S.A.L.M. and J.J.M. Bedaux. 1996. Some statistical properties of estimates of no-effects concentrations. *Water Research,* 30: 1724–1728.

Korthals, G.W., A.D. Alexiev, Th.M. Lexmond, J.E. Kammenga, and T. Bongers. 1996. Long-term effects of copper and pH on the terrestrial nematode community in an agroecosystem. *Environmental Toxicology and Chemistry,* 15: 979–985.

Kosalwat, P. and A.W. Knight. 1987. Chronic toxicity of copper to a partial life cycle of the midge, *Chironomus decorus. Archives of Environmental Contamination and Toxicology,* 16: 283–290.

Kovacs, T.G., P.H. Martel, R.H. Voss, P.E. Wrist, and R.F. Willes. 1993. Aquatic toxicity equivalency factors for chlorinated phenolic compounds present in pulp mill effluent. *Environmental Toxicology and Chemistry,* 12: 684–691.

Kros, J., G.J. Reinds, W. de Vries, J.B. Latour, and M. Bollen. 1995. Modelling the response of terrestrial ecosystems to acidification and desiccation scenarios. *Water, Air, and Soil Pollution,* 85: 1101–1106.

Kupper, L.L. and K.B. Hafner. 1989. How appropriate are popular sample size formulas? *American Statistician,* 43: 101–105.

Lackey, R.T. 1997. Ecological risk assessment: use, abuse and alternatives. *Environmental Management,* 21: 808–821.

Lamberson, J.O., T.H. DeWitt, and R.C. Swartz. 1992. Assessment of sediment toxicity to marine benthos. In Burton, G.A., Ed., *Sediment Toxicity Assessment,* Lewis Publishers, Boca Raton, FL, USA: pp. 183–211.

Landrum, P.F. and J.A. Robbins. 1990. Bioavailability of sediment-associated contaminants to benthic invertebrates. In Baudo, R., J.P. Giesy, and H. Muntau, Eds., *Sediments, Chemistry and Toxicity of In-Place Pollutants,* Lewis Publishers, Boca Raton, FL, USA: pp. 237–263.

Laskowski, R. 1995. Some good reasons to ban the use of NOEC, LOEC and related concepts in ecotoxicology. *Oikos,* 73: 140–144.

Latour, J.B. and R. Reiling. 1993. A multiple stress model for vegetation ("MOVE"), a tool for scenario studies and standard-setting. *Science of the Total Environment,* Suppl. 2: 1513–1526.

Latour, J.B., R. Reiling, and W. Slooff. 1994. Ecological standards for eutrophication and desiccation, perspectives for risk assessment. *Water, Air, and Soil Pollution,* 78: 265–278.

Lawton, J.H. 1994. What do species do in ecosystems? *Oikos,* 71: 367–374.

LeBlanc, G.A. 1984. Interspecies relationships in acute toxicity of chemicals to aquatic organisms. *Environmental Toxicology and Chemistry,* 3: 47–82.

Lee, G.F. and A. Jones-Lee. 1995. Appropriate use of numeric chemical concentration-based water quality criteria. *Human and Ecological Risk Assessment,* 1: 5–11.

Lee, G.F. and A. Jones-Lee. 1999. The single chemical probabilistic risk assessment approach is inadequate for OP pesticide aquatic life toxicity. *SETAC News,* 19: 20–21.

Lehtinen, K.-J., M. Notini, J. Mattsson, and L. Landner. 1988. Disappearance of bladderwrack (*Fucus vesiculosis* L.) in the Baltic Sea; Relation to pulp-mill chlorate. *Ambio,* 17: 387–393.

Leslie, J.K. and C.A. Timmins. 1998. Age 0+ fish occurrence in modified habitat in South-Western Ontario. Fisheries and Oceans report 2219, Burlington, Ontario, Canada.

Levin, S.A., M.A. Harwell, J.R. Kelly, and K.D. Kimball. 1989. *Ecotoxicology, Problems and Approaches,* Springer-Verlag, New York, USA.

Lewis, M.A. and V.T. Wee. 1983. Aquatic safety assessment for cationic surfactants. *Environmental Toxicology and Chemistry,* 2: 105–118.

Lexmond, T.M. and Th. Edelman. 1986. Reference lines and background concentrations for several heavy metals and arsene in the top soil of natural areas and agricultural soil. Report 1986-2, Section Soil Science and Plan Nutrition, Agricultural University, Wageningen, the Netherlands.

Lexmond, T.M. and Th. Edelman. 1992. Present background concentartions of several heavy metals and arsenic in soil. *Handbook for Environmental Protection and Soil Management.* D4110: 1–34. Samson Tjeenk Willink, Alphen aan de Rijn, the Netherlands [in Dutch].

Liber, K., K.R. Solomon, N.K. Kaushik, and J.H. Carey. 1994. Impact of 2,3,4,6-tetrachlorophenol (DIATOX) on plankton communities in limnocorrals. In Graney, R.L., J.L. Kennedy, and J.H. Rogers, Eds., *Aquatic Mesocosm Studies in Ecological Risk Assessment,* Lewis Publishers, Boca Raton, FL, USA: pp. 257–294.

Lloyd, R. 1992. *Pollution and Freshwater Fish,* Blackwell Scientific, Oxford, U.K.: 173 pp.

Løkke, H. 1989. Huor ren er ren jord? *Miljø and Teknologi,* 1: 10–13.

Løkke, H. 1994. Ecotoxicological extrapolation, Tool or toy? In Donker, M.H., H. Eijsackers, and F. Heimbach, Eds., *Ecotoxicology of Soil Organisms,* Lewis Publishers, Boca Raton, FL, USA: pp. 411–425.

Løkke, H. and C.A.M. van Gestel. 1998. *Handbook of Soil Invertebrate Toxicity Tests,* John Wiley & Sons, Chichester, U.K.

Long, E.R. 1992. Ranges in chemical concentrations in sediments associated with adverse biological effects. *Marine Pollution Bulletin,* 24: 38–45.

Long, E.R., and D.D. MacDonald. 1992. National Status and Trends Program approach. In Sediment Classification Methods Compendium, EPA 823-R-92-006, Office of Water, U.S. Environmental Protection Agency, Washington, D.C., USA.

Long, E.R. and L.G. Morgan. 1990. The potential for biological effects of sediment-sorbed contaminants tested in the National Status and Trends Program. National Oceanic and Atmospheric Administration Tech. Memo. NOS OMA 52. Seattle, WA, USA: 175 pp. + app.

Long, E.R., D.D. MacDonald, S.L. Smith, and F.D. Calder. 1995. Incidence of adverse biological effects within ranges of chemical concentrations in marine and estuarine sediments. *Environmental Management,* 19: 81–97.

Long, E.R., D.D. MacDonald, J.C. Cubbage, and C.G. Ingersoll. 1998. Predicting the toxicity of sediment-associated trace metals with simultaneously extracted trace metal: acid-volatile sulfide concentrations and dry weight-normalized concentrations, a critical comparison. *Environmental Toxicology and Chemistry,* 17: 972–974.

Lorenzato, S.G., A.J. Gunther, and J.M. O'Connor. 1991. Summary of a workshop concerning sediment quality assessment and development of sediment quality objectives. California State Water Resources Control Board, Sacramento, CA, USA: 32 pp.

Loring, D.H. and R.T.T. Rantala. 1992. Manual for the geochemical analysis of marine sediments and suspended particulate matter. *Earth-Science Reviews,* 32: 235.

Luttik, R. and T. Aldenberg. 1995. Extrapolation factors to be used in case of small samples of toxicity data (with a special focus on LD_{50} values for birds and mammals). RIVM report 679102 029. National Institute of Public Health and Environment, Bilthoven, the Netherlands.

Luttik, R. and T. Aldenberg. 1997. Extrapolation factors for small samples of pesticide toxicity data: special focus on LD_{50} values for birds and mammals. *Environmental Toxicology and Chemistry,* 16: 1785–1788.

Luttik, R., C.A.F.M. Romijn, and J.H. Canton. 1993. Presentation of a general algorithm to include secondary poisoning in effect assessment. *Science of the Total Environment,* Suppl. 1993: 1491–1500.

Luttik, R., T.P. Traas, and H. Mensink. 1997. Mapping the potential affected fraction of avian and mammalian indicator species in the notional ecological network. RIVM report 607504 002. National Institute of Public Health and the Environment, Bilthoven, the Netherlands.

Ma, W.C. 1982. The influence of soil properties and worm related factors on the concentration of heavy metals in earthworms. *Pedobiologia,* 24: 109–119.

Ma, W.C., Th. Edelman, I. van Beersum, and Th. Jans. 1983. Uptake of cadmium, zinc, lead and copper by earthworms near a zinc-smelting complex: influence of soil pH and organic matter. *Bulletin of Environmental Contamination and Toxicology,* 30: 424–427.

Ma, W.C., W. Denneman, and J. Faber. 1991. Hazardous exposure of ground-living small mammals to cadmium and lead in contaminated terrestrial ecosystems. *Archives of Environmental Contamination and Toxicology,* 20: 266–270.

MacArthur, R.H. and E.O. Wilson. 1967. *The Theory of Island Biogeography.* Princeton University Press, Princeton, NJ, USA.

MacDonald, D.D. 1993. Development of an Approach to the Assessment of Sediment Quality in Florida Coastal Waters. Prepared for the Florida Department of Environmental Protection. MacDonald Environmental Sciences, Ltd., Ladysmith, British Columbia, Canada: Vol. 1, 128 pp.; Vol. 2, 117 pp.

MacDonald, D.D. 1994. Approach to the Assessment of Sediment Quality in Florida Coastal Waters. Prepared for the Florida Department of Environmental Protection. MacDonald Environmental Sciences, Ltd, Ladysmith, British Columbia, Canada: Vol. 1, 123 pp.

MacDonald, D.D, S.L. Smith, M.P. Wong, and P. Mudroch. 1992. The Development of Canadian Marine Environmental Quality Guidelines. Report prepared for the Interdepartmental Working Group on Marine Environmental Quality Guidelines and the Canadian Council of Ministers of the Environment. Environment Canada, Ottawa, Canada: 50 pp. + app.

MacDonald, D.D., R.S. Carr, F.D. Calder, E.R. Long, and C.G. Ingersoll. 1996. Development and evaluation of sediment quality guidelines for Florida coastal waters. *Ecotoxicology,* 5: 253–278.

Mackay, D. 1991. *Multimedia Environmental Models. The Fugacity Approach*, Lewis Publishers, Chelsea, MI, USA.

Mackay, D. and S. Peterson. 1982. Fugacity revisited. *Environmental Science and Technology,* 16: 654A–660A.

Mage, D.T. 1982. An objective graphical method for testing normal distributional assumptions using probability plots. *The American Statistician,* 36: 116–120.

Malins, D.C., B.B. McCain, D.W. Brown, S.L. Chan, M.S. Myers, J.T. Landahl, P.G. Prohaska, A.J. Friedman, L.D. Rhodes, D.G. Burrows, W.D. Gronlund, and H.O. Hodgkins. 1984. Chemical pollutants in sediments and diseases of bottom-dwelling fish in Puget Sound, Washington. *Environmental Science and Technology,* 18: 705–713.

Malins, D.C., M.M. Krahn, M.S. Myers, L.D. Rhodes, D.W. Brown, C.A. Krone, B.B. McCain, and S.L. Chan. 1985. Toxic chemicals in sediments and biota from a creosote-polluted harbour, Relationships with hepatic neoplasms and other hepatic lesions in English sole (*Parophrys vetulus*). *Carcinogenesis,* 6: 1463–1469.

Manly, B.F.J. 1992. Bootstrapping for determining sample sizes in biological studies. *Journal of Experimental Marine Biology and Ecology,* 158: 189–196.

Manz, M., L. Weissflog, and G. Schüürman. 1999. Ecotoxicological hazard and risk assessment of heavy metal contents in agricultural soils of central Germany. *Ecotoxicology and Environmental Safety,* 42: 191–201.

Martin, E. and J. Novak. 1999. *Mathematica® — 4 Standard Add-on Packages*. Wolfram Research, Champaign, IL, USA: 535 pp.

Maund, S.J., K.Z. Travis, P. Hendley, J.M. Giddings and K.R. Solomon. 2001. Probabilistic risk assessment of cotton pyrethroids: V. Combining landscape-level exposures and ecotoxicological effects data to characterize risks. *Environmental Toxicology and Chemistry,* 20: 687–692.

Mayer, F.L. and M.R. Ellersieck. 1986. Manual of aquatic toxicity: interpretation and data base for 410 chemicals and 66 species of freshwater animals. Resource Publication 160, U.S. Department of the Interior, Fish and Wildlife Service, Washington, D.C., USA.

Mayer, L.M., Z. Chen, R.H. Findlay, J. Fang, S. Sampson, R.F.L. Self, P.A. Jumars, C. Quetel, and O.F.X. Donard. 1996. Bioavailability of sedimentary contaminants subject to deposit-feeder digestion. *Environmental Science and Technology,* 30: 2641–2645.

McCarty, L.S. and D. Mackay, 1993. Enhancing ecotoxicological modeling and assessment. *Environmental Science and Technology,* 27: 1719–1728.

McDaniel, M. and T.W. Snell. 1999. Probability distributions of toxicant sensitivity for freshwater rotifer species. *Environmental Toxicology,* 14: 361–366.

McGee, B.L., C.E. Schlekat, and E. Reinharz. 1993. Assessing sublethal levels of sediment contamination using the estuarine amphipod *Leptocheirus plumulosus*. *Environmental Toxicology and Chemistry,* 12: 577–587.

McGreer, E.R. 1982. Factors affecting the distribution of the bivalve *Macoma balthica* (L.) on a mudflat receiving sewage effluent, Fraser River Estuary, British Columbia. *Marine Environmental Research,* 7: 131–149.

McKone, T.E. 1993. CalTOX, a multimedia total exposure model for hazardous-waste sites. UCRL-CR-111456 Pt I-IV, U.S. Department of Energy, Lawrence Livermore National Laboratory, U.S. Government Printing Office, Washington, D.C., USA.

McLaughlin, S.B. and G.E. Taylor. 1985. SO$_2$ effects on dicot crops, some issues, mechanisms and indicators. In Winner W.E., H.A. Mooney, and R.A. Goldstein, Eds., *Sulfur Dioxide and Vegetation,* Stanford University Press, Stanford, CA, USA: pp. 227–249.

MEDCHEM. 1992. MEDCHEM database. Database and calculation method for K_{ow} values. Developed at Pomona College, Claremont, CA, USA.

Mennes, W., K.D. van den Hout, and E.J. van de Plassche. 1995. Characterization of human exposure patterns to environmental contaminants, possibilities of the USES approach. RIVM report 679101 021. National Institute of Public Health and the Environment, Bilthoven, the Netherlands.

Mennes, W., M. van Apeldoorn, M. Meijerinck, and T. Crommentuijn. 1998. The incorporation of human toxicity criteria into Integrated Environmental Quality Standards. RIVM report 601501 004. National Institute of Public Health and the Environment, Bilthoven, the Netherlands.

Michael, J.R. and W.R. Schucany. 1986. Analysis of data from censored samples. In D'Agostino, R.B. and M.A. Stephens, Eds., *Goodness-of-Fit Techniques,* Marcel Dekker, New York, USA: pp. 461–496.

Millard, S.P. and N.K. Neerchal. 2001. *Environmental Statistics with S-PLUS.* CRC Press, Boca Raton, FL, USA: 830 pp.

Mineau, P., B.T. Collins, and A. Baril. 1996. On the use of scaling factors to improve interspecies extrapolation of acute toxicity in birds. *Regulatory Toxicology and Pharmacology,* 24: 24–29.

Mineau, P., A. Baril, B.T. Collins, J. Duffe, G. Joerman, and R. Luttik. 2001. Pesticide acute toxicity reference values for birds. *Reviews in Environmental Contamination and Toxicology,* 170: 13–74.

Montgomery, D.C. 1997. *Design and Analysis of Experiments,* John Wiley & Sons, New York, USA.

Mood, A.M., F.A. Graybill, and D.C. Boes. 1974. *Introduction to the Theory of Statistics.* 3rd ed. McGraw-Hill, Tokyo, Japan: 564 pp.

Moore, D.R.J. and B.J. Elliott. 1996. Should uncertainty be quantified in human and ecological risk assessments used for decision-making? *Human and Ecological Risk Assessment,* 2: 11–24.

Morgan, M.G. 1998. Uncertainty analysis in risk assessment. *Human and Ecological Risk Assessment,* 4: 25–39.

Morton, M.G., K.L. Dickson, W.T. Waller, M.F. Acevedo, F.L. Mayer, Jr., and M. Ablan. 2000. Methodology for the evaluation of cumulative episodic exposure to chemical stressors in aquatic risk assessment. *Environmental Toxicology and Chemistry,* 4: 1213–1221.

Mount, D.I. 1982. Aquatic surrogates. In Surrogate Species Workshop Report, TR-507-36B. U.S. Environmental Protection Agency, Washington, D.C., USA: pp. A6-2–A6-4.

Mount, D.I. and T. Norberg. 1985. A seven-day life-cycle toxicity test. *Environmental Toxicology and Chemistry,* 3: 425–434.

Mount, D.I. and C.E. Stephan. 1967. A method of establishing acceptable toxicant limits for fish — malathion and the butoxyethanol ester of 2,4-D. *Transactions of the American Fisheries Society,* 96: 185–193.

Mullins, J.A., R.F. Carsel, J.E. Scarbrough, and A.M. Ivery. 1993. PRZM-2 a Model for Predicting Pesticide Fate in the Crop Root Zone and Unsaturated Soil Zones: Program and User's Manual for Release 2.0. Report EPA/600/R-93/046. U.S. Environmental Protection Agency, Athens, GA, USA.

Munro, I.C. 1990. Safety assessment procedures for indirect food additives: an overview. *Regulatory Toxicology and Pharmacology,* 12: 2–12.

Murphy, B.L. 1998. Dealing with uncertainty in risk assessment. *Human and Ecological Risk Assessment,* 4: 685–699.

Naeem, S. and S. Li. 1997. Biodiversity enhances ecosystem reliability. *Nature,* 390: 507–509.

Nagy, K.A., I.A. Girard, and T.K. Brown. 1999. Energetics of free-ranging mammals, reptiles, and birds. *Annual Review of Nutrition,* 19: 237–277.

Neter, J., W. Wasserman, and M.H. Kutner. 1990. *Applied Linear Statistical Models. Regression, Analysis of Variance, and Experimental Designs.* Richard D. Irwin, Homewood, MA, USA.

Newman, M.C. and M.G. Heagler. 1991. Allometry of metal bioaccumulation and toxicity. In M.C. Newman and A.W. McIntosh, Eds., *Metal Ecotoxicology, Concepts and Applications,* Lewis Publishers, Chelsea, MI, USA: pp. 91–130.

Newman, M.C., D.R. Ownby, L.C.A. Mézin, D.C. Powell, T.R.L. Christensen, S.B. Lerberg, and B.-A. Anderson. 2000. Applying species sensitivity distributions in ecological risk assessment, assumptions of distribution type and sufficient numbers of species. *Environmental Toxicology and Chemistry,* 19: 508–515.

Nichols, P., M. Hauschild, J. Potting, and P. White. 1996. Impact assessment of non toxic pollution in life cycle assessment. In Udo de Haes, H.A., Ed., *Towards a Methodology for Life Cycle Impact Assessment,* Society of Environmental Toxicology and Chemistry-Europe, Brussels, Belgium.

Niederlehner, B.R., J.R. Pratt, A.L. Buikema, and J. Cairns, Jr. 1986. Comparison of estimated hazard derived at three levels of complexity. In Cairns, J., Ed., *Community Toxicity Testing,* ASTM STP 920, American Society for Testing and Materials, Philadelphia, PA, USA: pp. 30–48.

Niethammer, J. and F. Krapp, Eds. 1978–1994. *Handbuch der Säugetiere Europas.* Aula-Verlag, Wiesbaden, Germany.

Norberg, T.J. and D.I. Mount. 1985. A new fathead minnow (*Pimephales promelas*) subchronic toxicity test. *Environmental Toxicology and Chemistry,* 4: 711–718.

Norberg-King, T.J. and D.I. Mount. 1986. Validity of effluent and ambient toxicity tests for predicting biological impact, Skeleton Creek, Enid, Oklahoma. EPA/600/30-85/044. Duluth Environmental Research Laboratory, Duluth, MN, USA.

Notenboom, J., M.A. Vaal, and J.A. Hoekstra. 1995. Using comparative ecotoxicology to develop quantitative species sensitivity relationships (QSSR). *Environmental Science and Pollution Research,* 2: 242–243.

NRC. 1993. National Research Council. Issues in Risk Assessment. National Academy Press, Washington, D.C., USA.

OECD. 1981. *Guidelines for Testing of Chemicals.* Organization for Economic Cooperation and Development, Paris, France.

OECD. 1984. *Guidelines for Testing of Chemicals,* Section 2, *Effects on Biotic Systems.* Organization for Economic Cooperation and Development, Paris, France.

OECD. 1989. *Report of the OECD Workshop on Ecological Effects Assessment.* OECD Environment Monographs No. 26, Organization for Economic Cooperation and Development, Paris, France.

OECD. 1992. *Report of the OECD Workshop on the Extrapolation of Laboratory Aquatic Toxicity Data to the Real Environment.* OECD Environment Monograph No. 59, Organization for Economic Cooperation and Development, Paris, France.

OECD. 1995. *Guidance Document for Aquatic Effects Assessment.* OECD Environment Monograph No. 92, Organization for Economic Cooperation and Development, Paris, France.

Okkerman, P.C., E.J. van de Plassche, W. Slooff, C.J. van Leeuwen, and J.H. Canton. 1991. Ecotoxicological effects assessment, a comparison of several extrapolation procedures. *Ecotoxicology and Environmental Safety,* 21: 182–193.

Okkerman, P.C., E.J. van de Plassche, H.J.B. Emans, and J.H. Canton. 1993. Validation of some extrapolation methods with toxicity data derived from multiple species experiments. *Ecotoxicology and Environmental Safety,* 25: 341–359.

O'Neill, R.V., R.H. Gardner, L.W. Barnthouse, G.W. Suter II, S.G. Hildebrand, and C.W. Gehrs. 1982. Ecosystem risk analysis, a new methodology. *Environmental Toxicology and Chemistry,* 1: 167–177.

Oreskes, N., K. Shrader-Frechette, and K. Belitz. 1994. Verification, validation, and confirmation of numerical models in the earth sciences. *Science,* 263: 641–646.

Ott, F.S. 1986. Amphipod Sediment Bioassays, Effects on Response of Methodology, Grain Size, Organic Content, and Cadmium. Ph.D. thesis, University of Washington, Seattle, WA, USA: 81–135 (cited in Long and Morgan, 1990).

Otte, J.-G., J.J.M. van Grinsven, W.J.G.M. Peijnenburg, and A. Tiktak. 1999. Determination of field-based sorption isotherms for Cd, Cu, Pb and Zn in Dutch soils. RIVM report nr. 711401007. National Institute of Public Health and the Environment, Bilthoven, the Netherlands: 99 pp. [in Dutch].

Owen, D.B. 1968. A survey of properties and applications of the non-central t-distribution. *Technometrics,* 10: 445–478.

Papoulis, A. 1965. *Probability, Random Variables, and Stochastic Processes.* McGraw-Hill, Tokyo, Japan: p. 583.

Parker, R. 1999. GENEEC Version 1.3. Environmental Fate and Effects Division, Office of Pesticide Programs, U.S. Environmental Protection Agency, Washington, D.C., USA.

Parkhurst, B.R., W. Warren-Hicks, R.D. Cardwell, J. Volosin, T. Etchison, J.B. Butcher, and S.M. Covington. 1996. Methodology for Aquatic Ecological Risk Assessment. RP91-AER-1. Water Environment Research Foundation, Alexandria, VA, USA.

Parkhurst, D. 1998. Arithmetic versus geometric means for environmental concentration data. *Environmental Science and Technology,* 92A–98A.

Parrott, J.L., P.V. Hodson, M.R. Servos, S.L. Huestis, and D.G. Dixon. 1995. Relative potencies of polychlorinated dibenzo-*p*-dioxins and dibenzofurans for inducing mixed function oxygenase activity in rainbow trout. *Environmental Toxicology and Chemistry,* 14: 1041–1050.

Pastorok, R.A. and D.S. Becker. 1990. Comparative sensitivity of sediment toxicity bioassays at three Superfund sites in Puget Sound. In Landis, W.G. and W.H. van der Schalie, Eds., *Aquatic Toxicology and Risk Assessment,* 13th vol., ASTM STP 1096, American Society for Testing and Materials, Philadelphia, PA, USA: pp. 123–139.

Paustenbach, D.J. 1995. The practice of health risk assessment in the United States (1975–1995): How the U.S. and other countries can benefit from that experience. *Human and Ecological Risk Assessment,* 1: 29–62.

Pavlou, S.P. and D.P. Weston. 1984. Initial evaluation of alternatives for development of sediment related criteria for toxic contaminants in marine waters (Phase II); Report prepared for U.S. Environmental Protection Agency, Washington, D.C., USA.

Payne, R.W. and P.W. Lane. 1987. *Genstat 5, Reference Manual.* Clarendon Press, Oxford, U.K.

Pedersen, F. 1994. Økotoksikologiske kvalitetskriterier for overfladevand. Miljøprojekt 250. Danish Environmental Protection Agency, Copenhagen, Denmark [in Danish].

Pedersen, F. and M.B. Pedersen. 1993. Principper for fastsættelse af jordkvalitetskriterier. Miljøprojekt 247, Miljøstyrelsen, Copenhagen, Denmark [in Danish].

Pedersen F. and G.I. Petersen. 1996. Variability of species sensitivity to complex mixtures. *Water Science and Technology,* 33: 109–119.

Peijnenburg, W.J.G.M., L. Posthuma, H.J.P. Eijsackers, and H.E. Allen. 1997. A conceptual framework for implementation of bioavailability of metals for environmental management purposes. *Ecotoxicology and Environmental Safety,* 37: 163–172.

Peijnenburg, W.J.G.M., R. Baerselman, A.C. de Groot, D.T. Jager, L. Posthuma, and R.P.M. van Veen. 1999a. Relating environmental availability to bioavailability: soil type dependent metal accumulation in the oligochaete *Eisenia andrei. Ecotoxicology and Environmental Safety,* 44: 294–310.

Peijnenburg, W.J.G.M., L. Posthuma, P.G.P.C. Zweers, R. Baerselman, A.C. de Groot, R.P.M. van Veen, and D.T. Jager. 1999b. Prediction of metal availability in Dutch field soils for the oligochaete *Enchytraeus crypticus*. *Ecotoxicology and Environmental Safety,* 43: 170–186.

Pittinger, C.A., D.A. Woltering, and J.A. Masters. 1989. Bioavailability of sediment-sorbed and aqueous surfactants to *Chironomus riparius* (midge). *Environmental Toxicology and Chemistry,* 8: 1023–1033.

Plackett, R.L. and P.S. Hewlett. 1952. Quantal responses to mixtures of poisons. *Journal of the Royal Statistical Society,* B14: 141–163.

Posthuma, L. 1997. Effects of toxicants on population and community parameters in field conditions, and their potential use in the validation of risk assessment methods. In Van Straalen, N.M. and H. Løkke, Eds., *Ecological Risk Assessment of Contaminants in Soil,* Chapman & Hall, London, U.K.: pp. 85–123.

Posthuma, L., R.F. Hogervorst, E.N.G. Joosse, and N.M. van Straalen. 1993. Genetic variation and covariation for characteristics associated with cadmium tolerance in natural populations of the springtail *Orchesella cincta* (L.). *Evolution,* 47: 619–631.

Posthuma, L., T. Aldenberg, R. Luttik, T.P. Traas, M.A. Vaal, and A. Willemsen. 1995. Extrapolation methods for laboratory toxicity data to indicator or wildlife species. RIVM report 719102 047. National Institute of Public Health and the Environment, Bilthoven, the Netherlands [in Dutch].

Posthuma, L., C.A.M. van Gestel, C.E. Smit, D.J. Bakker, and J.W. Vonk. 1998a. Validation of toxicity data and risk limits for soils: final report. RIVM report 607505 004. National Institute of Public Health and the Environment, Bilthoven, the Netherlands.

Posthuma L., J. Notenboom, A.C. de Groot, and W.J.G.M. Peijnenburg. 1998b. Soil acidity as major determinant of zinc partitioning and zinc uptake in two oligochaete worms (*Eisenia andrei* and *Enchytraeus cryptics*) exposed in contaminated field soils. In S. Sheppard, J.J.D. Bembridge, M. Holmstrup, and L. Posthuma, Eds., *Advances in Earthworm Ecotoxicology,* SETAC Publishers, Pensacola, FL, USA: pp. 111–127.

Posthuma, L., A.J. Schouten, P. van Beelen, and M. Rutgers. 2001. Forecasting effects of toxicants at the community level. Four case studies comparing observed community effects of zinc with forecasts from a generic ecotoxicological risk assessment method. In Rainbow, P.S., S.P. Hopkin, and M. Crane, Eds. *Forecasting the Environmental Fate and Effects of Chemicals,* John Wiley, Chichester, U.K.: pp. 151–175.

Power, M. and L.S. McCarty. 1996. Probabilistic risk assessment: betting on its future. *Human and Ecological Risk Assessment,* 2: 30–34.

Power, M. and L.S. McCarthy. 1997. Fallacies in ecological risk assessment practices. *Environmental Science and Technology,* 31: 370A–374A.

Racke, K.D. 1993. Environmental fate of chlorpyrifos. *Reviews of Environmental Contamination and Toxicology,* 131: 1–151.

Rademaker, M.C.J. and C.A.M. van Gestel. 1993. Pilot study to the effects of non-acidifying outdoor air-pollutants on arthropods. Report D93003 from the Vrije Universiteit, Amsterdam, the Netherlands [in Dutch].

Rand, G.M. and S.R. Petrocelli. 1985. *Fundamentals of Aquatic Toxicology.* Hemisphere, Washington, D.C., USA.

Reinert, K.H. 1987. Parameterization of predictive fate models: a case study. *Environmental Toxicology and Chemistry,* 6: 99–104.

Resampling Stats. 1995. *Resampling Stats User Guide.* Resampling Stats, Inc., Arlington, VA, USA.

Reuber, B., D. Mackay, S. Paterson, and P. Stokes. 1987. A discussion of chemical equilibria and transport at the sediment-water interface. *Environmental Toxicology and Chemistry,* 6: 731–739.

Reuther, C., T. Crommentuijn, and E.J. van de Plassche. 1998. Maximum permissible concentrations and negligible concentrations for aniline derivatives. RIVM report 601501 003. National Institute of Public Health and the Environment, Bilthoven, the Netherlands.

Richardson, G.M. 1996. Deterministic versus probabilistic risk assessment: strengths and weaknesses in a regulatory context. *Human and Ecological Risk Assessment,* 2: 44–54.

Ritter, A.M., J.L. Shaw, W.M. Williams, and K.Z. Travis. 2000. Characterizing aquatic ecotoxicological risks from pesticides using a diquat dibromide case study. I. Probabilistic exposure estimates. *Environmental Toxicology and Chemistry,* 19: 749–759.

RIVM. 1997a. National Environmental Outlook 4. Samson-Tjeenk Willink bv, Alphen aan den Rijn, the Netherlands.

RIVM. 1997b. Achtergronden bij de Milieubalans 97. Samson H.D. Tjeenk Willink bv, Alphen aan den Rijn, the Netherlands.

RIVM. 1998. National Environmental Balance 1998. Samson-Tjeenk Willink bv, Alphen aan den Rijn, the Netherlands.

RIVM. 2000. National Environmental Outlook 5. Samson-Tjeenk Willink bv, Alphen aan den Rijn, the Netherlands.

RIVM/TNO. 1999. EU Risk assessment on zinc and zinc compounds (EC Regulation 793/93). Draft version of 21-12-1999. Dutch National Institute of Public Health and the Environment (RIVM) and Netherlands Organization for Applied Scientific Research (TNO), Bilthoven, the Netherlands.

Roberts, S.M. 1999. Practical issues in the use of probabilistic risk assessment. *Human and Ecological Risk Assessment,* 5: 729–736.

Roghair, C.J., A. Buijze, and N.H.P. Schoon. 1992. Ecotoxicological risk evaluation of the cationic fabric softener DHDTMAC. I. Ecotoxicological effects. *Chemosphere,* 24: 599–609.

Roman, G., P. Isnard, and J.M. Jouany. 1999. Critical analysis of methods for assessment of predicted no-effect concentration. *Ecotoxicology and Environmental Safety,* 43: 117–125.

Romijn, C.F.A.M., R. Luttik, D. van de Meent, W. Slooff, and J.H. Canton. 1993. Presentation of a general algorithm to include effect assessment on secondary poisoning in the derivation of environmental quality criteria. Part 1. Aquatic food chains. *Ecotoxicology and Environmental Safety,* 26: 61–85.

Romijn, C.F.A.M., R. Luttik, and J.H. Canton. 1994. Presentation of a general algorithm to include effect assessment on secondary poisoning in the derivation of environmental quality criteria. Part 2. Terrestrial food chains. *Ecotoxicology and Environmental Safety,* 27: 107–127.

Roux, D.J., S.H.J. Jooste, and H.M. MacKay. 1996. Substance-specific water quality criteria for the protection of South African freshwater ecosystems: methods for derivation and initial results for some inorganic toxic substances. *South African Journal of Science,* 92: 198–206.

Royal Society of Chemistry. 1994. *Agrochemicals Handbook,* 3rd ed. Royal Society of Chemistry, Cambridge, U.K.

Russom, C.L., S.P. Bradbury, and S.J. Broderius. 1997. Predicting modes of toxic action from chemical structure, acute toxicity in the fathead minnow (*Pimephales promelas*). *Environmental Toxicology and Chemistry,* 16: 948–967.

Rutgers, M., I.M. van'tVerlaat, B. Wind, L. Posthuma, and A.M. Breure. 1998. Rapid method for assessing pollution-induced community tolerance in contaminated soil. *Environmental Toxicology and Chemistry,* 17: 2210–2213.

Sabljic, A., H. Güsten, H. Verhaar, and J. Hermens. 1995. QSAR modelling of soil sorption. Improvements and systematics of log K_{oc} vs. log K_{ow} correlations. *Chemosphere,* 31: 4489–4514.

Safe, S. 1998. Hazard and risk assessment of chemical mixtures using the toxic equivalency factor (TEF) approach. *Environmental Health Perspectives,* 106: 1051–1058.

Sample, B.E. and C.A. Arenal. 1999. Allometric models for interspecies extrapolation for wildlife toxicity data. *Bulletin of Environmental Contamination and Toxicology,* 62: 653–663.

Sample, B.E., G.W. Suter II, J.J. Beauchamp, and R.A. Efroymson. 1999. Literature-derived bioaccumulation models for earthworms: development and validation. *Environmental Toxicology and Chemistry,* 18: 2110–2120.

SAS. 1988. *SAS/STAT User Guide, Version 6.03.* SAS Institute, Cary, NC, USA.

Saterbak, A., R.J. Toy, D.C.L. Wong, B.J. McMain, M.P. Williams, P.B. Dorn, L.P. Brzuzy, E.Y. Chai, and J.P. Salanitro. 1999. Ecotoxicological and analytical assessments of hydrocarbon-contaminated soils and application to ecological risk assessment. *Environmental Toxicology and Chemistry,* 18: 1591–1607.

Saterbak, A., R.J. Toy, B.J. McMain, M.P. Williams, and P.B. Dorn. 2000. Ecotoxicological and analytical assessment of the effects of bioremediation of hydrocarbon-containing soils. *Environmental Toxicology and Chemistry,* 19: 2643–2652.

Schwarz, R.C., D.W. Schults, R.W. Ozretich, J.O. Lamberson, F.A. Cole, T.H. DeWitt, M.S. Redmond, and S.P. Ferraro. 1995. Sigma PAH: a model to predict the toxicity of polynuclear aromatic hydrocarbon mixtures in field-collected sediments. *Environmental Toxicology and Chemistry,* 14: 1977–1978.

Scott-Fordsmand, J.J. and M. Bruus Pedersen. 1995. Soil quality criteria for selected inorganic compounds. Working report 48, Danish Environmental Protection Agency, Copenhagen, Denmark.

Scott-Fordsmand, J.J., M. Bruus Pedersen, and J. Jensen. 1996. Setting a soil quality criterion. *Toxicology and Ecotoxicology News,* 3: 20–24.

Scott-Fordsmand, J.J., P.H. Krogh, and J.M. Weeks. 2000. Responses of *Folsomia fimetaria* (Collembola, Isotomidae) to copper under different soil copper contamination histories, in relation to risk assessment. *Environmental Toxicology and Chemistry,* 19: 1297–1303.

Shao, Q. 2000. Estimation of hazardous concentrations based on NOEC toxicity data: an alternative approach. *Environmetrics,* 11: 583–595.

Shapiro, S.S. and R.S. Francia. 1972. An approximate analysis of variance test for normality. *Journal of the American Statistical Association,* 67: 215–216.

Shapiro, S.S. and M.B. Wilk. 1965. An analysis of variance test for normality (complete samples). *Biometrika,* 52: 591–611.

Shea, D. 1988. Developing national sediment quality criteria. *Environmental Science and Technology,* 22: 1256–1261.

Shore, R.F. and P.E.T. Douben. 1994. The ecotoxicological significance of cadmium intake and residues in terrestrial small mammals. *Ecotoxicology and Environmental Safety,* 29: 101–112.

Slooff, W. 1983. Benthic macroinvertebrates and water quality assessment, some toxicological considerations. *Aquatic Toxicology,* 4: 73–82.

Slooff, W. 1992. Ecotoxicological risk assessment: deriving maximum tolerable concentrations (MTC) from single species toxicity data. RIVM report 719102 018. National Institute for Public Health and the Environment, Bilthoven, the Netherlands.

Slooff, W. and J.H. Canton. 1983. Comparison of the susceptibility of 11 freshwater species to 8 chemical compounds. II. (Semi) chronic toxicity tests. *Aquatic Toxicology,* 4: 271–282.

Slooff, W. and D. de Zwart. 1984. Bioindiactors and chemical pollution of surface waters. In Best, E.P.H. and J. Haeck, Eds., *Ecological Indicators for Quality Assessment of Air, Water, Soil and Ecosystems,* Pudoc, Wageningen, the Netherlands: pp. 39–50 [in Dutch].

Slooff, W. and D. de Zwart. 1991. The pT-value as environmental policy indicator for the exposure to toxic substances. RIVM report 719102 003. National Institute for Public Health and the Environment, Bilthoven, the Netherlands.

Slooff, W., J.H. Canton, and J.L.M. Hermens. 1983. Comparison of the susceptibility of 22 freshwater species to 15 chemical compounds. I. (Sub)acute toxicity tests. *Aquatic Toxicology*, 4: 113–128.

Slooff, W., J.A.M. van Oers, and D. de Zwart. 1986. Margins of uncertainty in ecotoxicological hazard assessment. *Environmental Toxicology and Chemistry*, 5: 841–852.

Slooff, W., P. van Beelen, J.A. Annema, and J.A. Janus. 1995. Integrated criteria document mercury. RIVM report 601014 008. National Institute of Public Health and the Environment, Bilthoven, the Netherlands.

Slovic, P. 1987. Perception of risk. *Science*, 236: 280–285.

Smissaert, H.R. and A.A.M. Jansen. 1984. On the variation of toxic effects over species, its cause, and analysis by "structure-selectivity relations." *Ecotoxicology and Environmental Safety*, 8: 294–302.

Smit, C.E. 1997. Field Relevance of the *Folsomia candida* Soil Toxicity Test. Ph.D. thesis, Free University Amsterdam, the Netherlands.

Smit, C.E. and C.A.M. van Gestel. 1998. Effects of soil type, prepercolation, and ageing on bioaccumulation and toxicity of zinc for the springtail *Folsomia candida*. *Environmental Toxicology and Chemistry*, 17: 1132–1141.

Smit, C.E., A.J. Schouten, P.J. van den Brink, M.L.P. van Esbroek, and L. Posthuma. in press. Effects of zinc contamination on the natural nematode community in outdoor mesocosms. *Archives of Environmental Contamination and Toxicology*, 42.

Smith, E.P. and J. Cairns, Jr. 1993. Extrapolation methods for setting ecological standards for water quality, statistical and ecological concerns. *Ecotoxicology*, 2: 203–219.

Snel, J.F.H., J.H. Vos, R. Gylstra, and T.C.M. Brock. 1998. Inhibition of photosystem II (PSII) electron transport as a convenient endpoint to assess stress of the herbicide linuron on freshwater plants. *Aquatic Ecology*, 32: 113–123.

Snell, T.W. and G. Persoone. 1989. Acute toxicity bioassays using rotifers. II. A freshwater test with *Brachionus rubens*. *Aquatic Toxicology*, 14: 81–92.

Snell, T.W., B.D. Moffat, C. Janssen, and G. Persoone. 1991. Acute toxicity bioassays using rotifers. IV. Effects of cyst age, temperature and salinity on the sensitivity of *Brachionus calyciflorus*. *Ecotoxicology and Environmental Safety*, 21: 308–317.

Solomon, K.R. 1996. Overview of recent developments in ecotoxicological risk assessment. *Risk Analysis*, 16: 627–633.

Solomon, K.R. 1999. Integrating environmental fate and effects information: the keys to ecotoxicological risk assessment for pesticides. In Brooks, G.T. and Roberts, T.R., Eds. *Pesticide Chemistry and Bioscience: The Food-Environment Challenge*, Royal Society of Chemistry, London, U.K.: pp. 313–326.

Solomon, K.R. and M.J. Chappel. 1998. Triazine herbicides: ecological risk assessment in surface waters. In Ballantine, L., J. McFarland, and D. Hackett, Eds., *Triazine Risk Assessment*, ACS Symposium Series, Vol. 683, American Chemical Society, Washington, D.C., USA: pp. 357–368.

Solomon, K.R. and J.M. Giddings. 2000. Understanding probabilistic risk assessment. *SETAC Globe*, 1: 34–35.

Solomon, K.R., M.K. Baker, H. Heyne, and J. van Kleef. 1979. The use of frequency diagrams in the survey of resistance to pesticides in ticks in Southern Africa. *Onderstepoort Journal of Veterinary Research*, 46: 171–177.

Solomon, K.R., D.B. Baker, R.P. Richards, K.R. Dixon, S.J. Klaine, T.W. La Point, R.J. Kendall, C.P. Weisskopf, J.M. Giddings, J.P. Giesy, L.W. Hall, and W.M. Williams. 1996. Ecological risk assessment for atrazine in North American surface waters. *Environmental Toxicology and Chemistry,* 15: 31–76.

Solomon, K., J. Giesy, and P. Jones. 2000. Probabilistic risk assessment of agrochemicals in the environment. *Crop Protection,* 19: 649–655.

Solomon, K.R., J.M. Giddings, and S.J. Maund. 2001. Probabilistic risk assessment of cotton pyrethroids: I. Distributional analyses of laboratory aquatic toxicity. *Environmental Toxicology and Chemistry,* 20: 652–659.

Southwood, J.M., R.C. Harris, and D. Mackay. 1989. Modeling the fate of chemicals in an aquatic environment: the use of computer spreadsheet and graphics software. *Environmental Toxicology and Chemistry,* 8: 987–996.

SPSS, 2000. *SigmaPlot for Windows.* Version 6.0. SPSS Incorporated, Chicago, IL, USA.

Sprague, J.B. 1970. Measurement of pollutant toxicity to fish. II Utilizing and applying bioassay results. *Water Research,* 4: 3–32.

Spurgeon, D.J., S.P. Hopkin, and D.T. Jones. 1994. Effects of cadmium, copper, lead and zinc on growth, reproduction and survival of the earthworm *Eisenia fetida* (Savigny), assessing the environmental impact of point-source metal contamination in terrestrial ecosystems. *Environmental Pollution,* 2: 213–234.

Stebbing, A.R.D. 1982. Hormesis — the stimulation of growth by low levels of inhibitors. *Science of the Total Environment,* 2: 213–234.

Steen, R.J.C.A., P.E.G. Leonards, U.A.T. Brinkman, D. Barcelo, T. Tronzynski, T.A. Albanis, and W.P. Cofino. 1999. Ecological risk assessment of agrochemicals in European estuaries. *Environmental Toxicology and Chemistry,* 18: 1574–1581.

Stephan, C.E. 1985. Are the "Guidelines for Deriving Numerical National Water Quality Criteria for the Protection of Aquatic Life and Its Uses" based on sound judgments? In Cardwell, R.D., R. Purdy, and R.C. Bahner, Eds., *Aquatic Toxicology and Hazard Assessment,* Seventh Symposium, American Society for Testing and Materials, Philadelphia, PA, USA: pp. 515–526.

Stephens, M.A. 1974. EDF statistics for goodness of fit and some comparisons. *Journal of the American Statistical Association,* 69: 730–737.

Stephens, M.A. 1982. Anderson-Darling test for goodness of fit. In Kotz, S. and N.L. Johnson, Eds., *Encyclopedia of Statistical Sciences,* Vol. 1, Wiley, New York, USA: pp. 81–85.

Stephens, M.A. 1986a. Tests based on EDF statistics. In D'Agostino, R.B. and M.A. Stephens, Eds., *Goodness-of-Fit Techniques,* Marcel Dekker, New York, USA: pp. 97–193.

Stephens, M.A. 1986b. Tests based on regression and correlation. In D'Agostino, R.B. and M.A. Stephens, Eds., *Goodness-of-Fit Techniques,* Marcel Dekker, New York, USA: pp. 195–233.

Stephenson, G.L., N.K. Kaushik, K.R. Solomon, K.E. Day, and P. Hamilton. 1986. Impact of methoxychlor on freshwater plankton communities in limnocorrals. *Environmental Toxicology and Chemistry,* 5: 587–603.

Struijs, J., D. van de Meent, W.J.G.M. Peijnenburg, M.A.G.T. van den Hoop, and T. Crommentuijn. 1997. Added risk approach to derive maximum permissible concentrations for heavy metals, how to take into account the natural background levels? *Ecotoxicology and Environmental Safety,* 37: 112–118.

Struijs, J., R. van de Kamp, and E.A. Hogendoorn. 1998. Isolating organic micropollutants from water samples by means of XAD resins and supercritical fluid extraction. RIVM report 607602 001. National Institute of Public Health and the Environment, Bilthoven, the Netherlands [in Dutch].

Stuijfzand, S.C., S. Engels, E. van Ammelrooy, and M. Jonker. 1999. Caddisflies (Trichoptera: Hydropsychidae) used for evaluating water quality of large European rivers. *Archives of Environmental Contamination and Toxicology,* 36: 186–192.

Suter, G.W. II. 1989. Ecological endpoints. In Warren-Hicks,W., B.R. Parkhurst, and J.S.S. Baker, Eds., Ecological Assessment of Hazardous Waste Sites: A Field and Laboratory Reference Document, EPA 600/3-89/013. Environmental Research Laboratory, Corvallis, OR, USA: pp. 1–28.

Suter, G.W. II, Ed. 1993. *Ecological Risk Assessment.* Lewis Publishers, Boca Raton, FL: 538 pp.

Suter, G.W. II. 1996. Abuse of hypothesis testing statistics in ecological risk assessment. *Human and Ecological Risk Assessment,* 2: 331–349.

Suter, G.W. II. 1998a. Comments on the interpretation of distributions in "Overview of recent developments in ecological risk assessment." *Risk Analysis,* 18: 3–4.

Suter, G.W. II. 1998b. An overview perspective of uncertainty. In Warren-Hicks, W.J. and D.R.J. Moore, Eds., *Uncertainty Analysis in Ecological Risk Assessment,* SETAC Press, Pensacola, FL, USA: pp. 121–130.

Suter, G.W. II. 1998c. Ecotoxicological effects extrapolation models. In M.C. Newman and C.L. Strojan, Eds., *Risk Assessment: Logic and Measurement.* Ann Arbor Press, Ann Arbor, MI, USA: pp. 167–185.

Suter, G.W. II and A.E. Rosen. 1988. Comparative toxicology for risk assessment of marine fishes and crustaceans. *Environmental Science and Technology,* 22: 548–556.

Suter, G.W. II, D.S. Vaughan, and R.H. Gardner. 1983. Risk assessment by analysis of extrapolation error, a demonstration for effects of pollutants on fish. *Environmental Toxicology and Chemistry,* 2: 369–378.

Suter, G.W. II, A.E. Rosen, E. Linder, and D.F. Parkhurst. 1987. Endpoints for responses of fish to chronic toxic exposures. *Environmental Toxicology and Chemistry,* 6: 793–809.

Suter, G.W. II, L.W. Barnthouse, R.E. Efroymson, and H. Jager. 1999. Ecological risk assessment of a large river-reservoir, 2. Fish community. *Environmental Toxicology and Chemistry,* 18: 589–598.

Swain, L.G. and R.A. Nijman. 1991. An approach to the development of sediment quality objectives for Burrard Inlet. In Chapman, P., F. Bishay, E. Power, K. Hall, L. Harding, D. McLeay, M. Nassichuk, and W. Knapp, Eds., *Proc. 17th Annual Aquatic Toxicity Workshop,* November 5–7, 1990, Vancouver, BC, Canada, Canadian Technical Report of Fisheries and Aquatic Sciences, 1774, vol. 2: pp. 12.

Swartjes, F.A. 1999. Risk-based assessments of soil and groundwater quality in the Netherlands, standards and remediation urgency. *Risk Analysis,* 19: 1235–1249.

Swartz, R.C., G.R. Ditsworth, D.W. Schults, and J.O. Lamberson. 1985. Sediment toxicity to a marine infaunal amphipod, cadmium and its interaction with sewage sludge. *Marine Environmental Research,* 18: 133–153.

Swartz, R.C., D.W. Schults, T.H. DeWitt, G.R. Ditsworth, and J.O. Lamberson. 1990. Toxicity of fluoranthene in sediment to marine amphipods, a test of the equilibrium partitioning approach to sediment quality criteria. *Environmental Toxicology and Chemistry,* 9: 1071–1080.

Ter Braak, C.J.F. 1995. Ordination. In Jongman, R.G.H., C.J.F. Ter Braak, and O.F.R. van Tongeren, Eds., *Data Analysis in Community and Landscape Ecology,* Cambridge University Press, Cambridge, U.K.: pp. 91–173.

The Cadmus Group, Inc. 1996a. *Aquatic Ecological Risk Assessment,* Version 1.1. Water Environment Research Foundation, Alexandria, VA, USA.

The Cadmus Group, Inc. 1996b. *Aquatic Ecological Risk Assessment Software User's Manual,* Version 2.0. RP91-AER1. Water Environment Research Foundation, Alexandria, VA, USA.

Thomas, J.M., J.R. Skalski, J.F. Cline, M.C. McShane, J.C. Simpson, W.E. Miller, S.A. Peterson, C.A. Callahan, and J.C. Greene. 1986. Characterization of chemical waste site contamination and determination of its extent using bioassays. *Environmental Toxicology and Chemistry*, 5: 487–501.

Tiktak, A., J.R.M. Alkemade, J.J.M. van Grinsven, and G.B. Makaske. 1998. Modelling cadmium accumulation at a regional scale in the Netherlands. *Nutrient Cycling in Agroecosystems*, 50: 209–222.

Tilman, D. 1996. Biodiversity: population versus ecosystem stability. *Ecology*, 77: 350–363.

Tilman, D., D. Wedlin, and J. Knops. 1996. Productivity and sustainability influenced by biodiversity in grassland ecosystems. *Nature*, 379: 718–720.

Tomlin, C.D.S., Ed. 1997. *Pesticide Manual*. British Crop Protection Council, Farnham, U.K.

Tong, Z., J. Hongjun, and Z. Huailan. 1996. Quality criteria for acrylonitrile for the protection of aquatic life in China. *Chemosphere*, 32: 2083–2093.

Traas, T.P. 2001. Guidance document on deriving Environmental Risk Limits. RIVM report 601501 012, National Institute of Public Health and the Environment, Bilthoven, the Netherlands.

Traas, T.P., R. Luttik, and R.H. Jongbloed. 1996. A probabilistic model for deriving soil quality criteria based on secondary poisoning of top predators. I. Model description and uncertainty analysis. *Ecotoxicology and Environmental Safety*, 34: 264–278

Traas, T.P., J.H. Janse, T. Aldenberg, and T.C.M. Brock. 1998a. A food web model for fate and direct and indirect effects of Dursban® 4E (active ingredient chlorpyrifos) in freshwater microcosms. *Aquatic Toxicology*, 32: 179–190.

Traas, T.P., R. Luttik, and R. Posthumus. 1998b. The potentially affected fraction for target species: additional data and calculations. RIVM report 607504 005. National Institute of Public Health and the Environment, Bilthoven, the Netherlands.

Traas, T.P., A.M. Breure, and A.J. Schouten. 1999. Evaluation report: toxicants and biodiversity — toxic pressure expressed as probability of presence of species. RIVM report 607601 005. National Institute of Public Health and the Environment, Bilthoven, the Netherlands [in Dutch].

Travis, K.Z. and P. Hendley. 2001. Probabilistic aquatic risk assessment of pyrethroids: IV. Landscape-level exposure characterization. *Environmental Toxicology and Chemistry*, 20: 679–686.

Troyer, M.E. and M.S. Brody. 1994. Managing ecological risks at EPA: issues and recommendations for progress. EPA/600/R-94/183. U.S. Environmental Protection Agency, Washington, D.C., USA.

Tsvetnenko, Y. 1998. Derivation of Australian tropical marine water quality criteria for the protection of aquatic life from adverse effects of petroleum hydrocarbons. *Environmental Toxicology and Water Quality*, 13: 273–284.

Tubbing, G.M.J., E.D. de Ruyter van Steveninck, and W. Admiraal. 1993. Sensitivity of planctonic photosynthesis to various toxicants in the river Rhine. *Environmental Toxicology and Water Quality*, 8: 51–62.

Tukey, J.W. 1962. The future of data analysis. *Annals of Mathematical Statistics*, 33: 21–24.

Twining, J.R. and R.F. Cameron. 1997. Decision-making processes in ecological risk assessment using copper pollution of Macquarie Harbor from Mt. Lyell, Tasmania, as a case study. *Hydrobiology*, 352: 207–218.

Twining, J., J. Perera, V. Nguyen, P. Brown, B. Ellis, and K. Wilde. 2000. AQUARISK: A computer code for aquatic ecological risk assessment. ANSTO/M-127. Australian Nuclear Science and Technology Organization, Lucas Heights, Australia.

Tyler, G. 1984. The impact of heavy metal pollution on forests: a case study of Gusum, Sweden. *Ambio*, 13, 18–24.

Udo de Haes, H.A. 1996. Discussion of general principles and guidelines for practical use. In Udo de Haes, H.A., Ed., *Towards a Methodology for Life Cycle Impact Assessment.* Society of Environmental Toxicology and Chemistry — Europe, Brussels, Belgium.

Udo de Haes, H.A., O. Jolliet, G. Finnveden, M. Hauschild, W. Krewitt, and R. Müller-Wenk. 1999. Best available practice regarding impact categories and category indicators in life cycle impact assessment. *International Journal of Life Cycle Analysis,* 4: 66–74.

Urban, D.J. and N.J. Cook. 1986. Standard Evaluation Procedure for Ecological Risk Assessment. EPA/540/09-86/167. Hazard Evaluation Division, Office of Pesticide Programs, U.S. Environmental Protection Agency, Washington, D.C., USA.

U.S. EPA. 1976. Quality Criteria for Water. EPA-440/9-76-023 or PB263943. U.S. Environmental Protection Agency, Washington, D.C., USA.

U.S. EPA. 1978a. *Federal Register,* 43: 21506–21518. May 18. U.S. Environmental Protection Agency, Washington, D.C., USA.

U.S. EPA. 1978b. *Federal Register,* 43: 29028. July 5. U.S. Environmental Protection Agency, Washington, D.C., USA.

U.S. EPA. 1979. *Federal Register,* 44: 15926–15981. March 15. [The aquatic life guidelines are on pages 15970–15974.] U.S. Environmental Protection Agency, Washington, D.C., USA.

U.S. EPA. 1980. *Federal Register,* 45: 79318–79379. November 28. [The aquatic life guidelines are on pages 79341–79347.] U.S. Environmental Protection Agency, Washington, D.C., USA.

U.S. EPA. 1984a. AQUIRE, Aquatic Information Retrieval Toxicity Database. Project Description, Guidelines and Procedures, by R.C. Russo and A. Pilli, EPA 600/8-84-021, U.S. Environmental Protection Agency, Office of Research and Development, Environmental Research Laboratory, Duluth, MN, USA.

U.S. EPA. 1984b. Estimating "concern levels" for concentrations of chemical substances in the environment. Report, Environmental Health Branch, Health and Environmental Review Division, U.S. Environmental Protection Agency, Washington, D.C., USA.

U.S. EPA. 1984c. *Federal Register,* 49: 4551–4554. February 7. [The guidelines are not in the *Federal Register,* but a summary of the revisions is in the *Federal Register.*] U.S. Environmental Protection Agency, Washington, D.C., USA.

U.S. EPA. 1985a. Guidelines for Deriving Numerical National Water Quality Criteria for the Protection of Aquatic Organisms and Their Uses. PB85-227049. U.S. Environmental Protection Agency, National Technical Information Service, Springfield, VA, USA.

U.S. EPA. 1985b. *Federal Register,* 50: 30784–30796. July 29. U.S. Environmental Protection Agency, Washington, D.C., USA.

U.S. EPA. 1985c. Ambient Water Quality Criteria for Copper — 1984. EPA 440/5-84-031. U.S. Environmental Protection Agency, Office of Water Regulations and Standards, Washington, D.C., USA: 141 pp.

U.S. EPA. 1987a. Ambient Aquatic Life Water Quality Criteria for Silver (Draft). EPA-440/5-87-011. U.S. Environmental Protection Agency, Office of Research and Development, Environmental Research Laboratory, Duluth, MN, USA.

U.S. EPA. 1987b. Ambient Aquatic Life Water Quality Criteria Document for Zinc, 1987. NTIS No. PB87-153581. U.S. Environmental Protection Agency, Environmental Research Laboratory, Duluth, MN, USA: 207 pp.

U.S. EPA. 1989. Risk assessment guidance for Superfund, Vol. II, Environmental Evaluation Manual. EPA/540-1-89/001. U.S. Environmental Protection Agency, Washington, D.C., USA.

U.S. EPA. 1991. Technical support document for water quality based toxics control. EPA/505/2-90-001. U.S. Environmental Protection Agency, Office of Water, Washington, D.C., USA.

U.S. EPA. 1992. Framework for ecological risk assessment. EPA/630/R-92/001. Risk Assessment Forum, U.S. Environmental Protection Agency, Washington, D.C., USA.

U.S. EPA. 1993a. Technical basis for deriving sediment quality criteria for nonionic organic contaminants for the protection of benthic organisms by using equilibrium partitioning. EPA-822-R-93-001. Office of Water, U.S. Environmental Protection Agency, Washington, D.C., USA.

U.S. EPA. 1993b. Wildlife exposure factors handbook. EPA/600/R-93/187a. U.S. Environmental Protection Agency, Washington, D.C., USA.

U.S. EPA. 1994a. Interim Guidance on Determination and Use of Water-Effect Ratios for Metals. EPA-823-B-94-001 or PB94-140951. U.S. Environmental Protection Agency, Washington, D.C., USA: 90 pp.

U.S. EPA. 1994b. Use of the Water-Effect Ratio in Water Quality Standards. EPA-823-B-94-001. U.S. Environmental Protection Agency, Office of Science and Technology, Washington, D.C., USA: 153 pp.

U.S. EPA. 1994c. Interim Guidance on Interpretation and Implementation of Aquatic Life Criteria for Metals. U.S. Environmental Protection Agency, Health and Ecological Criteria Division, Washington, D.C., USA.

U.S. EPA. 1994d. Pesticides registration rejection rate analysis. Ecological effects. EPA 738-R-94-035, U.S. Environmental Protection Agency, Washington, D.C., USA.

U.S. EPA. 1995a. *Federal Register,* 60: 15366–15425. March 23. [The methodology for deriving aquatic life Tier I criteria is on pages 15393–15399, whereas the methodology for deriving aquatic life Tier II values is on pages 15399–15400.] U.S. Environmental Protection Agency, Washington, D.C., USA.

U.S. EPA. 1995b. Water Quality Guidance for the Great Lakes System, Supplementary Information Document (SID). EPA-820-B-95-001. U.S. Environmental Protection Agency, Washington, D.C., USA: pp. 114–115.

U.S. EPA. 1995c. Final water quality guidance for the Great Lakes system; final rule. *Federal Register,* 60: 15366–15425. U.S. Environmental Protection Agency, Washington, D.C., USA.

U.S. EPA. 1995d. *Federal Register,* 60: 22228–22237. May 4, 1995. U.S. Environmental Protection Agency, Washington, D.C., USA.

U.S. EPA. 1996. 1995 Updates: Water Quality Criteria Documents for the Protection of Aquatic Life in Ambient Water. EPA-820-B-96-001. U.S. Environmental Protection Agency, Office of Water, Washington, D.C., USA.

U.S. EPA. 1997. AQUIRE. AQUatic toxicity Information REtrival Database. U.S. Environmental Protection Agency, Washington, D.C., Office of Research and Development, National Health and Environmental Effects Research Laboratory, Mid-Continent Ecology Division, Duluth, MN, USA, available at http://www.epa.gov/med/databases/aquire.html.

U.S. EPA. 1998a. 1998 Update of Ambient Water Quality Criteria for Ammonia. EPA-822-R-98-008. U.S. Environmental Protection Agency, Office of Water, Washington, D.C., USA.

U.S. EPA. 1998b. Ambient Aquatic Life Water Quality Criteria, Diazinon, Draft, 9/28/98. U.S. Environmental Protection Agency, Office of Water, Washington, D.C., USA.

U.S. EPA. 1998c. Guidelines for ecological risk assessment. EPA/630/R-95/002F. U.S. Environmental Protection Agency, National Center for Environmental Assessment, Washington, D.C., USA.

U.S. EPA. 1999a. National Recommended Water Quality Criteria — Correction. EPA 822-Z-99-001. U.S. Environmental Protection Agency, Office of Water, Washington, D.C., USA.

U.S. EPA. 1999b. Water Quality Criteria and Standards Newsletter, Spring/Summer 1999. EPA-823-N-99-001. U.S. Environmental Protection Agency, Office of Water, Washington, D.C., USA.

U.S. EPA. 1999c. Technical Basis for Deriving Equilibrium-Partitioning Sediment Guidelines (ESGs) for the Protection of Benthic Organisms, Nonionic Organics. EPA-822-R-93-011. U.S. Environmental Protection Agency, Washington, D.C., USA.

Vaal, M., J.T. van der Wal, J. Hermens, and J. Hoekstra. 1997a. Pattern analysis of the variation in the sensitivity of aquatic species to toxicants. *Chemosphere,* 35: 1291–1309.

Vaal, M., J.T. van der Wal, J. Hoekstra, and J. Hermens. 1997b. Variation in the sensitivity of aquatic species in relation to the classification of environmental pollutants. *Chemosphere,* 35: 1311–1327.

Vaal, M.A., C.J. van Leeuwen, J.A. Hoekstra, and J.L.M. Hermens. 2000. Variation in sensitivity of aquatic species to toxicants: practical consequences for effect assessment of chemical substances. *Environmental Management,* 25: 415–423.

Van Beelen, P. and P. Doelman. 1996. Significance and application of microbial toxicity tests in assessing ecotoxicological risks of contaminants in soil and sediment. RIVM report 719102 051. National Institute of Public Health and the Environment, Bilthoven, the Netherlands.

Van Beelen, P. and A.K. Fleuren-Kemilä. 1999. A comparison between toxicity tests using single species and a microbial process. *Chemosphere,* 38: 3277–3290.

Van Beelen, P., A.K. Fleuren-Kemilä, and T. Aldenberg. 2001. The relation between extrapolated risk, expressed as the potentially affected fraction, and community effects, expressed as pollution-induced community tolerance. *Environmental Toxicology and Chemistry,* 20: 1133–1140.

Van de Meent, D. 1993. Simplebox, a generic multimedia fate evaluation model. RIVM report 72720 001. National Institute of Public Health and the Environment, Bilthoven, the Netherlands.

Van de Meent, D. 1999. Potentially affected fraction as a rule for toxic stress on ecosystems. RIVM report 607504 007. National Institute of Public Health and the Environment, Bilthoven, the Netherlands [in Dutch].

Van de Meent, D. and J.H.M. de Bruijn. 1995. A modelling procedure to evaluate the coherence of independently derived environmental quality objectives for air, water and soil. *Environmental Toxicology and Chemistry,* 14: 177–186.

Van de Meent, D. and D. Toet. 1992. Dutch priority setting system for existing chemicals. RIVM report 679120 001. National Institute for Public Health and Environmental Protection, Bilthoven, the Netherlands.

Van de Meent, D., T. Aldenberg, J.H. Canton, C.A.M. van Gestel, and W. Slooff. 1990a. Desire for levels. Background study for the policy document "Setting environmental quality standards for water and soil." RIVM report 670101 002. National Institute of Public Health and the Environment, Bilthoven, the Netherlands.

Van de Meent, D., T. Aldenberg, J.H. Canton, C.A.M. van Gestel, and W. Slooff. 1990b. Background study for the policy document "Setting Environmental Quality Standards for Water and Soil." RIVM report 718922 001. National Institute of Public Health and Environmental Protection, Bilthoven, the Netherlands.

Van de Plassche, E.J. 1994. Towards integrated environmental quality objectives for several compounds with a potential for secondary poisoning. RIVM report 679101 012. National Institute of Public Health and the Environment, Bilthoven, the Netherlands.

Van de Plassche, E.J. and G.J.M. Bockting. 1993. Towards integrated environmental quality objectives for several volatile compounds. RIVM report 679101 011. National Institute of Public Health and the Environment, Bilthoven, the Netherlands.

Van de Plassche, E.J. and J.H.M. de Bruijn. 1992. Towards integrated environmental quality objectives for surface water, groundwater, sediment and soil for nine trace metals. RIVM report 679101 005. National Institute of Public Health and the Environment, Bilthoven, the Netherlands.

Van de Plassche, E.J., J.H.M. De Bruijn, R. Stephenson, S.J. Marschall, T.C.J. Feijtel, and S.E. Belanger. 1999a. Predicted no-effect concentrations and risk characterization of four surfactants, linear alkyl benzene sulfonate, alcohol ethoxylates, alcohol ethoxylated sulfates, and soap. *Environmental Toxicology and Chemistry*, 18: 2653–2663.

Van de Plassche, E.J., M.A.G.T. van de. Hoop, R. Posthumus, and T. Crommentuijn. 1999b. Risk limits for boron, silver, titanium, tellurium, uranium and organosilicon compounds in the framework of EU Directive 76/464/EEC. RIVM report 601501 005. National Institute of Public Health and the Environment, Bilthoven, the Netherlands.

Van den Berg, G.A., J.P.G. Loch, L.M. van der Heijdt, and J.J.G. Zwolsman. 1998. Vertical distribution of acid-volatile sulfide and simultaneously extracted metals in a recent sedimentation area of the river Meuse in the Netherlands. *Environmental Toxicology and Chemistry*, 17: 758–763.

Van den Berg, M., L. Birnbaum, A.T.C. Bosveld, B. Brunström, P. Cook, M. Feeley, J.P. Giesy, A. Hanberg, R. Hasegawa, S.W. Kennedy, T. Kubiak, J.C. Larsen, F.X.R van Leeuwen, A.K. Djien Liem, C. Nolt, R.E. Peterson, L. Poellinger, S. Safe, D. Schrenk, D. Tillitt, M. Tysklind, M. Younes, F. Wærn, and T. Zacharewski. 1998. Toxic equivalency factors (TEFs) for PCBs, PCDDs, PCDFs for humans and wildlife. *Environmental Health Perspectives*, 106: 775–792.

Van den Berg, R. 1994. Human exposure to soil contamination, a qualitative and quantitative analysis towards proposals for human toxicological intervention values (partly revised edition). RIVM report 725201 011. National Institute of Public Health and the Environment, Bilthoven, the Netherlands.

Van den Berg, R. and J.M. Roels. 1991. Human and environmental risk assessment in case of soil contamination. An integration of results. RIVM report 725201 007. National Institute of Public Health and the Environment, Bilthoven, the Netherlands [in Dutch].

Van den Brink, P.J. and C.J.F. Ter Braak. 1999. Principal response curves: analysis of time-dependent multivariate responses of a biological community to stress. *Environmental Toxicology and Chemistry*, 18: 138–148.

Van den Brink, P.J., R.P.A. van Wijngaarden, W.G.H. Lucassen, T.C.M. Brock, and P. Leewangh. 1996. Effects of the insecticide Dursban 4E (active ingredient chlorpyrifos) in outdoor experimental ditches: II. Invertebrate community responses and recovery. *Environmental Toxicology and Chemistry*, 15: 1143–1153.

Van den Brink, P.J., E.M. Hartgers, U. Fettweis, S.J.H. Crum, E. van Donk, and T.C.M. Brock. 1997. Sensitivity of macrophyte-dominated freshwater microcosms to chronic levels of the herbicide linuron: I. Primary producers. *Ecotoxicology and Environmental Safety*, 38: 13–24.

Van den Brink, P.J., J. Hattink, F. Bransen, E. van Donk, and T.C.M. Brock. 2000. Impact of the fungicide carbendazim in freshwater microcosms. II. Zooplankton, primary producers and final conclusions. *Aquatic Toxicology*, 48: 251–264.

Van der Hoeven, N. 1994. Statistical aspects of NOEC and ECx estimates. In Noppert, F., N. van der Hoeven, and A. Leopold, Eds., *How to Measure No Effect*, Netherlands Working Group on Statistics and Ecotoxicology, Delft, the Netherlands: pp. 11–17.

Van der Hoeven, N. 1998. Power analysis for the NOEC: what is the probability of detecting small toxic effects on three different species using the appropriate standardized test protocols? *Ecotoxicology*, 7: 355–361.

Van der Hoeven, N. 2001. Estimating the 5-percentile of the species sensitivity distributions without any assumptions about the distribution. *Ecotoxicology,* 10: 25–34.

Van der Hoeven, N., F. Noppert, and A. Leopold. 1997. How to measure no effect. Part I. Towards a new measure of chronic toxicity in ecotoxicology. Introduction and workshop results. *Environmetrics,* 8: 241–248.

Van der Kooy, L.A., D. van de Meent, C.J. van Leeuwen, and W.A. Bruggeman. 1991. Deriving quality criteria for water and sediment from the results of aquatic toxicity tests and product standards, application of the equilibrium partitioning method. *Water Research,* 25: 697–705.

Van der Valk, H.C.H.G. 1997. Community structure and dynamics in desert ecosystems: potential implications for insecticide risk assessment. *Archives of Environmental Contamination and Toxicology,* 32: 11–21.

Van Drecht, G., L.J.M. Boumans, D. Fraters, H.F.R. Reijnders, and W. van Duijvenboden. 1996. Diffuse metal contamination of soil, metal concentrations in topsoil and the relation between the two. RIVM report 714801 006. National Institute of Public Health and the Environment, Bilthoven, the Netherlands [in Dutch].

Van Eck, G.T.M., M.R.L. Ouboter, and B.J. Kater. 1995. A model for the behaviour of some trace metals in the Scheldt Estuary, S.W. Netherlands. In R.D. Wilken, U. Föstner, and A. Knöchel, Eds., *Proceedings of the 10th International Conference on Heavy Metals in the Environment,* Hamburg, CEP Consultants, Edinburgh, Scotland: Vol. 1, 57–60.

Van Geest, G.J., N.G. Zwaardemaker, R.P.A. van Wijngaarden, and J.G.M. Cuppen. 1999. Effects of a pulsed treatment with the herbicide afalon (active ingredient linuron) on macrophyte-dominated mesocosms. II. Structural responses. *Environmental Toxicology and Chemistry,* 18: 2866–2874.

Van Gestel, C.A.M., M.C.J. Rademaker, and N.M. van Straalen. 1995. Capacity controlling parameters and their impact on metal toxicity in soil invertebrates. In Salomons, W. and W.M. Stigliani, Eds., *Biogeodynamics of Pollutants in Soils and Sediments,* Springer-Verlag, Berlin, Germany: pp. 171–192.

Van Leeuwen, C.J. 1989. Environmental effects of fabric softeners. $H_2O,$ 22: 296–299.

Van Leeuwen, C.J. 1990. Ecotoxicological effects assessment in the Netherlands, recent developments. *Environmental Management,* 14: 779–792.

Van Leeuwen, C.J. and J.L.M. Hermens. 1995. *Risk Assessment of Chemicals. An Introduction,* Kluwer Academic Publishers, Dordrecht, the Netherlands.

Van Leeuwen, C.J., P.J.T. van der Zandt, T. Aldenberg, H.J.M. Verhaar, and J.L.M. Hermens. 1991. The application of QSARs, extrapolation and equilibrium partitioning in aquatic effects assessment for narcotic pollutants. *Science of the Total Environment,* 109/110: 681–690.

Van Leeuwen, C.J., C. Roghair, T. de Nijs, and J. de Greef. 1992a. Ecotoxicological risk evaluation of the cationic fabric softener DHDTMAC. III. Risk assessment. *Chemosphere,* 24: 629–639.

Van Leeuwen, C.J., P.J.T. van der Zandt, T. Aldenberg, H.J.M. Verhaar, and J.L.M. Hermens. 1992b. Application of QSARs, extrapolation and equilibrium partitioning in aquatic effects assessment. I. Narcotic industrial pollutants. *Ecotoxicology and Environmental Safety,* 11: 267–282.

Van Leeuwen, C.J., H.-J. Emans, E.J. van De Plassche, and J.H. Canton. 1994. The role of field tests in hazard assessment. In Hill, I.R., F. Heimbach, P. Leeuwangh, and P. Matthiessen, Eds., *Freshwater Field Tests for Hazard Assessment of Chemicals,* Lewis Publishers, London, U.K.: pp. 425–437.

Van Straalen, N.M. 1987. Stofgehalten in de bodem- (geen) effecten op bodemdieren. In *Symposium Bodemkwaliteit,* Ede. Rapport VTCB M86/44, Leidschendam, the Netherlands: pp. 75–84 [in Dutch].

Van Straalen, N.M. 1990. New methodologies for estimating the ecological risk of chemicals in the environment. In Price, D.G., Ed., *Proceedings of the 6th Congress of the International Association of Engineering Geology,* A.A. Balkema, Rotterdam, the Netherlands: pp. 165–173.

Van Straalen, N.M. 1993a. An ecotoxicologist in politics. *Oikos,* 66: 142–143.

Van Straalen, N.M. 1993b. Open problems in the derivation of soil quality criteria from ecotoxicity experiments. In Arendt, F., G.J. Annokkée, R. Bosman, W.J. van den Brink, Eds., *Contaminated Soil '93,* Kluwer Academic Publishers, Dordrecht, the Netherlands: pp. 315–326.

Van Straalen, N.M. 1994. Biodiversity of ecotoxicological responses in animals. *Netherlands Journal of Zoology,* 44: 112–129.

Van Straalen, N.M. and W.F. Bergema. 1995. Ecological risks of increased bioavailability of metals under soil acidification. *Pedobiologia,* 39: 1–9.

Van Straalen, N.M. and C.A.J. Denneman. 1989. Ecotoxicological evaluation of soil quality criteria. *Ecotoxicology and Environmental Safety,* 18: 241–251.

Van Straalen, N.M. and J.P. van Rijn. 1998. Ecotoxicological risk assessment of soil fauna recovery from pesticide application. *Reviews of Environmental Contamination and Toxicology,* 154: 83–141.

Van Straalen, N.M., J.H.M. Schobben, and R.G.M. de Goede. 1989. Population consequences of cadmium toxicity in soil microarthropods. *Ecotoxicology and Environmental Safety,* 17:190–204.

Van Straalen, N.M., J.H.M. Schobben, and T.P. Traas. 1992. The use of ecotoxicological risk assessment in deriving maximum permissible half-lives of pesticides. *Pesticide Science,* 34: 227–231.

Van Straalen, N.M., P. Leeuwangh, and P.B.M. Stortelder. 1994. Progressing limits for soil ecotoxicological risk assessment. In Donker, M.H., H. Eijsackers, and F. Heimbach, Eds., *Ecotoxicology of Soil Organisms,* Lewis Publishers, Boca Raton, FL, USA: pp. 397–409.

Van Wezel, A.P. 1998. Chemical and biological aspects of ecotoxicological risk assessment of ionizable and neutral organic compounds in fresh and marine waters: a review. *Environmental Reviews,* 6: 123–137.

Van Wezel, A.P., T.P. Traas, M.E.J. van der Weiden, T.H. Crommentuijn, and D.T.H.M. Sijm. 2000a. Environmental risk limits for polychlorinated biphenyls in the Netherlands, derivation with probabilistic food chain modeling. *Environmental Toxicology and Chemistry,* 19: 2140–2153.

Van Wezel, A.P., P. van Vlaardingen, R. Posthumus, T.H. Crommentuijn, and D.T.H.M. Sijm. 2000b. Environmental risk limits for two phthalates, with special emphasis on endocrine disruptive properties. *Ecotoxicology and Environmental Safety,* 46: 305–321.

Van Wijngaarden, R.P.A. and T.C.M. Brock. 1999. Population and community responses in pesticides-stressed freshwater ecosystems. In Del Re, A.A.M. et al., Eds., *Human and Environmental Exposure to Xenobiotics, Proc. of the XI Symposium Pesticide Chemistry,* Cremona, Italy.

Van Wijngaarden, R.P.A., P.J. Van den Brink, J.H. Oude Voshaar, and P. Leeuwangh. 1995. Ordination techniques for analyzing response of biological communities to toxic stress in experimental ecosystems. *Ecotoxicology,* 4: 61–77.

Van Wijngaarden, R.P.A., P.J. van den Brink, S.J.H. Crum, J.H. Oude-Voshaar, T.C.M. Brock, and P. Leewangh. 1996. Effects of the insecticide Dursban 4E (active ingredient chlorpyrifos) in outdoor experimental ditches: I. Comparison of short-term toxicity between laboratory and the field. *Environmental Toxicology and Chemistry,* 15: 1133–1142.

Van Wijngaarden, R.P.A., S.J.H. Crum, K. Decraene, J. Hattink, and A. van Kammen. 1998. Toxicity of derosal (active ingredient carbendazim) to aquatic invertebrates. *Chemosphere*, 37: 673–683.

Vega, M.M., A. Urzelai, and E. Angulo. 1997. Regression study of environmental quality objectives for soil, fresh water, and marine water, derived independently. *Ecotoxicology and Environmental Safety*, 38: 210–223.

Vega, M.M., A. Urzelai, and E. Angulo. 1999. Minimum data required for deriving soil quality criteria from invertebrate ecotoxicity experiments. *Environmental Toxicology and Chemistry*, 18: 1304–1310.

Verbruggen, E.M.J., R. Posthumus, and A.P. Van Wezel. 2001. Ecotoxicological Serious Risk Concentrations for soil, sediment and (ground) water: updated proposals for first series of compounds. RIVM report 711701 020. National Institute for Public Health and the Environment, Bilthoven, the Netherlands.

Verdonschot, P.F.M. and C.J.F. ter Braak. 1994. An experimental manipulation of oligochaete communities in mesocosms treated with chlorpyrifos or nutrient additions: multivariate analyses with Monte Carlo permutation tests. *Hydrobiologia*, 287: 251–266.

Verhaar, H.J.M., C.J. van Leeuwen, and J.L.M. Hermens. 1992. Classifying environmental pollutants. 1, Structure–activity relationships for prediction of aquatic toxicity. *Chemosphere*, 25: 471–491.

Verhaar, H.J.M., F. Busser, and J.L.M. Hermens. 1995. A surrogate parameter for the baseline toxicity content of contaminated water: simulating bioconcentration and counting molecules. *Environmental Science and Technology*, 29: 726–734.

Verhallen, E. and W.C. Ma. 1997. Bio-availability of cadmium, copper, lead and zinc for earthworms in relation to soil and worm properties. IBN-DLO, preliminary report used for the Fourth National Environmental Outlook, 1997. RIVM, Bilthoven, the Netherlands [in Dutch].

Versteeg, D.J., T. Feijtel, C. Cowan, T. Ward, and R. Rapaport. 1992. An environmental risk assessment for DHDTMAC in the Netherlands. *Chemosphere*, 24: 641–662.

Versteeg, D.J., S.E. Belanger, and G.J. Carr. 1999. Understanding single species and model ecosystem sensitivity, a data based comparison. *Environmental Toxicology and Chemistry*, 18: 1329–1346.

Vijver, M., D.T. Jager, L. Posthuma, and W.J.G.M. Peijnenburg. 2001. Impact of metal pools and soil properties on metal accumulation in *Folsomia candida* (Collembola). *Environmental Toxicology and Chemistry*, 20: 712–720.

Vonk, M., J.R.M. Alkemade, M. Bakkenes, D.J. van der Hoek, F.G. Wortelboer, S. van Dijk, C.J. Roghair, and D. van de Meent. 2000. Calculation of environmental effects on nature for the fifth national environmental survey. RIVM report 408129 033. National Institute of Public Health and the Environment, Bilthoven, the Netherlands [in Dutch].

VROM. 1989a. National Environmental Policy Plan. To choose or lose. Ministry of Housing, Spatial Planning and the Environment (VROM), Second chamber, session 1988–1989, 21137, no. 1–2, The Hague, the Netherlands.

VROM. 1989b. Premises for risk management. Risk limits in the context of environmental policy. Ministry of Housing, Spatial Planning and the Environment (VROM), Second chamber, session 1988–1989, 21137, no 5, The Hague, the Netherlands.

VROM. 1990. Environmental quality objectives for soil and water. Ministry of Housing, Spatial Planning and the Environment (VROM), Directoraat Generaal voor Milieubeheer. Tweede Kamer der Staten-Generaal. Session 1990–1991, 21990, no. 1, The Hague, the Netherlands.

VROM. 1991. National Environmental Policy Plan. Ministry of Housing, Spatial Planning and the Environment (VROM). Tweede Kamer der Staten-Generaal, session 1990–1991, 21137, no. 74, The Hague, the Netherlands: 7 pp.

VROM. 1994. Environmental Quality Objectives in the Netherlands. Ministry of Housing, Spatial Planning and Environment (VROM), The Hague, the Netherlands.

VROM. 1998. Third National Environmental Policy Plan. Ministry of Housing, Spatial Planning and the Environment (VROM), The Hague, the Netherlands.

VROM. 1999. Compounds and quality criteria. Directorate General for the Environment, Ministry of Housing, Spatial Planning and the Environment (VROM), Samson Publishers, Alphen aan de Rijn, the Netherlands [in Dutch].

VROM. 2000. Circular on target values and intervention values for soil clean-up. Ministry of Housing, Spatial Planning and the Environment (VROM), Staatscourant, The Hague, the Netherlands, 39: 8–16 [in Dutch].

VROM/NVZ. 1992. Environmental risk assessment of detergents. Proceedings of a Dutch Government-Industry workshop on April 9th 1992 in Leidschendam, the Netherlands. VROM/NVZ, Leidschendam, the Netherlands.

Wagner, C. and H. Løkke. 1991. Estimation of ecotoxicological protection levels from NOEC toxicity data. *Water Research,* 25: 1237–1242.

Walker, B. 1992. Biodiversity and ecological redundancy. *Conservation Biology,* 6: 18–23.

Walker, B. 1995. Conserving biological diversity through ecosystem resilience. *Conservation Biology,* 9: 747–752.

Walker, C.H. 1978. Species differences in microsomal monooxygenase activity and their relationship to biological half-lives. *Drug Metabolism Review,* 7: 295–323.

Walthall, W.K. and J.D. Stark. 1997. A comparison of acute mortality and population growth rate as endpoints of toxicological effect. *Ecotoxicology and Environmental Safety,* 37: 45–57.

Wardle, D.A., M.A. Huston, J.P. Grime, F. Berendse, E. Garnier, W.K. Lauenroth, H. Setala, and S.D. Wilson. 2000. Biodiversity and ecosystem function: an issue in ecology. *Bulletin of the Ecological Society of America,* July: 235–239.

Warne, M.S.J. and D.W. Hawker. 1995. The number of components in a mixture determines whether synergistic and antagonistic or additive toxicity predominate: the funnel hypothesis. *Ecotoxicology and Environmental Safety,* 17: 190–204.

Weber, C.I., W.H. Peltier, T.J. Norberg-King, W.B. Horning II, F. Kessler, J. Menkedick, T.W. Neiheisel, P.A. Lewis, D.J. Klemm, W.H. Pickering, E.L. Robinson, J. Lazorchak, L.J. Wymer, and R.W. Freyberg. 1989. Short-term methods for estimating the chronic toxicity of effluents and receiving waters to freshwater organisms. EPA-600/4-89/001. U.S. Environmental Protection Agency, Cincinnati, OH, USA.

Weltje, L. 1998. Mixture toxicity and tissue interactions of Cd, Cu, Pb and Zn in earthworms (Oligochaeta) in laboratory and field soils, a critical evaluation of data. *Chemosphere,* 36: 2643–2660.

Wiles, J.A. and G.K. Frampton. 1996. A field bioassay approach to assess the toxicity of insecticide residues on soil to Collembola. *Pesticide Science,* 47: 273–285.

Wilk, M.B. and R. Gnanadesikan. 1968. Probability plotting methods for the analysis of data. *Biometrika,* 55: 1–17.

Wong, D.C.L., E.Y. Chai, K.K. Chu, and P.B. Dorn. 1999. Prediction of ecotoxicity of hydrocarbon-contaminated soils using physicochemical parameters. *Environmental Toxicology and Chemistry,* 18: 2611–2621.

Word, J.Q. and A.J. Mearns. 1979. 60-meter control survey off southern California. TM 229. El Segundo, CA, USA, Southern California Coastal Water Research Project: pp. 27–31 [cited in Long and Morgan, 1990].

Zwolsman, J.J.G., G.T.M. van Eck, and C.H. van der Weijden. 1997. Geochemistry of dissolved trace metals (cadmium, copper, zinc) in the Scheldt estuary, southwestern Netherlands: impact of seasonal variability. *Geochimica et Cosmochimica Acta,* 61: 1635–1652.

Glossary

Acute toxicity The harmful effects of a chemical or mixture of chemicals occurring after a brief exposure. *See also* chronic toxicity.

Acute Occurring within a short period in relation to the life span of the organism (usually 4 days for fish). It can be used to define either the exposure (acute test) or the response to an exposure (acute effect).

Analysis of effects A phase in an ecological risk assessment in which the relationship between exposure to contaminants and effects on endpoint entities and properties are estimated along with associated uncertainties.

Analysis of exposure A phase in an ecological risk assessment in which the spatial and temporal distribution of the intensity of the contact of endpoint entities with contaminants are estimated along with associated uncertainties.

Assessment endpoint An explicit expression of the environmental value to be protected. An assessment endpoint must include an entity and specific property of that entity.

Background concentration The concentration of a substance in environmental media that are not contaminated by the sources being assessed or any other local sources. Background concentrations are due to natural occurrence or regional contamination.

Battery toxicity testing The parallel application of a range of different toxicity tests.

Bioaccumulation The net accumulation of a substance by an organism as a result of uptake from all environmental media.

Bioassay Commonly used as synonymous with toxicity test.

Bioavailability The extent to which the form of a chemical is susceptible to being taken up by an organism. A chemical is said to be bioavailable if it is in a form that is readily taken up (e.g., dissolved) rather than a less available form (e.g., sorbed to solids or to dissolved organic matter).

Bioconcentration The net accumulation of a substance by an organism due to uptake from an aqueous solution.

Chronic Occurring after an extended time relative to the life span of an organism (conventionally taken to include at least one tenth of the life span). Long-term effects are related to changes in metabolism, growth, reproduction, or the ability to survive.

Chronic toxicity The harmful effects of a chemical or mixture of chemicals occurring after an extended exposure. *See also* acute toxicity.

Community The biotic community consists of all plants, animals, and microbes occupying the same area at the same time. However, the term is commonly used to refer to a subset of the community such as the fish community or the benthic macroinvertebrate community.

Conceptual model   A representation of the hypothesized causal relationship between the source of contamination and the response of the endpoint entities.

Contaminant   A substance that is present in the environment due to release from an anthropogenic source and that is believed to be potentially harmful.

Cumulative distribution function (CDF)   A function expressing the probability that a random variable is less than or equal to a certain value. The CDF is obtained by integration of the PDF for a continuous random variable, or summation of the PDF in the case of a discrete random variable. *See also* probability density function.

Direct effect   An effect resulting from an agent acting on the assessment endpoint or other ecological component of interest itself, not through effects on other components of the ecosystem. *See also* indirect effect.

EC(D)$_n$ Concentration (dose) that affects designated effect criterion (e.g., a behavioral trait) in $n\%$ of the population observed. The EC(D)$_{50}$ is known as the median effective concentration (dose). The EC values and their 95% confidence limits are usually derived by statistical analysis of effects in several test concentrations, after a fixed period of exposure. The duration of exposure must be specified (e.g., 96-h EC$_{50}$). *See also* LC(D)$_n$.

Ecological risk assessment (ERA)   A process that evaluates the likelihood that adverse ecological effects may occur or are occurring as a result of exposure to one or more agents.

Ecosystem   A collection of populations (microorganisms, plants, and animals) that occur in the same place at the same time and that can therefore potentially interact with each other as well as their physical and chemical environment, thus forming a functional entity.

Ecosystem function   A biological, chemical, or biochemical processes taking place in an ecosystem.

Ecosystem structure   The composition of the biological community in an ecosystem and the interrelationships between the individual populations of species (e.g., food web structure).

Ecotoxicity   The property of a compound to produce adverse effects in an ecosystem or one of its components.

Ecotoxicology   The study of toxic effects of chemical and physical agents in living organisms, especially on populations and communities within defined ecosystems; it includes transfer pathways of these agents and their interaction with the environment.

Effect criterion   The type of effect observed in a toxicity test (e.g., immobility).

Endpoint entity   An organism, population, species, community, or ecosystem that has been chosen for protection. The endpoint entity is one component of the definition of an assessment endpoint.

Environmental quality criterion (EQC)   The concentration of a potentially toxic substance that can be allowed in an environmental medium over a defined period. The term is used in this book as a general term, for which also EQO (objective) and EQS (standard) are used in different contexts.

Exposure   The contact or co-occurrence of a contaminant with a receptor organism, population, or community.

Exposure assessment   The component of an ecological risk assessment that estimates the exposure resulting from a release or occurrence in a medium of a chemical, physical, or biological agent. It includes estimation of transport, fate, and uptake.

Exposure pathway   The physical route by which a contaminant moves from a source to a biological receptor. A pathway may involve exchange among multiple media and may include transformation of the contaminant.

Exposure route   The means by which a contaminant enters an organism (e.g., inhalation, stomatal uptake, ingestion).

Extrapolation   (1) An estimation of a numerical value of an empirical (measured) function at a point outside the range of data that were used to calibrate the function or (2) the use of data derived from observations to estimate values for unobserved entities or conditions.

Frequency distribution   The organization of data to show how often certain values or ranges of values occur.

Geographic Information System (GIS)   Systems that alow for the interrelation of quality data (as well as other information) from a diversity of sources based on multilayered geographic information processing techniques.

Hazard (toxic)   The set of inherent properties of a chemical substance or mixture that makes it capable of causing adverse effects in humans or the environment when a particular degree of exposure occurs. *See also* risk.

Hazard assessment   Comparison of the intrinsic ability to cause harm with expected environmental concentration. In Europe, it is typically a comparison of PEC with PNEC. It is sometimes loosely referred to as risk assessment.

HC_p (HC_5)   Hazardous concentration for $p\%$ (5%) of the species.

$HC_{Percentage}^{Endpoint}$ General notation to identify that an environmental quality criterion is a percentile derived from an SSD, explicitly stating the endpoint and the chosen percentage p as cutoff value, e.g., HC_5^{NOEC} to identify in general notation what is known as the HC_5 based on NOEC toxicity data. Optionally, one could add a superscript prefix to show the number of toxicity data from which the model was derived, such as $^4HC_{Percentage}^{Endpoint}$ for an HC_p based on four data points.

$HD_{Percentage}^{Endpoint}$ General notation to identify that an environmental quality criterion is based on a dose (e.g., μg/kg body weight/day). Optionally, one could add a superscript prefix to show the number of toxicity data from which the model was derived, such as $^4HD_{Percentage}^{Endpoint}$ for an HD_p based on four data points.

Indirect effect   An effect resulting from the action of an agent on components of the ecosystem, which in turn affect the assessment endpoint or other ecological component of interest. Indirect effects of chemical contaminants include reduced abundance due to toxic effects on food species or on plants that provide habitat structure. *See also* direct effect.

Intervention value   A screening criterion (Netherlands) based on risks to human health and ecological receptors and processes. The ecotoxicological component of the intervention value is the hazardous concentration 50 (HC_{50}), the concentration at which 50% of species are assumed to be protected.

Joint action   Two or more chemicals exerting their effects simultaneously.

LC(D)$_n$ The concentration/dose of a substance in water that is estimated to be lethal to $n\%$ of the test organisms. The LC_{50} is known as the median lethal concentration. The LC values and their 95% confidence limits are usually derived by statistical analysis of mortalities in several test concentrations, after a fixed period of exposure. The duration of exposure must be specified (e.g., 96-h LC_{50}).

Life cycle assessment   A method for determining the relative environmental impacts of alternative products and technologies based on the consequences of their life cycle, from extraction of raw materials to disposal of the product following use.

Line of evidence   A set of data and associated analysis that can be used, alone or in combination with other lines of evidence, to estimate risks. Each line of evidence is qualitatively different from any others used in the risk characterization. In ecotoxicological assessments, the most commonly used lines of evidence are based on (1) biological surveys, (2) toxicity tests of contaminated media, and (3) toxicity tests of individual chemicals.

Lowest-observed-adverse-effect level (LOAEL)   The lowest level of exposure to a chemical in a test that causes statistically significant differences from the controls in a measured negative response.

Lowest-observed-effect concentration (level) (LOEC)   The lowest concentration of a material used in a toxicity test that has a statistically significant effect on the exposed population of test organisms as compared with the controls.

Measure of effect   A measurable ecological characteristic that is related to the valued characteristic chosen as the assessment endpoint (equivalent to the earlier term *measurement endpoint*).

Measure of exposure   A measurable characteristic of a contaminant or other agent that is used to quantify exposure.

Mechanism of action   The process by which an effect is induced. It is often used interchangeably with "toxic mode of action" but is usually more specific. For example, the mode of action of an agent on a population may be lethality and its mechanism of action may be crushing, acute narcosis, cholinesterase inhibition, or burning.

Mechanistic model   A mathematical model that simulates the underlying processes of a system, rather than using simple empirical relationships.

Median lethal concentration   A statistically or graphically estimated concentration that is expected to be lethal to 50% of a group of organisms under specified conditions.

Model   A formal representation of some component of the world or a mathematical function with parameters that can be adjusted so that the function closely describes a set of empirical data.

Model uncertainty   The component of the uncertainty concerning an estimated value that is due to possible misspecification of a model used for the estimation. It may be due to the choice of the form of the model, its parameters, or its bounds.

Monte Carlo simulation   A resampling technique frequently used in uncertainty analysis in risk assessments to estimate the distribution of the output parameter of a model.

No-observed-adverse-effect level (NOAEL)   The highest level of exposure to a chemical in a test that does not cause statistically significant differences from the controls in any measured negative response.

No-observed-effect concentration (NOEC)   The highest concentration of a test substance to which organisms are exposed that does not cause any observed and statistically significant effects on the organism as compared with the controls. For example, the NOEC might be the highest tested concentration at which an observed variable such as growth did not differ significantly from growth in the control. The NOEC customarily refers to the most sensitive effect unless otherwise specified. NEL, NOAEL, NEC, and NOEC are used as equivalent terms.

Normal distribution   The classical statistical bell-shaped distribution, which is symmetric and parametrically simple in that it can be fully characterized by two parameters: its mean and variance. The normal distribution is observed in situations where many independent additive effects are influencing the values of the variates.

Normalization   Alteration of a chemical concentration or other property (usually by dividing by a factor) to reduce variance due to some characteristic of an organism or its environment (e.g., division of the body burden of a chemical by the organism's lipid content to generate a lipid-normalized concentration).

Octanol–water partition coefficient   The quotient of the concentration of an organic chemical dissolved in octanol divided by the concentration dissolved in water if the chemical is in equilibrium between the two solvents.

PAF^{Endpoint}   General notation to identify that the potentially affected fraction (PAF) of species is based on an SSD with a specific type of endpoint, e.g., PAF^{NOEC} to identify a PAF based on SSD^{NOEC}. Optionally, one could add a superscript prefix to show the number of toxicity data from which the model was derived, such as ^{4}PAF^{Endpoint} for a PAF based on four data points.

Percentile   Same as quantile, but with the proportion expressed as a percentage. The median is the 50th percentile.

Population   An aggregate of interbreeding individuals of a species, occupying a specific location in space and time.

Predicted environmental concentration (PEC)   The concentration in the environment of a chemical calculated from the available information on certain of its properties, its use and discharge patterns, and the associated quantities.

Predicted-no-effect concentration/level (PNEC)   The maximum level (dose or concentration) that, on the basis of current knowledge, is likely to be tolerated by an organism without producing any adverse effect.

Probability   According to the frequentist view, the probability is the frequency of an event in an infinite repetition of identical and independent trials. In the Bayesian view, probability is a measure for the degree of belief in possible values of a random variable. In both views, probability is a measure of uncertainty of some outcome of an experiment or a prediction.

Probability density function (PDF)   For a continuous random variable, the PDF expresses the probability that the random variable belongs to some very small interval. For a discrete random variable, the PDF expresses the probability that the random variable is equal to a specific (discrete) value.

Problem formulation   The phase in an ecological risk assessment in which the goals of the assessment are defined and the methods for achieving those goals are specified.

Quantile   The value of a random variable that corresponds to a specified proportion of the PDF of that random variable. Quantiles can be determined from the inverse CDF. The median is the 0.5th quantile. The quartiles are the 0.25th, 0.50th, and 0.75th quantiles.

Random variable   A probabilistic (i.e., uncertain) quantity that may assume different possible values, either in a continuous or a discrete way.

Receptor   An organism, population, or community that is exposed to contaminants. Receptors may or may not be assessment endpoint entities.

Recovery   The extent of return of a population or community to a condition that existed before contamination. Because of the complex and dynamic nature of ecological systems, the attributes of a "recovered" system must be carefully defined.

Risk (toxic)   The predicted or actual probability of occurrence of an adverse effect on humans or the environment of exposure to a chemical substance or mixture. *See also* hazard.

Risk assessment   A process that entails some or all of the following elements: hazard identification, effects assessment, exposure assessment, risk characterization. It is the identification and quantification of the risk resulting from a specific use or occurrence of a chemical compound including the establishment of dose–response relationships and target populations. When quantitative data on dose–response relationships for different types of population, including sensitive groups, are unavailable, such considerations may have to be expressed in more qualitative terms.

Risk characterization   A phase of ecological risk assessment that integrates the exposure and stressor response profiles to evaluate the likelihood of adverse ecological effects associated with exposure to the contaminants.

Risk management   The process of deciding what regulatory or remedial actions to take, justifying the decision, and implementing the decision.

Safety factor   A factor applied to an observed or estimated toxic concentration or dose to arrive at a criterion or standard that is considered safe. The terms *safety factor* and *uncertainty factor* are often used synonymously. *See also* uncertainty factor.

Species sensitivity distribution (SSD)   A PDF or CDF of the toxicity of a certain compound or mixture to a set of species that may be defined as a taxon, assemblage, or community. Empirically, a PDF or CDF is estimated from a sample of toxicity data for the specified species set.

SSD^{Endpoint}   General notation for an SSD based on a specific type of input data, e.g., NOEC (SSD^{NOEC}), EC$_{50}$ (SSD^{EC50}), or LC$_{50}$ values (SSD^{LC50}). A superscript prefix can be added to the SSD^{Endpoint} to show the number of toxicity data from which the model was derived, such as ^{4}SSD^{Endpoint} for an SSD based on four data points.

Susceptibility   The condition of an organism or other ecological system lacking the power to resist a particular disease, infection, or intoxication. It is inversely proportional to the magnitude of the exposure required to cause the response.

Test endpoint   A response measure in a toxicity test, i.e., the value(s) derived from a toxicity test that characterize the results of the test (e.g., NOEC or LC$_{50}$).

Toxic mode of action (TMoA)   A phenomenological description of how an effect is induced. *See also* mechanism of action.

Toxic unit   The concentration of a chemical expressed as a fraction or proportion of its effective concentration (measured in the same units). It may be calculated as follows: toxic unit = actual concentration of chemical in solution/LC$_{50}$. If this number is greater than 1.0, more than half of a group of aquatic organisms will be killed by the chemical. If it is less than 1.0, less than half the organisms will be killed.

Toxicity test   The determination of the effect of a substance on a group of selected organisms under defined conditions. A toxicity test usually measures either (a) the proportions of organisms affected (quantal) or (b) the degree of effect shown (graded or quantitative), after exposure to specific levels of a chemical or mixture of chemicals.

Uncertainty   Imperfect knowledge concerning the present or future state of the system under consideration; a component of risk resulting from imperfect knowledge of the degree of hazard or of its spatial and temporal pattern of expression.

Uncertainty factor   A factor applied to an exposure or effect concentration or dose to correct for identified sources of uncertainty. *See also* safety factor.

Acronyms

ACR	acute-to-chronic ratio
ACF	algae, crustaceans, fish
AE	alcohol ethoxylates
AERA	aquatic ecological risk assessment
AES	alcohol ethoxylated sulfates
AF	application factor
AIS	Association Internationale de la Savonnerie et de la Detergence
AQUIRE	AQUatic Information REtrieval database on toxicity
ARAMDG	Aquatic Risk Assessment and Mitigation Dialog Group (U.S.)
ASTER	ASsessment Tool for Evaluation of Risk
AUC	area under the curve
AVS	acid-volatile sulfide
AWQC	ambient water quality criterion
BAF	bioaccumulation factor
BCF	bioconcentration factor
BSAF	biota-to-sediment(or soil) accumulation factor
BW	body weight
CA	concentration addition
Cb	background concentration
CC	caloric content
CCC	criterion continuous concentration
CCME	Canadian Council of Ministers of the Environment
CCREM	Canadian Council of Resource and Environment Ministers
CDF	cumulative distribution function
CEC	Commission of the European Communities
CI	confidence interval
CMC	criterion maximum concentration
COA	co-occurrence analysis
COPC	chemical of potential concern
CSTE	Committee on Toxicity and Ecotoxicity of Chemicals (European Union)
CV	chronic value
DDT	dichlorodiphenyltrichloroethane
DEEDMAC	diethyl ester dimethyl ammonium chloride
DHTDMAC	dihydrogenated-tallow dimethyl ammonium chloride
DIBAEX	DIstribution BAsed EXtrapolation
DODMAC	dioctadecyl dimethyl ammonium chloride
DSDMAC	distearyl dimethyl ammonium chloride
DTDMAC	ditallow dimethyl ammonium chloride
EAC	ecologically acceptable concentration
EC	exposure concentration

EC_{50}	median effective concentration
ECD	exposure concentration distribution
ECDF	empirical cumulative distribution function
ECOFRAM	Ecological Committee on FIFRA Risk Assessment Methods (U.S.)
EEC	expected environmental concentration
EF	extrapolation factor
EINECS	European INventory of Existing Chemical Substances
EMR	existence metabolic rate
EPP	exceedence profile plot
EQC	environmental quality criterion
EqP	equilibrium partitioning method
EQS	environmental quality standard
ERA	ecological risk assessment
ERC	ecological risk criterion
ER-L	effects range–low (U.S.)
ERL	environmental risk limit
ER-M	effects range–median (U.S.)
ESG	equilibrium-partitioning sediment guidelines
ESQC	ecotoxicological soil quality criterion
ETR	expected total risk
EU	European Union
EUSES	European Uniform System for the Evaluation of Substances
EXAMS	EXposure Analysis Modeling System
EXF	exceedence function
FA	fraction affected
FAE	food assimilation efficiency
FAME	FActorial application MEthod
FAV	final acute value (U.S.)
FCV	final chronic value (U.S.)
FIFRA	Federal Insecticide, Fungicide and Rodenticide Act
FMR	field metabolic rate
FW	fresh water
GENEEC	GENeric Expected Environmental Concentration model
GIS	Geographic Information System
GLP	good laboratory practice
GMAV	genus mean acute value
HC_5	hazardous concentration to 5% of species
HC_p	hazardous concentration to some percentage p of species
HCS	hazardous concentration for sensitive species
HD_5	hazardous dose for 5% of species
HD_p	hazardous dose for some percentage p of species
HQ	hazard quotient
HRS	hazard ranking system (U.S. EPA)
HU	hazard unit
ICES	International Council for Exploration of the Seas
IJC	International Joint Commission

ISO	International Standardization Organization
IV	intervention value
JPC	joint probability curve
K_{oc}	partition coefficient or sorption coefficient to organic carbon
K_{ow}	octanol–water partitioning coefficient
K_p	partition coefficient or sorption coefficient
K_p	protection concentration for fraction of soil organisms (Denmark)
LAS	linear alkyl sulfonates
LC_{50}	median lethal concentration
LCA	life cycle assessment
LCIA	life cycle impact assessment
LOC	level of concern
LOD	limit of detection
LOEC	lowest-observed-effect concentration
LOQ	limit of quantitation
MATC	maximum acceptable toxicant concentration
MAV	mean acute value
MDL	method detection limit
MDR	minimum data requirement
MLE	maximum likelihood estimator
MOVE	MOdel for the VEgetation
MPA	maximum permissible addition
MPC	maximum permissible risk concentration
MPR	human toxicological maximum permissible risk
MR	metabolic rate
msPAF	multisubstance potentially affected fraction
MTTMAC	mono-tallow trimethyl ammonium chloride
MUSCRAT	MUltiple SCenario Risk Assessment Tool
MVUE	minimum variance unbiased estimator
NA	negligible addition
NAWQC	national ambient water quality criteria (U.S.)
NC	negligible risk concentration
NC	no concordance indication in data tables
NCT	non-central t-distribution
NE	no effect indication in data tables
NEC	no-effect concentration
NEN	National Ecological Network (the Netherlands)
NERI	National Environmental Research Institute (Denmark)
NG	no gradient indication in data tables
NOAA	National Oceanic and Atmospheric Administration (U.S.)
NOAEC	no-observed-adverse-effect concentration
NOEC	no-observed-effect concentration
NSTP	National Status and Trends Program (Canada)
OC	organic carbon
OECD	Organisation for Economic Co-operation and Development
OTS	Office of Toxic Substances (U.S. EPA)

PAF	potentially affected fraction of species
PAH	polycyclic aromatic hydrocarbon
PCA	principal component analysis
PCB	polychlorinated biphenyl compounds
PDF	potentially disappeared fraction of species
PDF	probability density function
PEC	predicted environmental concentration
PEL	probable effect level (Canada, sediment guideline derivation)
PERA	probabilistic ecological risk assessment
PES	probability of effects on a species
PNEC	predicted no-effect concentration
PRESS	prediction sum of squares
PRZM	pesticide root zone model
PSL	Priority Substance List
pT	toxic potency
QA	quality assurance
Q-Q plot	quantile–quantile plot
QSAR	quantitative structure–activity relationship
QSSR	quantitative species sensitivity relationship
RA	response addition
RADAR	Risk Assessment tool to evaluate Duration And Recovery
RIKZ	Dutch National Institute for Coastal and Marine Management
RIVM	National Institute of Public Health and the Environment of the Netherlands
RQ	risk quotient
RV	random variable
SAF	secondary acute factor
SAV	secondary acute value
SDU	sensitivity distribution units
SEF	standard extrapolation factor
SEM	simultaneously extracted metals
SEM	standard error of the mean
SETAC	Society of Environmental Toxicology and Chemistry
SF	safety factor
SG	small gradient indication in data tables
SLRA	screening-level risk assessment
SMAV	species mean acute value
SQG	sediment quality guideline
SR	sensitivity ratio
SRC	serious risk concentration
SS	species sensitivity
SSD	species sensitivity distribution
$SSD^{Endpoint}$	species sensitivity distribution for a specific ecotoxicological endpoint (e.g., SSD^{NOEC}, SSD^{EC50}, and SSD^{LC50})
SS_{error}	error sum of squares
SSTT	spiked sediment toxicity test
SW	salt water

TDI	tolerable daily intake
TE	toxic equivalent
TEF	toxic equivalence factor
TEL	Threshold Effect Level (Canada, sediment guideline derivation)
TEL	Threshold Effects Level (U.S.)
TGD	Technical Guidance Document
TMoA	toxic mode of action
TOC	total organic carbon
TSS	total suspended solids
TU	toxic unit
TWMC	time weighted mean concentration
U.S. EPA	Environmental Protection Agency of the United States of America
USDOD	U.S. Department of Defense
USES	Uniform System for the Evaluation of Substances
USES-LCA	Uniform System for the Evaluation of Substances in Life Cycle Assessment
VKI	Water Quality Institute (Denmark)
VROM	Ministry of Housing, Spatial Planning and the Environment (the Netherlands)
VTCB or TCB	Committee on Soil Protection in the Netherlands
WERF	Water Environment Research Foundation
WWTP	wastewater treatment plant

Index

T - #0301 - 071024 - C616 - 234/156/27 - PB - 9780367396480 - Gloss Lamination